CISM COURSES AND LECTURES

Series Editors:

The Rectors
Manuel Garcia Velarde - Madrid
Mahir Sayir - Zurich
Wilhelm Schneider - Wien

The Secretary General
Bernhard Schrefler - Padua

Former Secretary General
Giovanni Bianchi - Milan

Executive Editor
Carlo Tasso - Udine

The series presents lecture notes, monographs, edited works and proceedings in the field of Mechanics, Engineering, Computer Science and Applied Mathematics.
Purpose of the series is to make known in the international scientific and technical community results obtained in some of the activities organized by CISM, the International Centre for Mechanical Sciences.

INTERNATIONAL CENTRE FOR MECHANICAL SCIENCES

COURSES AND LECTURES - No. 442

THEORIES OF TURBULENCE

EDITED BY

MARTIN OBERLACK
DARMSTAT UNIVERSITY OF TECHOLOGY

FRIEDRICH H. BUSSE
UNIVERSITY OF BAYREUTH

Springer-Verlag Wien GmbH

This volume contains 147 illustrations

Originally published by Springer-Verlag Wien New York in 2002
SPIN 10890855

In order to make this volume available as economically and as rapidly as possible the authors' typescripts have been reproduced in their original forms. This method unfortunately has its typographical limitations but it is hoped that they in no way distract the reader.

ISBN 978-3-211-83694-1 ISBN 978-3-7091-2564-9 (eBook)
DOI 10.1007/978-3-7091-2564-9

Preface

The present book evolved from the summer course on *Theories of Turbulence* held at the *International Centre For Mechanical Sciences (CISM)* in Udine/Italy from 17.-21. September 2001. The original idea for the course was born in 1997 during a discussion with Prof. Dr. P. D. Panagiotopolous, a former member of the scientific council of CISM. Sadly, Prof. Panagiotopolous passed away in 1998 at the early age of 48. In recognition of his deep contributions to the mechanical sciences and as a mark of our gratitude to him for the inspiration of this course we dedicate the present book to him.

It addresses a variety of analytical and numerical approaches to problems of turbulent fluid flow with the main goal of providing insights into the nature of turbulence. The topics of the six chapters range from higher bifurcations and coherent structures through symmetry principles and renormalization theories to one- and two-point models and sub-grid scale modelling for large eddy simulations. Mathematical tools such as group theory and statistics are presented as needed for the theoretical developments. Applications to examples of turbulent flows range from isotropic turbulence and homogeneous turbulence with a constant mean velocity gradient to "semi-complex" flows with one or two homogeneous directions such as turbulent pipe, channel and alike flows. The six chapters in this book are organized as follows:

In the first chapter by R. Benzi and L. Biferale, which primarily deals with isotropic turbulence, emphasis is placed on turbulent scaling laws of two- and multi-point structure functions. Intermittency and the multifractal aspects of turbulence are emphasized.

In chapter two by F.H. Busse the evolution of dynamical structures and phase turbulence in weakly nonlinear fluid systems are discussed. The possibilities for the description of coherent structures through the sequence-of-bifurcation approach are then explored at higher Rayleigh and Reynolds numbers. Finally the theory of upper bounds for turbulent transports at asymptotic values of the control parameters is outlined.

The third chapter by W.D. McComb interprets turbulence as a many-body problem, along with renormalized perturbation theory. Then a review and assessment of the two-point, two-time closure which arises from renormalization is given. The relevance of both these and of renormalization group theory for sub-grid scale models, as well as their extensions to shear flows, is discussed.

Chapter four by C. Cambon gives an overview on non-local turbulence theories and models. A classical spectral description is introduced for multi-point correlations. Applications to stably-stratified and rotating turbulence as well as "wave-turbulence" and developments towards inhomogeneous turbulence are discussed.

The fifth chapter by A.V. Johannson discusses one-point turbulence models from a somewhat analytical point of view. These range from eddy-viscosity based two-equation models, with particular attention to explicit algebraic Reynolds stress models, to differential Reynolds stress and scalar flux models. Realizability and modelling concepts are discussed in detail.

The last chapter by M. Oberlack gives a short introduction to symmetries of differential equations. Symmetry methods are applied to the two- and multi-point correlation equations for the derivation of invariant solutions (scaling laws). These solutions comprise classical and new results for isotropic and homogeneous turbulence as well as for wall-bounded shear flows such as the logarithmic-law-of-the-wall. Finally it is shown that symmetries provide necessary conditions which are crucial in the development of physically sound Reynolds stress and sub-grid scale models.

For the readers' convenience and because the topics are based on similar equations, physical quantities have been assigned the same names (as far as possible) in the last three chapters.

The book is intended for advanced students and scientists from engineering and applied sciences, as well as for physicists and mathematicians interested in the fundamentals of the field. It is particularly aimed at readers who wish to be brought up to date with details of recent advances in statistical turbulence theory from a fundamental point of view, as well as those who are more concerned with the practical issues of turbulence and, in particular, turbulence modelling. Those readers will find it very helpful to learn more about modern developments in turbulence theory which can immediately be applied to turbulence modelling. For many of the recent results it has already been shown that they provide an effective framework to significantly improve turbulence models and their predictive capabilities. This is true for both Reynolds averaged equations and filtered equations in large-eddy simulation.

We sincerely thank all authors for their highly professional and instructive contributions to this book. Prof. W.D. McComb deserves our particular gratitude for his very valuable comments which helped to improve the English of several contributions. Last but not least we are very grateful to Jeroen Pieterse for integrating all contributions, commenting on language issues and giving the book the optical finish using LaTeX 2_ε.

Friedrich H. Busse and Martin Oberlack, June 2002

CONTENTS

Intermittency in Turbulence

Roberto Benzi and Luca Biferale

*Dept. of Physics and INFM, University of Rome, Tor Vergata,
Via della Ricerca Scientfica 1, 00133 Rome, Italy*

Abstract

We present a detailed review of recent developments in the statistical approach to fully developed turbulence. We address both ideal situations such as "homogeneous and isotropic turbulence" as well as problems of real anisotropic and wall bounded flows. We also discuss a set of theoretical questions connected to the calculation of anomalous exponents in the Navier-Stokes equations and in a class of shell models for the turbulent energy cascade.

1 Introduction: the Kolomogorov Equation

In these lectures we review some elements of small scale statistical properties of turbulent flows. We have tried to discuss all the problems and their possible solutions in a self consistent way, i.e. without assuming previous, detailed knowledge of the reader. However, in many cases we did not enter into the necessary mathematical and/or experimental details without which, we feel, it could be difficult to clearly follow our discussions. At any rate, we hope to give to the reader a systematic picture of current ideas on intermittency and open problems.

We start by reviewing two major concepts in turbulence. One is related to the so called Kolmogorov 4/5 equation while the other is related to the physical picture known as the Richardson cascade. The Kolmogorov equation is derived from the Navier-Stokes equations for homogeneous and isotropic turbulence. By denoting $S_n(r)$ the n-order longitudinal structure functions at r, namely $S_n(r) = \langle(\delta u(r))^n\rangle$ and $\delta u(r) = (\vec{u}(\vec{x}+\vec{r}) - \vec{u}(\vec{x}))\cdot\frac{\vec{r}}{r}$, ϵ the mean rate of energy dissipation, and ν the kinematic viscosity, the Kolmogorov equation reads for three-dimensional turbulence:

$$S_3(r) = -\frac{4}{5}\epsilon r + 6\nu\frac{dS_2(r)}{dr} \qquad (1)$$

The Richardson cascade (see Figure 1) is somewhat more difficult to describe. We would like to point out that the ideas introduced by Richardson simply tell us that for large Reynolds numbers the statistical properties of small scale fluctuations are due to dynamical instabilities occurring on many scales. These instabilities are responsible for an energy flux from large to small scales, where eventually the energy is dissipated. The Richardson cascade is a simple physical picture which suggests that many statistical properties of small scale fluctuations in turbulence, for large Reynolds numbers, should not depend on the details of forcing and dissipation. Such a supposed independence is known as "universality hypothesis" and it is one of the major open challenges to be understood from first principles by using the Navier-Stokes equations. Anisotropies and inhomogeneities are introduced in the Navier-Stokes equations only through boundary and forcing

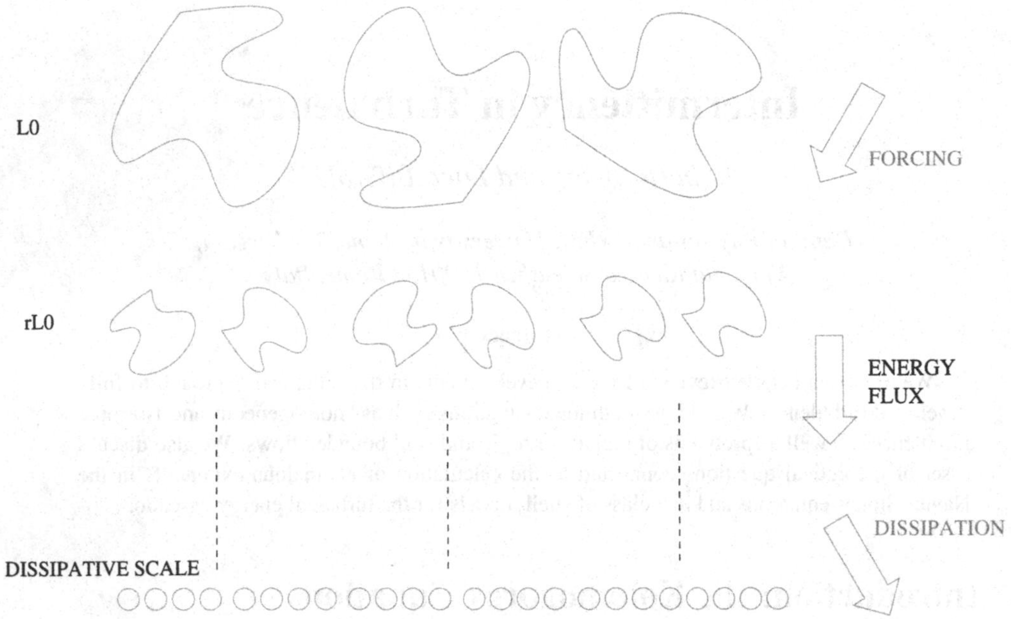

Figure 1: Schematic picture of the Richardson cascade.

terms. The universality hypothesis, as indicated in the Richardson picture, suggests that turbulent fluctuations, at small scales, should become more and more isotropic and homogeneous. Thus we arrive at the celebrated Kolmogorov 41 theory which tells us that the statistical properties of small scales depend only on ϵ and r, in agreement with equation (1). We do not want to discuss the details of the Kolmogorov theory (see the references in the bibliography for a complete and clear introduction to this problem). We remark, however, that there have been, and still are, many discussions on whether small scale fluctuations are isotropic or not. We will spend some time in clarifying the actual theoretical and experimental understanding of this important issue at the end of section 6 where important realizations of anisotropic and inhomogeneous turbulent flows are discussed (boundary layer and channel flows). In the following we shall restrict ourselves to isotropic and homogeneous turbulence unless otherwise specified.

Besides isotropy, the Kolmogorov theory also assumes that ϵ does not depend on Re for large Reynolds numbers. This is an important assumption which has been tested experimentally with rather good accuracy. Under the above assumptions, the Kolmogorov theory predicts that for large enough Reynolds numbers the dissipative contribution in equation (1) is vanishing and therefore we are left with:

$$S_n(3) \sim (\epsilon r)^{n/3} \tag{2}$$

The Kolmogorov theory implies that the statistical properties of turbulence, at small scales, are universal i.e. independent of the forcing and of the form of dissipation. If this is the case, the nonlinear (inertial) term of the Navier-Stokes equations are playing the major role in determining the statistical properties of turbulence. The universal properties of (isotropic) turbulence have been extensively investigated in the last 20 years and are the central questions addressed in these

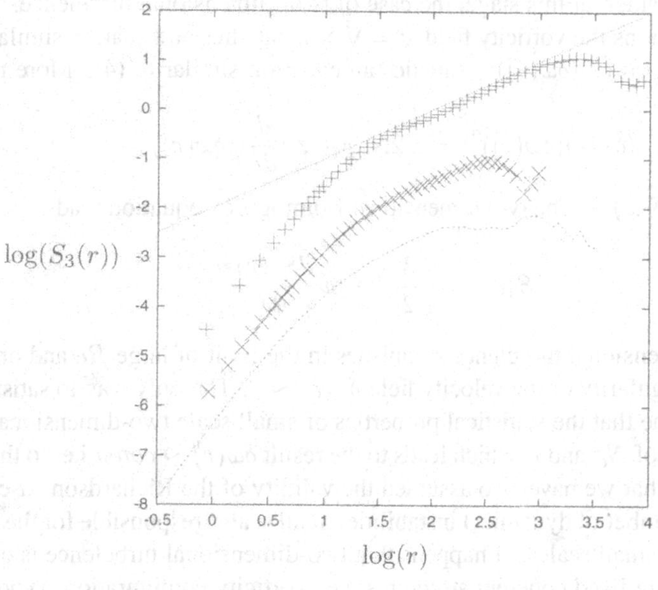

Figure 2: Log-log plot of the scale behaviour of the third order structure function. The straight line corresponds to slope 1. Only at large *Re* one can observe agreement with the scaling prediction of the Kolmogorov equation.

lectures. Although we do not yet posses any rigorous results, recently there have been substantial advances in our understanding of the problem. As we shall see, the question of universality is complicated by intermittency effects (see section 2) and the fact that any experimental test suffers from finite size effects for finite Reynolds numbers. To have a preliminary test of the difficulties, let us consider the results shown in Figure 2, where we plot, on a log-log scale, the quantity $S_3(r)$ versus r for different Reynolds number. Only in one case, corresponding to the largest Reynolds number, $Re = 8 \times 10^5$, a reasonable scaling range of $S_3(r)$ is observed, as predicted by equation (1) (the dashed line in the figure corresponds to the slope 1).

For low or moderate Reynolds number, the agreement with the Kolmogorov prediction is rather poor. As a consequence, it is difficult to use direct numerical simulation of the Navier-Stokes equations (the lowest Reynolds number data set shown in Figure 2), to study the universal properties of turbulence, because of limitations of our present computer power in reaching high enough Reynolds numbers.

The Kolmogorov equation can be generalized for any dimensions and for passive scalars. By denoting θ as a passive scalar field satisfying the equation:

$$\partial_t \theta + \vec{u} \cdot \vec{\nabla} \theta = \chi \Delta \theta \tag{3}$$

where χ is the kinematic diffusivity and N the mean rate of passive dissipation, we can obtain the analogue of the Kolmogorov equation, namely:

$$\langle \delta u(r) (\delta \theta(r))^2 \rangle = -\frac{4}{3} N r + 6\chi \frac{d}{dr} \langle (\delta \theta(r))^2 \rangle \tag{4}$$

It is interesting to consider, at this stage, the case of two- dimensional turbulence. As it is well known, in two dimensions the vorticity field $\omega = \vec{\nabla} \times \vec{u}$ satisfies an equation similar to (3) and, therefore the quantity $\langle \delta u(r)(\delta\omega(r))^2 \rangle$ satisfies an equation similar to (4). More precisely we have:

$$\langle \delta u(r)(\delta\omega(r))^2 \rangle = -2N_\omega r + 2\nu \frac{d}{dr} \langle (\delta\omega(r))^2 \rangle \tag{5}$$

where now $N_\omega = \nu\langle(\nabla\omega)^2\rangle$. The two-dimensional Kolmogorov equation reads:

$$S_3(r) = -\frac{3}{2}\epsilon r + 6\nu \frac{dS_2(r)}{dr} \tag{6}$$

Moreover, in two-dimensional turbulence ϵ vanishes in the limit of large Re and one can prove mathematically the regularity of the velocity field $\delta u(r) \sim r$. The only way to satisfy equations (5) and (6) is to assume that the statistical properties of small scale two-dimensional turbulence are described in terms of N_ω and r, which leads to the result $\delta\omega(r) \sim const$ i.e. to the Kraichnan theory. Let us notice that we have also assumed the validity of the Richardson cascade, i.e. the existence of a large number of dynamical instabilities at all scales responsible for the "enstrophy" cascade from large to small scales. It happens that two-dimensional turbulence is characterized by the existence of long lived coherent structures, i.e. vorticity configurations (vortices) which are stable solutions of the two-dimensional Euler equation. These vortices, which are often observed in numerical simulations of two-dimensional turbulence and whose sizes and numbers strongly depend on the characteristics of the large scale forcing, are at variance with the idea of a Richardson cascade. Actually, one is led to think that two-dimensional turbulence should be considered as a collection of coherent structures whose interactions and dynamics control all the statistical features of the flow.

The above considerations on two-dimensional turbulence raise many questions on the relevance of coherent structures in turbulent flows. In three dimensional turbulence, flow visualization of numerical simulations and laboratory experiments have shown the existence of well defined structures in the flow where a relevant part of vorticity is concentrated. However, in contrast to two-dimensional turbulence, coherent structures in three-dimensional turbulence are very unstable and their characteristic life time is not long compared to the eddy turnover time $r/\delta u(r)$ of their scale r. Therefore, the Richardson scenario may not be affected. It is important to remember, however, that the existence of coherent structures may lead to the conclusion that a statistical description similar to the one proposed by Kolmogorov, might be irrelevant for the understanding of the physical properties of turbulence. We are tempted to say that the crucial point in all physical theories is to build up a coherent scenario of the observed phenomena which leads to quantitative comparisons against experimental data and, in some cases, to non trivial predictions. Certainly the Kolmogorov theory of fully developed turbulence belongs to the above category.

Before closing this section, we want to describe how to use the Kolmogorov equation to derive interesting conjectures on the statistical properties of turbulent convection. Let us consider a flow heated from below in the so called Boussinesq approximation. In this case the Navier-Stokes equations become:

$$\partial_t \vec{u} + \vec{u} \cdot \vec{\nabla}\vec{u} = -\frac{1}{\rho}\vec{\nabla}p + \vec{g}\beta\theta + \nu\Delta\vec{u} \tag{7}$$

where $\vec{g}$ is the gravity vector and β is a constant. The temperature θ (more precisely the fluctuation of temperature with respect to a given reference temperature) satisfies the equation:

$$\partial_t \theta + \vec{u} \cdot \vec{\nabla} \theta = \chi \Delta \theta \tag{8}$$

Following the procedure leading to equation (1) and assuming isotropy (!) we obtain:

$$S_3(r) = -\frac{4}{5}\epsilon r + \frac{gC\beta}{r^4} \int x^4 \langle \delta u(r) \delta\theta(r) \rangle dx + 6\nu \frac{dS_2(r)}{dr} \tag{9}$$

$$\langle \delta u(r) (\delta\theta(r))^2 \rangle = -\frac{4}{3} N r + 6\chi \frac{d}{dr} \langle (\delta\theta(r))^2 \rangle \tag{10}$$

where C is a numerical constant. Let us remark that a new term appears in the Kolmogorov equation due to the buoyancy effect and that, rigorously speaking, this new term is proportional to $\delta u_z(r)\delta\theta r$, $\delta u_z(r)$ being the velocity difference along the direction selected by the gravity force $\vec{g}$. The assumption of isotropy implies that the statistical properties of $\delta u_z(r) \sim \delta u(r)$, which is of course a very strong assumption in this case. At any rate, by using equations (9) and (10) one can reach the following scenario: the new term in the Kolmogorov equation introduces a new length scale in the problem, namely the Bolgiano scale L_B defined as:

$$L_B = \frac{\epsilon^{5/4}}{N^{3/4} (g\beta)^{3/2}} \tag{11}$$

For r smaller than L_B, the new term in the Kolmogorov equation can be disregarded leading to the standard Kolmogorov prediction $S_n(r) \sim r^{n/3}$, i.e. the temperature field is acting as a passive scalar. On the other hand, for r larger than L_B the new term dominates and, by combining equations (9) and (10) we obtain:

$$S_n(r) \sim r^{3n/5} \tag{12}$$

$$\langle (\delta\theta(r))^n \rangle \sim r^{r/5} \tag{13}$$

This prediction has been subject of a number of experimental and numerical tests which seem to indicate the validity of equations (12) and (13) although with intermittency corrections. We remark that this is a non trivial "success" of the Kolmogorov scenario and we shall use the same kind of arguments in chapter 6 in order to understand the statistical properties of turbulence in boundary layers and shear flows.

References for section 1

A detailed description of all issues connected to the Kolmogorov theory can be found in the following list of books:

- L. D. Landau and E. M. Lifshitz, *Fluid Mechanics*, (Pergamon Press, New York, 1997)
- G. K. Batchelor, *The Theory of Homogeneous Turbulence*, (Cambridge University Press, 1971)

- A. S. Monin and A. M. Yaglom, *Stat. Fluid Mechanics*, (MIT Press, Cambridge, 1975)
- U. Frisch, *Turbulence*, (Cambridge University Press, Cambridge, 1995)
- T. Bohr, M. H. Jensen, G. Paladin and A. Vulpiani, *Dynamical System Approach to Turbulence*, (Cambridge University Press, Cambridge, 1997)
- J. O. Hinze, *Turbulence*, (McGraw-Hill, 1959)
- L. F. Richardson, *Weather Prediction by Numerical Process*, Cambridge U.P., Cambridge (1922)
- A. N. Kolmogorov, *CR. Acad. Sci. USSR* **30**, 301 (1941); *CR. Acad. Sci. USSR* **31**, 538 (1941); *CR. Acad. Sci. USSR* **32**, 16 (1941)

2 Intermittency in Fully Developed Turbulence

Let us consider the following simple equation:

$$x_{m+1} = 0.99x_m + \eta_m \tag{14}$$

where η_m is a Gaussian random variable $\delta - correlated$, i.e. $\langle \eta_n \eta_m \rangle = \delta_{n,m}$. The dynamical process x_m is clearly a Gaussian random process, as the reader can easily check directly. Let us now consider the variable $w_m = (x_m)^3$ which is no longer a Gaussian variable. A plot of w_m versus m shows a very intermittent behavior characterized by large fluctuations. This is a qualitative picture of an "intermittent" random process. A more quantitative description can be gained by computing the moments of the probability distribution and in particular the non dimensional quantities (generalized Kurtosis):

$$K_{2n} = \frac{\langle (w_m)^{2n} \rangle}{\langle (w_m)^2 \rangle^n} \tag{15}$$

By denoting K_{2n}^g the values of K_{2n} for a Gaussian random process, an intermittent signal is characterized by the relation $K_{2n} > K_{2n}^g$. The above example tells us that it is not difficult to define intermittency and to exhibit an explicit example of an intermittent process. As we shall see, the situation is much more complex in turbulence.

The Kolmogorov theory does not tell us anything about the value of K_{2n}. By using equation (2) we can compute K_{2n} as a function of r for the variable δu:

$$K_{2n}(r) = \frac{S_{2n}(r)}{(S_2(r))^n} \sim const \tag{16}$$

Thus, in principle, the Kolmogorov theory states that the "degree" of intermittency is kept constant at all scales. It turns out that this prediction is wrong for all Reynolds numbers. In Figure 3 we show the quantity $K_4(r)$, normalized to 1 for large scales, as a function of r obtained by a direct numerical simulation of homogeneous and isotropic turbulence. As we can see, intermittency is systematically growing going from large to small scales.

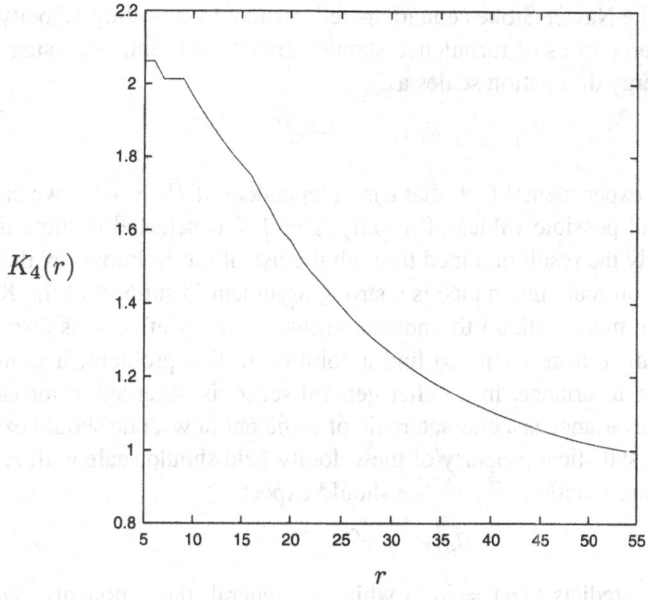

Figure 3: Behaviour of the flatness $K_4(r) \equiv \langle(\delta u(r))^4\rangle / \langle(\delta u(r))^2\rangle^2$ as a function of r, obtained by a direct numerical simulation of homogeneous and isotropic turbulence. The flatness has been divided by its gaussian value ($K_4^g = 3$). Intermittency is growing from large to small scales.

Near the dissipation scale, intermittency reaches a plateau and remains constant. Although the result shown in Figure 3 refers to a low Reynolds number, the same picture is observed for all three-dimensional turbulent flows and for all Reynolds numbers investigated so far. In particular by denoting the Kolmogorov dissipation scale with η given by:

$$\eta = \left(\frac{\nu}{\epsilon}\right)^{3/4} \tag{17}$$

and by L the (large) scale characterizing the turbulent flow, we find that $K_4(\eta)$ is an increasing function of Re. It seems, therefore, that the Kolmogorov theory fails to predict the correct scale behavior of fully developed turbulence. Even if the Kolmogorov theory is poorly representing the scale dependent probability distribution of the velocity field, we may hope to recover an important aspect of the theory which is scale invariance, i.e. a (simple) set of physical rules which identify scale behavior of turbulent flows. More precisely, one can hope that quantities like $K_{2n}(r)$ are functions of r and other (few) parameters independent of the Reynolds number for sufficiently large values of Re and for $\eta < r < L$ (the inertial range). In order to understand how this can happen, let us reconsider the Kolmogorov theory from a different point of view. As one can check directly, the Euler equations are invariant with respect to the time-scale transformation:

$$r \to \lambda r \quad u \to \lambda^h u \quad t \to \lambda t^{1-h} \tag{18}$$

where h is an arbitrary number. The Navier-Stokes equations are also scale invariant with respect to (18) if we let the kinematic viscosity change as:

$$\nu \to \lambda^{1+h} \nu \tag{19}$$

The scale invariance of the Navier-Stokes equations tells us that by rescaling velocity, space and viscosity the statistical properties of turbulence should remain invariant. By using (18,19) we find the mean rate of energy dissipation scales as:

$$\epsilon \rightarrow \lambda^{3h-1}\epsilon \sim \nu^{\frac{(3h-1)}{(1+h)}} \tag{20}$$

Because we know, as an experimental fact, that ϵ is independent of $Re \sim \nu^{-1}$, we can reach the conclusion that among all possible values of h, only $h = 1/3$ is selected in three dimensional turbulence. This is exactly the result obtained through the use of the Kolmogorov theory.

The above argument on scale invariance is a strong argument in support of the Kolmogorov theory which makes even more difficult to understand intermittency effects, as discussed at the beginning of this section. Before trying to find a solution to this problem, it is necessary to understand whether scale invariance, in a rather general sense, is observed in turbulence. It is quite clear that, if scale invariance is a characteristic of turbulent flows, one should expect to find that any scale dependent statistical property of the velocity field should scale with r. Therefore, if we consider the structure functions $S_n(r)$, we should expect:

$$S_n(r) \sim r^{\zeta(n)} \tag{21}$$

The Kolmogorov theory predicts $\zeta(n) = n/3$ while, in general, the exponents $\zeta(n)$ can be a non-linear function of n satisfying the general constraint $d^2\zeta(n)/dn^2 \leq 0$ and $\zeta(3) = 1$ because of (1).

In the last two decades there have been a number of laboratory experiments devoted to test the scaling hypothesis (21) and to estimate the values of the exponents $\zeta(n)$. In Figure 4 we show a log-log plot of $S_6(r)$ versus r for the same set of data already discussed in (2). We note that for large enough *Re* a scaling region is observed with a scaling exponent $\zeta(6) = 1.78$, i.e. significantly smaller than what is predicted by the Kolmogorov theory. The same analysis (not shown here) gives the following set of numbers: $\zeta(2) = 0.7$, $\zeta(4) = 1.28$, $\zeta(8) = 2.2$. Similar results have also been reported by many authors for homogeneous and isotropic turbulence at large *Re*. By using this information, we can easily obtain:

$$K_4(r) = \frac{S_4(r)}{(S_2(r))^2} \sim r^{\zeta(4)-2\zeta(2)} \sim r^{-.12} \tag{22}$$

which clearly indicates that intermittency is growing as one proceeds from large to small scales. We thus reach the conclusion that intermittency could be consistent with scale invariance although not in the form predicted by Kolmogorov theory. We want to remark, however, that intermittency seems to be independent of any scaling argument because it is observed even at low *Re*.

In an attempt to solve the issue raised by intermittency, one can introduce, following Kolmogorov, a scale dependent energy dissipation defined as:

$$\epsilon_r = \frac{1}{B(r)} \int_{B(r)} \epsilon(x)dx \tag{23}$$

where $B(r)$ is a "box" of side r and $\epsilon(x)$ is the local rate in x of energy dissipation. The quantity ϵ_r is a fluctuating quantity which can be used to generalize the relation $S_3(r) \sim \epsilon r$ as follows:

$$S_n(r) \sim \langle(\epsilon_r)^{n/3}\rangle r^{n/3} \tag{24}$$

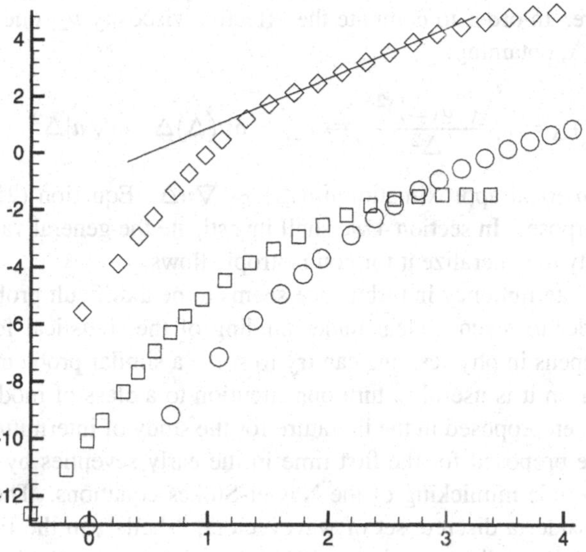

Figure 4: Scaling behaviour of $S_6(r)$ for three different values of Re (same set of experiments shown in Figure 2). Only at large Re a scaling region is observed. The scaling exponent is about 1.8 (straight line).

Equation (24) is known as the refined Kolmogorov similarity hypothesis (RKSH). It represents a simple and brilliant idea to explain intermittency effects (due to fluctuation of energy dissipation) in the framework of the Kolmogorov theory. Unfortunately, (24) does not tell us anything about the fluctuation of ϵ_r and its scale dependence. We want to remark that a number of experimental investigations have been carried out to check the validity of (24) and we shall discuss this point later on in section 5. Also, there have been a lot of discussions trying to understand whether (24) represents a new "physical" law or a trivial consequence of intermittency. Let us explain this point in more detail. The quantity $\epsilon(x)$ can be directly computed using the expression $\epsilon(x) = \nu \partial_i u_j \partial_i u_j$, where the indices i and j refer to spatial components and summation over repeated indices is assumed. Thus the local rate of energy dissipation can be computed directly from the velocity field which also enters in the definition of $S_n(r)$. Therefore, one may think that (24) is a kinematic constraint holding for most (in some mathematical sense) functions satisfying the constraint $S_3(r) \sim r$. In the following sections, we will present many results indicating that (24) is a non trivial statement deserving an important role in turbulence theory.

Equation (24) can be used as a general guideline for modeling turbulent parameterization in numerical simulations. If we want to numerically solve the Navier-Stokes equations at large Re, we are faced with the problem that the computational grid grows as $(L/\eta)^3 \sim Re^{9/4}$. Thus current computer power is useless for a direct numerical simulation. One may think to overcome the problem by introducing an effective viscosity, at scale Δ which is tuned such as to correctly transfer the energy from large to small scale without simulating the dynamic instabilities for scales smaller than Δ. This idea is a consequence of the Kolmogorov theory and the Richardson picture. In order to compute the effective viscosity ν_Δ one can use (24) in the form $(\delta u(\Delta))^3 \sim \epsilon_\Delta \Delta$, obtaining:

$$\epsilon_\Delta = \nu_\Delta \frac{(\delta u(\Delta))^2}{\Delta^2} \quad \rightarrow \nu_\Delta = \delta u(\Delta)\Delta \sim |\nabla u|\Delta^2 \tag{25}$$

where we use the numerical approximation $\delta u(\Delta) \sim \nabla u \Delta$. Equation (25) has been used in literature for many purposes. In section 4 we shall investigate the general validity of (25) and in section 6 the possibility to generalize it for non isotropic flows.

As we have seen, intermittency in turbulence seems to be a difficult problem which deserve much attention in order to reach a clear understanding of the statistical features of turbulent flows. As it often happens in physics, one can try to solve a similar problem in a much simpler situation. For this reason it is useful to turn our attention to a class of models, known as shell models, which have been proposed in the literature for the study of intermittency.

Shell models were proposed for the first time in the early seventies by the Russian school and are built upon a crude mimicking of the Navier-Stokes equations. The basic idea behind shell models is to consider a discrete set of wavevectors, "shells", in the Fourier k-space, and to construct a set of ordinary differential equations taking into account only a few variables per shell. The shells are chosen equispaced on logarithmic scale (the ratio between two successive wavenumbers, λ, is usually taken to be 2)

$$k_n = k_0 \lambda^n \tag{26}$$

This allows one to simulate flows with very high Reynolds numbers, while keeping only a few degrees of freedom. The amplitude $|u_n|$ of the velocity variable u_n can be thought of as the typical energy of the nth wavenumber shell

$$|u_n(t)| \sim \sqrt{\int_{k_n}^{k_{n+1}} 2E(k,t)\, dk}$$

and can also be regarded as the velocity increment $|u(x+l) - u(x)|$ on an eddy of scale $l \sim k_n^{-1}$. As the shell variable depends only of the shell index n, we necessarily have 0-dimensional shell models. If the number of variables per shell grows as 2^n we can have also 1-dimensional shell models (see section 4), the advantage of the latter being the possibility of resolving the spatial energy dissipation field.

Despite the fact that shell models are simplified versions of the Navier-Stokes equations they are not simple. They are less demanding in numerical simulations and because of the reduced number of degrees of freedom they allow simulations at very high *Re* numbers. Still they retain almost all of the complexity of the full Navier-Stokes equations when one tries to attack them analytically. The GOY model is perhaps the most popular shell model of turbulence. Equations (28) for the complex variable $u_n(t)$ are dictated by the following criteria:

(a) the linear term for $u_n(t)$ is given by $-\nu k_n^2 u_n$;

(b) the non linear terms for $u_n(t)$ are quadratic combinations of the form $k_n u_{n'} u_{n''}$;

(c) in absence of forcing and damping the energy $\frac{1}{2}\sum_n |u_n|^2$ is conserved;

(d) the interactions among shells are local in k-space (i.e. n' and n'' are close to n).

These criteria stem directly from an analysis of the Navier-Stokes equations, apart from (d) which, at this level, is like a sort of closure approximation (somehow supported by recent calculations using field theory techniques).
The dynamical GOY equations are the following:

$$\frac{d}{dt}u_n(t) + \nu k_n^2 u_n(t) = ik_n \left(a_n u_{n+1}u_{n+2} + b_n u_{n-1}u_{n+1} + c_n u_{n-1}u_{n-2}\right)^* + f_n\delta_{n,n_0} \quad (28)$$

The forcing term $f_n\delta_{n,n_0}$ acts only on shell n_0, usually concentrated at a large scale close to the first shell.

The following constraint has to be imposed if one wants the model to conserve the energy in the unforced and inviscid limit ($f = \nu = 0$):

$$a_n = 1; \quad b_n = -\frac{\delta}{\lambda}; \quad c_n = -\frac{1-\delta}{\lambda^2} \quad (29)$$

The constraint leaves a free parameter $\delta \in [0, 2]$ which plays an important role in determining the static and dynamical properties of the model as it is related to a second invariant of the model which, in the case $\delta = 1/2$, has the dimension of an helicity : $H = \sum(-1)^n k_n |u_n|^2$.

The model is originally defined with an infinite number of shells, but, because of the damping effect of the viscous terms, it can be truncated to a finite number of shells, N. The number of shells has to be chosen, in order to avoid an explosion of the numerical integration algorithm, about $4-5$ shells larger than the shell number at which viscosity starts to act: $n_d \sim -3/4 \log_\lambda \nu$. Usually the numerical integration is performed with a fourth order Adams-Bashfort slaved scheme. The constraints (29) have to be supplemented with the following boundary conditions: $b_1 = b_N = c_1 = c_2 = a_{N-1} = a_N = 0$.

The shell model has exact static (time-independent) solutions satisfying the Kolmogorov scaling at those scales where the forcing and the viscosity are negligible (indeed the solution $u_n = Ck_n^{-1/3}$ makes the nonlinear term vanish). It is possible to define shell number structure functions $S_p(n)$ and their scaling exponents ζ_p, are defined by means of the relation:

$$S_p(n) = \langle |u_n|^p \rangle \sim k_n^{-\zeta(p)} \quad (30)$$

The quality of the scaling is rather good and the scaling exponents can be computed for high order to very good accuracy. Let us remark that for any shell model it is possible to find the analogue of the Kolmogorov equation and one can prove that $\zeta(3) = 1$. It is a striking and yet not understood property that for ($\delta = 1/2$) the GOY models have ζ_p exponents equal (within statistical errors) to those of 3D Navier-Stokes equations. We remark that this result is rather important because it tells us that intermittency effects in the real Navier-Stokes equations can be studied without disregarding the scenario originally proposed by Kolmogorov and Richardson.

The GOY model suffers from a spurious symmetry: for any n the equations are invariant with respect to change of variables

$$\phi_{3n} \rightarrow \phi_{3n} + 2\alpha \quad \phi_{3n-1} \rightarrow \phi_{3n-1} - \alpha \quad \phi_{3n+1} \rightarrow \phi_{3n+1} - \alpha \quad (31)$$

where ϕ_n is the phase of the shell variable u_n and α is any real number.

Such a $U(1)$ symmetry produces small annoying oscillations in the scaling behavior of the structure functions. Recently an improved version of the GOY model has been introduced, named the Sabra model defined by:

$$\left(\frac{d}{dt}+\nu k_n^2\right)u_n = i(k_n u_{n+1}^* u_{n+2} + b k_{n-1} u_{n+1} u_{n-1}^* + (1+b) k_{n-2} u_{n-2} u_{n-1}) + f_n \quad (32)$$

The Sabra model does not suffer from any unwanted symmetry and displays a much better scaling behavior of the structure functions. In Figure 5 we give an example of the scaling behavior for the structure functions S_3 and S_6. The lines correspond to the best fit of the scaling in the inertial range, where we measure $\zeta(3) = 1$ and $\zeta(6) = 1.69$.

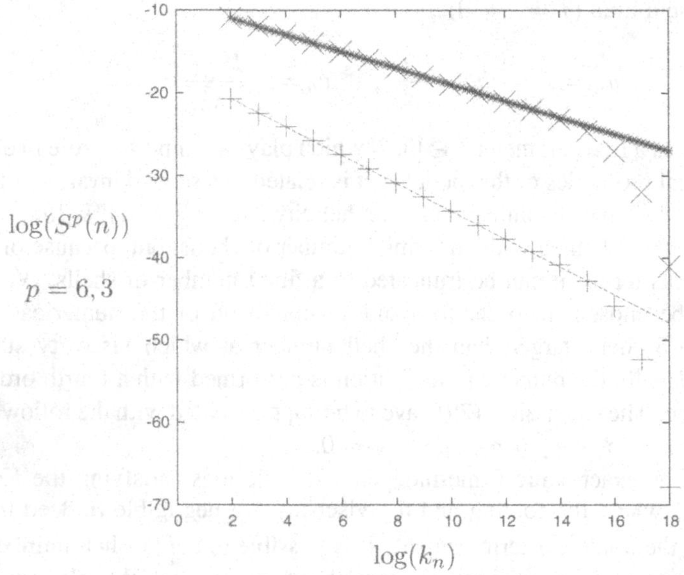

Figure 5: Scaling properties of the shell model (Sabra). With × we indicate the third order structure function (slope -1), while + indicates the sixth order structure function (slope -1.69).

3 The Multifractal Theory of Fully Developed Turbulence

In the last section we saw how intermittency and scale invariance can simultaneously hold in turbulent flows. Our task is now to set up a general framework which explicitly takes into account both scale invariance and intermittency. Let us come back to the scale transformation (18). In principle any number h is allowed for the scale transformation while the Kolmogorov equation (1) is fixed with the special value $h = 1/3$. On the other hand, it is quite clear that a unique value of h is not consistent with intermittency. One can solve this apparent paradox by assuming that many possible different values of h are allowed, with different probabilities, such that (1) is satisfied on average. This is precisely the idea behind the multifractal theory of fully developed turbulence.

The physical key idea behind the multifractal theory is to assume that at scale r the fluctuations of the velocity field $\delta u(r)$ are labelled in terms of h, each one with a different probability distribution. More precisely let us write $\delta u(r) \sim r^h$, then the probability distribution $P_h(r)$ to observe a fluctuation at scale r of size r^h is given by

$$P_h(r) \sim r^{3-D(h)} \tag{33}$$

There is a simple geometrical meaning for the function $D(h)$. Indeed, the quantity $r^{3-D(h)}$ is proportional to the number of "boxes" covering a set of fractal dimensions $D(h)$, divided by the number of boxes covering the three-dimensional space, which is r^{-3}. Thus, the idea which is underlying (33) is that for each possible fluctuation of size r^h there is only a subset of the full three-dimensional space which is dynamically "responsible" for that fluctuation.

By using equation (33) we can deduce that

$$S_n(r) = \int d\mu(h) r^{nh} r^{3-D(h)} \sim r^{\zeta(n)}, \qquad \zeta(n) = inf_h(nh + 3 - D(h)) \tag{34}$$

where we have used the saddle point technique to estimate $\zeta(n)$. It is now clear that a suitable function $D(h)$ can satisfy the constraint (1) and, at the same time, can provide values of $\zeta(n)$ in agreement with laboratory measurements. On the other hand, it is possible to estimate the function $D(h)$ knowing the values $\zeta(n)$. Defining h_n by the value of h where the minimum of $nh + 3 - D(h)$ is reached, one can write $\zeta(n) = nh_n + 3 - D(h_n)$, where h_n is the solution of the equation $n = dD(h)/dh$. We formally obtain:

$$\frac{d\zeta(n)}{dn} = h_n + n\frac{dh_n}{dn} - \frac{dD(h_n)}{dn} = h_n \tag{35}$$

Thus we finally obtain $3 - D(h_n) = \zeta(n) - nd\zeta(n)/dn$. To give an explicit example of the formalism, let us consider a particular form of $3 - D(h)$, namely

$$3 - D(h) = a(h - h_0)^2 \tag{36}$$

Using (34) we find

$$\zeta(n) = nh_0 - \frac{n^2}{4a} \quad h_n = h_0 - \frac{n^2}{2a} \tag{37}$$

Because $\zeta(3) = 1$, we can write

$$h_0 = \frac{1}{3} + \frac{3}{4a} \tag{38}$$

The final expression for $\zeta(n)$ reads:

$$\zeta(n) = \frac{n}{3} + \frac{3n - n^2}{4a} \tag{39}$$

Actually for a suitable value of $a \sim 20$, the values of $\zeta(n)$ obtained by (39) are in good agreement with the experimental numbers, although we have strong reason to consider (39) wrong for large n. Using the variable $h = \log(\delta u(r))/\log(r)$, one can rewrite $P_h(r)$ in terms of $\delta u(r)$:

$$P_h(r) \sim \exp\left[-\frac{(\log(\delta u(r)) - \mu(r))^2}{(2\sigma(r)^2)}\right] \tag{40}$$

where $\mu(r) = h_0 \log(r)$ and $2\sigma(r)^2 = -\log(r)/a$ (note that $\log(r) < 0$). The above expression makes it clear that in the example we are assuming the probability distribution to be log-normal. In the limit of vanishing fluctuations ($\sigma(r) \to 0$) we recover the Kolmogorov prediction.

With the knowledge of the function $D(h)$ it is possible to derive all information concerning the probability distributions of $\delta u(r)$ as a function of large scale fluctuations $\delta u(L)$. Let us define $P(\delta u(r))$ and $P(\delta u(L))$ the probability distribution of $\delta u(r)$ and $\delta u(L)$, where L is supposed to be the energy injection scale of turbulence. For simplicity we shall assume, as it is often observed in real turbulence, that $P(\delta u(L))$ is Gaussian, i.e.

$$P(\delta u(L)) = N \exp\left[-\left(\frac{(\delta u(L))^2}{2\sigma^2}\right)\right] \tag{41}$$

The multifractal theory tells us that

$$\delta u(r) = \left(\frac{r}{L}\right)^h \delta u(L) \tag{42}$$

with probability $P_h(r) \sim r^{3-D(h)}$. It follows that, for any h the probability $P_h(\delta u(r))$ is given by

$$P_h(\delta u(r)) = N \left|\frac{d\delta u(L)}{d\delta u(r)}\right| \exp\left[-\left(\frac{(\delta u(L))^2}{2\sigma^2}\right)\right] = N \left(\frac{r}{L}\right)^{-h} \exp\left[-\left(\frac{(\delta u(r))^2}{2\sigma^2 (r/L)^{2h}}\right)\right] \tag{43}$$

The final expression for $P(\delta u(r))$ is given by integrating $P_h(\delta u(r))$ over all possible h with probability $P_h(r) \sim r^{3-D(h)}$:

$$P(\delta u(r)) = N \int d\mu(h) \left(\frac{r}{L}\right)^{3-h-D(h)} \exp\left[-\left(\frac{(\delta u(r))^2}{2\sigma^2 (r/L)^{2h}}\right)\right] \tag{44}$$

To show that this expression is in agreement with our previous findings, we compute $S_n(r)$ as follows:

$$S_n(r) = \int d\delta u(r) (\delta u(r))^n P(\delta u(r)) = \int\int d\mu(h) dx x^n \left(\frac{r}{L}\right)^{(nh+3-D(h))} \exp\left[-\left(\frac{x^2}{2\sigma^2}\right)\right] \tag{45}$$

where $x = \delta u(r)/r^h$. Expression (45) is equivalent to (34). A plot of the probability distribution $P(\delta u(r))$ shows, for small r, long tails which can be fitted by a stretched exponential. Expression (44) shows that the stretched exponential is just the superposition of different Gaussian probability distributions with suitable variances. So, a direct inspection of the probability distribution of $\delta u(r)$ does not give us more information with respect to the scaling exponents $\zeta(n)$.

We can use the same argument leading to (44) to estimate the probability distribution of ∇u. The computation, however, requires some care in the definition of the scale where gradients should be computed. We have already noted in section 1 that at very small scales fluctuations are suppressed because of dissipation effects. There is no general physical law which tells us the characteristic scale at which dissipation is acting in turbulence. Following Kolmogorov, we assume that dissipation becomes relevant when the following condition holds:

$$\frac{\delta u(r) r}{\nu} \sim 1 \tag{46}$$

In the Kolmogorov theory condition (46) is reached for $r = \eta$ as defined by (17). In the multifractal theory we have

$$\delta u(L) L \left(\frac{r_d}{L}\right)^{h+1} \sim \nu \ \rightarrow \ \frac{r_d}{L} \sim \left(\frac{\nu}{\delta u(L) L}\right)^{1/(1+h)} \tag{47}$$

where we have defined r_d as the dissipation scale and we have estimated $\delta u(r_d)$ using (42). Equation (47) implies that the computation of gradients depends on the fluctuation exponent h for the multifractal theory, i.e. r_d is a function of h. The estimate of the gradient is now given by:

$$\nabla u \sim \frac{\delta u(r_d)}{r_d} \sim \left(\frac{\delta u(L)}{L}\right) \left(\frac{\nu}{\delta u(L) L}\right)^{(-(1-h)/(1+h))} \tag{48}$$

with probability $(r_d/L)^{(3-D(h))}$. By assuming (41), we can compute the probability distribution of ∇u using the formula

$$P(\nabla u) \sim P(\delta u(L)) \left| \frac{d(\delta u(L))}{d(\nabla u)} \right| \tag{49}$$

we finally obtain using $t = |\nabla u|$ and putting $L = 1$:

$$P(t) \sim \int d\mu(h) \left(\frac{\nu}{t}\right)^{2-(h+D(h))/2} \exp\left[-\left(\frac{\nu^{1-h} t^{1+h}}{2\sigma^2}\right)\right] \tag{50}$$

Expression (50) cannot be handled in an analytical way as we did for (34).

A consequence of (46) and (47) is that $\langle (\nabla u)^2 \rangle \sim \nu^{-1}$. A direct computation gives:

$$\langle (\nabla u)^2 \rangle \sim \langle r_d^{2(h-1)} \rangle \sim \langle \nu^{2(h-1)/(h+1)} \rangle \sim \int d\mu(h) \nu^{(1+2h-D(h))/(1+h)} \tag{51}$$

The saddle point technique gives the following equation:

$$(1+h)\left(2 - \frac{dD(h)}{dh}\right) - (1 - D(h) + 2h) = 0 \tag{52}$$

Let us now call h_3 the value of h used to compute $\zeta(3)$. Employing the definition of h_3 we know that the following relations must be satisfied,

$$\frac{dD(h_3)}{dh} = 3, \quad 3h_3 + 3 - D(h_3) = 1 \tag{53}$$

One can directly check that, by using (53) h_3 is a solution of (52). Finally inserting (53) into (51) we obtain

$$\langle (\nabla u)^2 \rangle \sim \nu^{-1} \tag{54}$$

Expression (51) can be generalized for all moments of the gradient as follows:

$$\langle (\nabla u)^{2n} \rangle \sim \int d\mu(h) \nu^{(2n(h-1)+3-D(h))/(1+h)} \tag{55}$$

The above expression gives non trivial estimates for the scaling of $\langle (\nabla u)^{2n} \rangle$ as a function of $Re \sim \nu^{-1}$ which are not simply related to the exponents $\zeta(n)$. This prediction is based on the multifractal theory and the assumption (46). An explicit set of numbers can be obtained only if we know an explicit form of $D(h)$.

Although the multifractal theory is providing many interesting interpretations and predictions on intermittency, we would like to understand more clearly what a multifractal field is. This is a non-trivial question which we want to address in detail in the next section. Here we would like to reformulate the multifractal theory in a somewhat simpler way using the so called multiplicative process. Our starting point is the basic definition of the multifractal hypothesis which, for any $\eta < r < R < L$, can be written as:

$$\delta u(r) = \left(\frac{r}{R}\right)^h \delta u(R) \quad P_h(r,R) \sim \left(\frac{r}{R}\right)^{(3-D(h))} \tag{56}$$

In order to clarify the meaning of (56), let us consider a set of scales $l_m = a^m L$ with $a < 1$ and let us define the corresponding velocity differences as $u_m = \delta u(l_m)$ with $u_0 = \delta u(L)$. For each m we can write

$$u_m = a^h u_{m-1} = \Pi_{k=1..m} \; a^{h(k)} u_0 \tag{57}$$

where each $h(k)$ is chosen with the probability $a^{3-D(h)}$. Expression (57) can be rewritten by saying that u_m is the product of m random variables with the same probability distribution. Using a more general formalism, we can say that

$$u_m = \Pi_{k=1..m} \; A_k \tag{58}$$

where the variables A_k are distributed according to the same probability distribution $P(A)$. Let us now make the (strong) assumption that the variables A_k are also independently distributed. Then, one can immediately write:

$$\langle (u_m)^n \rangle \sim \langle A^n \rangle^m \sim a^{m \log\langle A^n \rangle / \log a} \sim l_m^{\zeta(n)} \quad \zeta(n) = -\frac{\log(\langle A^n \rangle)}{\log(a)} \tag{59}$$

Although (57) is one of the possible interpretations of (56), it is interesting to observe that it gives us a simple way to "build" a multifractal variable.
Expression (57) defines a random multiplicative process. This way of interpreting the multifractal theory is somewhat reminiscent of the Richardson cascade: during the process of energy transfer from one scale to the other, the energy transfer is fluctuating in a self similar way, i.e. with the same scale-independent probability distribution. By dimensional analysis we can even estimate the flux of energy transfer as $u_m^3/l_m \sim a^m \Pi_{k=1..m} \; A_k^3$. The requirement of the probability

distribution $P(A)$ is such that $\zeta(3) = 1$. One can easily define a simple model for $P(A)$ as follows:

$$a = \frac{1}{2} \quad P(A) = p\delta(A - a^{0.2}) + (1-p)\delta(A - a^{0.8}) \quad p = 0.66 \tag{60}$$

which gives values of $\zeta(n)$ in reasonable agreement with experimental values for $n \leq 10$.

The random multiplicative process provides us with an interesting interpretation for the function $D(h)$ based on large deviation theory. Let us consider expression (59) again. We may write $\langle u_m^n \rangle$ as:

$$\langle u_m^n \rangle \sim \Pi_{k=1..m} \; A_k^n \sim \exp(\Sigma_{k=1,..m} n \log(A_k)) \sim \exp(mn\Phi_m) \quad \Phi_m = \frac{1}{m}\Sigma_{k=1,..m} \log(A_k) \tag{61}$$

One can naively think that Φ_m, being the average of $\log(A_k)$, must be independent of m. This is clearly in contradiction to (59). The reason of this apparent paradox is that the variable Φ_m is a random variable with exponentially small probability to exhibit large deviations with respect to the average value of $\langle \log(A_k) \rangle$. More precisely one can write the general estimate:

$$Prob(\Phi_m = x) \sim \exp(-mS(x)) \tag{62}$$

where the function $S(x)$, called the Cramer entropy, is nonnegative and becomes 0 for $x = \langle \log(A_k) \rangle$. By using (62) we obtain:

$$\langle u_m^n \rangle \sim \langle \exp(mnx - mS(x)) \rangle \sim \exp(mH(n)) \quad H(n) = inf_x(nx - S(x)) \tag{63}$$

which is formally equivalent to (34). The above discussion highlights a possible interpretation of the multifractal theory as "large deviation" theory for turbulence fluctuations. In this case, it is not strictly demanded that the function $D(h)$ should be interpreted as a geometrical (fractal) dimension.

References for sections 2 and 3

A pedagogical introduction to intermittency in fully developed turbulence and to shell models can be found in the two books:

- U. Frisch, *Turbulence*, Cambridge University Press, Cambridge, 1995.
- T. Bohr, M. H. Jensen, G. Paladin and A. Vulpiani, *Dynamical System Approach to Turbulence*, Cambridge University Press, Cambridge, 1997.

A more detailed discussion of all the issues described in this section can be found in the following list of papers:

- A. N. Kolmogorov, " A refinement of previous hypothesis concerning the local structure of turbulence in a viscous incompressible fluid at high Reynolds number" *J. Fluid Mech.* **62**, 82 (1962).
- G. Parisi and U. Frisch, "On the singularity structure of fully developed turbulence" in *Turbulence and predictability of geophysical fluid dynamics*, ed. by M. Ghil, R. Benzi and G. Parisi, North-Holland, Amsterdam, 1985, p. 84.

- R. Benzi, G. Paladin, G. Parisi and A. Vulpiani, "On the multifractal nature of fully developed turbulence and chaotic systems" *J. Phys. A* **17**, 3521 (1984).
- R. Benzi, L. Biferale, G. Paladin, M. Vergassola and A. Vulpiani, Multifractality in the statistics of the velocity gradients in turbulence *Phys. Rev. Lett.* **67**, 2299 (1991).
- U. Frisch and M. Vergassola, Intermediate dissipative range in turbulence. *Europhys. Lett.* **14**, 439 (1991).
- C. M. Meneveau and K. R. Sreenivasan, The multifractal spectrum of the dissipation field in turbulent flows *Nucl. Phys. B Proc. Suppl.* **2**, 49 (1987).
- M.H. Jensen, G. Paladin and A. Vulpiani, Intermittency in a cascade model for 3-dimensional turbulence *Phys. Rev. A* **43**, 798-805 (1991).
- L. Biferale, A. Lambert, R. Lima and G. Paladin, Transition to chaos in a shell model of turbulence *Physica D* **80**, 105 (1995).
- L. Kadanoff, D. Lohse, J. Wang, R. Benzi, Scaling and dissipation in the GOY shell model *Phys. Fluids* **7**, 617 (1995).
- V. L'vov, E. Podivilov, A. Pomyalov, I. Procaccia and D. Vandembroucq, Improved shell model of turbulence. Phys. Rev. E. **58**, 1811 (1998).

4 Multifractal Fields and the Navier-Stokes Equations

This is the most difficult section of our lecture. Our aim in this section is to understand how the multifractal theory of turbulence can be "fixed" by using the deterministic equations of motion, i.e. how to compute $D(h)$ by using the equation of motion.

We start by generalizing the random multiplicative process in order to define a multifractal field. We will answer this question for a one dimensional field, in order to keep the notation as simple as possible. Our technical device to define a one dimensional "random" multifractal field is the wavelet decomposition. It is not necessary to have a deep mathematical knowledge of wavelets in order to follow our technique. In a nutshell, wavelet decomposition allows us to write any function $f(x)$ in the interval $[0,1]$ as the superposition of infinite basic functions $\psi_{j,k}(x)$ which satisfy the following properties:

(a) the function $\psi_{j,k}(x)$ is concentrated near the point $x \sim k/2^j$;

(b) the Fourier transform of the function $\psi_{j,k}(x)$ is concentrated near wavenumber $q \sim 2^j$;

(c) the wavelets obey the orthogonality conditions $\int \psi_{j,k}(x)\psi_{j',k'}(x)dx = \delta_{j,j'}\delta_{k,k'}$.

The general theory of wavelet representation tells us that for any $f(x)$ we can write:

$$f(x) = \Sigma_j \Sigma_k \alpha_{j,k} \psi_{j,k}(x) \tag{64}$$

where

$$\psi_{j,k}(x) = 2^{j/2}\psi(2^j x - k) \tag{65}$$

and $\psi(x)$ is a (suitable) basis function with zero mean in $[0, 1]$. For our purpose we do not need any particular form of $\psi(x)$ and we do not even need, in most cases, the orthogonality condition. Note, however, that the orthogonality condition is mandatory if one wants to decompose a function $f(x)$ by using (64).

We can think of the wavelet as a gaussian shaped function whose spread is proportional to 2^{-j} and located at the points $x \sim k/2^j$. For instance, one can choose the function

$$\psi(x) = -\frac{d^2}{dx^2} \exp\left(-\frac{x^2}{2s^2}\right) \quad s \sim 2^j \tag{66}$$

The coefficients $\alpha_{j,k}$ form a dyadic structure. On the top there is $\alpha_{0,0}$, the next line is formed by $\alpha_{1,0}$ and $\alpha_{1,1}$, the third line is formed by 4 numbers $\alpha_{2,i}$ $i = 0, 1, 2, 3$ and so on.

We can now generalize the random multiplicative process in the following way. Let us denote with A a random variable with probability distribution $P(A)$ and let us define another random variable, B, which assumes with equal probability the values ± 1. Next we define:

$$\alpha_{0,0} = 1 \quad \alpha_{1,0} = B_{1,0} A_{0,0} \alpha_{0,0} \quad \alpha_{1,1} = B_{1,1} A_{1,1} \alpha_{0,0} \tag{67}$$

$$\alpha_{2,0} = B_{2,0} A_{2,0} \alpha_{1,0} \quad \alpha_{2,1} = B_{2,2} A_{2,1} \alpha_{1,0} \quad \alpha_{2,2} = B_{2,1} A_{2,2} \alpha_{1,1} \quad \alpha_{2,3} = B_{2,3} A_{2,3} \alpha_{1,1} \tag{68}$$

At each level j in the dyadic structure we have:

$$\langle |\alpha_{j,k}|^n \rangle = \langle A^n \rangle \langle |\alpha_{j-1,k}|^n \rangle = 2^{j \log_2(\langle A^n \rangle)} \langle |\alpha_{0,0}|^n \rangle \tag{69}$$

which is the estimate given by the multiplicative random process. Our aim is to show that the function $f(x)$ satisfies

$$\langle |f(x+r) - f(x)|^n \rangle \sim r^{\zeta(n)}, \; \zeta(n) = -\frac{1}{2} n - \log_2(\langle A^n \rangle) \tag{70}$$

In order to obtain our result, let us consider the second order structure function:

$$S_2(r) = \langle (f(x+r) - f(x))^2 \rangle \tag{71}$$

where $\langle \bullet \rangle$ represents the space average. Using the wavelet decomposition we can write:

$$S_2(r) = \langle \Sigma_{j,k} (\alpha_{j,k} 2^{j/2} (\psi(2^j x + 2^j r - k) - \psi(2^j x - k)))^2 \rangle \tag{72}$$

Next we observe that the spatial average can be changed into an average over the random numbers $\alpha_{j,k}$, which are uncorrelated with zero mean. This observation allows us to write:

$$S_2(r) = \Sigma_{j,k} 2^j \langle \alpha_{j,k}^2 \rangle \langle (\psi(2^j x + 2^j r - k) - \psi(2^j x - k))^2 \rangle \tag{73}$$

We now introduce the function $G_2(r)$ defined as

$$G_2(r) = \int dx (\psi(x+r) - \psi(x))^2 \tag{74}$$

Using G_2 we can write S_2 in the form

$$S_2(r) = \Sigma_j 2^j \langle \alpha_{j,k}^2 \rangle G_2(2^j r) \tag{75}$$

where we use the fact that $\langle\alpha_{j,k}^2\rangle$ are independent of k and there are 2^j different values of k for a fixed j. The proof of the scaling exponents for $S_2(r)$ can now be completed. We have:

$$S_2(2r) = \Sigma_j 2^j \langle\alpha_{j,k}^2\rangle G_2(2^{j+1}r) = \Sigma_j 2^{j(1+\log_2(\langle A^2\rangle))} G_2(2^{j+1}r) \tag{76}$$

which gives:

$$S_2(2r) = 2^{-(1+\log_2(\langle A^2\rangle))} \Sigma_j 2^{(j+1)(\log_2(\langle A^2\rangle+1))} G_2(2^{j+1}r) = 2^{-(1+\log_2(\langle A^2\rangle))} S(r) \tag{77}$$

which implies (70) for $n = 2$. In the same way we can compute the structure function $S_4(r)$. After a long but simple computation we obtain:

$$S_4(r) = \Sigma_j 2^j \langle\alpha_{j,k}^4\rangle G_4(2^j r) + 3S_2^2(r) \tag{78}$$

where $G_4(r) = \int (\psi(x+r) - \psi(x))^4 dx$. After some computations, we finally get:

$$S_4(r) = c_4 r^{\zeta(4)} + c_2 r^{2\zeta(2)} \tag{79}$$

Because r is much smaller than 1, the first term on the r.h.s of (79) dominates the scaling behavior. The above argument can be generalized leading to the proof for the scaling exponents (70). In order to check that our mathematical theory is correct, we report a numerical simulation performed by using the probability distribution $P(A) = 0.125\delta(A - 2^{-1/2}) + 0.875\delta(A - 2^{-5/6})$. We use the theory outlined before with $N = 2^{16}$ level in the dyadic structure (i.e. the maximum number of j is 15), and as a wavelet function the one given in (66).

In Figure 6 we report the one realization of $f(x)$ in the interval $[0, 0.2]$, in Figure 7 the scaling behavior of $S_4(r)$ including the theoretical prediction, and finally in Figure 8 the exponents $\zeta(n)$ computed numerically for $n \leq 10$ also compared with the theoretical prediction. Altogether we see that the mathematical theory so far defined works rather well.

We want to stress two important points obtained by having a constructive (numerical) example of a multifractal field. First of all, one can check the quality of the scaling behavior against the statistics which is used either experimentally or numerically. Even in our well controlled case, we see that the estimate of $\zeta(10)$ is not good enough. Second, one can introduce dissipation effects, using (46), to simulate finite values of *Re* and to understand the effect of finite size scaling in the data analysis. This is an important point because no laboratory measurements can show exact scaling behavior, and any discussion of the relevance of scaling in fully developed theory might be very frustrating without a quantitative assessment of finite size effects. We shall return to this last remark in the next section. Starting with the results obtained for the multifractal field, we can generalize the shell model for the same dyadic structure used in the wavelet representation. This "tree" model can be regarded as a description of evolution of the coefficients of an orthonormal wavelet expansion of a one-dimensional projection of the velocity field $u(x,t)$:

$$u(x,t) = \sum_{n,k} u_{n,k}(t)\psi_{n,k}(x). \tag{80}$$

Each dynamical variable $u_{n,k}$ describes fluctuations in a box of length $l_n = 2^{-n}$, centered in the interval ranging from $(k-1)l_n$ to kl_n. At each scale n there are 2^{n-1} boxes, covering a total length $\Lambda_T = 2^{n-1}l_n = 1/2$ (see Figure 9). For the sake of convenience we define the tree model

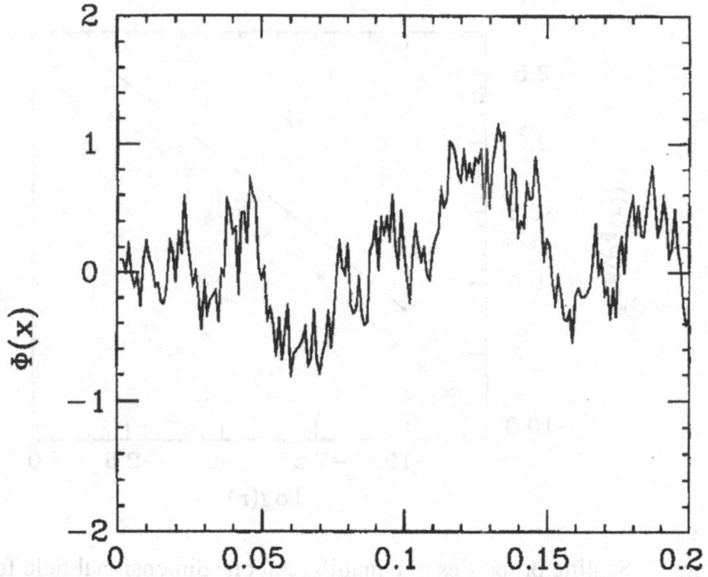

Figure 6: An example of realization of a multifractal one dimensional field obtained by a random multiplicative process.

in terms of *density* variables, $u_{n,k}$, which would correspond to $2^{n/2}u_{n,k}$ in a wavelet expansion. In this notation, $|u_{n,k}|^2$ represents the energy density in a flow structure of length $l_n = 2^{-n}$ and is spatially labeled by the index k. In this tree structure, each variable $u_{n,k}$ continues to interact with the nearest and next nearest levels, as in equation (28); however, a lot of possibilities are now opened by the presence of many horizontal degrees of freedom localized on each shell. The simplest choice is depicted in Figure 10 , where a portion of the tree structure is shown and the variables $u_{n,j}$ are represented by a black ball. In the figure, solid lines connect interacting balls (variables). The dynamical tree equations are as follows:

$$\begin{aligned}\dot{u}_{n,k} &= -D_n u_{n,j} + \delta_{n,n_0} F + \\ &+ik_n \left\{ \tfrac{a}{4} \left[u_{n+1,2k-1} \left(u_{n+2,4k-3} + u_{n+2,4k-2} \right) + u_{n+1,2k} \left(u_{n+2,4k-1} + u_{n+2,4k} \right) \right] + \right. \\ &\left. + \tfrac{b}{2} \left[u_{n-1,\bar{k}} \left(u_{n+1,2k-1} + u_{n+1,2k} \right) \right] + c \left[u_{n-2,\bar{\bar{k}}}\, u_{n-1,\bar{k}} \right] \right\}^*\end{aligned}$$

where, in the indices, $\bar{\bar{k}}$ is the integer part of $\left(\frac{k+3}{4}\right)$ and $\bar{k}$ is the integer part of $\left(\frac{k+1}{2}\right)$.

The dissipation acts through the term $D_n = \nu k_n^2$. The interaction terms with coefficients $a/4$, $b/2$ and c are depicted in Figure 10,a,b,c respectively. The numerical values of a, b and c are the same as in the original shell models (see section 2).

In a tree structure we can define a 1-dimensional surrogate of the spatial energy dissipation field. The first step in constructing the energy dissipation field in any tree model is to consider

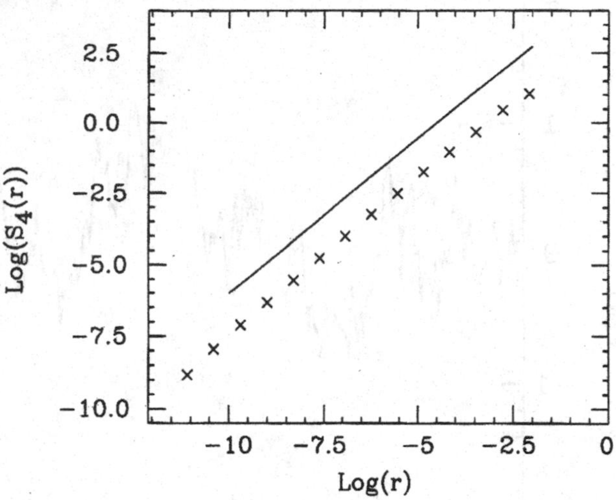

Figure 7: Scaling properties of a multifractal one dimensional field for S_4.

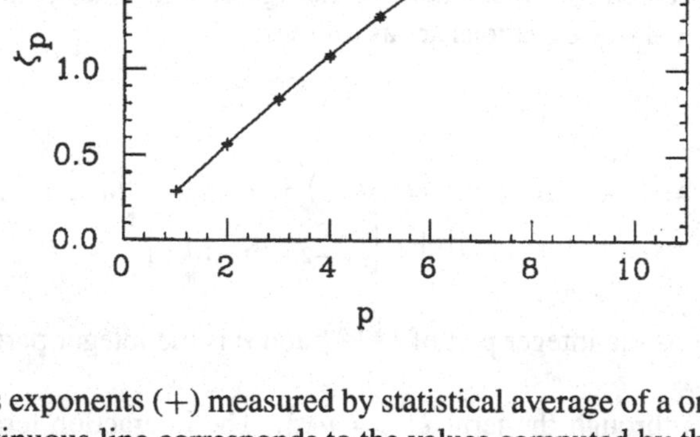

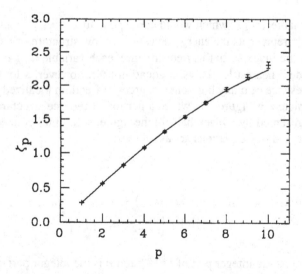

Figure 8: Anomalous exponents (+) measured by statistical average of a one dimensional multifractal field. The continuous line corresponds to the values computed by the random multiplicative process.

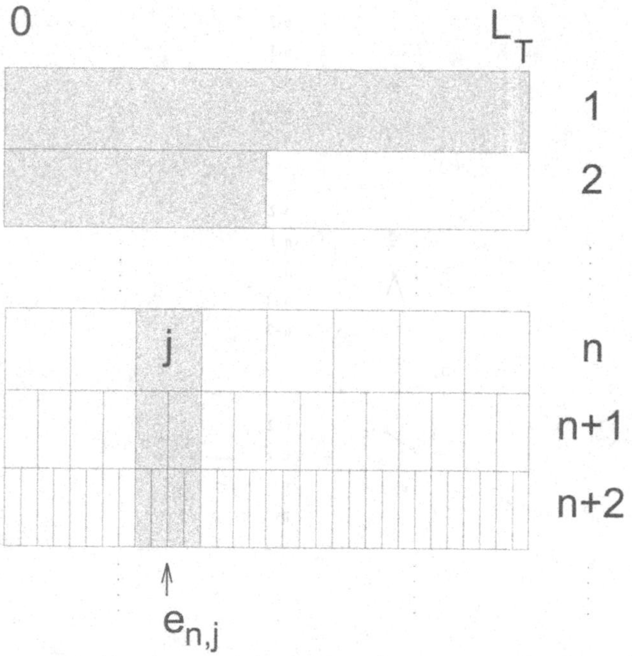

Figure 9: Definition of the space scale properties of the energy dissipation in a tree model.

the energy dissipation *density*, $\eta_{n,k}$, in the structure covering the region $\Lambda_k(n)$ of length 2^{-n}, centered in the spatial site labeled by k. These structures are represented by boxes in Figure 9:

$$\eta_{n,k} = D_{n,k}\left(|u_{n,k}|^2 + |u_{n,k}|^2\right). \tag{82}$$

The total energy dissipation density, $\epsilon = (1/\Lambda_T)\int_{\Lambda_T}\epsilon(x)\,dx$, where Λ_T is the total space length, is, by definition, the sum of all these contributions (sum over boxes at all scales in Figure 9):

$$\epsilon = \sum_{n,j} 2^{-n}\eta_{n,j}. \tag{83}$$

On the other hand, in order to define a local energy dissipation field, one has to disentangle in ϵ the contributions coming from the coarse-grained energy dissipation field ϵ_l. We can then rewrite our formulation:

$$\epsilon = \frac{1}{\Lambda_T}\int_{\Lambda_T}\epsilon(x)\,dx = \frac{1}{2^{n-1}}\sum_{k=1}^{2^{n-1}}\left(\frac{1}{2^{-n}}\int_{\Lambda_k(n)}\epsilon(x)\,dx\right) = \frac{1}{2^{n-1}}\sum_{k=1}^{2^{n-1}}\epsilon_{n,k}, \tag{84}$$

where the last expression is independent of n and the $\epsilon_{n,k}$'s are the coarse-grained energy dissipation densities, obtained as averages over spatial regions of length 2^{-n}. Note that the average

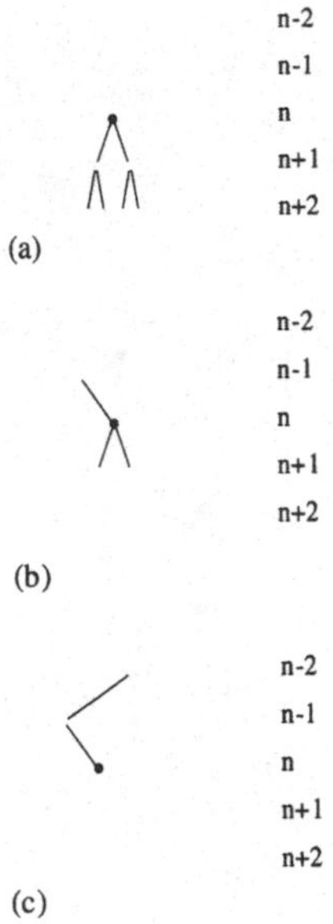

Figure 10: An example of three possible scale-scale interactions for a tree model.

density $\epsilon_{n,k}$ over $\Lambda_k(n)$ does not coincide simply with the density $\eta_{n,k}$ of the structure living in $\Lambda_k(n)$, namely:

$$\epsilon_{n,k} = \eta_{n,k} + \sum_{m \langle n} \eta_{m,K(m)} + \sum_{m \rangle n} \langle \eta_{m,K(m)} \rangle_{I(m)} \,. \tag{85}$$

Here, in the second (third) term of the right hand side we take into account density contributions coming from larger (smaller) scale structures (as an example, all regions contributing to the definition of $\epsilon_{n,k}$ are represented as shadowed boxes in Figure 9).

The index $K(m)$ in the second term on the right hand side labels the location of larger scale structures containing the region $\Lambda_k(n)$ under consideration (shadowed boxes with $m < n$ in Figure 9). In the third term, an average is performed over $K(m) \in I(m)$, where $I(m)$ labels the set of structures contained in $\Lambda_k(n)$, for any $m \rangle n$ (in Figure 9 $I(m)$ labels the two boxes at $n+1$, the four boxes at $n+2$, and so on).

In order to be as clear as possible, let us consider the simplest case of a three level model. In this case we have

$$\epsilon = \epsilon_{1,1} = \frac{1}{2}(\epsilon_{2,1} + \epsilon_{2,2}) = \frac{1}{4}(\epsilon_{3,1} + \dots + \epsilon_{3,4}) \tag{86}$$

By using (85), we have:

$$\epsilon_{2,1} = \eta_{2,1} + \eta_{1,1} + \frac{1}{2}(\eta_{3,1} + \eta_{3,2}) \tag{87}$$

$$\epsilon_{2,2} = \eta_{2,2} + \eta_{1,2} + \frac{1}{2}(\eta_{3,3} + \eta_{3,4}) \tag{88}$$

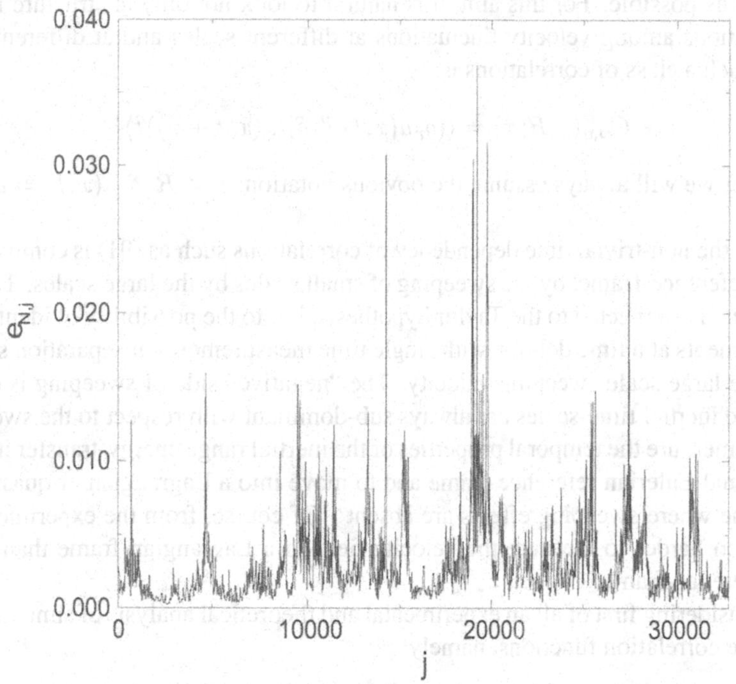

Figure 11: Instantaneous behavior of the energy dissipation in the tree shell model of turbulence

Finally, by using (86) we obtain:

$$\frac{1}{2}(\epsilon_{2,1} + \epsilon_{2,2}) = \eta_{1,1} + \frac{1}{2}(\eta_{2,1} + \eta_{2,2}) + \frac{1}{4}(\eta_{3,1} + \ldots + \eta_{3,4}) \equiv \epsilon_{1,1} \quad (89)$$

in agreement with (84).

The best spatially resolved energy dissipation field is for $n = N$:

$$\epsilon_{N,j} = \sum_{m \leq N} \eta_{m,K(m)}; \quad k = 1, \ldots, 2^{N-1}. \quad (90)$$

In Figure 11 an example of the instantaneous values assumed by $\epsilon_{N,k}$ in the $N_T/2 = 32768$ locations of the last level is shown.

The chaotic, intermittent character of this spatial signal is evident.

The tree shell model enables us to verify the validity of the refined Kolmogorov similarity (24). Let us remark that such a relationship is build neither for the multifractal signal nor for the tree shell model. A direct test of the structure function of order 9 gives an extremely good result with deviations of less than 1 *percent* against the theoretical prediction made by (24).

We have seen how to define a multifractal field in a constructive way, at least in one dimension. Now we want to exploit our understanding of intermittency in terms of the multifractal field in order to characterize the statistical properties of small scale velocity fluctuations in turbulent flows as much as possible. For this aim, it is natural to look not only at structure functions but also at correlations among velocity fluctuations at different scales and at different times. The prototype of such a class of correlations is:

$$C_{p,q}(r,R;\tau) = \langle (\delta_r u(x,t))^p (\delta_R u(x,t+\tau))^q \rangle \tag{91}$$

where hereafter we will always assume the obvious notation: $r < R$, $\delta_r u(x,t) = u(x+r,t) - u(x,t)$.
Unfortunately, the non-trivial time dependency of correlations such as (91) is completely hidden, in a Eulerian reference-frame, by the sweeping of small scales by the large scales. The "positive" side of sweeping is connected to the Taylor hypothesis, i.e. to the possibility of identifying single point measurements at a time delay τ with single time measurements at separation scale $r \sim \tau\bar{V}$ where $\bar{V}$ is the large scale sweeping velocity. The "negative" side of sweeping is connected to the fact that the inertial time-scales are always sub-dominant with respect to the sweeping time.

In order to measure the temporal properties of the inertial range energy-transfer it is necessary to leave the usual Eulerian reference-frame and to move into a Lagrangian or quasi-Lagrangian reference-frame where sweeping effects are absent . Of course, from the experimental point of view, it is much harder to measure the velocity field in a Lagrangian frame than in the usual laboratory reference-frame.
We start by considering first of all an experimental and theoretical analysis of simultaneous single time multiscale correlation functions, namely

$$F_{p,q}(r,R) = \langle (\delta_r u(x,t))^p (\delta_R u(x,t)^q \rangle \tag{92}$$

with $\eta < r < R < L$. In order to simplify our discussion, we will confine our analysis to the case of longitudinal collinear velocity differences.

Our starting point is, as before, the multifractal theory of turbulence as discussed so far. In particular within the framework of random multiplicative processes we can write:

$$\delta_r u(x) = W(r,R) \cdot \delta_R u(x) \tag{93}$$

where, $W(r,R)$ is the random multiplier. Requiring homogeneity along the cascade process, the random function W should depend only on the ratio r/R. Structure functions are then described in terms of the W process:

$$S_n(r) \sim \langle [W(r/L)]^n \rangle,$$

if the stochastic multiplier may be considered almost uncorrelated with the large-scale velocity field. Pure power laws arise in the high Reynolds regime: in this limit we must have

$$\langle (W(r,R)^n \rangle \sim (r/R)^{\zeta(n)}.$$

In the same framework, it is straightforward to compute the leading term of the multiscale correlation functions (92):

$$F_{p,q}(r,R) \sim \langle W(r,L)^p W(R,L)^q \rangle \sim C_{p,q} \langle W(r,R)^p W(R,L)^p W(R,L)^q \rangle \tag{94}$$

which, in the hypothesis of negligible correlations among multipliers, becomes:

$$F_{p,q}(r,R) = C_{p,q}\langle W(r,R)^p\rangle\langle W(R,L)^{p+q}\rangle \sim \frac{S_p(r)}{S_p(R)}\cdot S_{p+q}(R) \quad (95)$$

These expressions are named "fusion rules". We should emphasize that fusion rules give the leading behavior of (92) when $r/R \to 0$. Let us notice that expression (95) is also the zero-*th* order prediction starting from any multiplicative uncorrelated random cascade satisfying $S_p(r) = \langle [W(r,R)]^p\rangle S_p(R)$. Let us also stress that the fusion rules prediction as stated in (95) does not necessarily requires any scaling property of the underlying structure functions, a fact which suggests that the validity of the statement should be almost Reynolds independent.
We also want to stress that fusion rules require some care if applied to any p, q. For instance, let us consider the following correlation:

$$F_{1,q}(r,R) = \langle (\delta_r u(x,t))^1 (\delta_R u(x,t)^q\rangle \quad (96)$$

The multiplicative prediction gives:

$$F_{1,q}(r,R) = \frac{S_1(r)}{S_1(R)}\cdot S_{1+q}(R).$$

Such a prediction is wrong because, if homogeneity can be assumed, $S_1(r) \equiv 0$ for all scales r. In this case prediction (95) cannot represent the leading contribution.

In order to check the validity of the fusion rules we can analyze experimental data. Moreover, the quality of the results should be compared against the same data analysis performed on a multifractal signal, obtained by the mathematical procedure discussed at the beginning of this section. We refer to the multifractal signal with the name "synthetic turbulence".

Experimental data sets have been obtained in a wind tunnel (Modane) with $Re_\lambda = 2000$. The integral scale was $L_0 \sim 20\,m$ and the dissipative scale was $\eta = 0.31\,mm$. A second data set comes from a recirculating wind tunnel (ENS de Lyon) with a working section $3m$ long with a cross section of $50cm \times 50cm$. Values of Re_λ realized in the experiments were 400 (wake behind a cylinder) and 800 (jet turbulence). Integral scales were $0.1m$ and $0.125m$, respectively, whereas the dissipative scales were $0.15mm$ and $0.1mm$.
We proceed with a simple but basic observation.
Notice that for any 1-dimensional string of numbers (such as the typical outcome of laboratory experiments in turbulence) the multiscale correlations (92) feel unavoidable strong geometrical constraints. In particular, we can always write down a set of identities which will be referred to as "Ward-Identities" (WI):

$$S_p(R-r) \sim \langle (((u(x+R)-u(x)) - (u(x+r)-u(x)))^p\rangle = \sum_{k=0}^{p}(-1)^k\binom{p}{k}F_{k,p-k}(r,R), \quad (97)$$

For example, for $p = 2$ we have

$$2F_{1,1}(r,R) \sim S_2(r) + S_2(R) - S_2(R-r) \sim \left(\left(\frac{r}{R}\right)^{\zeta(2)} + O\left(\frac{r}{R}\right)\right)\cdot S_2(R) \quad (98)$$

where the latter expression has been obtained by expanding $S_2(R-r)$ in the limit $r/R \to 0$. For $p=3$ we have

$$S_3(R-r) = S_3(R) - S_3(r) + 3F_{2,1}(r,R) - 3F_{1,2}(r,R)$$

"Ward-Identities" can be useful for understanding sub-leading corrections to the multiplicative cascade process. One may argue that in a geometrical set-up different from the one specified in (92) the same kind of constraint will appear with eventually different weights among different terms.

The most important result one can extract from (97) is that the multiscale correlation functions, as stated in (92), may not be perfect scaling functions even in the limit of very high Reynolds numbers.

Our major result is that all multiscale correlations functions are well reproduced in their leading term, $r/R \to 0$, by a simple uncorrelated random cascade (95). The recipe for calculating multiscale correlations is the following: first, apply the multiplicative guess for the leading contribution and look for geometrical constraints in order to find out sub-leading terms. Second, in all cases where the leading multiplicative contribution vanishes because of underlying symmetries, look directly for the geometrical constraints and find out what the leading contribution applying the multiplicative random approximation to all, non-vanishing, terms in the WI is.

Let us check the fusion rules prediction (95) for even moments $p, q = 2, 4, \ldots$.

In order to better highlight the scaling properties we will often use in the following the compensated fusion rule $\tilde{F}_{p,q}(r,R)$ defined as:

$$\tilde{F}_{p,q}(r,R) = \frac{F_{p,q}(r,R) \cdot S_p(R)}{S_p(r) \cdot S_{p+q}(R)} \tag{99}$$

In order to compare experiments with different Reynolds numbers we may use as independent variable in our plot the quantity: $x(R) \equiv \frac{R-r}{L-r}$, where with L we denote the integral scale of each different experiment. In this way, by fixing the small scale $r = 5\eta$ and by changing $r \leq R \leq L$ for each set of data we have a variation of $0 \leq x(R) \leq 1$.

In Figure 12 one can see the compensated correlation functions for two different sets of moments. In the limit of large separation $R \to L_0$ at fixed r, we indeed see a tendency toward a plateau. On the other hand, there are clear deviations for $r/R \sim \mathcal{O}(1)$.

In order to understand the physical meaning of the observed deviations to the fusion rules (95), we compare, in Figure 13, the experimental data against the equivalent quantities measured by using a 1-dimensional synthetic signal.

We notice an almost perfect superposition of the two data sets, indicating that the deviations observed in real data may not be considered a "dynamical effect".

The result so far obtained, i.e. that both the experimental data and the synthetic signal show the same quantitative behavior, is a strong indication that multiscale correlation functions, at least for even order moments, are in good agreement with the random multiplicative model.

An even stronger proof of this statement comes from the analysis of multiscale correlations in terms of the coefficients obtained by a wavelet analysis of the experimental signal.

The wavelet coefficient $\alpha_{j,k}$ may be seen as the representative of a velocity fluctuation at scale

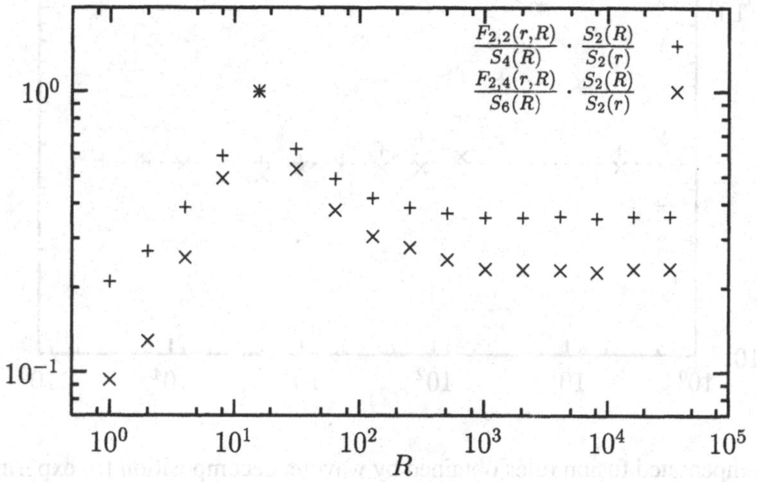

Figure 12: Compensated fusion rules obtained for the data analysis discussed in the text.

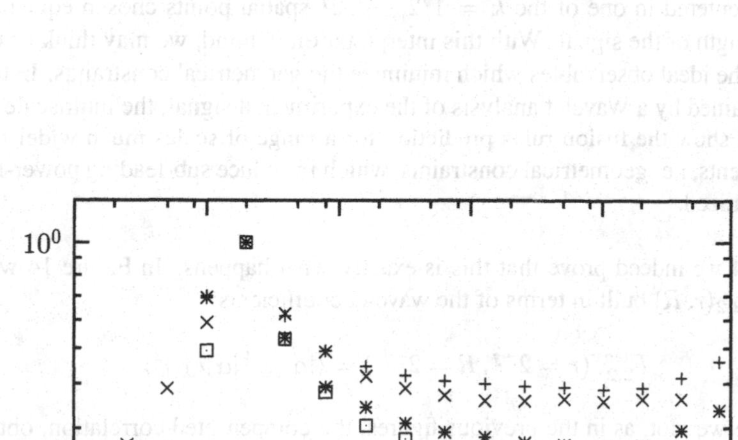

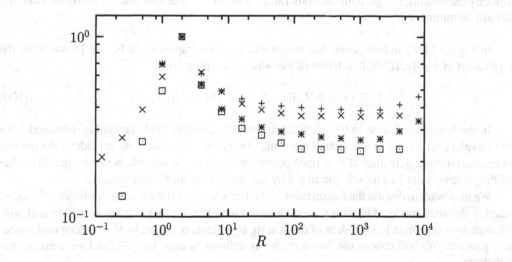

Figure 13: Same as in Figure 12 for experimental data and for a synthetic one dimensional multifractal field.

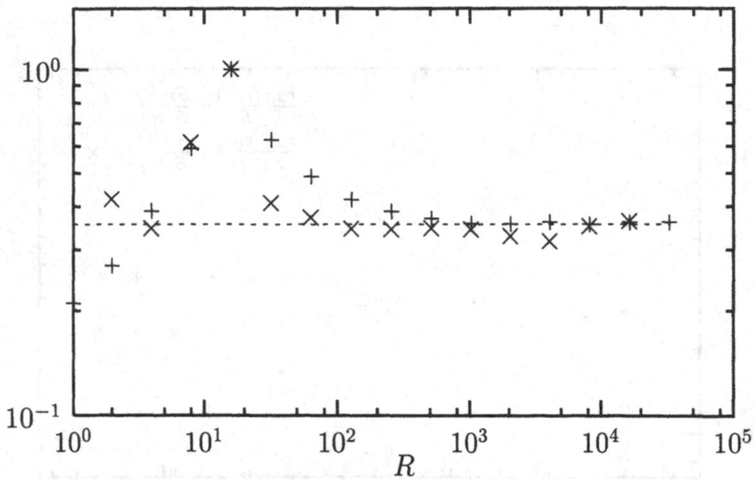

Figure 14: Compensated fusion rules obtained by wavelet decomposition for experimental data.

$r = 2^{-j}$ and centered in one of the $k = 1, 2, \cdots, 2^j$ spatial points chosen equispaced in the original total length of the signal. With this interpretation in mind, we may think of the wavelet coefficients as the ideal observables which minimize the geometrical constraints. In terms of the coefficients obtained by a Wavelet analysis of the experimental signal, the multiscale correlation function should show the fusion rules prediction for a range of scales much wider than for the velocity increments, i.e. geometrical constraints, which introduce sub-leading power-law decays, should be minimized.

In Figure 14 we indeed prove that this is exactly what happens. In Figure 14 we show the equivalent of $F_{2,2}(r, R)$ built in terms of the wavelet coefficients:

$$F_{2,2}^{wav}(r = 2^{-j}, R = 2^{-j'}) = \langle |\alpha_{j,k}|^2 \, |\alpha_{j',k'}|^2 \rangle \tag{100}$$

In the figure we plot, as in the previous figures, the compensated correlation, obtained from the wavelet coefficients, at fixed small scale and at varying large scales. As is evident, the plateau is reached immediately after almost one fragmentation step. We remark, however, that the value of the plateau is not 1 as naively predicted by the random multiplicative process.

We now want to discuss the fusion rules in the framework of the simplest "turbulence" model, namely the shell model. Our aim is to understand whether the fusion rules are consistent with the equation of motion independent of the scaling arguments provided by the random multiplicative process. We will choose the Sabra model as defined by equation (32) and we consider the equations:

$$\frac{d}{dt} \langle |u_n|^2 |u_m|^2 \rangle = 0 \tag{101}$$

in the limit of zero viscosity, i.e. neglecting dissipation effects. Let us introduce the variables $Q_n = u_n^2$ and $Z_n = Im(u_{n+1}^* u_n u_{n-1})$. Inserting (32) into equation (101), we obtain after a long but simple computation:

$$f(n,m) + f(m,n) = 0 \tag{102}$$

where

$$f(n,m) = k_{n+1}\langle Q_m Z_{n+1}\rangle + bk_n\langle Q_m Z_n\rangle - (1+b)k_{n-1}\langle Q_m Z_{n-1}\rangle \tag{103}$$

Notice that $f(n,m) = \langle Q_m dQ_n/dt\rangle$. Let us now consider $n = m + p$ with $p \gg 1$. Because $k_n = \lambda^n$ with $\lambda = 2$, we can neglect all terms proportional to $k_m \ll k_n$. In this limit, the equations (101) become $f(n,m) = 0$. We warn the reader that such a "naive" result is a first order approximation. Indeed, subtle dynamical properties must introduce cancellation between the three terms contributing to $f(n,m)$ when $n \gg m$. Each of the three terms has a leading contribution given by its fusion rule prediction. The sum of the three leading contributions must add to zero if one wants to satisfy (102). Therefore, the most useful way to use this dynamical information is to find a parameterization of the $\langle Q_m Z_n\rangle$ correlation function such that its asymptotic behavior satisfies its fusion rules prediction but with a detailed dependency on the scales such that the constraint $f(n,m) = 0$ when $n \gg m$ is satisfied.

We next consider the quantity:

$$R_n = \frac{\langle Q_m Z_n\rangle}{\langle Q_m Z_{n-1}\rangle} \tag{104}$$

In terms of R_n the equations $f(n,m) = 0$ reads:

$$k_{n+1}R_{n+1}R_n + bk_nR_n - (1+b)k_{n-1} = 0 \tag{105}$$

which can be solved as an iterative equation for R_n giving the fixed point solution $R_* = \lambda^{-1}$. Thus we have shown that

$$\langle Q_m Z_{m+p}\rangle \sim_{p\gg 1} k_p^{-1}\langle Q_m Z_m\rangle. \tag{106}$$

Equation (106) is nothing but the fusion rules applied to the quantity $\langle Q_m Z_{m+p}\rangle$. One could think that this result is a consequence of the Kolmogorov equation, i.e. $\zeta(3) = 1$. However, by considering more complex quantities like $d\langle Q_m Q_n Q_p ...\rangle/dt$ the final result does not seem to be different, i.e. we always recover the fusion rules as an exact result when the scale separation, among different quantities, becomes very large. The situation becomes much more complex and interesting when we consider equation (101) for small scale separation. To explain this point, let us consider the case $m = n$. In this case we obtain:

$$k_{n+1}\langle Q_n Z_{n+1}\rangle + bk_n\langle Q_n Z_n\rangle - (1+b)k_{n-1}\langle Q_n Z_{n-1}\rangle = 0 \tag{107}$$

If we apply the fusion rules to this equation we obtain:

$$k_{n+1}C_1\lambda^{-1}S_5(n) + bk_nS_5(n) - (1+b)k_{n-1}\lambda^{-\zeta(2)}D_1\lambda^{\zeta(5)}S_5(n) = 0 \tag{108}$$

where we have defined:

$$S_5(n) = \langle Q_n Z_n\rangle \tag{109}$$

$$\langle Q_{n+p}Z_n\rangle = D_p\lambda^{-\zeta(2)}S_5(n) \quad \langle Z_{n+p}Q_n\rangle = C_p\lambda^{-1}S_5(n) \tag{110}$$

where the terms C_p and D_p take into account deviations from a perfect fusion rule like the one observed in the data analysis. By looking at equation (108), we immediately understand that anomalous exponents (intermittency) may be present only if $C_1 \neq 1$ and $D_1 \neq 1$. If $C_1 = 1$ and $D_1 = 1$ we get:

$$1 + b - (1+b)\lambda^{-1-\zeta(2)+\zeta(5)} = 0 \quad \rightarrow \zeta(5) = \zeta(2) + 1 \tag{111}$$

which implies $\zeta(n) = n/3$ as in the Kolmogorov theory. Thus we reach the important result that the detailed way by which the fusion rules are satisfied (i.e. in our case D_p and C_p) is the key issue when intermittency is described, in a quantitative way, by the equation of motion. We can also say that the multifractal theory must be "dressed" by some dynamical information (i.e. the equation of motion) if we want to take into account intermittency in a more quantitative way. Notice that the random multiplicative process, at least in the way it has been defined so far, does not tell us anything about D_p or C_p. We now want to obtain more information on the way the dynamics affect the fusion rules. In order to do that we need to understand the difficult subject of correlations in space *and in time*, i.e. correlations like (91).

We start with the simplest case: two velocity fluctuations, both at scale R but separated by a time delay t. As long as the time separation t is smaller than the "instantaneous" eddy-turn-over time of that scale, τ_R, we expect, in the language of random multiplicative process, that $W_{R,L_0}(t) \sim W_{R,L_0}(0)$. On the other hand, we also expect that the two scale fluctuations should be almost uncorrelated for time larger than τ_R. Considering that the eddy-turn-over time at scale R is itself a fluctuating quantity $\tau_R \sim R/(\delta_R u) \sim R^{1-h}$ we may write down:

$$C^{p,q}(R,R|t) = \langle \delta_R u(0) \delta_R u(t) \rangle \sim \int d\mu(h) \left(\frac{R}{L_0}\right)^{h(p+q)} E_{p,q}\left(\frac{t}{\tau_R}\right) \tag{112}$$

where the time-dependency is hidden in the function $E_{p,q}(x)$ which must be a smooth function of its argument (for example a decreasing exponential).
From (112) it is straightforward to realize that at zero-time separation we recover the usual structure function representation. It is much more interesting to notice that (112) is also in agreement with the constraints imposed by the non-linear part of the Navier-Stokes equations. Indeed, we may say that only under the hypothesis that non-linear terms are dominated by local interactions in the Fourier space we can safely assume that, as far as power-law counting is concerned, the inertial terms of Navier-Stokes equations for the velocity difference $\delta_R u$ can be estimated to be of the form:

$$\partial_t \delta_R u(t) \sim \mathcal{O}\left[\frac{(\delta_R u(t))^2}{R}\right] \tag{113}$$

and therefore we may check that:

$$\partial_t C^{p,q}(R,R|t) \sim \int d\mu(h) \left(\frac{R}{L_0}\right)^{h(p+q)} (\tau_R)^{-1} E'_{p,q}\left(\frac{t}{\tau_R}\right) \sim \frac{C^{p+1,q}(R,R|t)}{R} \tag{114}$$

where of course in the last relation there is hidden the famous closure-problem of turbulence, now restated in term of the relation : $\frac{d}{dt} E_{p,q}(t) \sim E_{p+1,q}(t)$. Expression (112) has been obtained by using simple scaling arguments. We now want to generalize the idea of a multiplicative process in order to take into account time behavior. Our generalization will tell us whether (112) is consistent or not with the random multiplicative process. Let us, for the sake of simplicity,

introduce a hierarchical set of scales, $l_n = 2^{-n}L_0$ with $n = 0,\ldots,n_d$, which span the whole inertial range and let us simplify the notation by writing $u_n = \delta_r u$ in order to refer to a velocity fluctuation at scale $r = l_n$.
More precisely, we can perform a wavelet decomposition of the field of velocity differences in a quasi-Lagrangian frame of reference: then u_n stands, as usual, for a representative of the wavelet coefficients at the octave n.

The picture which will allow us to generalize the time-multifractal representation to the multi-scale multi-time case goes as follows.
For time-delays, $t \sim \tau_m$, typical of the eddy-turn-over time of the m-th scale we may safely say that the two velocity fluctuations follow the same fragmentation process from the integral scale L_0 down to the scale l_m while they follow two uncorrelated processes from scale l_m down to the smallest scale in the process, l_n. In the multifractal language we must write that for time $t \sim \tau_m$ we have:

$$u_n(0) \sim W'_{n,m}(0)W_{m,0}(0)u_0(0) \sim \left(\frac{l_n}{l_m}\right)^{h'}\left(\frac{l_m}{L_0}\right)^{h} u_0(0) \tag{115}$$

$$u_n(t) \sim W''_{n,m}(t)W_{m,0}(t)u_0(t) \sim W''_{n,m}(t)W_{m,0}(0)u_0(0) \sim \left(\frac{l_n}{l_m}\right)^{h''}\left(\frac{l_m}{L_0}\right)^{h} u_0(0) \tag{116}$$

where with $W, W', W'', \ldots$ we mean different independent outcomes of the cascade process with exponents h, h', h'' and where we have used the fact that in this time-window $W_{m,0}(t) \sim W_{m,0}(0)$.
Apart from subtle further time dependencies we should therefore conclude that for time $t \sim \tau_m$ the correlation functions may be approximated as:

$$C^{p,q}_{n,n}(\tau_m) \sim \langle W^p_{n,m}\rangle\langle W^q_{n,m}\rangle\langle W^{p+q}_{m,0}\rangle \tag{117}$$

which must be considered the fusion-rules prediction for the time-dependent fragmentation process. The expression (117) summarizes the idea that for time delays larger than τ_m but smaller than τ_{m-1}, velocity components with support on scales $r > l_{m-1}$ did not have enough time to relax and therefore the local exponent, h, which describes fluctuations on those scales must be the same for both fields. On the other hand, components with support on scales $r < l_{m-1}$ have already decorrelated for $t > \tau_{m-1}$ and therefore we must consider two independent scaling exponents h', h'' for describing fluctuations at these scales. Adding up all these fluctuations, centered at different time-delays, we end up with the following multifractal representation for $C^{p,q}(l_n, l_n|t) \equiv C^{p,q}_{n,n}(t)$:

$$C^{p,q}_{n,n}(t) = \int d\mu_{n,0}(h)\left(\frac{l_n}{L_0}\right)^{(q+p)h} E_{p,q}\left(\frac{t}{\tau_n}\right)$$
$$+\sum_{m=1}^{n-1}\int d\mu_{m,0}(h)d\mu_{n,m}(h_1)d\mu_{n,m}(h_2)\left(\frac{l_m}{L_0}\right)^{(q+p)h}\left(\frac{l_n}{l_m}\right)^{qh_1}\left(\frac{l_n}{l_m}\right)^{ph_2} E_{p,q}\left(\frac{t}{\tau_m}\right) \tag{118}$$

Note that the first term of the right hand side of (118) is the leading contribution in the static limit ($t = 0$).

Let us now jump to the most general multi-scale multi-time correlation functions:

$$C^{p,q}(r,R|t) = \langle \delta_R u^q(0) \cdot \delta_r u^p(t) \rangle \tag{119}$$

where from now on we will always suppose that $\delta_r u$ describes the velocity fluctuation at the smallest of the two scales considered, i.e. $r < R$. The first is the slower, at large scale, $\delta_R u(0)$ and the second is the faster, at small scale and at a time delay t, $\delta_r u(t)$. We shall now use the same discrete notation already introduced for the equal time correlation. Following the same reasoning as before we may safely assume that from zero time delay up to time delays of the order of the slower component, $t \le \tau_n$, the velocity field at small scale feels the same transfer process of u_n up to scale n and then from scale n to scale N an uncorrelated transfer mechanism:

$$u_N(t) = W_{N,n}(t)u_n(t) \sim W_{N,n}(t)u_n(0) \sim W_{N,n}(t)W'_{n,0}(0)u_0 \quad \text{for } 0 \le t \le \tau_n \tag{120}$$

Similarly, for time delays within $\tau_n \le \tau_m < t < \tau_{m-1} \le \tau_0$ also the field at large scale n will start to see different transfer processes:

$$u_n(0) \sim W''_{n,m}(0)u_m(0) \sim W''_{n,m}(0)W'_{m,0}(0)u_0 \tag{121}$$

$$u_N(t) \sim W_{N,m}(t)u_m(t) \sim W_{N,m}u_m(0) \sim W_{N,m}(t)W'_{m,0}(0)u_0 \tag{122}$$

It is clear now, how we may write down the correlation for any time:

$$C^{p,q}_{N,n}(t) = \int d\mu_{n,0}(h)d\mu_{N,n}(h_1)\left(\frac{l_n}{L_0}\right)^{(q+p)h}\left(\frac{l_N}{l_n}\right)^{ph_1} E_{p,q}\left(\frac{t-\tau_{nN}}{\tau_n}\right) + \tag{123}$$

$$\sum_{m=1}^{n-1}\int d\mu_{m,0}(h)d\mu_{n,m}(h_1)d\mu_{N,m}(h_2)\left(\frac{l_m}{L_0}\right)^{(q+p)h}\left(\frac{l_n}{l_m}\right)^{qh_1}\left(\frac{l_N}{l_m}\right)^{ph_2} g_{p,q}\left(\frac{t}{\tau_m}\right)$$

where $\tau_{nN} \simeq \tau_n - \tau_N$ represents the time-delay needed for an energy burst to travel from shell n to shell N and the functions $E_{p,q}(x)$ and $g_{p,q}(x)$ are defined in the same way as done for (118).

Let us observe that the sum in the above expression goes only up to the index of the largest scale n: this is because only for time larger than τ_n the correlation is a true multi-time correlation. Indeed, for time-delays shorter than τ_n only the field at small scale, u_N, is changing but always under the same large scale configuration, u_n.

This ends our theoretical analysis which we will use in the following to improve our knowledge of the link between the multifractal theory, and more precisely the random multiplicative process, and the equation of motion. Let us recall that, while trying to answer this problem, we have found some difficulties arising from a direct application of the fusion rules to the equation of motion. We argued that time dynamics could play an important role (the equation of motion certainly does). In a nutshell our question is: how much does time dynamics affect the equal time fusion rules ?

Let us start with a simple but very interesting example by considering a random multiplicative process defined, as usual by now, on scales $l_n = 2^{-n}L$. We assume that the dynamics is somehow trivial: each eddy turnover time depends only on the scale and it does not depend on the random process:

$$\tau_n = l_n^H \tag{124}$$

where H is a constant to be defined later. At each scale, the random multiplier is taken constant for a time τ_n and changed according to a given probability distribution. Only when the multiplier is changed, the field at scale l_n is computed as the product of the new outcome of the probability distribution multiplied by the field at scale l_{n-1}. Our model can be interpreted as a "passive" scalar advected by some velocity flow with a given (non multifractal) power spectrum. For this reason, we will use the notation θ_n for the "passive" difference on scale l_n. In order to clarify the picture, we show in Figure 15 a numerical simulation performed by using as a random multiplicative process the one defined in (60) with $p = 0.4$ (in this case $\zeta(2) = 1$) and $H = -1$. As we can see, the time dynamics seems to have some delay from scale to scale.
The time dynamics does not change the scaling behavior of $\langle\theta_n^q\rangle$ as one can see from Figure 16 where we plot $\langle\theta_n^6\rangle$ versus n compared with the scaling predicted by our model.

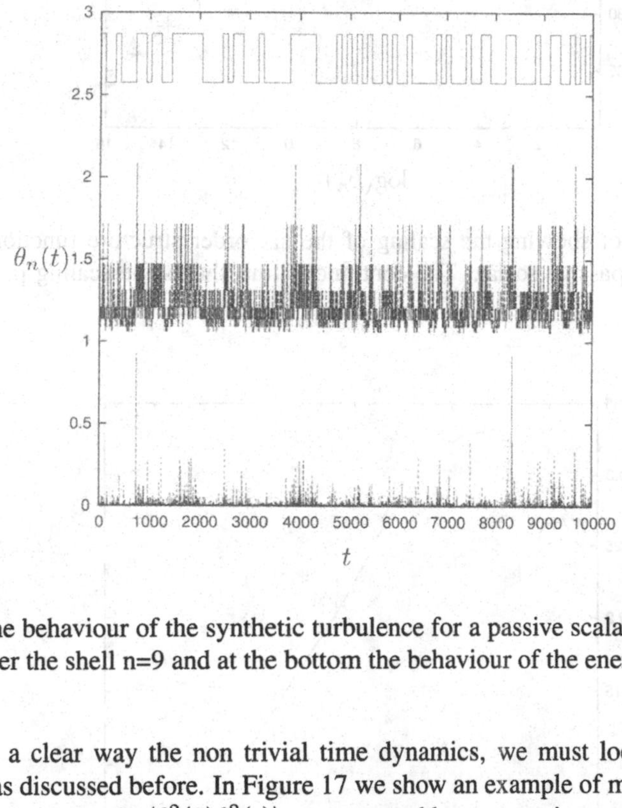

Figure 15: Time behaviour of the synthetic turbulence for a passive scalar. On the top the shell n=2, in the center the shell n=9 and at the bottom the behaviour of the energy dissipation.

To show in a clear way the non trivial time dynamics, we must look for the correlation $\langle\theta_n^p(t)\theta_m^q(0)\rangle$, as discussed before. In Figure 17 we show an example of multi-time, multi-space correlation function, namely $\langle\theta_3^2(0)\theta_9^2(t)\rangle$: as expected by our previous analysis, we observe an increase in the correlation up to time of the order τ_3.

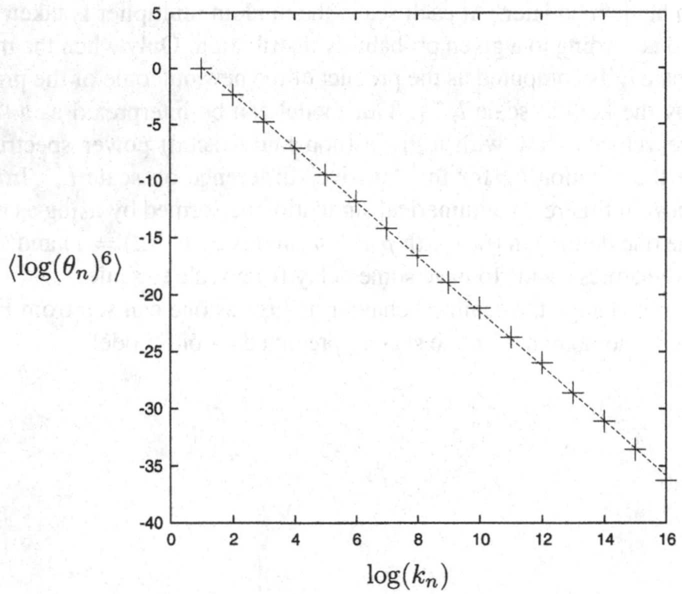

Figure 16: Log-log plot showing the scaling of the six order structure function for the synthetic turbulence for a passive scalar. The continuous line shows the scaling predicted by the multiplicative process.

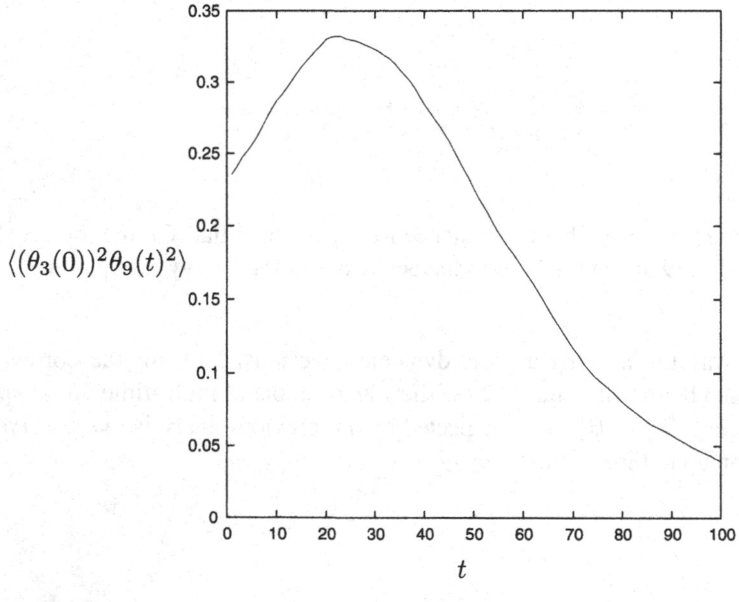

Figure 17: Time behaviour of the correlation $\langle \theta_3(0)^2 \theta_9(t)^2 \rangle$.

The non trivial time dynamics affects the fusion rules. In Figure 18, we show the compensated fusion rule for $\langle\theta_3^2\theta_n^2\rangle$. Although we still observe that a plateau is reached, the asymptotic offset changed. Why? The reason is quite simple and, as one can imagine, it has to do with the time delay among different scales.

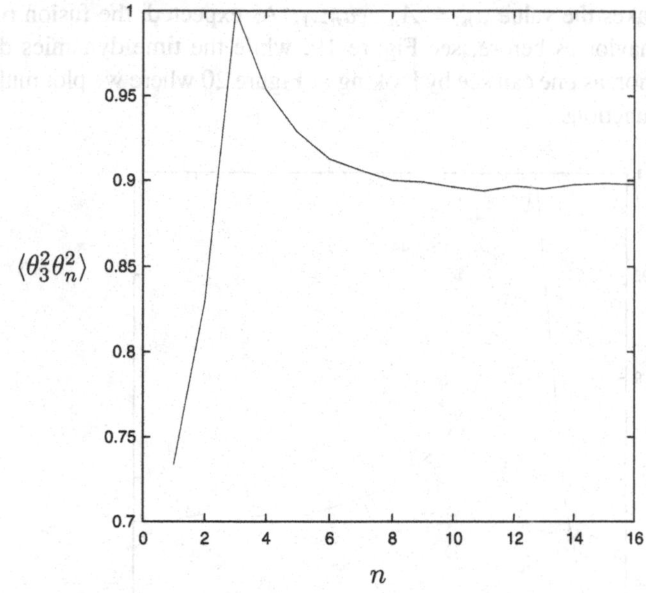

Figure 18: Compensated fusion rules for $\langle\theta_3^2\theta_n^2\rangle$ for different values of n.

Let us consider two scales l_n and l_{n+m}. For a fixed time, the quantities θ_n and θ_{n+m} are not necessarily products of the same random variables. They "feel" the same chain of multipliers for some larger scale $l_{n'}$ with $n'\langle n$. Thus we should have:

$$\langle\theta_n^q\theta_{n+m}^p\rangle = \langle\Pi_{k=1..n'}A_k^{q+p}\rangle\langle\Pi_{k=n'..n}A_k^q\rangle\langle\Pi_{k=n'..n+m}A_k^p\rangle \tag{125}$$

where A_k is the random multiplier acting between scale $k-1$ and scale k, i.e. $\theta_{k+1} = A_k\theta_k$. The above expression gives the following result for the compensated fusion rules:

$$\frac{\langle\theta_n^q\theta_{n+m}^p\rangle\langle\theta_n^p\rangle}{\langle\theta_{n+m}^p\rangle\langle\theta_n^{p+q}\rangle} \sim \frac{\langle A^{q+p}\rangle^{n'}\langle A^q\rangle^{n-n'}\langle A^p\rangle^{m+n-n'}\langle A^p\rangle^n}{\langle A^p\rangle^{n+m}\langle A^{p+q}\rangle^n} \tag{126}$$

that is:

$$\frac{\langle\theta_n^q\theta_{n+m}^p\rangle\langle\theta_n^p\rangle}{\langle\theta_{n+m}^p\rangle\langle\theta_n^{p+q}\rangle} = \left(\frac{\langle A^{q+p}\rangle}{\langle A^q\rangle\langle A^p\rangle}\right)^{n-n'} \leq 1 \tag{127}$$

Let us notice that for $n' = n$, which corresponds to a random multiplicative process without time dynamics, the r.h.s of the above expression is just 1. The above equation should be considered a qualitative explanation of our numerical findings. In general we expect n' to be a function of p and q.

We can generalize the above model for a more complex time dynamics, i.e for the case where the eddy turnover time does depend on the random multiplicative process. In order to highlight the important points with some graphics, we performed a numerical simulation for a random multiplicative process defined as in (60) ($\zeta(3) = 1$ in this case) and we let τ_n change randomly according to the rule $\tau_n = l_n/u_n$. For each scale l_n, the velocity u_n is kept unchanged for time τ_n after which, it takes the value $u_n = A_{n-1}u_{n-1}$. As expected, the fusion rules exhibit the same qualitative behavior as before, see Figure 19, while the time dynamics displays the correct qualitative behavior, as one can see by looking at Figure 20 where we plot multi-time and multi-scale correlation functions.

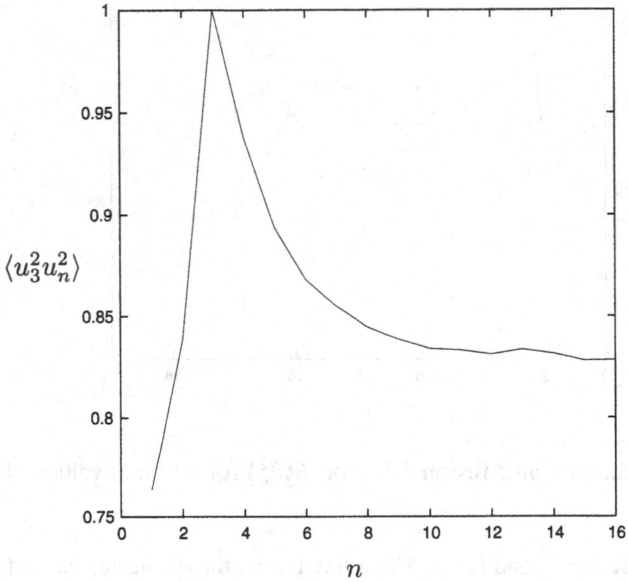

Figure 19: Compensated fusion rules for $\langle u_3^2 u_n^2 \rangle$ for different values of n.

However, at variance with the "passive" case discussed before, the scaling exponents of the structure functions $\langle u_n^p \rangle$ are now changed. In Figure 21 we show the scaling of $\langle u_n^3 \rangle$ which should be -1. A slope -1 is indeed plotted in the same picture indicating that the scaling behavior is clearly changed.

In order to understand the reason of such a (dramatic) change, we recall that the structure functions in this model are computed as time average over the dynamics, i.e.

$$\langle u_n^p \rangle = \int dt u_n^p(t) \tag{128}$$

We can write the integral in time as a sum over the eddy turnover time at scale l_n:

$$\langle u_n^p \rangle = \sum_k \tau_n(k) u_n^p(k) \tag{129}$$

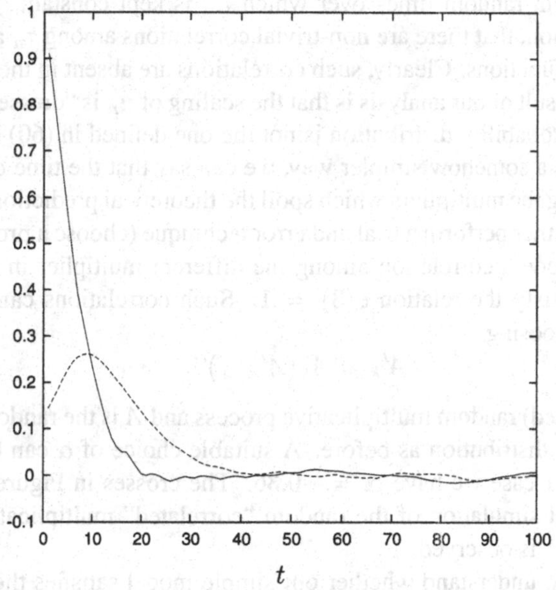

Figure 20: Time behaviour of the correlation $\langle u_3(0)^2 u_9(t)^2 \rangle$ (dashed line) and $\langle u_3(0)^2 u_3(t)^2 \rangle$ (solid line).

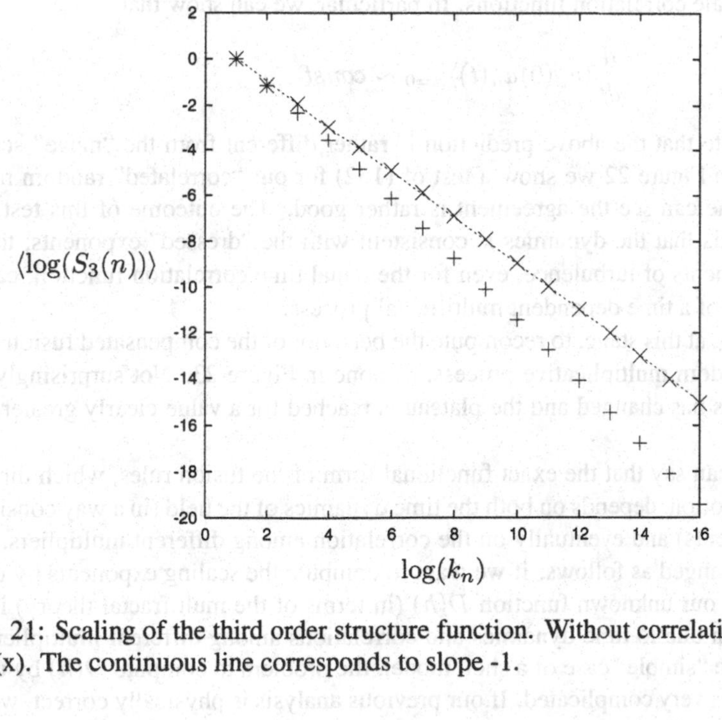

Figure 21: Scaling of the third order structure function. Without correlation(+) and with correlation (x). The continuous line corresponds to slope -1.

where $\tau_n(k)$ are different random times over which u_n is kept constant. We can deduce, by using the above expression, that there are non-trivial correlations among τ_n and u_n which fix the scaling of the structure functions. Clearly, such correlations are absent in the (previous) case of a constant τ_n. The final result of our analysis is that the scaling of u_n is "dressed" by the dynamical behavior, i.e. the true probability distribution is not the one defined in (60) but it is the result of a dynamical process. In a somehow simpler way, we can say that the time dynamics introduces some correlations among the multipliers which spoil the theoretical prediction. In order to control our simulation we can either perform a trial and error technique (choose a probability distribution until $\zeta(3) = 1$) or include a correlation among the different multiplier in such a way that the "dressed" exponents satisfy the relation $\zeta(3) = 1$. Such correlations can be introduced in a rather simple way by choosing

$$A'_k = A_k(A'_{k-1})^\alpha \tag{130}$$

where A' is the (correlated) random multiplicative process and A is the random number extracted by the same probability distribution as before. A suitable choice of α can be found in order to obtain $\zeta(3) = 1$. In our case we have $\alpha = -0.36$. The crosses in Figure 21 show the result obtained by a numerical simulation of the random "correlated" multiplicative process: a clear agreement with $\zeta(3) = 1$ is observed.

It is now tempting to understand whether our simple model satisfies the scaling constraints given by the dimensional relation:

$$\frac{du_n}{dt} \sim \frac{u_n^2}{l_n} \tag{131}$$

As previously discussed, one can translate such a constraint in a well defined prediction on the multi-time and multi-scale correlation functions. In particular, we can show that

$$\frac{d}{dt}\langle u_n(0)u_n(t)\rangle|_{t=0} \sim const \tag{132}$$

if $\zeta(3) = 1$. Let us note that the above prediction is rather different from the "naive" scaling proportional to $\langle u_n^2\rangle$. In Figure 22 we show a test of (132) for our "correlated" random multiplicative process: as one can see the agreement is rather good. The outcome of this test, and others not shown here, is that the dynamics is consistent with the "dressed" exponents, that is to say the scaling exponents of turbulence, even for the equal time correlation function, can be described only in terms of a time dependent multifractal process.

It is quite interesting, at this stage, to recompute the behavior of the compensated fusion rules for the "correlated" random multiplicative process, as done in Figure 23. Not surprisingly, the shape of the fusion rules has changed and the plateau is reached for a value clearly greater than before.

As a summary, we can say that the exact functional form of the fusion rules, which directly enters the equation of motion, depends on both the time dynamics of the field (in a way consistent with our scaling arguments) and eventually on the correlation among different multipliers. The last statement can be changed as follows: if we want to compute the scaling exponents by using the equation of motion, our unknown function $D(h)$ (in terms of the multifractal theory) is the result of renormalization due to time dynamics and correlations among different multipliers. It is clear that, even for the "simple" case of a shell model, the problem to compute $D(h)$ by using the equation of motion is very complicated. If our previous analysis is physically correct, we are

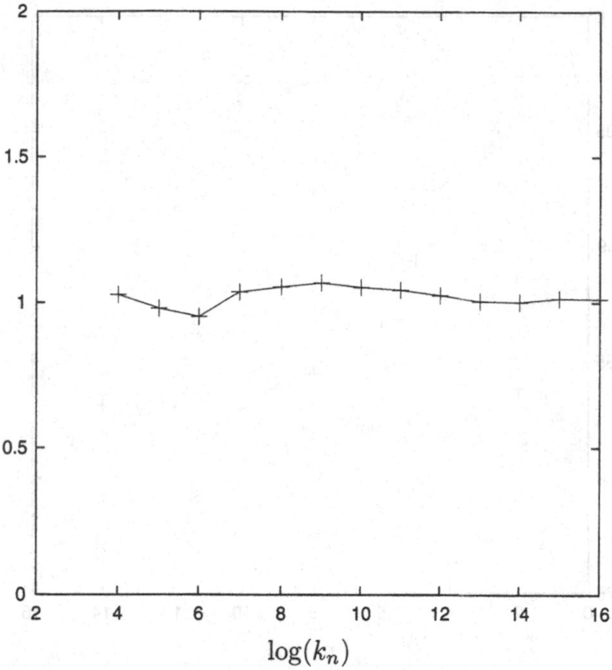

Figure 22: Scaling of $\frac{d\langle u_n(0)u_n(t)\rangle}{dt}|_{t=0}$

facing a complex problem in an (unknown) functional space of correlated random multiplicative processes. Although such a task may not be achieved in any exact way, nevertheless we argue that we have "grasped" all the physical features of intermittency, at least in relation to the multifractal theory. Now we will show that, in at least one non trivial case, we can exactly solve the problem.

The model we want to study, introduced by Kraichnan in 1968, is the advection of a passive scalar by a random, Gaussian velocity field, white-in-time and whose two-point velocity correlation function is given by: $\langle v_i(x,t)v_j(x',t')\rangle = \delta(t-t')D_{ij}(|x-x'|)$, with $D_{ij}(x) = D_{ij}(0) - \hat{D}_{ij}(x)$. Here, $\hat{D}$ is the d-dimensional velocity-field structure function: $\hat{D}_{ij}(x) = D_0|x|^{\xi}\left[(d-1+\xi)\delta_{ij} - \xi x_i x_j|x|^{-2}\right]$, where the scaling exponent of the second order velocity structure function, ξ $(0 < \xi < 2)$, is a free parameter. Higher order velocity-field correlation functions are fixed by the Gaussian assumption. Although such a choice is far from being realistic, many interesting analytical and phenomenological results have been obtained for this toy-model. Due to the delta correlation in time, moment equations are closed to all orders. For instance one can show that:

$$\tilde{S}_p(r) \equiv \langle|\theta(x) - \theta(x+r)|^p\rangle \sim r^{\zeta_p}. \tag{133}$$

$\theta(x)$ being the scalar field transported by the turbulent velocity field. In the inertial range $\tilde{S}_p(r) \sim r^{\zeta_p}$ and the set of scaling exponents ζ_p fully characterizes intermittency.

Because we want to keep our discussion as simple as possible, we consider the shell model version of the model discussed above, warning the reader that most of our results can also be

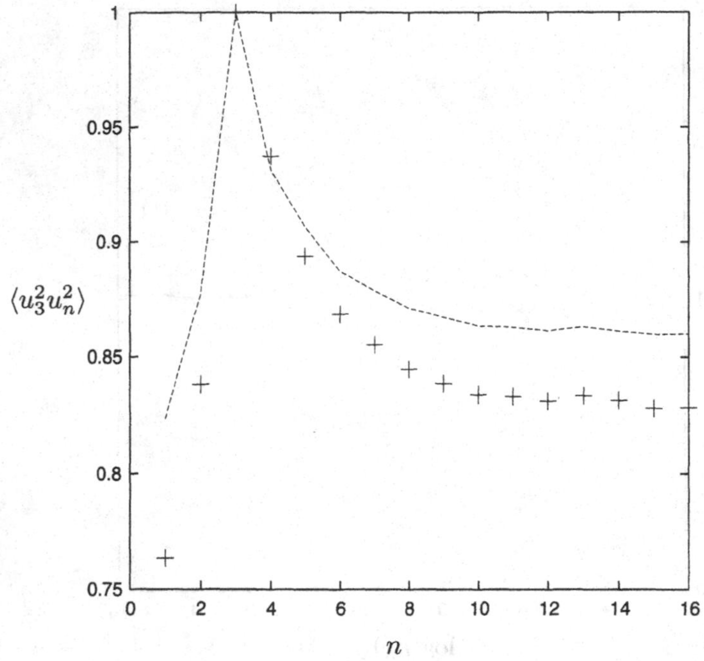

Figure 23: Compensated fusion rules for $\langle u_3^2 u_n^2 \rangle$ for different values of n with correlations (dashed line) and without correlation (+).

obtained for the continuous model:

$$\partial_t \theta + \vec{u} \cdot \nabla \theta = \kappa \Delta \theta \tag{134}$$

Our shell model version of (134) is constructed in the following way. Passive increments at scale $l_n = k_n^{-1}$ are described by a complex variable $\theta_n(t)$. The time evolution is obtained according to the following criteria: (i) the linear term is a purely diffusive term given by $-\kappa k_n^2 \theta_n$; (ii) the advection term is a combination of the form $k_n \theta_{n'} u_{n''}$; (iii) interacting shells are restricted to nearest neighbors of n; (iv) in the absence of forcing and damping the model preserves the volume in the phase-space and the passive-energy $E = \sum_n |\theta_n|^2$. Properties (i), (ii) and (iv) are valid also for the original equation of a passive scalar advected by a Navier-Stokes velocity field in the Fourier space, while property (iii) is an assumption of locality of interactions among modes. This assumption is rather well founded as long as $0 \ll \xi < 2$. The model is defined by the following equations ($m = 1, 2, \ldots$)

$$\left[\frac{d}{dt} + \kappa k_m^2\right] \theta_m(t) = i[c_m \theta_{m+1}^*(t) u_m^*(t) + b_m \theta_{m-1}^*(t) u_{m-1}^*(t)] + \delta_{1m} f(t) \tag{135}$$

where the star denotes complex conjugation and $b_m = -k_m, c_m = k_{m+1}$ for imposing energy conservation in the zero diffusivity limit. Boundary conditions are defined as: $u_0 = \theta_0 = 0$. The forcing term $\delta_{1m} f(t)$is Gaussian and delta correlated: $\langle f(t) f(t') \rangle = F_1 \delta(t - t')$ acts only on the first shell. In numerical implementations, the model is truncated to a finite number of shells N with the additional boundary conditions $\theta_{N+1} = 0$. We assume that the velocity variables $u_m(t)$ and the forcing term $f(t)$ are independent complex Gaussians and white-in-time, with the scaling law: $\langle u_m(t)u_n^*(t')\rangle = \delta(t-t')\delta_{nm}d_m$, $d_m = k_m^{-\xi}$. Due to the delta-correlation in time, we can close the equations of motion for all structure functions. In deriving the equations for the structure functions, we make use of an important relation (true only for Gaussian variables), namely:

$$\langle u_n Q(\theta_{n_1}, \theta_{n_2}, ..., \theta_{n_k})\rangle = d_n \frac{\delta}{\delta u_n^*}\langle Q(\theta_{n_1}, \theta_{n_2}, ..., \theta_{n_k})\rangle \qquad (136)$$

The functional derivate $\delta/\delta u_n^*$ is computed by observing that

$$\theta_m = i\int_0^t ds[c_m\theta_{m+1}^*(s)u_m^*(s) + b_m\theta_{m-1}^*(s)u_{m-1}^*(s)] + ... \qquad (137)$$

where dots indicate terms not containing u_m. Then we must use the relation:

$$\frac{\delta u_m(t)}{\delta u_n(s)} = \delta(t-s)\delta_{n,m} \qquad (138)$$

in order to finally compute the different quantities in (136). We concentrate on the non-perturbative analytic calculation of the fourth-order structure function $P_{mm} = \langle(\theta_m\theta_m^*)^2\rangle \propto k_m^{-\zeta_4}$(the lowest order with non-trivial anomalous scaling).
The closed equation satisfied by $P_{mq} = \langle(\theta_m\theta_m^*)(\theta_q\theta_q^*)\rangle$, is,

$$\begin{aligned}\dot{P}_{mq} &= (\delta_{1,m}E_m + \delta_{1,q}E_q)F_1 - \kappa(k_m^2 + k_q^2)P_{mq} \\ &+ [-P_{mq}c_m^2 d_m((1+\delta_{q,m+1}) + \lambda^{\xi-2}(1+\delta_{q,m-1})) \\ &+ P_{m+1,q}c_m^2 d_m(1+\delta_{q,m}) + P_{m-1,q}b_m^2 d_{m-1}(1+\delta_{q,m}) \\ &+ (q \leftrightarrow m)]\end{aligned} \qquad (139)$$

where $E_n = \langle\theta_n\theta_n^*\rangle$. We can symbolically represent equation (139) as:

$$\dot{P}_{mq} = \mathcal{I}_{mq,lp}P_{lp} + \kappa\mathcal{D}_{mq,lp}P_{lp} + \mathcal{F}_{mq} \qquad (140)$$

where $\mathcal{I}$ and $\mathcal{D}$ are the inertial and the diffusive 4-order tensor and $\mathcal{F}$ is the forcing term.
Our main result is derived by using the following ansatz: the *symmetric* matrix P_{mq}, which fully determines the scaling properties for any fourth-order quantity in the model, can be described as:

$$P_{n,n+l} = C_l P_{n,n}\ (l \geq 0), \quad P_{n,n-l} = D_l P_{n,n}\ (l \geq 0). \qquad (141)$$

The independence of C_l and D_l of n is equivalent to demanding the absence of strong boundary effects, i.e. the matrix is formally infinite-dimensional. Note that our ansatz is nothing else than the application of the fusion rules relations for our model. Using (141) we obtain:

$$\frac{C_{l+1}}{C_l C_1} = \frac{D_{l+1}}{D_l D_1}$$

which is equivalent to: $P_{n+l,n+l} = k_l^{-\zeta_4} P_{n,n}$ where $C_l/D_l = k_l^{-\zeta_4}$ and $\zeta_4 = 2(2-\xi) - \rho_4$, where we indicate by ρ_4 the anomalous correction to the scaling exponent.

Notice that (141) does not force the solution to have global scaling invariance: only the diagonal part is required to have pure scaling. Let us proceed by analyzing (139) restricted to the inertial operator and for the diagonal ($m = q$) and sub-diagonal terms ($q = m - 1$):

$$\dot{P}_{m,m} = 2P_{m,m}c_m^2 d_m(-1 - x + 2(C_1 + D_1 x)) \tag{142}$$

$$\dot{P}_{m,m-1} = 2P_{m,m-1}c_m^2 d_m \left(-1 - 4x - x^2 + \frac{x}{D_1} + \frac{x + C_2 + \frac{x^2 C_2}{R}}{C_1}\right) \tag{143}$$

where we have introduced $x = \lambda^{\xi-2}$ and $R = C_1/D_1$. By plugging the scaling (141) into (143) one obtains two equations with three unknowns which can be taken to be C_1 and the ratios C_2/C_1 and R. Our knowledge of the fusion rules suggests the following relationship:

$$P_{n,n+l} = C_l P_{n,n}, \quad \text{with } C_l = C_1 k_{l-1}^{\xi-2} \tag{144}$$

and

$$P_{n,n-l} = D_l P_{n,n}, \quad \text{with } D_l = D_1 k_{l-1}^{-(\xi-2)-\rho_4} \tag{145}$$

By plugging this scaling in (143) we end up with two equations and two unknowns and we can calculate ρ_4. Let us anticipate that this assumption gives results in very good agreement with the numerical simulations, indicating that the true solution is not very far from having pure scaling behavior. In order to solve the full problem, without imposing any "pure scaling" behavior, we analyze the other entry of the matrix $P_{n,q}$ with $q \neq n$ and $q \neq n-1$. Let us put $\gamma_l = D_{l+1}/D_l, \delta_l = C_{l+1}/C_l$. It is then possible to show that for $l > 1$ by plugging the scaling (141) into the inertial part of (139) and studying the equation for $\dot{P}_{n,n\pm l}$ we obtain two recursion equations:

$$-(1+x)(1+x^l) + \gamma_l(R + x^{l+1}) + \frac{1}{\gamma_{l-1}}\left(x^l + \frac{x}{R}\right) = 0 \tag{146}$$

$$(1+x)(1+x^l) - \lambda^{\zeta_4}\delta_l(R + x^{l+1}) - \frac{\lambda^{-\zeta_4}}{\delta_{l-1}}\left(x^l + \frac{x}{R}\right) = 0 \tag{147}$$

These two relations can indeed be seen as two maps connecting successive values of γ_l and δ_l respectively. By iterating forward (backward) the map (146) we move from the diagonal (IR boundary) to the IR boundary (diagonal) along a row of the matrix $P_{l,n}$. By iterating forward (backward) the map (147) we move from the diagonal (UV boundary) to the UV boundary (diagonal) along a row of the matrix $P_{l,n}$.

Note that the two maps are not independent, i.e. they satisfy our scaling ansatz $\delta_l = R\gamma_l$ and therefore we are going to consider only one of the two in what follows. In order to test stability under weak perturbation of boundary conditions in the map (146) we are interested in the behavior by backward iterations, i.e. iterating from $l = \infty$ to $l = 0$. In the limit $l \to \infty$ and $\xi \neq 2$ the map (146) has only two fixed points corresponding to $\gamma_1^* = x/R$ and $\gamma_2^* = 1/R$. It turns out that γ_1^* is stable for back iterations, i.e. iterating from the IR boundary ($l \gg 1$) to the diagonal ($l = 0$). The global solution can now be obtained by a self-consistent method. First, let us take as initial value for R the value that one would have guessed from imposing "pure scaling"

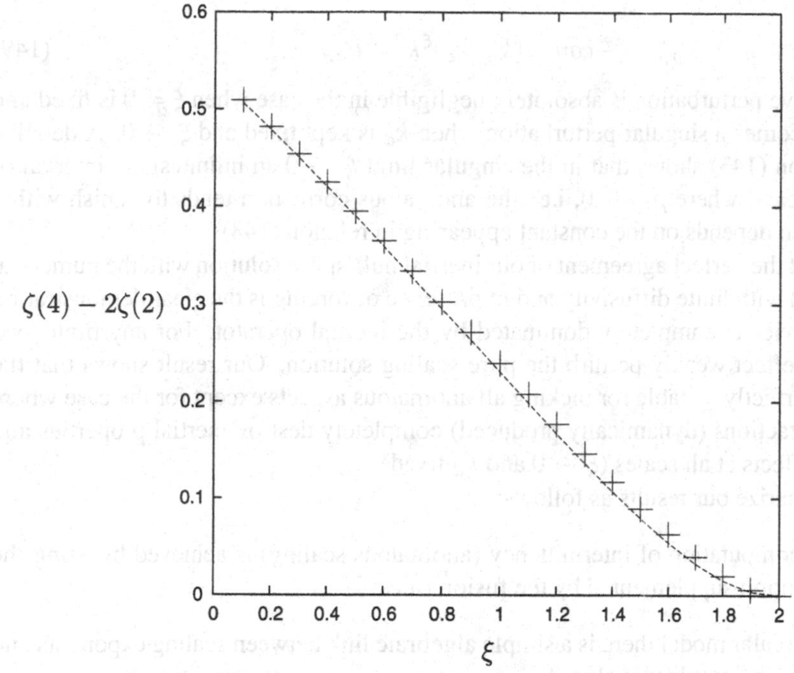

Figure 24: Prediction of the anomaly (vertical axes) in the shell of a passive scalar with random Gaussian velocity field with scaling exponent ξ (horizontal axes).

as discussed previously, then we can iterate (146) from the boundary toward the diagonal and find the value for C_2/C_1. This value can be used to close (143) exactly. Next, with the improved value for R, one can restart the full procedure getting a new improved value of R and so on up to the moment when the new value of R reaches its fixed point. In Figure 24 we show the computation of ρ_4 obtained by numerical integration of equation (135) as a function of ξ. In the same figure we plot the ρ_4 as a function of ξ obtained by the analytical solution previously discussed. As one can see, the agreement is perfect. Notice that it is impossible to reach values of ξ close to zero by numerical simulations because of strong diffusive effects which completely destroy scaling behavior. Note that it is the strong stability under UV and IR perturbations that allows us to consistently iterate the procedure. We have therefore proved that anomalous scaling comes only from the inertial operator and that it shows a very strong degree of universality as a function of the forcing and dissipative mechanisms, at least as far as the situation with $\xi = const. > 0$ and molecular diffusivity $\kappa \to 0$ is considered. New and interesting phenomena occur at the other asymptotic limit ($\xi \to 0$) at fixed molecular diffusivity. A simple dimensional analysis tells us that the following relation holds true:

$$\kappa = const. k_d^{-\xi}, \tag{148}$$

where k_d represents a preassigned diffusive scale such that all the inertial dynamics are at $k \ll k_d$. One can show that in such a situation the diffusive operator $\mathcal{D}$ in (140) gives a contribution of the form

$$\dot{P}_{n,n} = -const.(k_n/k_d)^{\xi} k_n^{2-\xi} P_{n,n} \tag{149}$$

Therefore, the diffusive perturbation is absolutely negligible in the case when $\xi \neq 0$ is fixed and $k_d \to \infty$ while it becomes a singular perturbation when k_d is kept fixed and $\xi \to 0$. A detailed analysis of the relation (143) shows that in the singular limit $\xi \to 0$ an infinitesimal interval of values of $\xi \sim 0$ appears, where $\rho_4 \to 0$, i.e. the anomalous correction tends to vanish with a particular shape which depends on the constant appearing in relation (148).

Let us remark that the perfect agreement of our inertial null-space solution with the numerical simulation performed with finite diffusivity and in presence of forcing is the clear demonstration that the scaling behavior is completely dominated by the inertial operator. For any finite system, the IR and UV effect weakly perturb the pure scaling solution. Our result shows that the inertial operator is perfectly suitable for picking all anomalous aspects except for the case where strong non-local interactions (dynamically produced) completely destroy inertial properties and introduce diffusive effects at all scales ($\xi \to 0$ and k_d fixed).

We want to summarize our results as follows:

(I) a quantitative computation of intermittency (anomalous scaling) is achieved by using the equations of motion supplemented by the fusion rules;

(II) for the passive scalar model there is a simple algebraic link between scaling exponents and fusion rules, i.e. the problem is closed;

(III) the analytical solution shows that the intermittency effects do not depend on the dissipation mechanism and on the forcing (universality), i.e. the analytical solution is stable with respect to change in the forcing and dissipation mechanism;

(IV) for the non linear case there is no simple link between the fusion rules and the scaling exponents; the scaling exponents are "renormalized" by the time dynamics and eventually by correlations among different multipliers.

Item (IV) is telling us that analytical computations of the scaling exponents for the non linear case may not be "physically" different from the computations we performed for the Kraichnan model. However, in the nonlinear case the computations are not closed and the solution cannot be found step by step by solving the equations for the correlation functions in increasing order of complexity. Rather, one should make a reasonable ansatz for the function $D(h)$ and for the correlation among different scales and, by using the equation of motion with the fusion rules, refining the ansatz up to the precision which is needed. Apart from the complexity of the computations to be performed, we believe that the statistical properties of the real turbulence are quite close to the picture outlined in this section.

The above discussion, among the many points which should deserve further investigations, provided us with a very important concept, analytically proved at least in one case, i.e. universality. Universality implies that intermittency should develop in a fluid system independent of the dissipation mechanism. This implies that there is no reason to resolve, in a numerical simulation for homogeneous and isotropic turbulence, small scale dynamics in the dissipation region to correctly obtain inertial physics and intermittency. We thus reach the conclusion that eddy viscosity, as explained in section 2, could be an effective tool to simulate large Reynolds number flows.

First we review some basic concepts concerning eddy viscosity and then we numerically check, in some well controlled cases, the application of eddy viscosity in numerical simulations.

By mere dimensional arguments, the effective eddy viscosity at scale r reads as follows

$$\nu_E(r) \sim r \cdot \delta v(r) \tag{150}$$

where $\delta v(r)$ is the velocity fluctuation across a distance r (vector indices are relaxed for simplicity). Equation (150) can also be deduced by using the Refined Kolmogorov Similarity Hypothesis (RKSH) as discussed in section 2. Let us define Δ as the scale at which we want to compute the eddy viscosity. Generally speaking, the common aim of eddy viscosity models is to incorporate the effects of unresolved scales $r < \Delta$ (Δ being a typical mesh size), on the resolved ones, $r > \Delta$.

One of the simplest and most popular sub-grid-model is due to Smagorinsky and it can be derived starting from equation (150).
The idea is to replace r with the mesh size Δ in the equation (150) and subsequently replace $\delta v(\Delta) \sim S\Delta$, where (we dispense with tensor indices for the sake of the argument) S is the strain tensor $S \sim \delta v/r$ evaluated at $r = \Delta$. The result is

$$\nu_{SGS} \sim \Delta^2 S \tag{151}$$

Although simple, this expression is less transparent than it looks. In fact, it is based on the assumption that the velocity field at the scale Δ is smooth enough to allow the definition of the space derivative S. This is in contrast to the fact that, if Δ belongs to the inertial range (as it should for the whole LES procedure to make sense), the velocity field is *known* not to be differentiable since δv scales like $r^{1/3}$. On account of this, one expects $\delta v(\Delta)/\Delta$ to be much larger than the corresponding ratio evaluated at $r = \eta$ (the only scale where this operation is conceptually allowed). This "inconsistency" is usually acknowledged by prefactoring the right hand side of the equation with an empirical coefficient C_S smaller than unity, typically $C_S \sim 0.12$.

On a more rigorous basis we can define a large scale velocity field $\overline{v}_i$, where with the averaging procedure only wavevectors smaller than a given threshold $1/\Delta$ are kept. Taking the average of the Navier-Stokes equations we end up with

$$\frac{\partial \overline{v}_i}{\partial t} + \frac{\partial}{\partial x_j}(\overline{v}_i \overline{v}_j) = -\frac{1}{\rho}\frac{\partial \overline{p}}{x_i} + \nu \frac{\partial^2 \overline{v}_i}{\partial x_i \partial x_j} - \frac{\partial \tau_{ij}}{\partial x_j}; \quad \frac{\partial \overline{v}_i}{\partial x_i} = 0 \tag{152}$$

where the Reynolds stress tensor, $\tau_{i,j}$, is the only term which involves sub-grid contributions (i.e. coming from scales which are not resolved).

$$\tau_{ij} - \frac{1}{3}\tau_{kk}\delta_{ij} = -2\nu_S \overline{S}_{ij} \tag{153}$$

In the Smagorinsky closure model the eddy viscosity is approximated as

$$\nu_S = (C_S \Delta)^2 \sqrt{2\overline{S}_{ij}\overline{S}_{ij}} \tag{154}$$

where $\overline{S}_{ij} = \frac{1}{2}(\partial_i \overline{v}_j + \partial_j \overline{v}_i)$ is the large-scale stress tensor. Since the discussion of the sophisticated developments of LES modeling is beyond our scope, we shall focus here exclusively on the specific question of the interrelation between dissipation and intermittency. Tackling this question within the true three-dimensional Navier-Stokes context is a very daunting task, in view of the enormous amount of data to be produced and carefully analyzed. It therefore makes sense to attack the problem within the context of simplified dynamical models sharing as much physics as possible with the Navier-Stokes equations while giving away most of its computational complexity.

Following the above discussion, we consider a shell model of turbulence, namely the model described by equation (28). However, in our idealized picture, at least mathematically, real turbulence consists of localized eddies of all sizes that interact, merge and subdivide locally: the physical picture is that of a large eddy which decays into smaller eddies. The number of degrees of freedom in such a field problem in d dimensions grows with the wave number as $N(k) \sim k^D$ ($D = 0$ in shell models). The first step in reproducing this kind of hierarchical structure is the one introduced in this section, namely the tree model.

To make contact with the issue of intermittency-dissipation interrelation, we shall replace the viscous coefficients D_n of the equations of motion with an "effective viscosity" term, $\mathcal{D}_n$, which now acquires both non-trivial dependencies of time and shell indeces. It reads for the two cases:

$$\mathcal{D}_n(t) \equiv \nu_S(\delta_{n,N} + \delta_{n,N-1})\frac{|u_n|}{k_n}k_n^2\,; \quad \mathcal{D}_{n,j}(t) \equiv \nu_S(\delta_{n,N} + \delta_{n,N-1})\frac{|u_{n,j}|}{k_n}k_n^2 \tag{155}$$

where ν_S is an empirical constant of order 1. This "sub-grid-scale" term is clearly patterned after the simplest Navier-Stokes effective viscosity model. The only difference is that due to the short range interactions of our shell models, the sub-grid-modeling is applied only to the last and last-but-one shell k_N, k_{N-1}.

Our sub-grid closure combines features of the classical Smagorinsky Large Eddy Simulation model and the so-called hyperviscosity models used in the direct spectral simulation of incompressible turbulence. This is consistent with the double-locality in real and momentum space of the wavelet basis functions.

The two methods are quite different in scope and formulation: Smagorinsky works in real space as a local, dynamic, effective viscosity responding to the local stress so as to mimic the effects of unresolved scales on the resolved ones. Hyperviscosity is local in k-space, static, and does *not* aim at representing the effects of unresolved scales, but simply at reducing the size of the dissipative region so as to take full advantage of the grid resolution.

In Figure 25 we show the energy spectra for the chain model with eddy-viscosity at three different resolutions $N = 16, 20, 24$. For the sake of comparison the case with normal viscosity is also reported for $N = 16$. As a first remark, we note that the presence of the eddy-viscosity considerably widens the inertial regime which extends deep down to the last-but one shell. Moreover, the slope of the spectrum is basically the same, independent of the number of shells used, which is exactly the property we were looking for (in particular this tell us that we are not dealing with finite size effects).

We note that this is not the case with normal viscosity, where in order to widen the inertial range it is necessary to lower the value of the viscosity so as to increase the Reynolds number. In that case, resolution must be increased accordingly in order to resolve the dissipative region and to prevent numerical problems. A similar result, not shown, holds true for the tree model. In order to gain a more quantitative assessment of the grid independence of our results, we shall

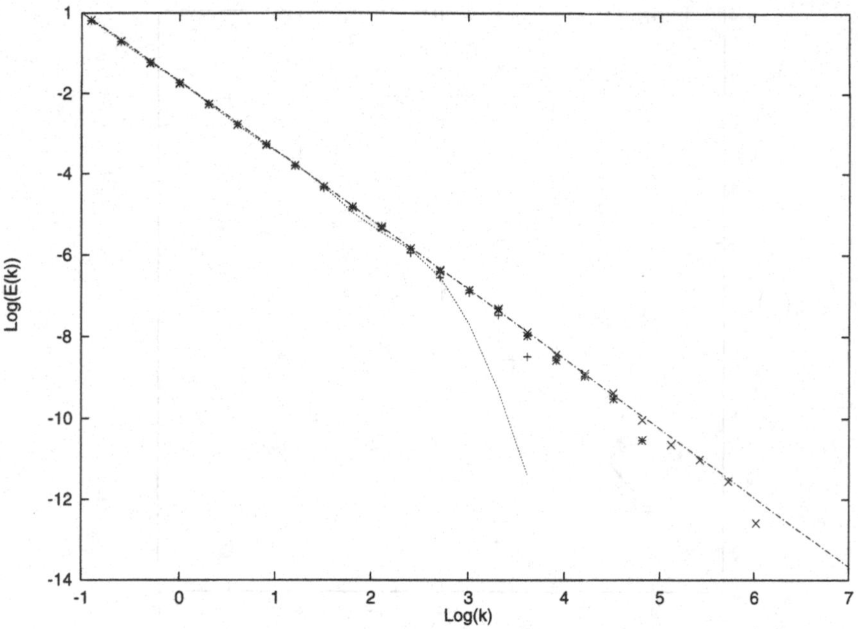

Figure 25: Energy spectrum for the shell model with and without eddy viscosity.

evaluate the scaling exponents $\zeta(n)$ up to $n = 12$. This is done in Figure 26 for the tree model (a similar picture can be obtained for the chain model).

The first remark is that in both cases a significant departure from Kolmogorov K41 law is observed, i.e. the sub-grid model does not destroy intermittency.
More quantitatively, the two sets of exponents coincide within statistical error, which means that intermittency survives and it is basically insensitive to eddy viscosity. Moreover, by using the tree model, we can check whether the introduction of eddy viscosity destroys the RKSH. For this purpose, we must use eddy viscosity in the computation of the energy dissipation and *not* the original dissipation term. The numerical agreement we found in the numerical simulation is excellent with deviations from the prediction of RKSH of the order of 2 percent.

Our detailed check confirms that inertial range dynamics is "stable" with respect to the dissipation mechanism. Thus eddy viscosity can be used safely in homogeneous and isotropic turbulence in order to reach large Reynolds number flows. The fact that intermittency is not affected by the details of the eddy-viscosity models indicates that fine-tuning of the coefficients in front of the eddy-viscosity term is probably not demanded. Nevertheless, oversimplified eddy-viscosity models based only on dimensional analysis would probably fail on the same goal, due to their inability to dissipate violent intermittent bursts.

References for section 4

The details of the more technical results presented in this section can be found in:

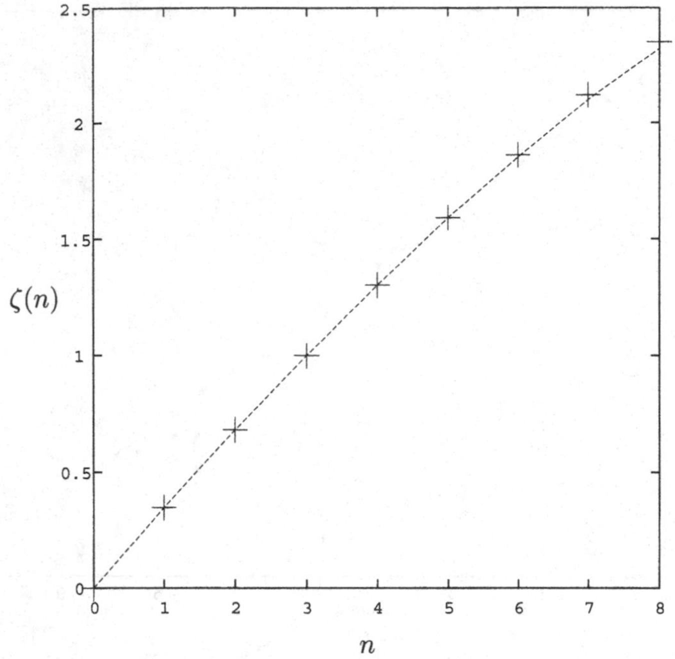

Figure 26: Anomalous exponents of the tree-shell model. The continuous line represents full resolution and the crosses the results obtained by using the eddy viscosity.

- R. Benzi, L. Biferale, A. Crisanti, G. Paladin, M. Vergassola, A. Vulpiani; A random process for the construction of multiaffine fields. *Physica D* **65**, 352, (1993).

- R. Benzi, L. Biferale, R. Tripiccione and E. Trovatore; (1+1)-dimensional turbulence *Phys. Fluids* **9**, 2355, (1996).

- V.S. L'vov and I. Procaccia;Towards a nonperturbative theory of hydrodynamic turbulence: Fusion rules, exact bridge relations, and anomalous viscous scaling functions *Phys. Rev. E* **54** 6268 (1996).

- R. Benzi, L. Biferale and F. Toschi. Multiscale correlation functions in turbulence. *Phys. Rev. Lett.* **80**, 3244, (1998).

- L. Biferale, G. Boffetta, A. Celani, F. Toschi ; Multi-time Multi-scale correlation function in turbulence and in turbulent models *Physica D* **127**, 187, 1999.

- K. Gawdezki and A. Kupiainen, Anomalous scaling of the passive scalar *Phys. Rev. Lett.* **75**, 3834 (1995).

- R.H. Kraichnan, Anomalous scaling of a randomly advected passive scalar *Phys. Rev. Lett.* **72**, 1016 (1994).

- M. Vergassola, Anomalous scaling for passively advected magnetic fields *Phys. Rev. E* **53** R 3021 (1996).
- R. Benzi, L. Biferale & A. Wirth ; Analytic calculation of anomalous scaling in random shell models for a passive scalar *Phys. Rev. Lett.* **78**, 4926, (1997).
- R. Benzi, L. Biferale, S. Succi, F. Toschi; Intermittency and eddy-viscosities in dynamical models of turbulence. *Phys. Fluids* **11**, 1221, (1999).

5 Extended Self Similarity

Most of the arguments discussed in the previous section are based on the scale invariance of the Navier-Stokes equation. We already discussed in section 1 that only at large Reynolds number the experimental data show a range of scales where scaling properties can be measured with reasonable accuracy. Thus, it seems that our theoretical analysis can be applied only in the limit $Re \to \infty$. This is a strong limitation to our present numerical simulations which can reach low to moderate values of Re. Moreover, high Reynolds number flows are difficult to obtain in laboratory experiments and at high Re the dissipation scale becomes so small that velocity gradients can hardly be measured. In summary, there is a clear gap between many theoretical statements and available experimental information.

Most of our theoretical knowledge has been discussed within the framework of the scaling hypothesis. Scaling is a powerful concept because it provides simple rules to compute in a simple way the scale dependence of the statistical properties of our fundamental variables. In a rather general way, let us consider an observable $Q(s)$ where s is the "scale" of the observable Q (for the time being the definition of the scale is arbitrary and does not play any role in what follows). We assume that Q is a random field with probability distribution $P_s(Q)$. Moments of Q are defined as $F_n(s) = \int P_s(Q)Q^n dQ$ (note that the integral is not made with respect to s). When $F_n(s)$ satisfies the scaling relation:

$$F_n(s) = A_n s^{\alpha(n)} \tag{156}$$

in the interval $s = (\eta_A, L_A)$, then the knowledge of $(A_n, \alpha(n))$ completely defines the probability distribution of Q and its scale dependence. Scaling laws like (1) have played an important role in our understanding of many complex physical phenomena as well as in fully developed turbulence. An important remark on scaling can be made by rewriting (156) in the following form:

$$F_n(s) = C_{n,m} F_m(s)^{\beta(n,m)} \tag{157}$$

where $C_{n,m} = A_n(1/A_m)^{\beta(m,n)}$ and $\beta(n,m) = \frac{\alpha(n)}{\alpha(m)}$. If $\beta(n,m) = n/m$, then scaling (2) satisfies dimensional counting. When dimensional counting is broken, then $\beta(n,m) \neq n/m$ are called anomalous exponents. Anomalous scaling implies that dimensionless quantities, like the skewness and the kurtosis for instance, have non trivial behavior in s. Because of the condition $\alpha(0) = 0$, which is satisfied for any random field with decent support in s, anomalous scaling also implies that $\alpha(n)$ is a convex non linear function of n. From a theoretical point of view, the scaling (157) can be considered a more general scaling than (156). In terms of equation (157), the probability distribution $P_s(Q)$ is determined by the set $(C_{m,n}, \beta(m,n))$ and the behavior of

one of the moments of Q, say m, is determined by the function $f(s) = F_m(s)$. The relevant remark is that (157) can be true even if (156) is not observed.
Now we want to investigate the relevance of (157) when analyzing turbulence data. In Figure 27 we plot the same data set of Figure 2 for S_6, but we use (157). More precisely, we look at the scaling properties in the form

$$S_6(r) \propto S_3(r)^{\beta(6)} \tag{158}$$

By inspecting Figure 27, an extremely good scaling behavior is observed for all Re. Moreover, a quantitative analysis shows that the exponent $\beta(6) < 2$. More precisely it turns out that $\beta(6) = \zeta(6)$, i.e. the scaling is anomalous and the scaling exponents are the same as observed at large Re.
Results similar to those present in Figure 27 have been obtained in a number of different (homogeneous and isotropic) turbulent flows at low and moderate Re. Even in numerical simulations, the generalized scaling (157) is observed. The quality of the scaling relation shown in Figure 27 can be better understood by looking at the so called local slope of the scaling, namely the quantity

$$z(6,r) = \frac{d\log S_6(r)}{d\log S_3(r)}$$

shown in Figure 28 for the DNS data.

As one can see, the anomalous local scaling extends from few grid points up to 60 grid points, while scaling with respect to r never holds. Let me remark that for small r, the scaling cannot be anomalous simply because in the dissipation region $\delta u(r) \propto r$. Figures 27 and 28 tell us that the generalized scaling (158) extends the usual concept of scaling both at low and moderate Re and at small scales. For this reason, scaling (158) has been called Extended Self Similarity (ESS in the following).
One obvious question of ESS is why it happens and what its physical meaning is. In order to answer these questions, we will use the framework of random multiplicative process discussed in details in the previous section.

Our starting point is to revise the concept of scaling in fully developed turbulence.

Let us consider three length scales $r_1 > r_2 > r_3$ and our basic variables to describe the statistical properties of turbulence, namely the velocity difference $\delta u(r_i)$.
We shall assume that there exists a statistical equivalence of the form:

$$\delta u(r_i) = a_{ij}\delta u(r_j) \tag{159}$$

where $r_i < r_j$ and a_{ij} is a random number with a prescribed probability distribution P_{ij}.
By definition, we have:

$$a_{13} = a_{12}a_{23} \tag{160}$$

Equation (160) is true no matter which value the ratios $\frac{r_1}{r_2}$ and $\frac{r_2}{r_3}$ have. Now we ask ourselves the following question: what is the probability distribution P_{ij} which is functionally invariant under the transformation (160)? This question can be answered by noting that equation (160) is equivalent to:

$$\log a_{13} = \log a_{12} + \log a_{23} \tag{161}$$

(we assume $a_{ij} > 0$). Thus our question is equivalent to ask what the probability distribution stable under convolution is. For independently distributed random variables a solution of this

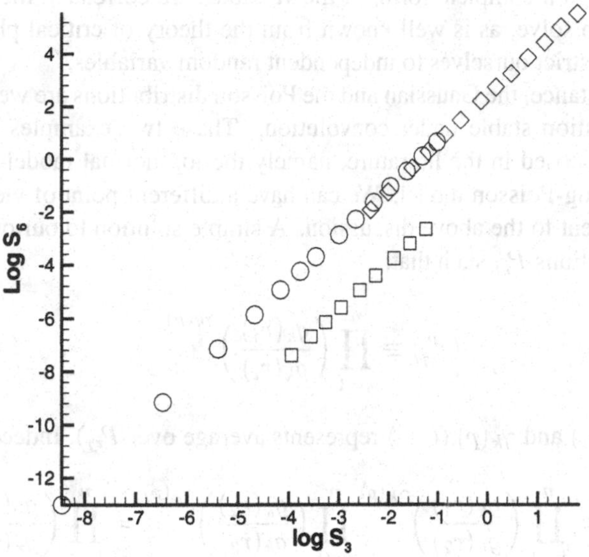

Figure 27: Log-log plot of S_6 versus S_3 for the same set of experimental data as discussed in section 1 and 2.

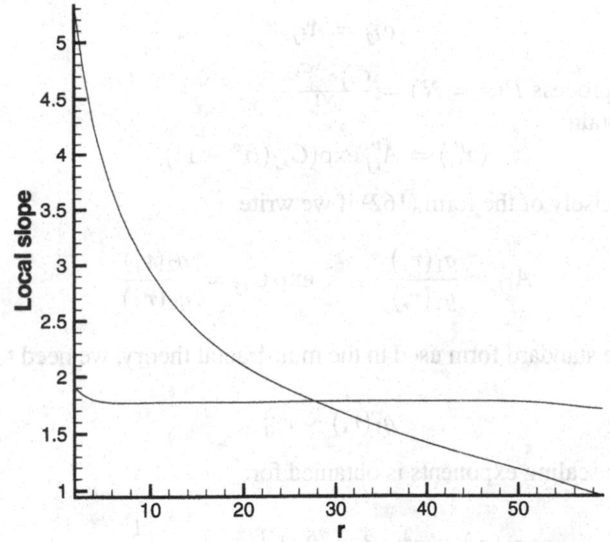

Figure 28: Local slopes of $dlog(S_6)/dlog(S_3)$ (almost straight line) and $dlog(S_6)/dlog(r)$ (continuous curve) for a direct numerical simulation of homogenous and isotropic turbulence (low Re).

problem can be given in a complete form. If the variables are correlated the situation becomes much more difficult to solve, as is well known from the theory of critical phenomena. For the time being we shall restrict ourselves to independent random variables.

In this case, for instance, the Gaussian and the Poisson distributions are well known examples of probability distribution stable under convolution. These two examples correspond to two turbulence models proposed in the literature, namely the log-normal model (already discussed in section 2) and the log-Poisson model. We can have a different point of view on our question which is fully equivalent to the above discussion. A simple solution to our question is given by all probability distributions P_{ij} such that:

$$\langle a_{ij}^p \rangle \equiv \prod_{k=1}^{n} \left(\frac{g_k(r_j)}{g_k(r_i)} \right)^{\gamma_k(p)} \tag{162}$$

for any functions $g_k(r_i)$ and $\gamma_k(p)$ ($\langle \cdots \rangle$ represents average over P_{ij}). Indeed we have:

$$\langle a_{13}^p \rangle = \langle a_{12}^p a_{23}^p \rangle = \prod_{k=1}^{n} \left(\frac{g_k(r_1)}{g_k(r_2)} \right)^{\gamma_k(p)} \prod_{k=1}^{n} \left(\frac{g_k(r_2)}{g_k(r_3)} \right)^{\gamma_k(p)} = \prod_{k=1}^{n} \left(\frac{g_k(r_1)}{g_k(r_3)} \right)^{\gamma_k(p)} \tag{163}$$

We want to remark that equation (162) represents the most general solution to our problem, independent of the scale ratio r_i/r_j.
Let us give a simple example in order to link equation (162) to the case of a probability distribution stable under convolution. Consider the case of a random log-Poisson multiplicative process, namely:

$$a_{ij} = A_{ij} \beta^x \tag{164}$$

where x is a Poisson process $P(x = N) = \frac{C_{ij}^N e^{-C_{ij}}}{N!}$.
By using (164) we obtain:

$$\langle a_{ij}^p \rangle = A_{ij}^p \exp(C_{ij}(\beta^p - 1)) \tag{165}$$

Equation (165) is precisely of the form (162) if we write

$$A_{ij} = \frac{g_1(r_j)}{g_1(r_i)} \qquad , \exp C_{ij} = \frac{g_2(r_i)}{g_2(r_j)} \tag{166}$$

In order to recover the standard form used in the multifractal theory, we need to assume that (see also 172):

$$g_1(r_i) \sim r_{i0}^{h} \tag{167}$$

while a good fit to the scaling exponents is obtained for:

$$g_2(r_i) \sim r_i^2 \quad \beta = (2/3)^{1/3} \quad h_0 = \frac{1}{9} \tag{168}$$

This example highlights one important point in our discussion, i.e. the general requirement of a scale invariant random multiplier (162) does not necessarily imply a simple power law scaling as expressed by the equations (167-168). Moreover, the general expression (162) is compatible only to an infinitively divisible distribution.

It is worthwhile to review the multifractal "language" in light of the previous discussion. In the multifractal language of turbulence, the two basic assumptions are:

I) The velocity difference at scale r shows local scaling law with exponent h, i.e. $\delta u(r) \sim r^h$;

II) the probability distribution to observe the scaling $\delta v(r) \sim r^h$ is given by $r^{3-D(h)}$.

Hence, using the multifractal language, there are two major ansatzes: one concerns power law scaling of the velocity difference (assumption I) and the other one concerns a geometrical interpretation (the fractal dimension $D(h)$) of the probability distribution to observe a local scaling with exponent h. How is it possible to generalize the multifractal language in order to take into account equation (162)?

As we shall see, the theory of infinitively divisible distribution is the tool we need to answer the previous questions. All published models of turbulence based on infinitively divisible distribution are equivalent to writing $D(h)$ in the form:

$$3 - D(h) = d_0 f\left[\frac{h - h_0}{d_0}\right] \tag{169}$$

where d_0 and h_0 are two free parameters while the function $f(x)$ depends only on the choice of the probability distribution. For instance for the log-normal distribution $f(x) = x^2$. Equation (169) allows us to write:

$$\langle (\delta u(r)^p \rangle = \int d\mu(h) r^{hp} r^{3-D(h)} = r^{h_0 p + d_0 H(p)} \tag{170}$$

where

$$H(p) = \inf_x (px + f(x)) \tag{171}$$

We can see that equation (170) is equivalent to a random multiplicative process given by:

$$\langle a_{ij}^p \rangle = \left(\frac{r_j}{r_i}\right)^{h_0 p} \left(\frac{r_j}{r_i}\right)^{d_0 H(p)} \tag{172}$$

Equation (172) can be generalized to the form (162) by allowing h_0 and d_0 to depend on r, i.e.

$$\langle a_{ij}^p \rangle = \frac{\left(r_j^{\overline{h}_0(r_j)}\right)^p}{\left(r_i^{\overline{h}_0(r_i)}\right)^p} \left[\frac{r_j^{\overline{d}_0(r_j)}}{r_i^{\overline{d}_0(r_j)}}\right]^{H(p)} \tag{173}$$

where:

$$\overline{h}_0(r) = h_0 s_h(r) \qquad \overline{d}_0(r) = d_0 s_d(r) \tag{174}$$

Equation (173) becomes equivalent to (162) through the use of:

$$g_1(r_i) = r_i^{\overline{h}_0(r_i)} \qquad g_2(r_i) = r_i^{\overline{d}_0(r_i)} \tag{175}$$

$$\gamma_1(p) = p \tag{176}$$

$$\gamma_2(p) = H(p) \tag{177}$$

The same results can be obtained by (170), i.e. we have

$$\langle \delta u(r)^p \rangle = r^{\overline{h}_0(r)p+\overline{d}_0(r)H(p)} \tag{178}$$

Note that the saddle point evaluation of (170) is not spoiled by the dependence of h_0 and d_0 on r.

We have seen that (162) can be reformulated in terms of multifractal language for infinitively divisible distribution whose function $D(h)$ can be rewritten as in (169). We can ask the following question: what is the physical meaning of (162) or its multifractal analogues (173- 178)? It is precisely the multifractal language which allows us to answer this question. Indeed, the two basic assumptions for the multifractal language can now be replaced in the following way:

I) the velocity difference on scale r behaves as

$$\delta v(r) \sim g_1(r) g_2(r)^x; \tag{179}$$

II) the probability distribution to observe I is $g_2(r)^{f(x)}$.

From this, we obtain

$$\langle \delta v(r)^p \rangle = \int d\mu(x) g_1(r)^p g_2(r)^{px+f(x)} = g_1(r)^p g_2(r)^{H(p)} \tag{180}$$

by employing a saddle point integration. The clearest physical interpretation of (180) is that the probability to observe a given fluctuation of the velocity difference has no longer a geometrical interpretation linked to the fractal dimension $D(h)$. The probability distributions are controlled by a dynamical variable $g_2(r)$ which at this stage we still need to understand. An insight into the dynamical meaning of $g_2(r)$ can be obtained from the following considerations. Let us define $\epsilon(r)$ as the average of the energy dissipation at a scale r. We can define the eddy turnover time $\tau(r)$ at scale r as:

$$\frac{\delta u^2(r)}{\tau(r)} \sim \epsilon(r) \tag{181}$$

In order to compute $\tau(r)$ we need some information linking velocity fluctuations $\delta u(r)$ and energy dissipation $\epsilon(r)$. In principle this information is obtained by using the Kolmogorov equation (1) or the refined hypothesis (24). The latter, however, assumes scaling properties with respect to r. In view or our theoretical analysis we will generalize (24) to the form:

$$\langle (\delta u(r))^{3p} \rangle = \frac{\langle (\epsilon(r))^p \rangle \langle (\delta u(r))^3 \rangle^p}{\langle \epsilon \rangle} \tag{182}$$

Equation (182) is equivalent to:

$$\frac{\epsilon(r)}{\langle \epsilon \rangle} =^s \frac{\delta v^3(r)}{\langle \delta v^3(r) \rangle} \tag{183}$$

where $=^s$ means that all moments on the r.h.s. are equal to the l.h.s. By using (181-183) we obtain the definition of length $L(r)$:

$$L(r) \equiv \delta v(r) \tau(r) = \frac{\langle \delta v^3(r) \rangle}{\epsilon} \tag{184}$$

$L(r)$ cannot be regarded as a real length scale in physical space. Rather, $L(r)$ should be considered as a dynamical variable entering into the statistical description of turbulence. This is precisely the idea behind ESS which reformulates the scaling properties of turbulence in terms of $L(r)$. Indeed in order to obtain ESS from (180) it is sufficient to state that, within the range of scales where ESS is observed, $g_1(r)^{1/h_0} \sim g_2(r)^{1/d_0} \sim L(r)$.

The physical meaning of ESS is strictly linked to (184) and in particular to (183) which is a generalization of the Kolmogorov Refined Similarity Hypothesis. In order to check this idea, we show in Figure 29 and 30 the scaling proposed in (182) for $p = 6$. A striking agreement with (182) is observed.

Let us summarize all our previous findings:
A) we have introduced the idea of a scale invariant random multiplier satisfying equation (162);
B) we have shown that infinitively divisible distributions are all compatible with (162);
C) we have shown that the multifractal language specialized for the case of infinitively divisible distribution gives equation (162) (with $n = 2$ and $\gamma_1(p)$ linear in p) and it is equivalent to a scale invariant random multiplier;
D) finally we have argued that the correct scaling parameter to describe the statistical properties of small scale turbulent flows is not directly linked to a simple geometrical interpretation, rather it should be considered a dynamical variable.

We have shown, in the theoretical framework so far exposed, that we recover the ESS when $g_1(r)^{1/h_0} \sim g_2^{1/d_0} \sim L(r)$.
We can also use (178) and (180) to simulate a synthetic signal according to a random multiplicative process satisfying (173). This can be done by using the algorithm introduced in section 4. Let us consider a wavelet decomposition of the function $\phi(x)$:

$$\phi(x) = \sum_{j,k=0}^{\infty} \alpha_{j,k}\psi_{j,k}(x) \tag{185}$$

In order to generalize this construction for a function showing ESS of the form (178), it is now sufficient to take a probability distribution, $P_l(\eta)$, for the random multiplier with the appropriate scale dependency (162). This will be implemented by allowing a dependency of $P(\eta_{jk})$ on the scale $r_j = 2^{-j}$.
According to the previous discussion, ESS corresponds to having only one seed-function defining the multiplicative process, i.e. $g_1(r)^{1/h_0} = g_2(r)^{1/d_0} = L(r)$ in the range of scales where ESS is valid. Following this recipe we define the signal such that :

$$\langle \delta u(r)^p \rangle = U_0^p L(r)^{p/3} L(r)^{\zeta(p)-p/3} \tag{186}$$

where

$$L(r) = \langle \delta u(r)^3 \rangle / U_0^3 \tag{187}$$

The function $\langle \delta u(r)^3 \rangle$ is always very well fitted by the Batchelor parameterization:

$$\langle \delta v(r)^3 \rangle = \frac{U_0^3}{L\eta_k^2} \frac{r^3}{(1 + (r/\eta_k)^2)}. \tag{188}$$

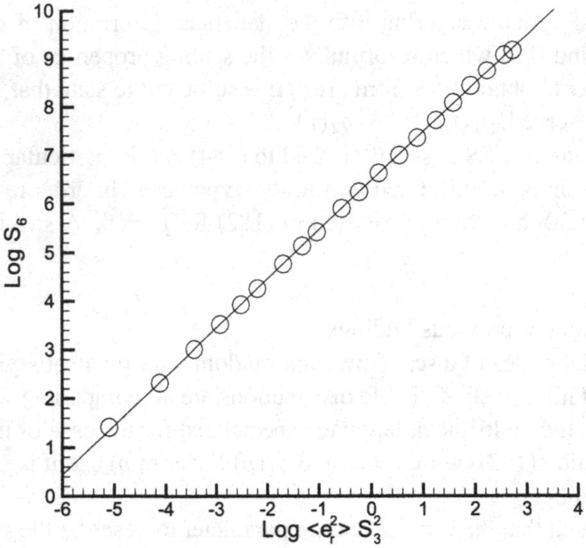

Figure 29: Check of the scaling properties in ESS form of the refined Kolmogorov similarity hypothesis (experimental data).

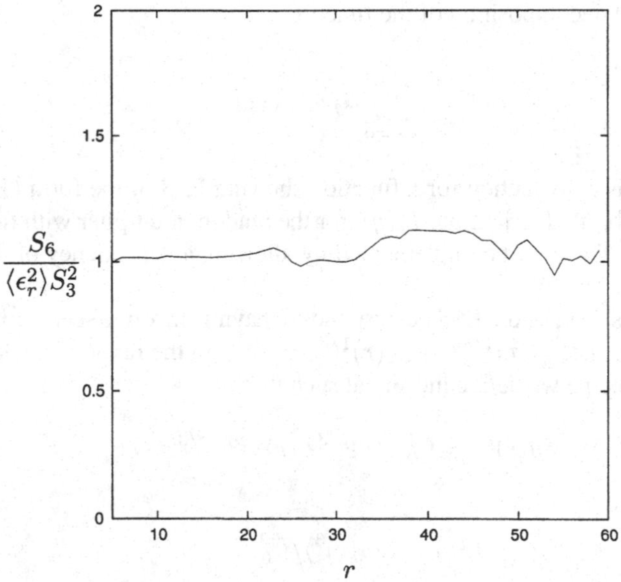

Figure 30: Ratio $S_6/\langle\epsilon_r^2\rangle S_3^2$ as a function of r for the same data set of Figure 29.

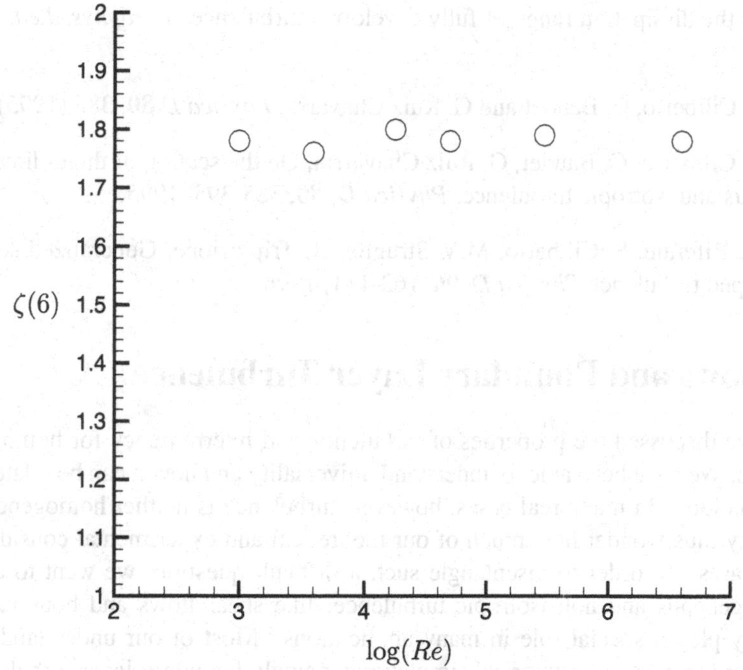

Figure 31: Values of $\zeta(6)$ computed from different data sets for homogenous and isotropic turbulence plotted against $log(Re)$.

Numerical simulations based on the above algorithm exhibit the scaling properties of ESS, i.e. there is no clear scaling in r while there is an exact scaling behaviour with respect to S_3. So we can safely reach the conclusion that ESS is a property of a class of anomalous (multifractal) random fields. More importantly, we reach the conclusion that ESS is strongly linked to the generalization of refined Kolmogorov similarity hypothesis (182), very well reproduced in laboratory experiments and in DNS.

Before closing this section, it is worthwhile to discuss briefly the laboratory findings on intermittency. In particular, we want to understand whether with increasing Re there is a tendency towards increasing or decreasing intermittency in turbulent flows. Using ESS, one can accurately measure $\zeta(n)$ even for low Re. In Figure 31 we show the outcome of such an analysis, i.e. we plot $\zeta(6)$ as a function of Re. No change in $\zeta(6)$ is observed over 4 order of magnitude. This is a strong and clear check of universal properties of intermittency in turbulent flows.

References for section 5

Extended self-similarity has been extensively discussed in:

- R. Benzi, S. Ciliberto, R. Tripiccione, C. Baudet, F. Massaiolie, S. Succi, Extended Self Similarity in turbulent flows. *Phys. Rev E*48, R29 (1993);

- R. Benzi, S. Ciliberto, C. Baudet, G. Ruiz Chavarria and R. Tripiccione, Extended self similarity in the dissipation range of fully developed turbulence. *Europhys. Lett.* **24**, 275 (1993);

- R. Benzi, S. Ciliberto, C. Baudet and G. Ruiz Chavarria *Physica D* **80**, 385 (1995);
- R. Benzi, S. Ciliberto, C. Baudet, G. Ruiz Chavarria, On the scaling of three-dimensional homogeneous and isotropic turbulence. *Physica D*, **80**, 385-398, 1995.
- R. Benzi, L. Biferale, S. Ciliberto, M.V. Struglia, R. Tripiccione, Generalized scaling in fully developed turbulence *Physica D*, **96**, 162-181, 1996.

6 Shear Flows and Boundary Layer Turbulence

Up to now we have discussed the properties of turbulence and intermittency for homogeneous and isotropic flows. We have been able to understand universality and how it can be related to the Navier-Stokes equations. In many real cases, however, turbulence is neither homogeneous nor isotropic. One may thus wonder how much of our theoretical and experimental considerations survive in those cases. In order to disentangle such a difficult question, we want to consider the case of homogeneous and non isotropic turbulence, like shear flows and boundary layer turbulence, as they play a special role in many applications. Most of our understanding will be obtained by using two sets of numerical simulations, namely boundary layer turbulence and homogeneous shear flow.

For boundary layer turbulence, we consider a channel flow as depicted in Figure 32 obtained through the use of a direct numerical simulation with high statistical accuracy (about 10^3 eddy turnover times where simulated, in time units U_0/h, where U_0 is the centerline velocity and h is the channel half-width).

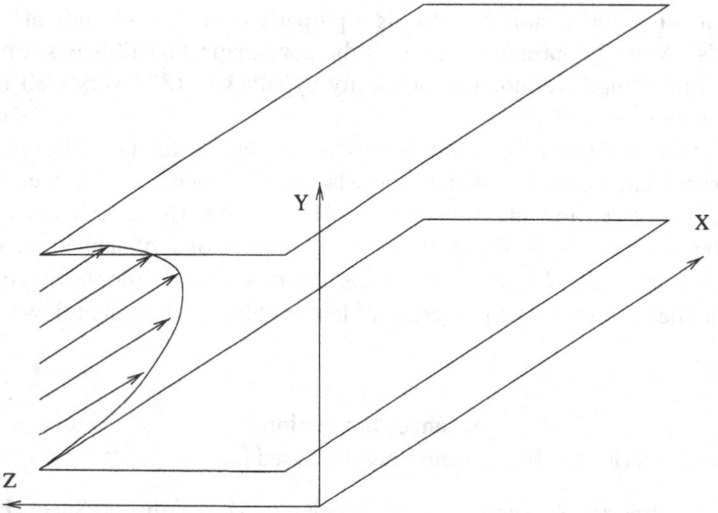

Figure 32: Schematic picture of the flow in a channel.

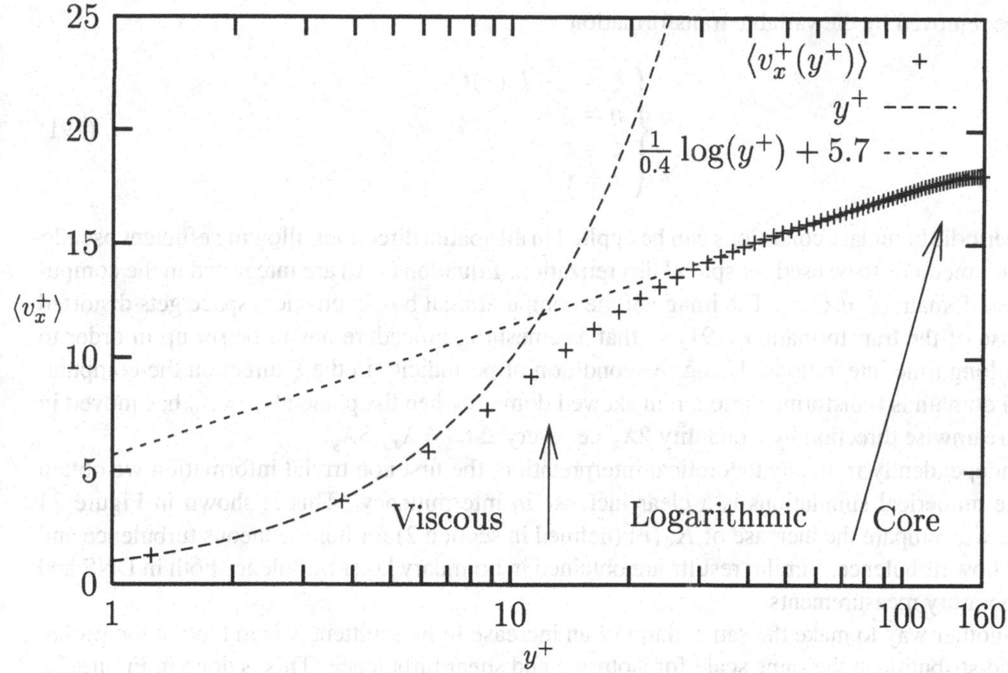

Figure 33: Mean velocity profile of the turbulent boundary layer in a channel flow.

In the following we use wall units defined as $y^+ = y/\delta$ and $v^+ = v/v^*$ where v^* is the friction velocity and $\delta = \nu/v^*$ is the typical boundary layer thickness. The mean velocity profile obtained near the wall is shown in Figure 33 and it is compared with well known boundary layer theories.

Concerning the homogeneous shear flow, we have considered a turbulent flow with an imposed mean velocity gradient S. The Navier-Stokes equations are written in terms of velocity fluctuations

$$\vec{u}(\vec{x},t) = U(y)\vec{e}_1 + \vec{v}(\vec{x},t) \tag{189}$$

where $\vec{x} \in V = [-\lambda_x, \lambda_x] \times [-\lambda_y, \lambda_y] \times [-\lambda_z, \lambda_z]$ identifies the computational box, $U(y) = S(y + \lambda_y)$ is the mean flow and $\vec{e}_1$ is the unit vector in the streamwise direction x. The mean gradient S is in the normal direction y, while z denotes the spanwise coordinate. The Navier-Stokes equations for the velocity fluctuations may be written as

$$\begin{cases} \vec{\nabla} \cdot \vec{u} = 0 \\ \dfrac{\partial \vec{u}}{\partial t} = (\vec{u} \times \vec{\zeta}) - \vec{\nabla}(p + \dfrac{u^2}{2}) + \nu\nabla^2\vec{u} - Sv\vec{e}_1 - U(y)\dfrac{\partial \vec{u}}{\partial x}. \end{cases} \tag{190}$$

Because of the mean flow a non periodic term $U(y)\partial\vec{u}/\partial x$ appears in the equations. This term can be removed by the variable transformation

$$\begin{cases} \xi = x - U(y)t \\ \eta = y \\ \zeta = z \\ \tau = t. \end{cases} \tag{191}$$

and periodic boundary conditions can be applied in all spatial directions allowing efficient pseudo-spectral methods to be used for spatial discretization. Equation (190) are integrated in the computational domain (ξ, η, ζ, τ). The image of the computational box in physical space gets distorted because of the transformation (191) so that a remeshing procedure has to be set up in order to allow long time integrations. Using the condition of periodicity in the ξ direction the computational domain is transformed into a non skewed domain when the plane at $y = \lambda_y$ has moved in the streamwise direction by a quantity $2\lambda_x$ i.e. every $\Delta t_r = \lambda_x / S\lambda_y$.

Independently from any theoretical interpretation, the first non trivial information we obtain by the numerical simulations is a clear increase in intermittency. This is shown in Figure 34 where we compare the increase of $K_4(r)$ (defined in section 2) for homogeneous turbulence and shear flow turbulence. Similar results are obtained in boundary layer turbulence both in DNS and in laboratory measurements.

Another way to make the same claim of an increase in intermittency, is to look at the probability distribution at the same scale for isotropic and shear turbulence. This is done in Figure 35.

The next step in our analysis is to search scaling relations in order to understand the increase of intermittency. In non-isotropic turbulence, scaling ideas may be very dangerous because the usual framework of the Kolmogorov scenario may be misleading. At any rate, by keeping in mind the above considerations, we shall investigate the scaling properties of longitudinal structure functions. We start with the boundary layer turbulence. To study intermittency in the channel, we introduce the following y-dependent longitudinal streamwise structure functions:

$$S_p(r^+, y^+) = \langle |(v_x(x^+ + r^+, y^+, z^+) - v_x(x^+, y^+, z^+)|^p \rangle \tag{192}$$

The average is taken at a fixed y^+ value (the normal to wall coordinate). The quantities $S_p(r^+, y^+)$ have been measured for each value of y. Due to the low Reynolds number, we use ESS in order to extract ζ_p values. For the computation of the scaling exponents $\zeta_p(y)$ it is necessary to analyze the ESS local slopes $D_{p,q}(r^+, y^+) = d\log(S_p(r^+, y^+))/d\log(S_q(r^+, y^+))$ for each value of the y^+ coordinate. A sample of the quantities $D_{6,3}(r^+, y^+)$ is shown in Figure 36.

By an accurate inspection of the local slopes, one reaches the conclusion that there are two regions in y^+, hereafter referred to as region H ("Homogeneous") and region B ("Boundary"), where well defined constant local slopes for the scaling exponents can be detected. Region H is close to the center of the channel ($y^+ \geq 100$) while region B is close to the viscous sublayer ($20 \leq y^+ \leq 50$). In region H, the scaling exponents $\zeta_p(H)$ are found to be approximately the same as the ones measured in homogeneous and isotropic turbulence. On the other hand, in region B the scaling exponents $\zeta_p(B)$ have been found to be much smaller than $\zeta_p(H)$. Moreover, while in region H the scaling range starts at $r^+ \geq 25$, in region B the scaling range starts at $r^+ \geq 50$. In the intermediate region between region H and region B, it is difficult to identify a range in r where a scaling exponent can be defined with enough confidence. In Figure 37 we

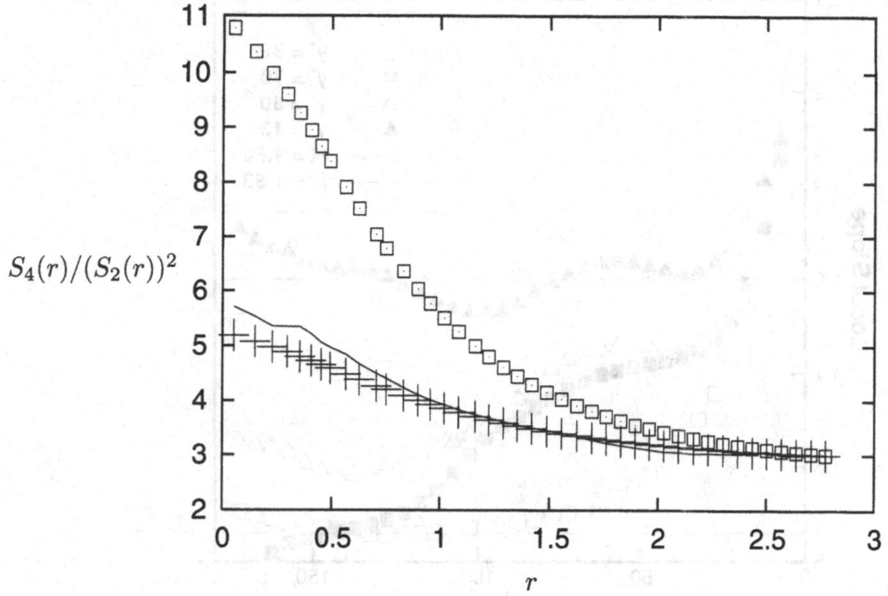

Figure 34: Observed values of the flatness for the velocity differences as a function of r, observed in homogenous and isotropic turbulence (+) and shear flows (squares).

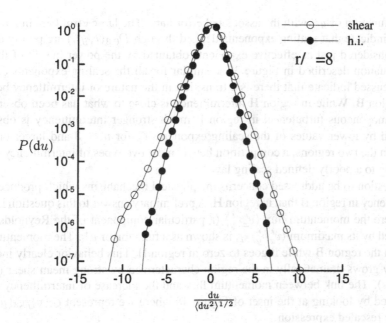

Figure 35: Probability density function at $r \sim 8\eta$ for homogenous and isotropic turbulence and shear flows.

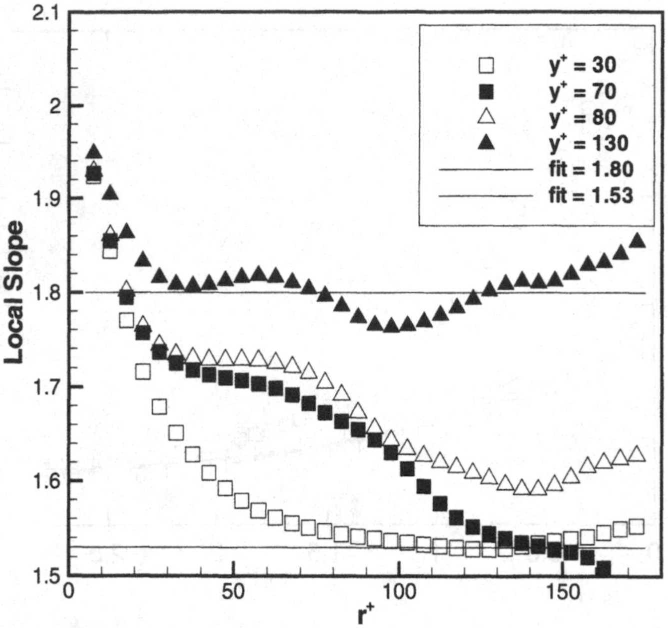

Figure 36: ESS local slopes of the longitudinal structure functions at different values of y^+ in a turbulent channel flow.

show ζ_6/ζ_3 as a function of y^+ with the associated error bars. The large error bars in the region $50 \leq y^+ \leq 100$ indicate that scaling exponents defined through $D_{6,3}(r, y^+)$ are poorly defined and should be considered just as effective exponents obtained by the power law fit of the ESS analysis. The situation described in Figure 37 is similar for all the scaling exponents ζ_p. The results so far discussed indicate that there is a transition in the nature of intermittency between region H and region B. While in region H intermittency is close to what has been observed in isotropic and homogeneous turbulence, in region B much stronger intermittency is observed, which is reflected by lower values of the scaling exponents ζ_p for $p > 3$ and larger ones for $p < 3$. In between the two regions, a competition between the two types of intermittency should take place, leading to a poorly defined scaling law.

An important question to be addressed concerns the physical mechanisms which produce much stronger intermittency in region B than in region H. A preliminary answer to this question is given in Figure 38 where the momentum flux $\langle v'_x v'_y \rangle$ (a particular component of the Reynolds stress tensor) normalized by its maximum $\langle v'_x v'_y \rangle_M$, is shown as a function of y^+. The momentum flux has a peak within the region B while it goes to zero in region H. This behavior clearly indicates that intermittency grows dramatically in the region characterized by strong mean shear (strong momentum fluxes). The link between momentum flux and the increase of intermittency can be further investigated by looking at the inset of Figure 38 where we represent $\zeta_6(y)/\zeta_3(y)$ from Figure 37 and the rescaled expression

$$\frac{\zeta_6(y^+)}{\zeta_3(y^+)} - \frac{\zeta_6(H)}{\zeta_3(H)} = \left(\frac{\zeta_6(B)}{\zeta_3(B)} - \frac{\zeta_6(H)}{\zeta_3(H)} \right) \frac{\langle v'_x v'_y \rangle}{\langle v'_x v'_y \rangle_M}. \tag{193}$$

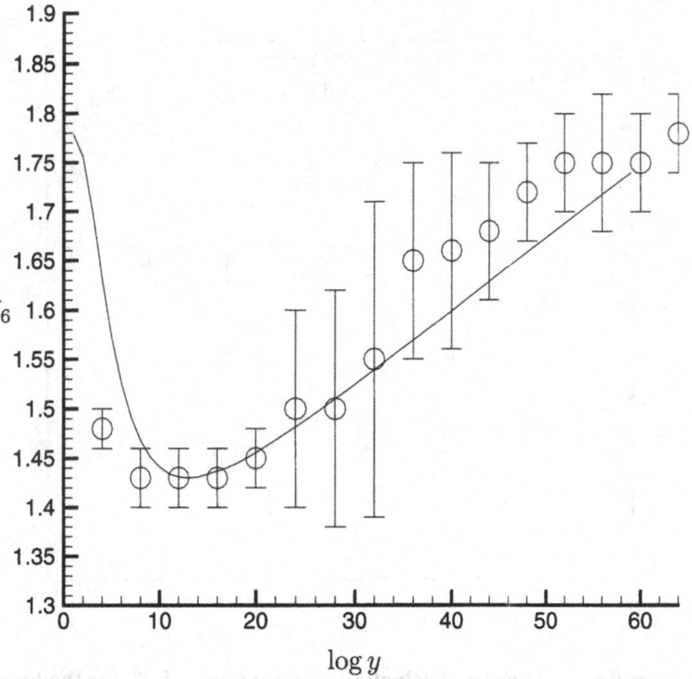

Figure 37: ESS exponent ζ_6 computed (with error bars) for the turbulent channel flow as a function of y^+. The continuous line represents the normalized value of the momentum flux (see the text).

We can therefore argue that the increase of intermittency should be related to the increase of momentum flux and, therefore, to the (local) mean shear.
A more quantitative way to investigate the increase of intermittency in region B, can be achieved by the following argument. One can generalize the Kolmogorov equation for homogeneous shear flows, in a way similar to what has been discussed in section 1 for thermal convection. Without doing a lot of mathematics, it is clear that the presence of a mean shear $S(y)$ introduces a new term in the Kolmogorov equations proportional to $S(y)$. Then, one can define, by dimensional consideration, a length scale $L_s(y)$ in terms of the mean energy dissipation $\epsilon(y)$ and the mean shear $S(y)$, as follows:

$$L_s(y) = \left(\frac{\langle \epsilon(y) \rangle}{S(y)^3} \right)^{1/2} . \tag{194}$$

In presence of mean shear $S(y)$, for any scale r we can define two characteristic timescales, namely the eddy turnover time $r/\delta v(r)$ and $1/S(y)$. We expect that when the mean shear is large enough, the eddy turnover time is not the relevant timescale for energy transfer from large to small scales. The inequality $r/\delta v(r) \langle 1/S(y)$ gives the range of scales r where the effect of shear should not be relevant to small scale statistics. By using the Kolmogorov estimate $\delta v(r) \sim \epsilon^{1/3} r^{1/3}$, we find that the above inequality can be written as $r \leq L_s$. Thus, for $r \leq L_s$ one expects that the scaling properties of turbulence are not affected by the mean shear. On the other hand, for $r \geq L_s$ one expects that the mean shear may significantly change the amount

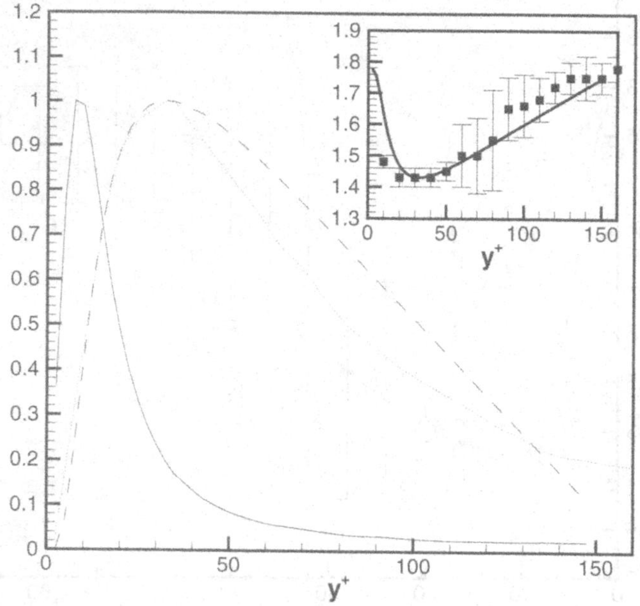

Figure 38: Momentum flux and energy dissipation as a function of y^+ for the turbulent channel flow.

of intermittency. In our numerical simulation $L_s(y)$ becomes small only in the region B, i.e. in the region where an increase of intermittency is observed. A computation of $L_s(y)$ is shown in Figure 39. This preliminary analysis seems to indicate that the change of scaling exponents cannot be represented as a perturbative effect in terms of the mean shear.

In principle, one may attempt to describe the increase of intermittency by employing the refined Kolmogorov similarity hypothesis (24) or its generalized version (182). The larger intermittency (smaller ζ_p) would be provided by an increase of intermittent fluctuations of ϵ_r (larger values of $|\tau(p)|$), due, for instance, to the quite ordered vortical structures, which are known to be active in the regions very close to the wall. In such conditions, the anomaly of the scaling exponents would strongly depend on local flow properties, loosing, thus, any trait of universality.

It is therefore interesting to investigate the validity of the refined Kolmogorov similarity hypothesis in shear dominated flows. We start by considering the boundary layer turbulence. In Figure 40 we plot on a logarithmic scale the structure function of order six versus $\langle \epsilon(r)^2 \rangle \langle \delta u(r)^3 \rangle^2$.

On the basis of the assumed validity of (182), the plot should result in a straight line of slope $s = 1$, independent of the distance from the wall. This behavior actually emerges near the center of the channel while in the wall region a quite clear, though small, violation manifests itself. Specifically, for $y^+ = 31$ two different scaling laws appear. One, characterized by the slope $s = 1$, trivially pertains to the dissipative range. The other, with slope $s = .88$, which doesn't satisfy (182), shows a first clear example of failure of the RKSH.

The previous discussion may suggest a relationship between the increase of intermittency, observed in the near wall region, and the simultaneous breaking of the RKSH. In this respect,

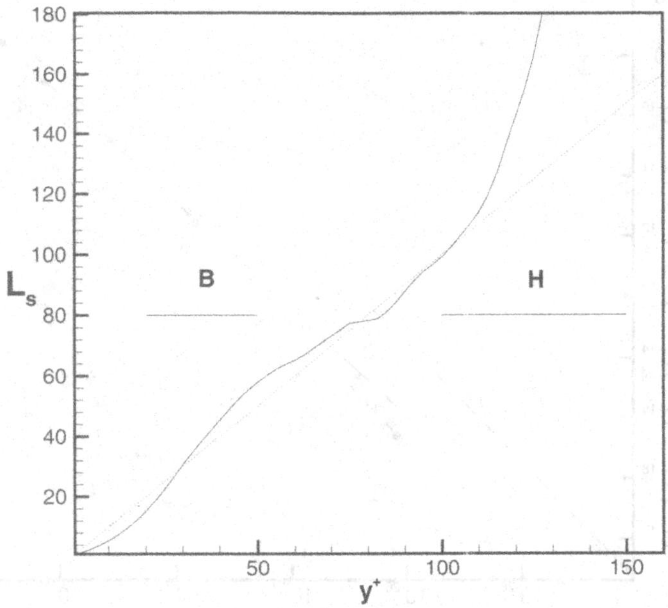

Figure 39: Plot of L_S as a function of y^+ for the turbulent channel flow.

it seems interesting to investigate the possible existence of a new form of RKSH valid in the near wall region. In fact RKSH, somehow suggested by the well known "4/5" Kolmogorov equation tells us, in physical terms, that the "energy flux" in the inertial range, represented by the term $(\delta u(r))^3$, fluctuates with a probability distribution which is the same as that of $\epsilon(r)r$. As discussed before, when a mean homogeneous shear is acting in the system, we expect that a new term is acting in the Kolmogorov equation, which by dimensional counting, should be proportional to $rS(y)\langle(\delta u(r))^2\rangle$. If this term becomes dominant, as it may occur for a very large shear, one is led to assume that the fluctuations of the energy flux in the inertial range are proportional to $(\delta u_r)^2$, i.e. $\epsilon(r) \propto A(r)(\delta u_r)^2$, with $A(r)$ a non fluctuating function of r. Hence, we may expect that a new form of the RKSH should hold which, in its generalized form, reads as

$$\langle(\delta u(r))^p\rangle \propto \frac{\langle\epsilon(r)^{p/2}\rangle}{\langle\epsilon\rangle^{p/2}}\langle\delta u(r)^2\rangle^{p/2}. \qquad (195)$$

The above expression of the new RKSH is given in terms of the structure function of order two, without explicit reference to the separation r, in the same way as the generalized RKSH (182). In the spirit of the extended self similarity, we assume the new form of RKSH to be valid also in the region very close to the wall, where the shear is certainly prevailing.

In order to verify this set of assumptions, we show in Figure 41 a log-log plot of equation (195) for $p = 4$ at $y^+ = 31$.

In the insert, we show for the same plane the compensated plot of both (195) for $p = 4$ and (182) for $p = 6$. A quite clear agreement of equation (195) with the numerical data is seen.

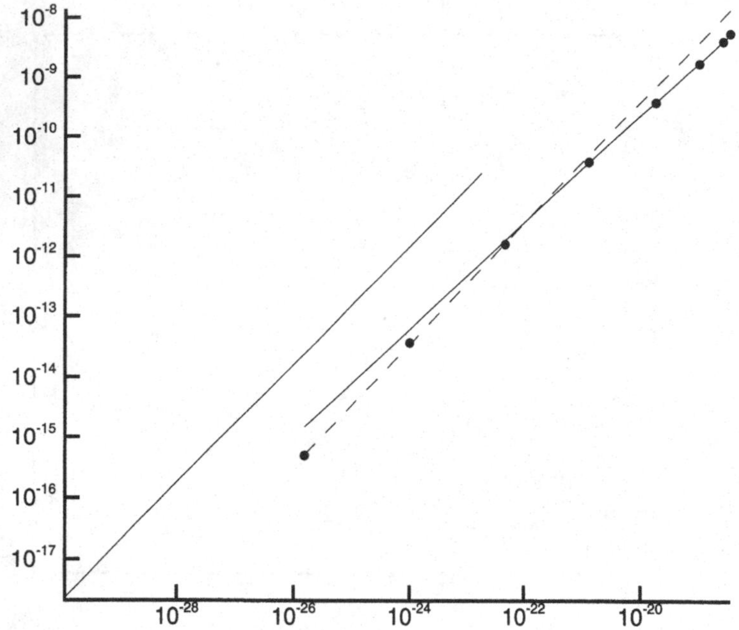

Figure 40: Log-log plot of $\langle(\delta u(r))^6\rangle$ versus $\langle\epsilon(r)^2\rangle\langle(\delta u(r))^3\rangle^2$ in the center of the channel (triangles) and in the buffer region (black circles). The best fit in the center of the channel has slope -1 which verifies the refined Kolmogorov similarity hypothesis. For the buffer region at large enough scales the slope -1 (dashed line) is no longer in agreement with the numerical findings, showing that the similarity hypothesis is broken.

In principle, the function $A(r)$ might be evaluated theoretically starting from the Kolmogorov equation for anisotropic shear flow.

The same analysis can be done for homogeneous shear flows and it gives exactly the same results, i.e. a breaking of (182) and the validity (for $r \geq L_s$) of (195). This result shows that the increase of intermittency and its possible explanation, is due to the presence of a mean shear and *not* due to the presence of the wall. Moreover, in the case of homogeneous shear flow the geometry is much simpler than in a turbulent channel flow. This allows us to compute more accurately the energy dissipation and its scale dependent properties. In Figure 42 we show a log-log plot of $\langle\epsilon(r)^3\rangle$ versus $\langle\epsilon(r)^2\rangle$, while in the insert we show a plot of $\langle\epsilon(r)^2\rangle$ versus r. These plots are produced for both homogeneous shear flow turbulence and homogeneous and isotropic turbulence. By inspecting Figure 42 we can safely deduce that the scale dependent properties of energy dissipation do not change.

Starting with the above discussion, we are now able to "predict" the quantitative increase of intermittency in shear dominated flows. Let us define the scaling exponents $\tau(q)$ by the relation

$$\langle(\epsilon(r))^q\rangle \sim \langle(\delta u(r))^3\rangle^{\tau(q)} \tag{196}$$

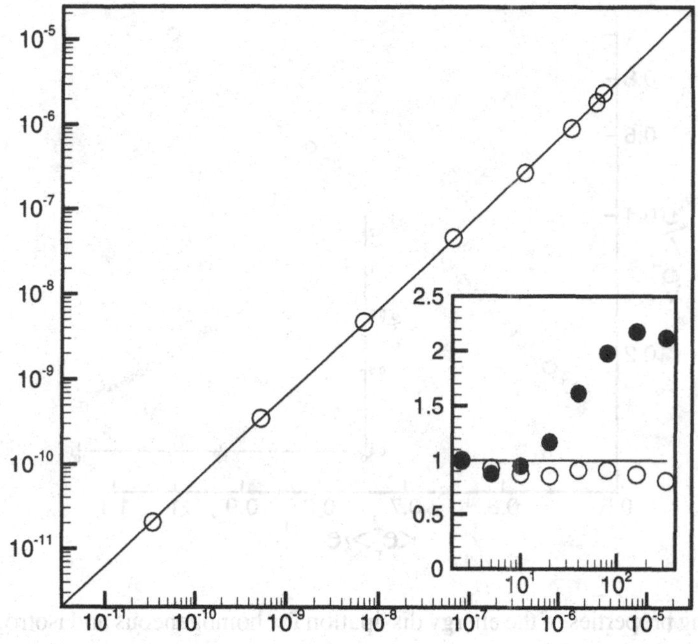

Figure 41: Log-log plot of $\langle(\delta u(r))^4\rangle$ versus $\langle\epsilon(r)^2\rangle\langle(\delta u(r))^2\rangle$ in buffer region (white circles). The best fit has slope -1 which verifies the new form of the refined Kolmogorov similarity hypothesis. In the insert, we show the ratio $\langle(\delta u(r))^4\rangle/\langle\epsilon(r)^2\rangle\langle(\delta u(r))^2\rangle$ (white circles) and $\langle(\delta u(r))^6\rangle/\langle\epsilon(r)^2\rangle\langle(\delta u(r))^3\rangle^2$ black circles, for the buffer region.

where we use the "ESS" language for this definition. In terms of (182) the relation among the $\tau(q)$ and the scaling exponents of the velocity structure functions $\zeta(q)$, defined by using ESS, is

$$\tau(q) = \zeta(3q) - q \tag{197}$$

On the other hand, equation (195) tells us that a new relation must be employed for shear dominated flows, namely:

$$\tau(q) = \zeta(2q) - q\zeta(2) \tag{198}$$

Knowing the values of $\zeta(q)$ for shear dominated flows, we can compute $\tau(q)$ by using (197) and (198) and compare the results with homogeneous and isotropic turbulence. This is done in Figure 43 where the black circles correspond to (197), the triangles to (198) and the white circles connected by a line to the values of $\tau(q)$ obtained for homogeneous and isotropic turbulence. Again, we reach the conclusion that the scaling properties of energy dissipation are not changed in shear dominated flows.

The increased intermittency of the velocity fluctuations for shear dominated flows may be estimated by considering how the flatness $K_4(r)$ grows with $r \to 0$, with

$$K_4(r) = \frac{\langle\delta u^4(r)\rangle}{\langle\delta u^2(r)\rangle^2}. \tag{199}$$

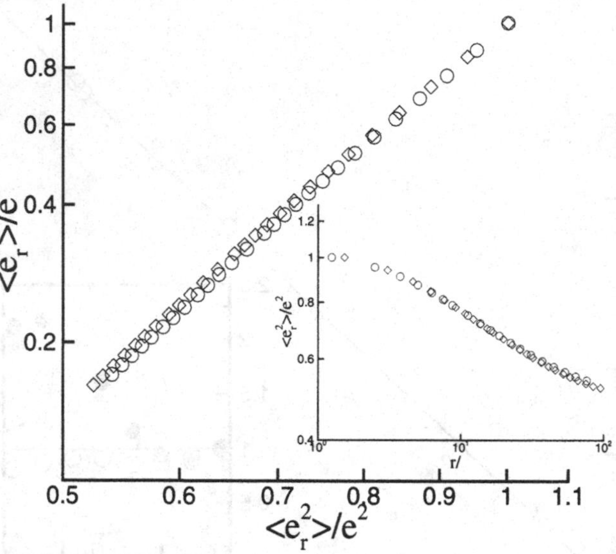

Figure 42: Scaling properties of the energy dissipation for homogeneous and isotropic turbulence and shear flow turbulence.

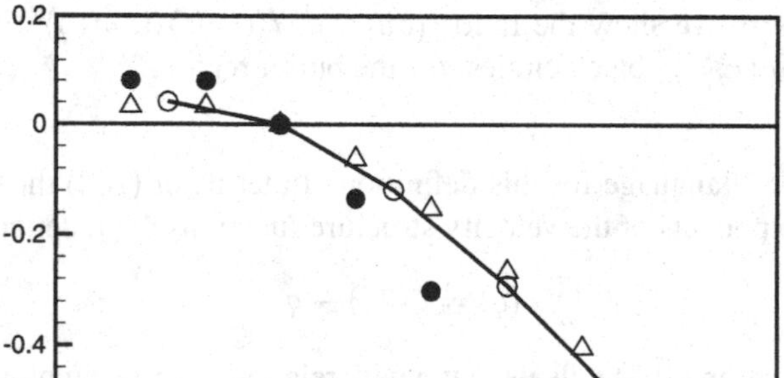

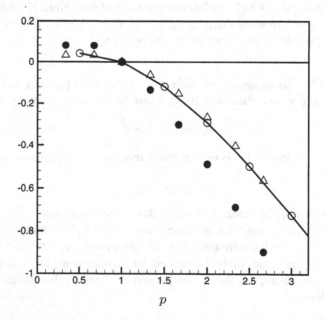

Figure 43: Computation of the scaling exponents $\tau(q)$ derived by the old RKSH (black circles) and new RKSH (triangles) as compared to the homogenous and isotropic values (white circles and continuous line)

By combining the definitions (196) and (198) with (199) we obtain the following expressions in terms of $\epsilon(r)$,

$$K_4^H(r) = \frac{\langle \epsilon_r^{4/3} \rangle}{\langle \epsilon_r^{2/3} \rangle^2} \qquad K_4^S(r) = \frac{\langle \epsilon_r^{4/2} \rangle}{\langle \epsilon_r^{2/2} \rangle^2}, \tag{200}$$

which are suitable for homogeneous and isotropic turbulence and shear dominated flows, respectively. In Figure 44 we show the quantities K_4^H and K_4^S as functions of r with the estimate $K_4^S(r)$. As we can see, there is rather good agreement between the computed intermittency growth and our theoretical analysis.

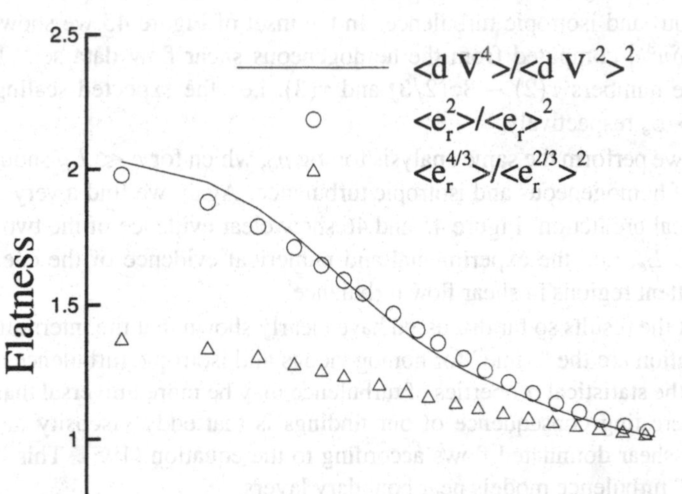

Figure 44: Computation of the flatness for the velocity increments for shear turbulence and comparison using old and new RKSH.

Following equations (182) and (195) and the above discussion, we can now compute the quantities

$$\sigma_p \equiv \langle \delta u^p \rangle / \langle \delta u^2 \rangle^{p/2} \quad \rho_p \equiv \langle \delta u^p \rangle / \langle \delta u^3 \rangle^{p/3}$$

which are expected to satisfy the relations:

$$\sigma_p \propto \begin{cases} \langle \delta u^3 \rangle^{\tau(p/2)} & r \gg L_s \\ \langle \delta u^3 \rangle^{\tau(p/3) - \tau(2/3)p/2} & r \ll L_s \end{cases} \tag{201}$$

and

$$\rho_p \propto \begin{cases} \langle \delta u^3 \rangle^{\tau(p/2)-\tau(3/2)p/3} & r \gg L_s \\ \\ \langle \delta u^3 \rangle^{\tau(p/3)} & r \ll L_s. \end{cases} \tag{202}$$

Equations (201) and (202) allows us to compare the ESS exponents of σ_p and ρ_p against the exponents predicted by equations (24) and (195). Let us remark that equations (201) and (202) are also based on the assumption that $\tau(q)$ is the same for both shear dominated flows and homogeneous and isotropic turbulence, as supported by the data analysis so far performed.

In Figure 45 we plot $\log \sigma_6$ against $\log\langle \delta u^3 \rangle$ for the data of the homogeneous shear flow and the turbulent boundary layer (experimental data).

The fits for $r \gg L_s$ are in close agreement with the value of $\tau(3) = -0.59$ expected from homogeneous and isotropic turbulence. In the inset of Figure 45 we show the local slope $d[\log \sigma_6]/d[\log\langle \delta u^3 \rangle]$ computed from the homogeneous shear flow data set. The two dashed lines indicate the numbers $\tau(2) - 3\tau(2/3)$ and $\tau(3)$, i.e. the expected scaling exponents for $r \ll L_s$ and $r \gg L_s$ respectively.

In Figure 46 we perform the same analysis for $\log \rho_6$, which for $r \ll L_S$ should scale according to the laws of homogeneous and isotropic turbulence. Again we find a very good agreement with the theoretical prediction. Figure 45 and 46 show clear evidence of the two scaling regions below and above L_s, i.e. the experimental and numerical evidence of the coexistence of two different intermittent regions in shear flow turbulence.

We think that the results so far discussed have clearly shown that the intermittent fluctuations of energy dissipation are the "same" for homogeneous and isotropic turbulence and shear dominated flows, i.e. the statistical properties of turbulence may be more universal than we previously thought. An interesting consequence of our findings is that eddy viscosity models should be reformulated for shear dominated flows according to the equation (195). This may explain the failure of "naive" turbulence models near boundary layers.

We finally want to investigate more precisely the effects of anisotropic contributions in the region $r < L_S$. In this region, we should expect that anisotropic contributions tend to decrease with some power of the ratio r/L. The idea stems from the simple observation that inertial and dissipative Navier-Stokes terms are invariant under rotation, i.e. anisotropies may be introduced only by either the boundary conditions or by the external forcing. Therefore, understanding how much anisotropies persist (or not) at small scales is a direct measure of a possible breaking of the universality hypothesis. The best way to disentangle isotropic from anisotropic fluctuations in a systematic way is to decompose any correlation function on the basis of the SO(3) group, the group of spatial rotation in three dimensions. A systematic way to do it has recently been proposed in the literature and a few first attempts to analyze anisotropies in this way have been recently proposed for both experimental and numerical data. The main results, still preliminary, can be summarized as follows.

In the range of scales where anisotropies do not directly influence the time evolution of velocity, i.e. for scales smaller than the shear lengths, L_s, the non-linear rotational invariant terms, $v\nabla v$, and ∇P dominate the velocity physics. In this region, turbulent fluctuations "foliate" with respect to the different sectors of the rotational group. This foliation mechanism leads to the conclusions that (i) scaling properties of any sector (isotropic or not) are universal, i.e. large-scale

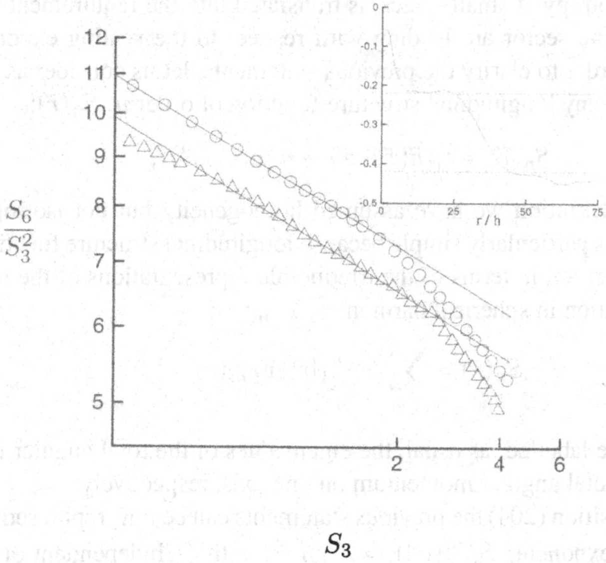

Figure 45: Anomalous scaling of S_6/S_3^2 for homogenous shear flow turbulence (circles) and boundary layer turbulence (triangles).

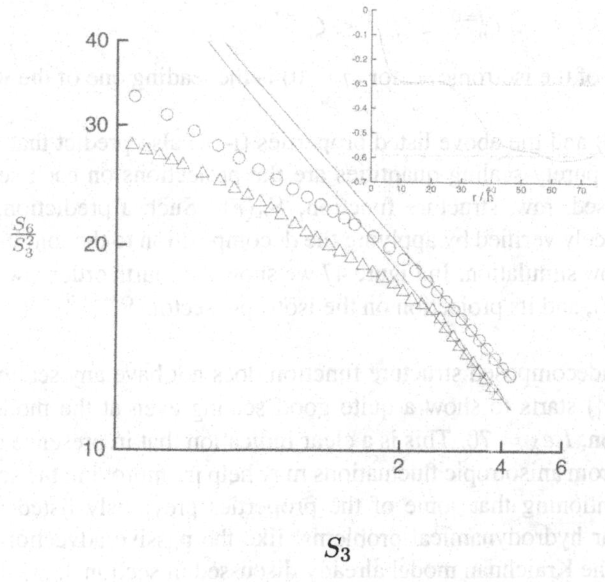

Figure 46: Anomalous scaling of S_6/S_2^3 for homogenous shear flow turbulence (circles) and boundary layer turbulence (triangles).

independent; (ii) different sectors are dominated by different scaling exponents; (iii) universality, i.e. recovering of isotropy at small scales, is translated into the requirement that the scaling exponents of the isotropic sector are leading with respect to the scaling exponents of all the anisotropic sectors. In order to clarify the previous statements, let us consider as an example the SO(3) decomposition of any longitudinal structure function of order n, $S_n(\vec{r})$:

$$S_n(\vec{r}) = \langle [(\vec{u}(\vec{x}+\vec{r}) - \vec{u}(\vec{x})) \cdot \hat{r}]^n \rangle \ , \tag{203}$$

where in the previous definition we have assumed homogeneity but not isotropy. The SO(3) decomposition of (203) is particularly simple because longitudinal structure functions are scalars, i.e. their exact decomposition in terms of the irreducible representations of the rotational group reduce to the decomposition in spherical harmonics, $Y_{j,m}(\hat{r})$:

$$S_n(\vec{r}) = \sum_{j,m} S_n^{j,m}(|r|) Y_{j,m}(\hat{r}) \ . \tag{204}$$

where with j, m we have labelled, as usual, the eigenvalues of the total angular momentum and of the projection of the total angular momentum on one axis, respectively.
In terms of the decomposition (204) the previous statements can be now rephrased: (i) each sector has a universal scaling exponent: $S_n^{j,m}(|r|) \sim a_{jm} r^{\zeta_n^j}$, with ζ_n^j independent of the large scale physics and (ii) in general $\zeta_n^j \neq \zeta_n^{j'}$. The most interesting remark is to translate the recovery of isotropy at scales small enough in terms of this SO(3) decomposition. Indeed, if we want that, independent of the kind of large-scales anisotropies introduced by the forcing and by the boundary conditions, small scale turbulent fluctuations become more and more isotropic it is necessary to require a hierarchical organization between all ζ_n^j:

$$\zeta_n^{j=0} < \zeta_n^{j=1} < \zeta_n^{j=2} < \cdots \tag{205}$$

i.e. the scaling exponent of the isotropic sector ($j = 0$) is the leading one of the whole hierarchy for any moment n.
The decomposition (204) and the above listed properties (i–iii) also predict that in the presence of anisotropies the only purely-scaling quantities are the projections on each sector, $S_n^{j,m}(|r|)$ and not the undecomposed, raw, structure function, $S_n(\vec{r})$. Such a prediction, together with many others, has been nicely verified by applying the decomposition to the longitudinal structure functions of a channel flow simulation. In Figure 47 we show the fourth order raw undecomposed structure functions, $S_4(\vec{r})$, and its projection on the isotropic sector, $S_4^{j=0,m=0}(|r|)$, measured at 1/4 of the channel height.

Clearly, while the undecomposed structure function does not have any scaling property, the projection, $S_4^{j=0,m=0}(|r|)$ starts to show a quite good scaling even at the moderate Reynolds numbers of the simulation, $Re_\lambda = 70$. This is a clear indication that in presence of anisotropies, disentangling isotropic from anisotropic fluctuations may help in improving the scaling behavior. Let us conclude by mentioning that some of the properties previously listed may be proved rigorously in some linear hydrodynamical problems, like the passive advection of scalars and vectors as described by the Kraichnan model already discussed in section 4.

References for section 6

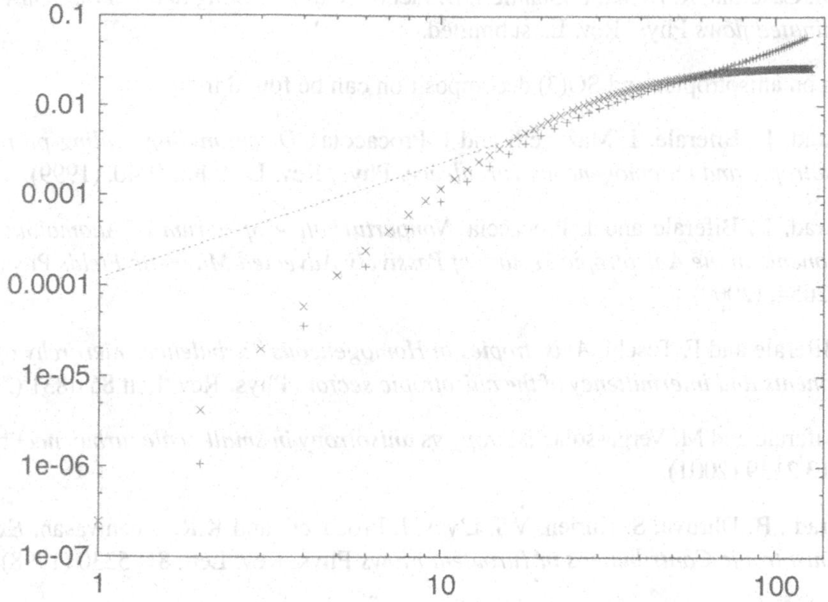

Figure 47: Fourth order raw undecomposed structure functions, $S_4(\vec{r})$ (x) and its projection on the isotropic sector, $S_4^{j=0,m=0}(|r|)$ (+), measured at 1/4 of the channel height.

The works on shear flows and channel flows can be found in :

- F. Toschi, G. Amati, S. Succi, R. Benzi, R. Piva, *Intermittency and structure functions in channel flow turbulence*, Phys. Rev. Lett, **82**(25), 5044-5049, 1999.
- A. Pumir, Phys. Fluids, **8**(11), 3112-3127, 1996.
- Benzi, Amati, Casciola, Toschi, Piva, *Intermittency and scaling laws for wall bounded turbulence*, Phys. Fluids, **11**(6), 1-3, 1999.
- Gualtieri, Casciola, Benzi, Amati, Piva, *Scaling laws and intermittency in homogeneous shear flow*, Submitted to Phys. Fluids., Nov. 2000.
- Jacob, Olivieri, Casciola, *Experimental assessment of a new form of scaling law for near wall turbulence*, Submitted to Phys. Fluids, Gen. 2001.
- Rogallo, *Numerical experiments in homogeneous turbulence*, NASA T.M., **81315**, 1981.
- Benzi, Casciola, Gualtieri, Piva, *Numerical evidence of a new similarity law in shear dominated flows* , to appear in Comp. and Math.
- Toschi, Leveque, Ruiz-Chavarria, *Shear effects in non-homogeneous turbulence*, Phys. Rev. Lett, **85**(7), 1436-1439, 2000.

- C. M. Casciola, R. Benzi, P. Gualtieri, B. Jacob, *Double scaling and intermittency in shear dominated flows* Phys. Rev. E., submitted.

The works on anisotropies and SO(3) decomposition can be found in

- I. Arad, L. Biferale, I. Mazzitelli and I. Procaccia ; *Disentangling scaling properties in anisotropic and inhomogeneous Turbulence.* Phys. Rev. Lett. **82**, 5040, (1999).
- I. Arad, L. Biferale and I. Procaccia *Nonperturbative Spectrum of Anomalous Scaling Exponents in the Anisotropic Sectors of Passively Advected Magnetic Fields* Phys. Rev. E **61**, 2654, (2000)
- L. Biferale and F. Toschi, *Anisotropies in Homogeneous Turbulence: hierarchy of scaling exponents and intermittency of the anisotropic sectors* Phys. Rev. Lett.**86** 4831 (2001)
- L. Biferale and M. Vergassola ; *Isotropy* vs *anisotropy in small-scale turbulence*Phys. Fluids **13** 2139 (2001).
- I. Arad , B. Dhruva, S. Kurien, V.S. L'vov, I. Procaccia and K.R. Sreenivasan, *Extraction of Anisotropic Contributions in Turbulent Flows* Phys. Rev. Lett, **81**, 5330 (1998)
- I. Arad, V. L'vov and I. Procaccia, *Correlation functions in isotropic and anisotropic turbulence: The role of the symmetry group* Phys. Rev. E **81** 6753 (1999).

7 Acknowledgments

In this short review we have tried to summarize in this short review the list of problems solved and unsolved which have attracted the attention of the community working on theoretical numerical and experimental turbulence during the last 10 years. We have, of course, presented our biased view on what must be considered important and interesting. Also, we have tried to summarize all the important contributions without paying the right credits to all the people, our colleagues and friends, who have contributed to them, directly or indirectly. The very short list of references added at the end of each section does not reflect correctly the many important contributions given by many scientists in the field. Among them we are very pleased to mention all our collaborators and friends, namely: G. Amati, I. Arad, G. Boffetta, C. Casciola, A. Celani, M. Cencini, S. Ciliberto, U. Frisch, P. Gualtieri, M.H. Jensen, L. Kadanoff, A. Lanotte, D. Lohse, F. Massaioli, C. M. Meneveau , I. Mazzitelli, G. Parisi, G. Paladin, R. Piva, I. Proccaccia, K. R. Sreenivasan, M.V. Struglia, G. Stolovitsky, S. Succi, F. Toschi, E. Trovatore, R. Tripiccione, M. Vergassola, D. Vergni, A. Vulpiani, A. Wirth.

The Problem of Turbulence and the Manifold of Asymptotic Solutions of the Navier-Stokes Equations

Friedrich H. Busse

Institute of Physics, University of Bayreuth,
95440 Bayreuth, Germany

Abstract

After a brief introduction to the concepts of energy stability, global stability and linear stability three theoretical approaches to different cases of hydrodynamic turbulence under stationary external conditions are discussed. Phase turbulence may occur in the weakly non-linear limit of the Navier-Stokes-equations of motion when the instability of the basic primary state occurs in the form of a highly degenerate bifurcation as, for example, in the case of convection in a fluid layer heated from below and rotating about a vertical axis. Another quite generally applicable approach towards understanding turbulent fluid flow and its coherent structures in particular is the sequence-of-bifurcations approach. Discrete transitions from simple to complex fluid flows can be analyzed in the form of successive bifurcations when the system exhibits a maximum of symmetries based on the assumption of homogeneity in two spatial dimensions. The spatially periodic solutions generated through the sequence-of-bifurcations approach may not be realized in experimental situations where inhomogeneous onsets of disturbances may lead to chaotic fluid flows long before all spatially periodic solutions become unstable. But the tertiary and quaternary solutions exhibit in the clearest way the dynamic mechanisms that operate in turbulent states of flow. Finally the theory of bounds for turbulent transports in the asymptotic range of high Reynolds and Rayleigh numbers is outlined.

1 Introduction: Turbulence as a Problem of Non-Equilibrium Physics

Continuous material systems such as fluids or elastic bodies are subject to the conservation laws of mass, of linear and of angular momentum, and of energy[1]. Since mechanical energy is not conserved there is always the tendency to convert mechanical energy into heat, whereby entropy increases. Since thermodynamic equilibrium is characterized by a maximum entropy, we expect that there will be a tendency for the mechanical energy to assume a minimum under the constraints of prescribed total momentum and/or angular momentum. Systems which have reached this state and are also in thermodynamic equilibrium are called equilibrium systems. Systems

[1] We shall stay within the realm of classical physics and shall not consider relativistic effects in this chapter.

subjected to dissipative processes which have not attained such a state of minimum mechanical energy are called non-equilibrium systems. A system can only be permanently maintained in a non-equilibrium state if an energy flux is applied. In a stationary non-equilibrium state the energy entering and leaving the system per unit time will be the same while entropy production occurs owing to dissipation.

Let's consider some simple examples: A fluid column of constant density $\varrho = 1$ in the shape of a cylinder is rotating about its axis with the angular velocity $\omega(r)$ which depends only on the distance r from the axis as shown in Figure 1a. Assuming that the mechanical energy is just given by the kinetic energy K per unit length we find the following expressions for K and the angular momentum A per unit length of the cylinder

$$K = 2\pi \int_0^{r_0} (\omega(r)r)^2 r\,dr, \quad A = 2\pi \int_0^{r_0} \omega(r)r^2 \; r\,dr \tag{1}$$

where r_0 is the radius of the cylinder. We now ask the question: What is the state of minimum energy K for a fixed value A_0 of A? This mathematical analysis of this question leads us to the formulation of the variational problem,

$$\delta\mathcal{F}_1 \equiv \delta \left\{ 2\pi \int_0^{r_0} \omega^2(r) r^3 dr/2 + \lambda(2\pi \int_0^{r_0} \omega(r) r^3 dr - A_0) \right\} = 0, \tag{2}$$

where δ indicates the variation and λ represents the Lagrange multiplier with which the constraint has been taken into account. The Euler-Lagrange equations providing the necessary condition for an extremal value of the functional are given by

$$\omega(r)r^3 + \lambda r^3 = 0 \quad \text{with} \quad \int_0^{r_0} \omega r^3 dr = A_0/2\pi, \tag{3}$$

which are solved by

$$\omega(r) = -\lambda = \text{ constant } \equiv \omega_0 \quad \text{with} \quad \omega_0 = \frac{2}{\pi r_0^4} A_0. \tag{4}$$

Rigid rotation thus represents the equilibrium state for a prescribed angular momentum. Similarly we can consider cylindrical liquid jets whose velocity parallel to the axis depends for simplicity only on the distance r from the axis as indicated in Figure 1b. Again, the minimization of mechanical energy, $\pi \int_0^{r_0} (v(r))^2 r dr$, for a given value M_0 of the momentum leads to a variational problem

$$\delta\mathcal{F}_2 \equiv \delta \left\{ \pi \int_0^{r_0} (v(r))^2 s ds + \lambda \left(2\pi \int_0^{r_0} v(r) r dr - M_0 \right) \right\} = 0 \tag{5}$$

which yields the Euler-Lagrange equations

$$v(r)r + \lambda r = 0 \quad \text{with} \quad \int_0^{r_0} v(r) r dr = M_0/2\pi. \tag{6}$$

The solution $v(r) = -\lambda = \text{constant} \equiv v_0$ with $v_0 = M_0/\pi r_0^2$ indicates that the equilibrium state is given by a constant velocity.

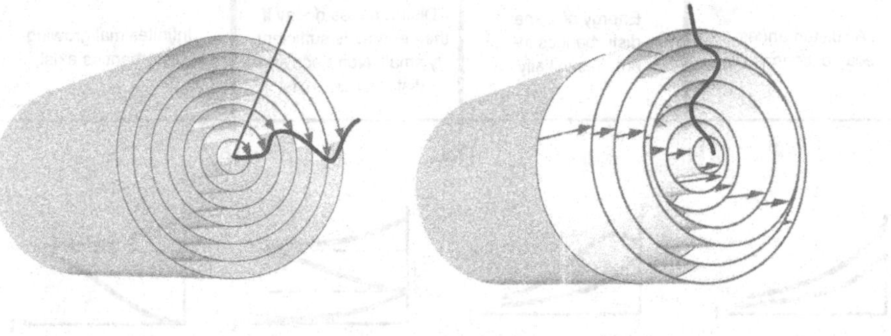

Figure 1: a) Sketch of differentially rotating fluid with angular velocity $\omega(r)$ depending only on distance r from the axis. b) Sketch of parallel flow with a velocity depending only on distance r from the axis.

Because they have reached states of minimum energy the above mentioned systems in the states of rigid rotation or constant velocity are absolutely stable, i.e. any disturbance with vanishing angular and linear momentum will decay or it will just change the values of ω_0 and v_0 if its linear and angular momentum do not vanish. Any system that does not have attained its equilibrium state must be regarded as unstable as long as there are processes available that convert mechanical energy into heat. The fact that this availability is restricted owing to dynamical and other constraints and the variety of energy fluxes that sustain non-equilibrium states give rise to the richness of theories of hydrodynamic instabilities and of turbulence. If, for instance, the cylindrical jet of Figure 1b is enclosed by a solid pipe then the state of constant velocity is not feasible. Instead a state will be realized with strong gradients of the velocity fields at the cylindrical wall and energy in the form of work done by an imposed pressure difference along the pipe must be supplied to sustain a constant momentum. The question of the conditions under which the non-equilibrium state in this case in the form of the laminar Hagen-Poiseuille flow is stable is a fundamental one and still not fully solved problem of hydrodynamic stability theory.

An example illustrating that the state of minimum mechanical energy may not be attainable even if the non-equilibrium state is not sustained by an applied energy flux can be visualized if surface tension is considered in the case of the liquid jet. Surface energy must then be included in the mechanical energy and instead of a cylindrical shape of the jet a spherical shape is found to minimize the latter. Indeed, as anybody knows after watching a water jet emerging from a faucet, the liquid jet breaks up into nearly spherical droplets of varying size. Unless these droplets coalesce, which is usually not possible, a state of minimum energy can not be achieved. But each droplet in isolation may be regarded as an equilibrium system since it corresponds to a state of minimum mechanical energy. In this respect this example differs from the case of a non-equilibrium system sustained by an energy flux mentioned above. In the following we restrict the attention to the latter kind of systems.

Let us consider a simple system like channel flow between two parallel plane walls with the distance d. Instead of using the momentum per unit length it is customary to use a dimension-

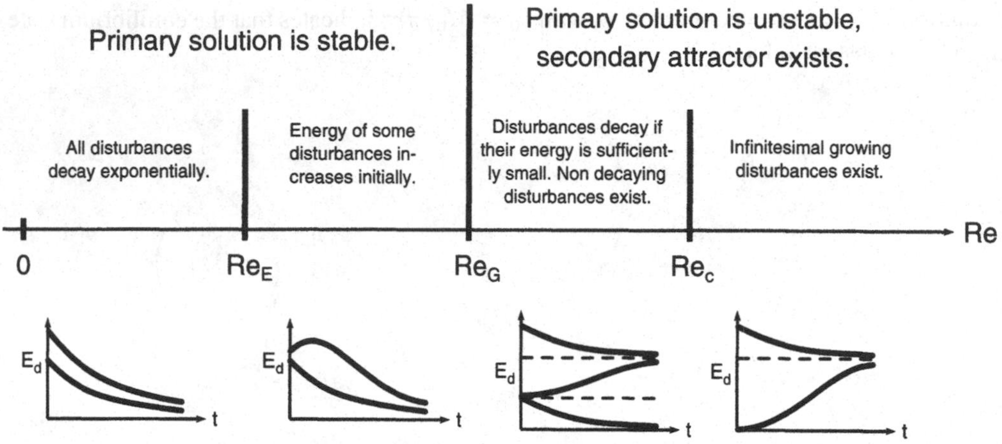

Figure 2: Regimes of stability and instability for a steady flow as a function of *Re*.

less parameter, namely the Reynolds number Re as control parameter. The latter is defined by $Re = Ud/\nu$ where U is the average velocity and ν is the kinematic viscosity of the fluid. Since any problem in physics in which the steady external conditions exhibit symmetries has at least one steady solution exhibiting the same symmetries, we expect that there exists a solution independent of the x, y-coordinates parallel to the walls and depending only on the z-coordinate perpendicular to the walls. This basic or primary solution is well known, $\boldsymbol{v} = (Re(\frac{1}{4} - z^2), 0, 0)$ and is called Poiseuille flow. But the particular form of the basic flow is unimportant at this point where we consider the general problem of stability. Four regimes of different stability properties can be distinguished as sketched in Figure 2:

A) Any disturbance of the primary state, large or small, decays monotonically in energy in the regime $0 \leq Re < Re_E$, where Re_E is the Energy-Stability-Reynolds number since it is determined by the Energy Method.

B) In the regime $Re_E \leq Re < Re_G$, where Re_G is the Global-Stability-Reynolds number, disturbances exist which can grow in energy initially, but asymptotically they all decay. This regime could be subdivided further into the regime $Re_E \leq Re < R_u$ where the primary state is still the unique solution of the problem and into the regime $Re_u \leq Re < Re_G$ where permanent (steady or oscillatory) states exist which are distinct from the primary state. But since these states are unstable and exhibit vanishing basins of attraction the system asymptotically always returns to still evolves into the primary state.

C) Infinitesimal disturbances still decay monotonically in the regime $Re_G \leq Re < Re_c$ where Re_c is the critical Reynolds number. But disturbances of finite amplitude exist. They do not decay, but instead give rise to the evolution of the system into a new state different from the primary one.

D) Finally there exists the regime $Re_c < Re \leq \infty$ where infinitesimal disturbances grow exponentially and where secondary solutions bifurcate from the primary solution.

Obviously the primary state is stable in the regimes A and B, but unstable in the regimes C and D. The different regimes are defined in close analogy to the case of a point mass in a potential field, except that no analogon to the case B exists. We also note that the above definitions can be readily generalized to the case of time periodic external conditions.

From the physical point of view the distinction between regimes A and B may not be important. But the value Re_E can be derived relatively easily for hydrodynamic problems since Re_E is determined as the eigenvalue of the linear boundary value problem. Re_c is also determined by a linear eigenvalue problem. But there are cases for which finite values of Re_c do not exist such as plane Couette flow and pipe flow. Re_G can be determined only under exceptional circumstances. For instance when it is found that Re_E and Re_c coincide. Then of course Re_G is given by $Re_E = Re_G = Re_c$.

One may be inclined to identify the primary state with what is usually called the laminar state and the new state mentioned in C) with the turbulent state. But this would not be correct since many examples demonstrate that the new states can be steady states corresponding to laminar flow. Instead the mathematical distinction to be made is between the uniqueness of the asymptotic state which is guaranteed in regimes A and B and the possibility of other asymptotic states which characterizes the regimes C and D. The turbulent nature of any other than the primary state is thus of secondary importance from the mathematical point of view. In the next section we shall discuss the methods for the determination of R_E and R_c.

2 Theory of Hydrodynamic Stability

2.1 Reynolds-Orr Energy Equations

For simplicity we restrict the attention to flows of homogeneous incompressible fluids with a constant kinematic viscosity ν in an arbitrary fixed domain $\mathcal{D}$. Using the diameter d of the domain as length scale and d^2/ν as time scale we can write the Navier-Stokes equations of motion in dimensionless form,

$$\frac{\partial}{\partial t}\boldsymbol{v} + \boldsymbol{v}\cdot\nabla\boldsymbol{v} = -\nabla p + \boldsymbol{f} + \nabla^2\boldsymbol{v} \tag{7}$$

$$\nabla\cdot\boldsymbol{v} = 0 \tag{8}$$

where $\boldsymbol{f}$ denotes some given steady force density distribution. On the boundary $\partial\mathcal{D}$ of the domain $\mathcal{D}$ steady velocities parallel to the boundary may be specified. We assume that the basic steady solution of the problem is given by $\boldsymbol{v}_s = Re\hat{\boldsymbol{v}}_s$ where the average of $(\hat{\boldsymbol{v}}_s)^2/2$ over the domain is unity,

$$\langle\hat{\boldsymbol{v}}_s\cdot\hat{\boldsymbol{v}}_s\rangle = 2. \tag{9}$$

Any velocity distribution $\boldsymbol{v}_t$ different from $\boldsymbol{v}_s$ must obey the equations

$$\frac{\partial}{\partial t}\boldsymbol{u} + \boldsymbol{v}_s\cdot\nabla\boldsymbol{u} + \boldsymbol{u}\cdot\nabla\boldsymbol{v}_s + \boldsymbol{u}\cdot\nabla\boldsymbol{u} = -\nabla\tilde{p} + \nabla^2\boldsymbol{u} \tag{10}$$

$$\nabla\cdot\boldsymbol{u} = 0 \tag{11}$$

for the difference velocity field $\boldsymbol{u} = \boldsymbol{v}_t - \boldsymbol{v}_s$ together with homogeneous boundary conditions for $\boldsymbol{u}$ on $\partial\mathcal{D}$. By multiplying equation (10) by $\boldsymbol{u}$ and averaging the results over the domain $\mathcal{D}$ we obtain the relationship

$$\frac{1}{2}\frac{d}{dt}\langle \boldsymbol{u}\cdot\boldsymbol{u}\rangle = -\langle \mid \nabla \boldsymbol{u}\mid^2\rangle - Re\langle \boldsymbol{u}\cdot(\boldsymbol{u}\cdot\nabla)\hat{\boldsymbol{v}}_s\rangle \tag{12}$$

where the vanishing of $\boldsymbol{u}$ on $\partial\mathcal{D}$ and equations such as $\langle \boldsymbol{u}\cdot(\boldsymbol{v}_s\cdot\nabla)\boldsymbol{u}\rangle = \frac{1}{2}\langle \boldsymbol{v}_s\cdot\nabla\boldsymbol{u}\cdot\boldsymbol{u}\rangle = \frac{1}{2}\langle \nabla\cdot(\boldsymbol{v}_s\boldsymbol{u}\cdot\boldsymbol{u})\rangle = 0$ have been used to prove that the terms $\boldsymbol{v}_s\cdot\nabla\boldsymbol{u}$, $\boldsymbol{u}\cdot\nabla\boldsymbol{u}$ and $\nabla\tilde{p}$ do not enter the balance (12). This balance is called the Reynolds-Orr energy equation and is the basis for the application of the energy method. Using general estimates for the terms on the right hand side Serrin (1959) has proven his famous theorem stating that a bound Re_B exists such that for $Re \le Re_B$ the solution $\boldsymbol{v}_s$ is unique and any disturbance energy, $\frac{1}{2}\langle \boldsymbol{u}\cdot\boldsymbol{u}\rangle$, decays exponentially in time. Many details on energy methods can be found in Joseph's book (1976).

Here we wish to remark that the Reynolds-Orr balance (12) is also valid if the hydrodynamical problem is considered in a rotating system with constant angular velocity $\boldsymbol{\Omega}_D$. The basic equations (7) and (8) must be supplemented by the Coriolis force term $2\boldsymbol{\Omega}\times\boldsymbol{v}$ on the left hand side of equation (7) where the dimensionless rotation vector $\boldsymbol{\Omega}$ is defined by $\boldsymbol{\Omega} = \boldsymbol{\Omega}_D d^2/\nu$. The centrifugal force does not have to be considered explicitly in this case since it can be combined with the body force $\boldsymbol{f}$. Although the term $2\boldsymbol{\Omega}\times\boldsymbol{u}$ must also be added on the left hand side of equation (10) in a rotating system, it does not contribute to the energy balance (12).

2.2 Energy Stability Limits for Plane and Circular Couette Flows

The energy Reynolds number Re_E can be determined as solutions of the variational problem:

For a given flow $\hat{\boldsymbol{v}}_s$ in $\mathcal{D}$ find the minimum Re_E of the functional

$$\mathcal{R}_E \equiv \frac{\langle \mid \nabla\check{\boldsymbol{u}}\mid^2\rangle}{\langle -\check{\boldsymbol{u}}\cdot(\check{\boldsymbol{u}}\cdot\nabla)\hat{\boldsymbol{v}}_s\rangle} \tag{13}$$

among all vector fields $\boldsymbol{u}$ which satisfy the conditions $\nabla\cdot\check{\boldsymbol{u}} = 0$ in $\mathcal{D}$, $\check{\boldsymbol{u}} = 0$ on $\partial\mathcal{D}$ and $\langle \check{\boldsymbol{u}}\cdot(\check{\boldsymbol{u}}\cdot\nabla)\hat{\boldsymbol{v}}_s\rangle < 0$.

A comparison of the functional (13) with the energy balance (12) demonstrates that all energies $\langle \boldsymbol{u}\cdot\boldsymbol{u}\rangle$ must decay for $Re < Re_E$. For $Re \ge Re_E$ at least one vector field $\boldsymbol{u}$ will exist, namely the minimizing solution $\check{\boldsymbol{u}}$ of the variational problem (13), the energy of which does not decrease, at least initially. The constraint $\nabla\cdot\check{\boldsymbol{u}} = 0$ can be taken into account through a Lagrange multiplying function $\check{\pi}$ when we derive the Euler-Lagrange equations for stationary values of the variational functional (13),

$$\frac{1}{2}M(\check{u}_\kappa\partial_\kappa\hat{v}_{si} + \check{u}_\kappa\partial_i\hat{v}_{s\kappa}) = -\partial_i\check{\pi} + \partial_\kappa\partial_\kappa\check{u}_i \tag{14}$$

$$\partial_\kappa\check{u}_\kappa = 0 \tag{15}$$

which must be considered together with the boundary condition $\check{u}_i = 0$ on $\partial\mathcal{D}$. M is the stationary value of the functional (13) and in general many of them exist. They are determined as eigenvalues of the linear boundary value problem (14) together with its boundary condition. Only the infinum of all M provides the energy Reynolds number Re_E.

As an example we want to consider the case of the simplest solution of the Navier-Stokes equations (7) and (8) which is the flow between two parallel plates moving in opposite direction with the velocity $\boldsymbol{U}_D$ relative to each other. Using the distance d of the plates the plane Couette flow solution of this problem can be written in dimensionless form

$$\boldsymbol{v}_s = -Rez\boldsymbol{i} \tag{16}$$

where a Cartesian system of coordinates (x, y, z) has been introduced together with corresponding unit vectors $\boldsymbol{i}, \boldsymbol{j}, \boldsymbol{k}$ such that $\boldsymbol{i}$ points in the direction of $\boldsymbol{U}_D$ and the plates are located at $z = \pm\frac{1}{2}$. The Reynolds number Re is defined by $Re = \mid \boldsymbol{U}_D \mid d/\nu$. For the solution of equations (14) and (15) it is convenient to introduce the general representation[2]

$$\check{\boldsymbol{u}} = \nabla \times (\nabla\varphi \times \boldsymbol{k}) + \nabla\psi \times \boldsymbol{k} \tag{17}$$

for a solenoidal vector field $\check{\boldsymbol{u}}$ and to take the z-components of the curl and of the (curl)^2 of equation (14). This procedure yields

$$\nabla^4 \Delta_2 \varphi = \frac{1}{2} M (2\partial_x \partial_z \Delta_2 \varphi + \partial_y \Delta_2 \psi) \tag{18}$$

$$\nabla^2 \Delta_2 \psi = \frac{1}{2} M \partial_y \Delta_2 \varphi \tag{19}$$

where the two-dimensional Laplacian, $\Delta_2 = \partial^2_{xx} + \partial^2_{yy}$, has been introduced. These equations must be solved subject to the boundary conditions

$$\varphi = \partial_z \varphi = \psi = 0. \tag{20}$$

If we restrict the attention to x-independent solutions, ψ can easily be eliminated and the problem can be reduced to

$$(\nabla^6 - \frac{1}{4} M_y^2 \partial^2_{yy}) \Delta_2 \varphi = 0 \text{ with } \varphi = \partial_z \varphi = \nabla^4 \varphi = 0 \text{ at } z = \pm\frac{1}{2}. \tag{21}$$

This eigenvalue problem is identical to the problem determining the critical Rayleigh number in a fluid layer heated from below with rigid boundaries. We thus obtain

$$\frac{1}{4} M_y^2 = 1708 \text{ corresponding to } \varphi = \cos \alpha y f(z) \text{ with } \alpha_c = 3.116 \tag{22}$$

as lowest eigenvalue. Of course, before we can claim that this value of M_y provides the absolute minimum Re_E we have to prove that more general solutions, φ, ψ, depending on x as well as y do not yield values of M lower than (22). Fortunately, such a proof is available (Busse, 1972) and we thus have as final result $Re_E = 2\sqrt{1708} \approx 82.6$ for plane Couette flow.

The values for Re_E that have been obtained for various shear flows in non-rotating systems are much too conservative when compared with the observed onset of turbulence, which we denote by Re_G. But the results obtained from the linear theory for the onset of infinitesimal

[2]If one requires that φ and ψ are bounded functions in the x, y-plane, a spatially constant vector $\boldsymbol{U}$ should be added on the right hand side (see proof of Schmitt and von Wahl, 1992). We shall neglect it since it is not needed in this and the next section.

disturbances are not much better, as can be seen from the values of Re_c given in Table 1. It should be noted that the energy stability limit Re_E holds in rotating systems independent of the angular velocity $\mathbf{\Omega}$ as long as the primary state of flow is given by (16). In particular, circular Couette flow obeys equation (16) relative to an appropriate rotating system in the small gap approximation as is discussed in more detail in the next subsection.

For simplicity we have identified the value of Re beyond which a sustained state of turbulent fluid flow can be found in experiments with the global Reynolds number Re_G. As Schmiegel and Eckhardt (1997,2000) have shown it is difficult to determine this value theoretically and Re_G may be defined only in a statistical sense. More definitive results may be obtained for the value of Re_u which is defined as the upper limit of Re for which the primary state (16) is the unique steady solution. Nagata (1990) and Busse and Clever (1992) have found secondary steady solutions for values of Re as low as 580. But these solutions are unstable and higher Reynolds numbers are required before the basin of attraction of the primary state becomes sufficiently small, such that the turbulent state can persist for sufficiently long time. The question of the nature of the transition to the turbulent state is a subject of intense current research and we refer to the recent comprehensive review of Grossmann (2000) for details.

Table 1: Reynolds Numbers for Shear Flows in Non-Rotating Systems

	Re_E	Re_G (from exp.)	Re_c
Plane Couette Flow	82.6	≈ 1300	∞
Poiseuille Flow (Channel Flow)	99.2*	≈ 2000*	5772*
Hagen-Poiseuille Flow (Pipe Flow)	81.5*	≈ 2100*	∞

* The maximum velocity and the channel width d (radius d in the case of pipe flow) have been used in definition of Re

2.3 Linear Stability Theory of Circular Couette Flow

While plane parallel shear flows in non-rotating systems exhibit a complex transition to turbulence, a gradual evolution from the primary state of flow through successive transitions to more complex flows can be observed in rotating systems. Here the energy method as well as the linear stability analysis describe the experimental observations quite well and for a particular case the values of Re_E and Re_c even coincide.

We consider the flow between coaxial cylinders with radii r_1 and r_2 $(r_2 > r_1)$ which rotate with the angular velocities Ω_1 and Ω_2, respectively. The basic solution of equation (7) for the azimuthal velocity v_φ is called circular Couette flow,

$$v_\varphi = \frac{(r_2^2\Omega_2 - r_1^2\Omega_1)}{r_2^2 - r_1^2} r - \frac{r_1^2 r_2^2(\Omega_2 - \Omega_1)}{(r_2^2 - r_1^2) r}. \tag{23}$$

In order to simplify the stability analysis we restrict the attention to the small gap limit, $d \equiv r_2 - r_1 \ll r_1$, and to the case $0 < \Omega_1 - \Omega_2 \ll \Omega_1$. In this limit the solution (23) assumes the form (16) of plane Couette flow relative to the system rotating with the angular velocity $\Omega_D = \frac{1}{2}(\Omega_1 + \Omega_2)$. In the dimensionless formulation of the problem we have introduced a

Cartesian coordinate system with the x-coordinate in the azimuthal direction, the y-coordinate in the axial direction and the z-coordinate in the radial direction. The Reynolds number Re in expression (16) is now defined by $Re = (\Omega_1 s_1 - \Omega_2 s_2)d/\nu$.

For the analysis of the stability of circular Couette flow with respect to infinitesimal disturbances we may use equations (10) and (11) after adding the Coriolis force term and neglecting the term $\tilde{\boldsymbol{u}} \cdot \nabla \tilde{\boldsymbol{u}}$,

$$\frac{\partial}{\partial t}\tilde{\boldsymbol{u}} + \boldsymbol{v}_s \cdot \nabla \tilde{\boldsymbol{u}} + \tilde{\boldsymbol{u}} \cdot \nabla \boldsymbol{v}_s + 2\boldsymbol{\Omega} \times \tilde{\boldsymbol{u}} = -\nabla \tilde{\pi} + \nabla^2 \tilde{\boldsymbol{u}} \tag{24}$$

$$\nabla \cdot \tilde{\boldsymbol{u}} = 0 \tag{25}$$

where the definition $\boldsymbol{\Omega} = \boldsymbol{\Omega}_D d^2/\nu$ has been used. Without losing generality an exponential dependence on time can be assumed, $\exp\{\sigma t\}$, since $\boldsymbol{v}_s$ describes a steady velocity field. The growth rate σ thus becomes the eigenvalue of the boundary value problem given by equations (24) and (25) together with the boundary condition $\boldsymbol{u} = 0$ at $z = \pm\frac{1}{2}$. Wherever there exists a growth rate σ with positive real part, the flow $\boldsymbol{v}_s$ is unstable; otherwise it is considered to be stable with respect to infinitesimal disturbances.

For the solution of equations (24) and (25) we again introduce the general representation as in (17)

$$\tilde{\boldsymbol{u}} = \nabla \times (\nabla \times \boldsymbol{k}\tilde{\varphi}) + \nabla \times \boldsymbol{k}\tilde{\psi} \tag{26}$$

and take the z-components of the (curl)^2 and of the curl of equation (24),

$$\nabla^4 \Delta_2 \tilde{\varphi} - 2\boldsymbol{\Omega} \cdot \nabla \Delta_2 \tilde{\psi} = \boldsymbol{v}_s \cdot \nabla \nabla^2 \Delta_2 \tilde{\varphi} + \sigma \nabla^2 \Delta_2 \tilde{\varphi} - \boldsymbol{v}_s'' \cdot \nabla \Delta_2 \tilde{\varphi} \tag{27}$$

$$\nabla^2 \Delta_2 \tilde{\psi} + 2\boldsymbol{\Omega} \cdot \nabla \Delta_2 \tilde{\varphi} = \boldsymbol{v}_s \cdot \nabla \nabla^2 \Delta_2 \tilde{\psi} + \sigma \Delta_2 \tilde{\psi} + \boldsymbol{k} \cdot (\nabla \Delta_2 \tilde{\varphi} \times \boldsymbol{v}_s') \tag{28}$$

where $\boldsymbol{v}_s'$ and $\boldsymbol{v}_s''$ denote the first and second derivatives of $\boldsymbol{v}_s$ with respect to z. As we shall prove later there is no need to obtain the general solution of the linear stability equations (27) and (28) together with the boundary conditions $\varphi = \partial_z \varphi = \psi = 0$ at $z = \pm\frac{1}{2}$. Instead we focus the attention on x-independent disturbances for which the principle of exchange of stabilities holds, i.e. the imaginary part of σ vanishes. In this case the critical disturbances correspond to $\sigma = 0$ and equations (27) and (28) reduce to

$$\nabla^4 \partial^2_{yy} \tilde{\varphi} - 2\Omega \partial_y \partial^2_{yy} \tilde{\psi} = 0 \tag{29}$$

$$\nabla^2 \partial^2_{yy} \tilde{\psi} - (Re - 2\Omega)\partial_y \partial^2_{yy} \tilde{\varphi} = 0 \tag{30}$$

where we have used the expression (16) for the circular Couette flow $\boldsymbol{v}_s$. A comparison with equations (18), (19) and (20) demonstrates that equations (29) and (30) are identical with the latter in the case of x-independent solutions. Since also the boundary conditions are the same we obtain expressions (22) as solution of equations (29) and (30) with $(Re - 2\Omega)2\Omega$ replacing $\frac{1}{4}M_y^2$, i.e.

$$Re_y = 2\Omega + \frac{1708}{2\Omega}. \tag{31}$$

The minimum value of Re_y as a function of Ω is given by

$$Re_c = 2\sqrt{1708} \text{ corresponding to } 2\Omega = \sqrt{1708} \tag{32}$$

The identity of expression (32) with the energy stability limit Re_E proves the following statements:

(i) The value of Re_y at $\Omega = \frac{1}{2}\sqrt{1708}$ must be equal to Re_c since x-dependent and oscillatory disturbances can not produce a lower value of Re.

(ii) The bifurcation at the point (32) must be supercritical since the basic state of circular Couette flow is unique for $Re < Re_E$.

(iii) The global stability limit Re_G is also given by expression (32) because of the relationship $Re_E \leq Re_G \leq Re_c$.

Although we can not offer a proof here that the x-independent stability equations (29) and (30) describe the onset of instability for other values but for $\Omega = \frac{1}{2}\sqrt{1708}$, this appears to be the case except when Ω becomes rather small. The growing disturbance corresponds to axisymmetric Taylor vortices which are observed in experiments over a wide range of the parameter space. The measured onsets of Taylor vortices therefore agree quite well with the theoretical curve of Figure 3 if a sufficiently small gap is used. For finite gap sizes the energy Reynolds number and the critical Reynolds number do not touch anymore as a function of Ω but the shape of the Re_c-curve remains the same. Asymptotically for large Ω it approaches the Rayleigh stability criterion for an arbitrary distribution of angular velocity ω as a function of the distance r from the axis,

$$\frac{d(\omega(r)r^2)^2}{dr} \geq 0 \quad \text{for stability with respect to axisymmetric disturbances.} \tag{33}$$

When applied to circular Couette flow this stability criterion which Rayleigh derived for an inviscid fluid assumes the form

$$|\, \Omega_1 r_1^2 \,|\leq|\, \Omega_2 r_2^2 \,| \tag{34}$$

and in the small gap limit gives rise to the condition $Re < 2\Omega$.

3 Phase Turbulence in the Limit of Low Reynolds Number

3.1 Rayleigh-Bénard Convection in a Fluid Layer Heated from Below

Besides the Taylor vortices between differentially rotating coaxial cylinders, convection flow in horizontal fluid layers heated from below and cooled from above has become the most popular example of a transition from a primary to a secondary state through a supercritical bifurcation. When the Boussinesq approximation is adopted, i.e. all material properties are assumed to be constant except for a linear dependence of the density on the temperature which is considered only in the gravity term, then the energy Rayleigh number Ra_E becomes identical to the critical value Ra_c just as in the case (32) of the Reynolds number. The Rayleigh number is a dimensionless measure of the temperature difference, $T_2 - T_1$, between the boundaries of the layer and is defined by

$$Ra = \frac{\gamma g(T_2 - T_1)d^3}{\nu\kappa} \tag{35}$$

where γ is the coefficient of thermal expansion, g is the acceleration of gravity, d is the height of the layer and κ is the thermal diffusivity of the fluid.

The onset of instability of the basic static state is due to the release of potential energy when cold fluid descends from above and hot fluid rises from below. Since the moving fluid loses

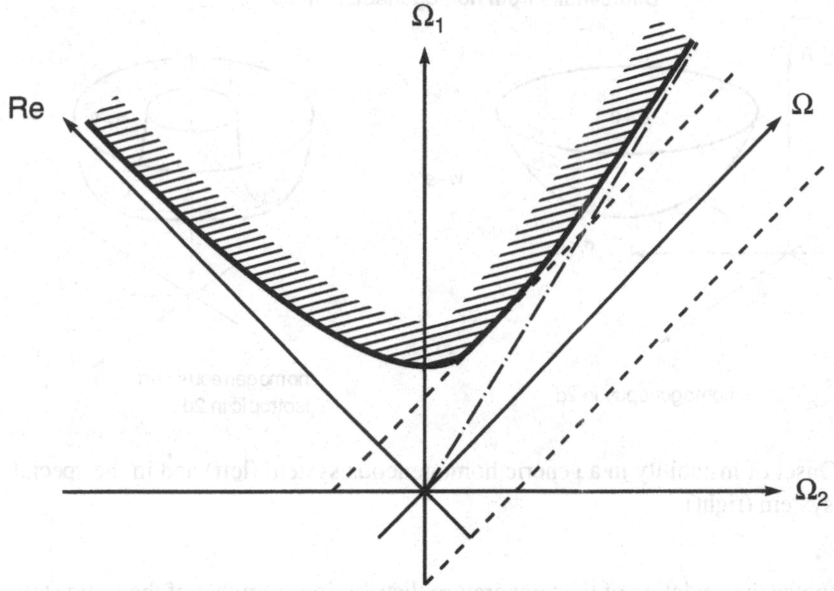

Figure 3: Energy stability boundaries (dashed lines) and onset of instability (shaded region) for circular Couette flow in the small gap limit. The Rayleigh stability criterion (34) corresponds to the dash-dotted line.

its buoyancy by thermal conduction and must overcome the effects of viscous friction a finite temperature gradient corresponding to a finite critical value Ra_c is needed for the onset of convection. This instability is the only hydrodynamic instability corresponding to a primary state which is not only homogeneous, but also isotropic in two dimensions. The basic difference between the generic case of a bifurcation as function of the control parameter R and the special case of an isotropic system such as the Rayleigh-Bénard layer with $R = Ra$ is visualized in Figure 4. Since there does not exist a preferred horizontal direction for the convection flow in the extended Rayleigh-Bénard layer, we have a bifurcation with infinite degeneracy. It is this property which makes the Rayleigh-Bénard problem of particular interest to the problem of fully developed turbulence since in both cases an essentially infinite number of competing solutions of the basic equations exists. It is thus not surprising that some kind of turbulence can exist in Rayleigh-Bénard convection even close to the critical value Ra_c of the Rayleigh number.

The basic equations governing Rayleigh-Bénard convection in the Boussinesq approximation do not differ much from equations (7) and (8),

$$\frac{\partial}{\partial t}\boldsymbol{v} + \boldsymbol{v} \cdot \nabla \boldsymbol{v} = -\nabla \pi + \Theta \boldsymbol{k} + \nabla^2 \boldsymbol{v} \tag{36}$$

$$0 = \nabla \cdot \boldsymbol{v} \tag{37}$$

$$Pr \left[\frac{\partial}{\partial t}\Theta + \boldsymbol{v} \cdot \nabla \Theta \right] = Ra\, \boldsymbol{k} \cdot \boldsymbol{v} + \nabla^2 \Theta \tag{38}$$

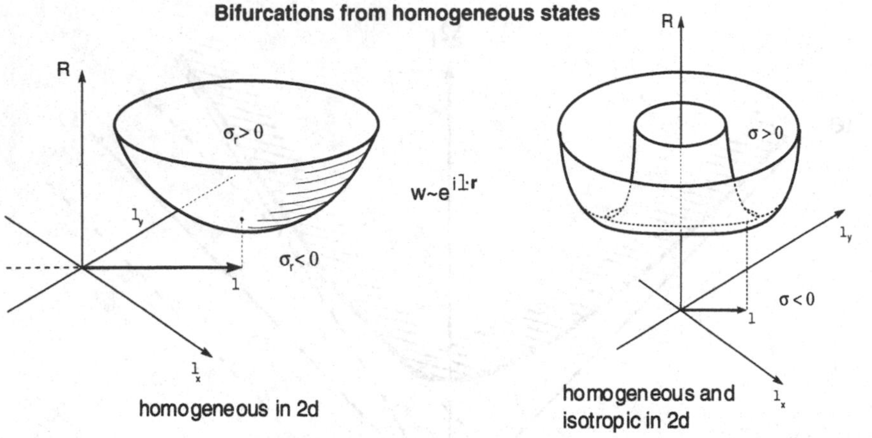

Figure 4: Onset of instability in a generic homogeneous system (left) and in the special case of a isotropic system (right).

where Θ denotes the deviation of the temperature distribution from that of the static state of pure conduction. It has been made dimensionless through the use of the temperature scale $(T_2 - T_1)/Ra$. $\boldsymbol{k}$ is the vertical unit vector of a Cartesian system of coordinates, and all terms that can be written as gradients in the equation of motion have been combined into $\nabla\pi$. The Prandtl number Pr appears as second dimensionless parameter in the heat equation (38),

$$Pr = \frac{\nu}{\kappa} \tag{39}$$

where κ is the thermal diffusivity of the fluid. Since the onset of convection occurs in the form of monotonously growing disturbances, it can be described by equations (36), (37) and (38) in the limit where the left hand sides are neglected. After taking the z-component of the (curl)^2 of equation (36) and eliminating Θ through the use of equation (38) we obtain

$$(\nabla^6 - Ra\Delta_2)v_z = 0 \tag{40}$$

which can be solved through $v_z = w(x,y)f(z)$ where $\Delta_2 w = -\alpha^2 w$ can be assumed. After the no-slip conditions at the boundaries with fixed temperatures have been used, $f = df/dz = (d^2/dz^2 - \alpha^2)^2 f = 0$, and Ra has been minimized with respect to the wavenumber α the result

$$Ra_c = 1708 \quad \text{with} \quad \alpha_c = 3.116 \tag{41}$$

is obtained to which we have referred in the preceding section. In contrast to the cases (29) and (30) of circular Couette flow which gives rise to the same mathematical equation, the present linear analysis leaves the form of the convection flow unspecified except for the absolute value α of the wavenumber, i.e. any solution of the form

$$w(x,y) = \sum_{n=-N}^{N} c_n \exp\{i\boldsymbol{q}_n \cdot \boldsymbol{r}\} \quad \text{with} \quad c_n = c_{-n}^{+} \text{ and } \boldsymbol{q}_{-n} = -\boldsymbol{q}_n \tag{42}$$

is admitted by equation (40) where the parameter N may tend to infinity. Here c_n^+ denotes the complex conjugate of c_n and the wavevectors $\boldsymbol{q}_n$ are arbitrary except for the conditions $\boldsymbol{q}_n \cdot \boldsymbol{k} = 0$ and $| \boldsymbol{q}_n | = \alpha$. As we shall see the nonlinear terms provide restrictions and the manifold of steady solutions of the nonlinear equations (36), (37) and (38) is much smaller in the limit of vanishing amplitudes than the manifold (42). In order to select physically realizable solutions among the still infinite manifold of steady solutions a stability analysis will be required. In order to demonstrate the mathematical structure of the problem we write the basic equations in the form

$$\underline{\boldsymbol{L}}\boldsymbol{X} + Ra\underline{\boldsymbol{K}}\boldsymbol{X} - \underline{\boldsymbol{M}}\partial\boldsymbol{X}/\partial t = \boldsymbol{N}(\boldsymbol{X}, \boldsymbol{X}) \tag{43}$$

where $\underline{\boldsymbol{L}}, \underline{\boldsymbol{K}}, \underline{\boldsymbol{M}}$ are linear matrix operators and $\boldsymbol{N}(\boldsymbol{X}, \boldsymbol{X})$ represents the nonlinear terms while the vector $\boldsymbol{X}$ represents the dependent variables (ϕ, ψ, Θ). Here the scalar functions φ and ψ describe poloidal and toroidal components of the velocity field

$$\boldsymbol{v} = \nabla \times (\nabla\varphi \times \boldsymbol{k}) + \nabla\psi \times \boldsymbol{k} \tag{44}$$

as in the case of expression (26).

Expanding $\boldsymbol{X}$ in the powers of an amplitude parameter ϵ and restricting the attention to steady solutions we write

$$\boldsymbol{X} = \epsilon\boldsymbol{X}_1 + \epsilon^2\boldsymbol{X}_2 + \ldots \qquad , Ra = R_0 + \epsilon R_1 + \epsilon^2 R_2 + \ldots \tag{45}$$

where the solution in the order ϵ of equation (43) is given by $\boldsymbol{X}_1 = \boldsymbol{F}(z)w(x, y)$. In the order ϵ^2 the linear equation

$$\boldsymbol{L}\boldsymbol{X}_2 + R_0\boldsymbol{K}\boldsymbol{X}_2 = \boldsymbol{N}(\boldsymbol{X}_1, X_1) - R_1\boldsymbol{K}\boldsymbol{X}_1 \tag{46}$$

is obtained from (43). Equation (46) can only be solved if the solvability conditions are satisfied since the homogeneous version already admits a solution. Accordingly the right hand side must be orthogonal to all solutions of the adjoint homogeneous linear problem. The latter solutions can be assumed in the form $\boldsymbol{X}_m^* = \exp\{i\boldsymbol{q}_m \cdot \boldsymbol{r}\}\boldsymbol{F}^*(z)$ with $m = -N, \ldots, -1, +1, \ldots, N$.

The solvability conditions in the order ϵ^2 do not yield restrictions on the choice of the coefficients c_n and can be satisfied with $R_1 = 0$ since the expression $\boldsymbol{N}(\boldsymbol{X}, \boldsymbol{X})$ is antisymmetric in z while $\boldsymbol{F}(z)$ and $\boldsymbol{F}^*(z)$ are symmetric. After the solution $\boldsymbol{X}_2$ has been obtained and included on the right side of equation (43), the solvability conditions in the order ϵ^3 can be derived, leading to equations of the form

$$0 = R_2 c_m^+ - \sum_{n=-N}^{N} B(\boldsymbol{q}_n \cdot \boldsymbol{q}_m) \mid c_n \mid^2 c_m^+ + \beta \sum_{n,i} c_n c_i \delta(\boldsymbol{q}_n + \boldsymbol{q}_i + \boldsymbol{q}_m) \text{ for } -N \le m \le N \tag{47}$$

In a slight relaxation of our formulation above we have taken into account small deviations from the Boussinesq approximation which lead to asymmetries of the properties of the convection layer and give rise to the term proportional to β in equation (47). This latter term contributes only when the three $\boldsymbol{q}$ vectors $\boldsymbol{q}_n, \boldsymbol{q}_i, \boldsymbol{q}_m$ form an equilateral triangle as indicated by the δ-function.

Simple solutions of equation (47) can be obtained when the angle between neighboring $\boldsymbol{q}$-vectors is given by π/N. For these regular $\boldsymbol{q}$-vector distributions

$$\mid c_1 \mid^2 = \mid c_2 \mid^2 = \ldots = \mid c_N \mid^2 = \frac{1}{2N} \tag{48}$$

is obtained from equations (47) where we have assumed the normalization condition $\sum_{n=-N}^{N} \mid c_n \mid^2 = 1$. In the cases $N = 1$ and $N = 2$ the solution of the form (42) is fully determined by expression (48) since the complex phases of the coefficients c_n just describe all the patterns that can be obtained from a given pattern through translations in the x, y-plane. The two-dimensional solution corresponding to $N = 1$ is usually called "rolls" and the solution with $N = 2$ describes a square pattern. In the case of a hexagonal pattern, $N = 3$, the phase of the third coefficient introduces a new parameter. There exists indeed a one-dimensional manifold of patterns on the hexagonal lattice, the end members of which are the l-hexagons and the g-hexagons corresponding to convection cells with up- and down-flow in the center of the cell, respectively. The letters l and g stand for liquid and gas since upflow in the center is usually observed for convection in liquids which have kinematic viscosities that decrease with increasing temperature, while the opposite property is found for gases. A visual impression of the convection flows corresponding to $N = 1, 2, 3$ can be gained from Figure 5. For a more general discussion of solutions on the hexagonal lattice including those which do not satisfy conditions (48) we refer to Golubitsky et al. (1984).

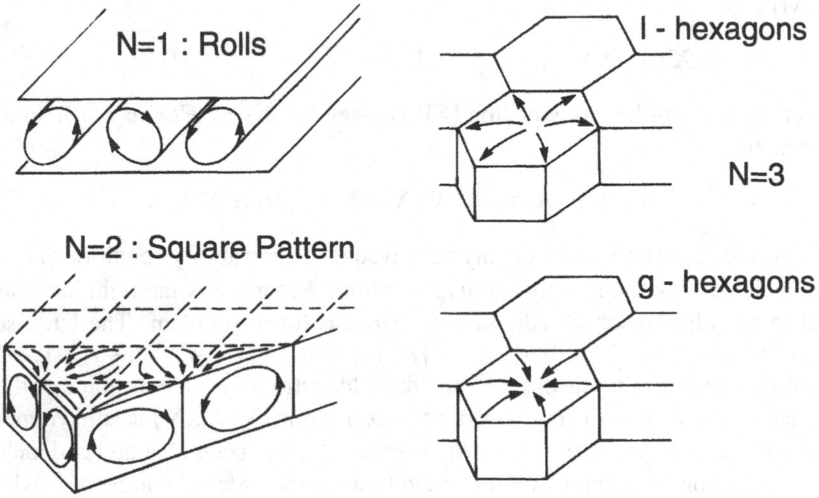

Figure 5: Pattern of convection flows in a fluid layer heated from below.

Equations (47) for the constant coefficients c_n of the steady solutions of equation (43) can be generalized through the inclusion of the influence of a weak time dependence. After replacing the coefficients c_n in expression (42) by $C_n(t)/\epsilon$ we obtain in place of equation (47)

$$M\frac{d}{dt}C_m^+(t) = (Ra - Ra_c - \sum_{n=-N}^{N} B(\boldsymbol{q}_m \cdot \boldsymbol{q}_n) \cdot \mid C_n \mid^2)C_m^+ + \beta\epsilon \sum_{n,i} C_n C_i \delta(\boldsymbol{q}_n + \boldsymbol{q}_i + \boldsymbol{q}_m) \quad (49)$$

for $m = 1, \ldots, N$ where we have set $R_0 = Ra_c$ and where terms of higher order in the expression (45) for Ra have been neglected. These equations have the remarkable property that the right hand sides can be derived as the negative derivatives with respect to C_m of a single functional

$$H(C_1,\ldots,C_N) \equiv \tfrac{1}{2}(Ra - Ra_c) \sum_{m=1}^{N} | C_m |^2 + \tfrac{1}{4} \sum_{n,m} B(\boldsymbol{q}_m \cdot \boldsymbol{q}_n) | C_m |^2 | C_n |^2 - \epsilon \tfrac{\beta}{3} \sum_{n,i,m} C_n C_i C_m \delta(\boldsymbol{q}_n + \boldsymbol{q}_i + \boldsymbol{q}_m). \qquad (50)$$

Equations (49) thus represent evolution equations for the amplitudes $C_n(t)$ of the Lyapunov type. Steady solutions of equations (49) correspond to stationary points of the functional $H(C_1, \ldots, C_N)$. Whenever these stationary points are minima, the steady solutions are stable. From the form of the functional it is evident that there exists at least one minimum and thus there is at least one stable steady solution. There can be more than one stable steady solution just as it occurs in the case of the region of coexistence of stable rolls and stable hexagons (Busse, 1967a; Walden and Ahlers, 1981). But any steady solution corresponding to a saddle point or a maximum of H is unstable.

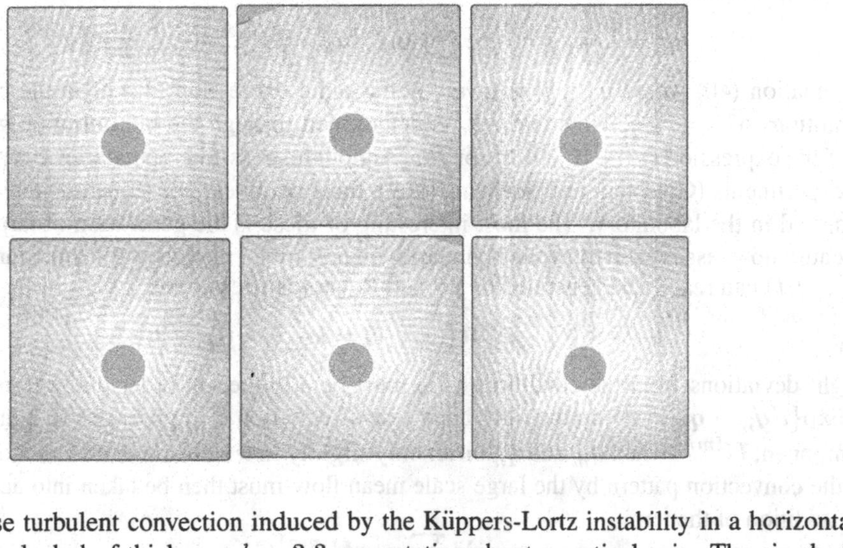

Figure 6: Phase turbulent convection induced by the Küppers-Lortz instability in a horizontal layer of methyl alcohol of thickness $d = 3.3mm$ rotating about a vertical axis. The six shadowgraph pictures have been taken 2 minutes apart in the clockwise sequence (upper row left to right, then lower row right to left). The central circle covers the outlet from the cooling water channel on top of the convection layer and does not interfere with the pattern dynamics. (For details see Busse and Heikes, 1980, and Heikes and Busse, 1980.)

The fact that a Lyapunov functional (50) does not exist in the general case becomes apparent when we consider a horizontal layer heated from below that is rotating about a vertical axis. In that case the problem is still isotropic with respect to the horizontal dimensions, since the centrifugal force can usually be neglected in comparison with gravity. But the term $B(\boldsymbol{q}_n \cdot \boldsymbol{q}_m)$ in expressions (47) and (49) must now be replaced by

$$B(\boldsymbol{q}_n \cdot \boldsymbol{q}_m) + \boldsymbol{k} \cdot \boldsymbol{q}_n \times \boldsymbol{q}_m D(\boldsymbol{q}_n \cdot \boldsymbol{q}_m) \qquad (51)$$

because the left-right symmetry is lost in the rotating system. No functional does exist whose derivative could reproduce the right hand side of equation (49) in this case. In fact, Küppers and Lortz (1969) have shown that all steady solutions do become unstable when the rotating parameter exceeds a critical value which depends on the Prandtl number (Clever and Busse, 1979). Because the resulting spatio-temporally chaotic convection exhibits little variation in its mean amplitude while the local orientation of the convection rolls changes in time, the phenomenon is called phase turbulence. An example of an experimental realisation of phase turbulent convection in a Rayleigh-Bénard layer rotating about a vertical axis is shown in Figure 6. The convection rolls seen in this time series of shadowgraph images are irregularly replaced by new rolls differing in their orientation by about 60° in the sense of rotation.

3.2 Phase Turbulent Convection in the Presence of Stress-Free Boundaries

There is another case, in which the absence of a Lyapunov functional (49) also leads to phase turbulence. Convection with stress-free instead of no-slip boundaries is the favorite configuration among mathematicians because of the simple analytical solution

$$v_z = w(x,y)\cos\pi z \quad \text{with} \quad Ra_c = 27\pi^4/4, \;\; \alpha_c = \pi/\sqrt{2} \tag{52}$$

of equation (40) where $w(x,y)$ is given by the same expression (42) as in the case of no-slip conditions of $z = \pm\frac{1}{2}$. Note that Ra_c is determined through the minimisation with respect to α of the expression $(\pi^2+\alpha^2)^3\alpha^{-2}$ for Ra. Although stress-free boundaries can be approached in experiments (Goldstein and Graham, 1969), their nonlinear consequences have not yet been explored in the laboratory. The most interesting of these is the generation of large scale flows. Because no stress is exerted by the boundaries on horizontal motion of the fluid, large scale flows $\boldsymbol{U}(x,y,t)$ can readily be generated by a small Reynolds stress,

$$\boldsymbol{k}\cdot\nabla\times(\partial_t-\Delta_2)\boldsymbol{U}(x,y,t) = \boldsymbol{k}\cdot\nabla\times\partial_z\left(\nabla\partial_z\varphi\Delta_2\varphi\right). \tag{53}$$

Slight deviations from the conditions (42) must be admitted in order that a term proportional to $\exp\{i(\boldsymbol{q}_n-\boldsymbol{q}_m)\cdot\boldsymbol{r}\}$ on the right hand side of equation (53) gives rise to a large scale flow component $\boldsymbol{U}^{(mn)}$ when $\boldsymbol{q}_m$ and $\boldsymbol{q}_n$ differ only slightly in length and in direction. The advection of the convection pattern by the large scale mean flow must then be taken into account through the addition of the term

$$iM\sum_n \boldsymbol{q}_n\cdot\boldsymbol{U}^{(mn)}C_n^+ \tag{54}$$

on the right hand side of equations (49). As a consequence they can no longer be written as derivatives with respect to C_m of a functional of the form (50). Indeed, it is found that all steady convection flows in the presence of stress-free boundaries are unstable at sufficiently low Prandtl numbers Pr (Zippelius and Siggia, 1982; Busse and Bolton, 1984) and are replaced by phase-turbulent convection. This latter phenomenon has been studied through numerical simulations by Busse (1986), Busse et al. (1992), Xi et al. (1997) and others. Because a finite horizontal periodicity interval must be used in the numerical treatments, Rayleigh numbers exceeding the critical value by a finite amount must be used. Owing to the finite horizontal periodicity interval the convection exhibits the properties of a dynamical system with a finite number of degrees of freedom. In dependence on the Rayleigh number intervals of chaotic convection alternate with intervals of time periodic convection as shown in Figure 7.

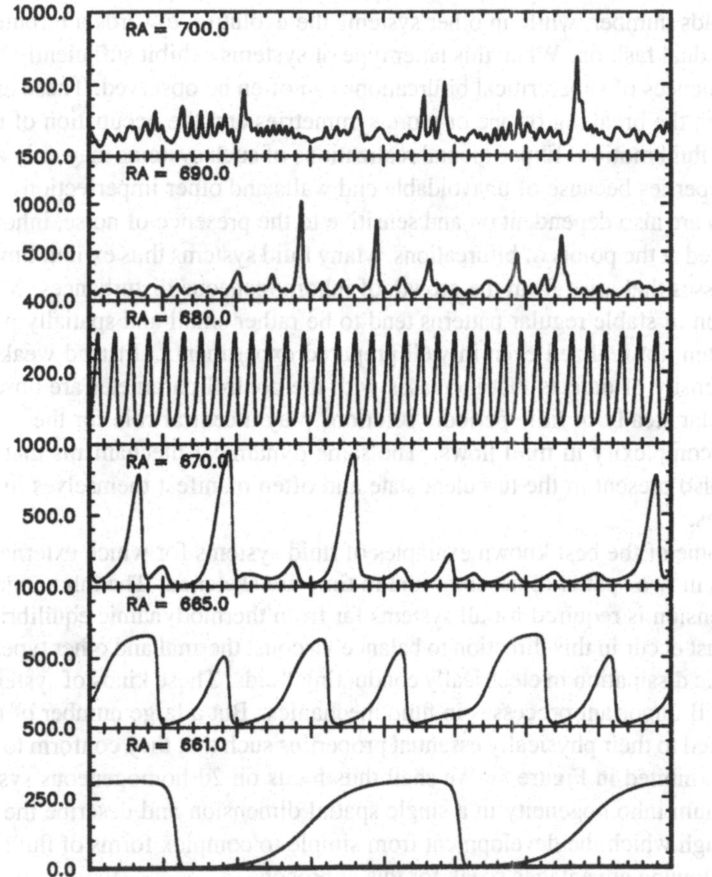

Figure 7: The convective heat transport in the presence of stress-free boundaries as a function of time for various Rayleigh numbers as indicated. The Prandtl number $Pr = 0.15$ and an aspect ratio of 8 for the horizontal periodicity have been used.

4 Transition from Simple to Complex Flows through Sequences of Bifurcations

4.1 The Sequence-of-Bifurcations Approach

We have already seen that some fluid systems experience an abrupt transition to turbulence with increasing Reynolds number, while in other systems the evolution towards a turbulent state occurs in a more gradual fashion. When this latter type of systems exhibit sufficiently high degrees of symmetry, sequences of supercritical bifurcations can often be observed. These discrete steps are associated with the breaking of one or more symmetries and the occupation of new degrees of freedom of the fluid motion. Experimental realizations of such systems may only approximate the symmetry properties because of unavoidable end walls and other imperfections. In addition, since bifurcations are also dependent on and sensitive to the presence of noise, inhomogeneities are often introduced at the points of bifurcations. Many fluid systems thus exhibit time dependent adjustment processes that arise from the growth of inhomogeneous disturbances. Moreover, the basins of attraction of stable regular patterns tend to be rather small and spatially periodic fluid flows are thus often not realized even in well prepared experiments. Instead weakly turbulent patterns with a density of defects that increases with the control parameter are observed. Nevertheless the regular steady or time periodic solutions play a central role for the understanding of the origins of complexity in fluid flows. The same dynamical mechanisms that give rise to bifurcations are also present in the turbulent state and often manifest themselves in the form of coherent structures.

In Figure 8 some of the best known examples of fluid systems for which external conditions are homogeneous in two spatial dimensions and in time are sketched. The inhomogeneity in the third spatial dimension is required for all systems far from thermodynamic equilibrium because an energy flux must occur in this direction to balance viscous, thermal and other types of dissipation, such as ohmic dissipation in electrically conducting fluids. These kinds of systems certainly do not represent all important processes in fluid mechanics. But a large number of them can be idealized or reduced to their physically essential properties such that they conform to symmetries similar to those exhibited in Figure 8. We shall thus focus on 2d-homogeneous systems which exhibit the minimum inhomogeneity in a single spatial dimension and describe the mathematical methods through which the development from simple to complex forms of fluid flow can be analyzed. The following advantages speak for this approach.

(i) In most cases the reduction of inhomogeneity to single dimension reduces a physical mechanism to its simplest form.

(ii) The homogeneity in two spatial dimensions and in time provides a maximum of symmetries, the breaking of which identifies the bifurcations in the manifold of solutions for the fluid flow.

(iii) The relative simplicity of the physical properties is reflected in the simplifications of the numerical analysis. Symmetries can be employed to reduce the numerical effort.

(iv) Although physically realized systems can only approximate homogeneity in two spatial dimensions, the bifurcations of the ideal system become only slightly imperfect bifurcations

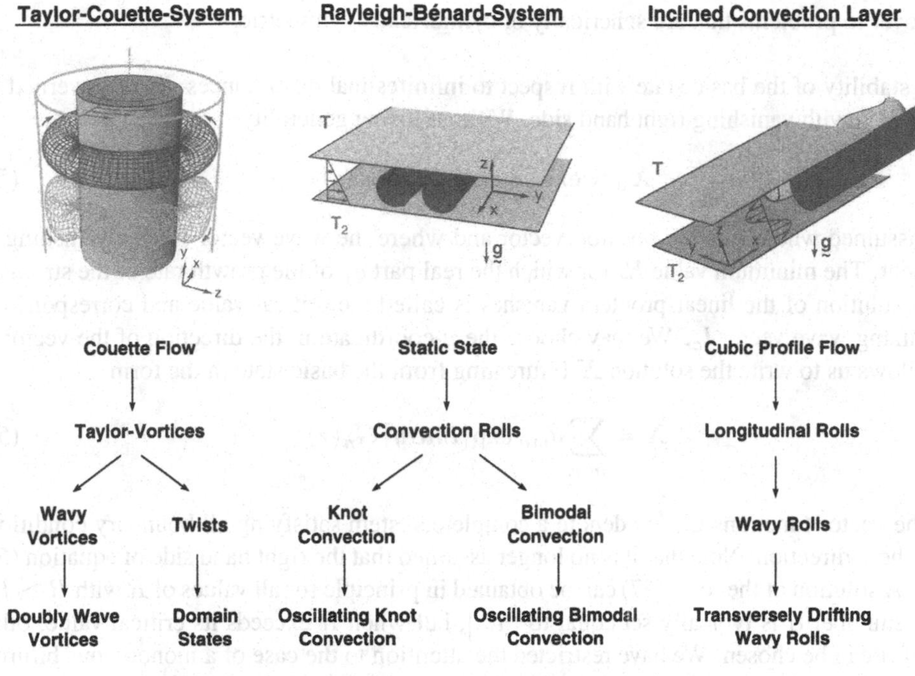

Figure 8: Examples for sequences of bifurcations in fluid dynamical systems.

in the real system as long as the typical wavelengths introduced by bifurcating solutions are small in comparison to the length scales associated with the deviations from homogeneity.

(v) The spatially and time periodic solutions that are obtained in the sequence-of-bifurcation approach represent only a minute manifold of the realizable solutions of the basic equations. Even if they are stable their basins of attraction decrease with increasing control parameter and solutions describing more irregular spatio-temporal flow structures are typically observed in experimental realizations. Nevertheless, the regular, spatially periodic solutions usually exhibit most clearly the dynamical properties and transport mechanisms of the fluid system as a function of the control parameter.

4.2 Secondary Solutions

In order to demonstrate the approach we use the same general formalism (43) except that the control parameter is now called R and only homogeneity, not isotropy in two spatial dimensions is assumed

$$\underline{L}\boldsymbol{X} + R\underline{K}\boldsymbol{X} - \underline{M}\frac{\partial}{\partial t}\boldsymbol{X} = \boldsymbol{N}(\boldsymbol{X}, \boldsymbol{X}) \tag{55}$$

Although the surfaces of homogeneity could be cylindrical or spherical surfaces, we focus on the planar case for simplicity and adopt a Cartesian system of coordinates x, y, z with the z-coordinate in the direction of inhomogeneity. Accordingly the operators of equations (55) may depend on z, but not on x, y or t. All considerations discussed in the following can be easily carried over to problems that are spherically or cylindrically symmetric.

The stability of the basic state with respect to infinitesimal disturbances $\boldsymbol{X}_0$ is governed by equation (55) with vanishing right hand side. Without losing generality

$$\boldsymbol{X}_0 \propto \exp\{i\boldsymbol{l} \cdot \boldsymbol{r} + \sigma t\}\boldsymbol{G}(\boldsymbol{l}, z) \tag{56}$$

can be assumed where $\boldsymbol{r}$ is the position vector and where the wave vector $\boldsymbol{l}$ has a vanishing z-component. The minimum value R_c for which the real part σ_r of the growth rate of the strongest growing solution of the linear problem vanishes is called the critical value and corresponds to a minimizing wave vector $\boldsymbol{l}_c$. We may choose the y-coordinate in the direction of the vector $\boldsymbol{l}_c$ which allows us to write the solution $\boldsymbol{X}$ bifurcating from the basic state in the form

$$\boldsymbol{X} = \sum_{m,n} a_{mn} \exp\{im\alpha y\}\boldsymbol{G}_n(z) \tag{57}$$

where the vector functions $\boldsymbol{G}_n(z)$ denote a complete system satisfying all boundary conditions of $\boldsymbol{X}$ in the z-direction. Note that it is no longer assumed that the right hand side of equation (55) is small. A solution of the form (57) can be obtained in principle for all values of R with $R > R_c$. The wavenumber α is typically set equal to $\mid \boldsymbol{l}_c \mid$, but when R exceeds its critical value, other values of α can be chosen. We have restricted the attention to the case of a monotonous bifurcation for which the imaginary part σ_i of the growth rate vanishes. But analogous representations can be used in the case of time periodic bifurcating solutions.

Since except for its sign a single wave vector $\boldsymbol{l}_c$ corresponds to the minimum R_c of the control parameter, solutions of the form (57) describing roll-like motions represent the generic secondary solution bifurcating from the homogeneous basic state. But in special cases the basic state exhibits symmetries such that two independent wave vectors, $\boldsymbol{l}_1$ and $\boldsymbol{l}_2$, correspond to the minimum value of R_c. In that case rectangular solutions corresponding to the superposition of roll solutions of the form (3) may also bifurcate at the point $R = R_c$. Moreover, in the case of the Rayleigh-Bénard problem the basic state is isotropic with respect to the x, y-plane which leads to an infinite degeneracy of the linear eigenvalue problem as we have discussed in section 3. But even in this case rolls of the form (57) represent the only stable solution (Schlüter et al., 1967) if deviations from the Boussinesq approximation are neglected. There is thus no need to distinguish between general homogeneous systems and the special case of the isotropic Rayleigh-Bénard system as far as the higher order bifurcations are concerned.

In Table 2 the symmetry properties of rolls are listed. They can be divided into the first three which are obeyed by all solutions of the form (57) and the remaining fourth and fifth properties which are satisfied only in more special cases. The fifth property requires the symmetry of the problem with respect to the midplane of the layer. But even in the case of Taylor vortices between cylinders rotating differentially about their common axis these symmetries are attained in the small gap limit with nearly co-rotating cylinders as we have seen in section 2.3.

For the determination of the coefficients a_{mn} in the representation (57) it is necessary to project the basic equation (55) onto the space of the expansion functions introduced in (57). The infinite system of nonlinear algebraic equations that is obtained in this way can be solved through a Newton-Raphson iteration method after a suitable truncation has been introduced. A typical procedure is to neglect all coefficients and corresponding equations whose subscripts satisfy the

Table 2: Symmetry properties of two-dimensional rolls

A translation in time:	$\partial\varphi/\partial t = 0$
B translation along roll axis:	$\partial\varphi/\partial x = 0$
C transverse periodicity:	$\varphi(y + 2\pi/\alpha, z) = \varphi(y, z)$
D transverse reflection:	$\varphi(-y, z) = \varphi(y, z)$ or $a_{-mn} = a_{mn}$
E inversion about roll axis:	$\varphi(y + \frac{\pi}{\alpha}, z) = -\varphi(y, -z)$ or $a_{mn} = 0$ for odd $m + n$

Table 3: Symmetries Broken by Bifurcations from Convection Rolls

Broken Symmetries		A	B	C	D	E
Properties of disturbances		$\sigma_i \neq 0$	$b \neq 0$	$d \neq 0$	$\tilde{a}_{mn} \neq \tilde{a}_{-mn}$	$\tilde{a}_{mn} \neq 0$ for $m + n =$ odd
Eckhaus Instab.				X	X	
Crossroll Instab.	CR		X			X
Knot-Instability	KN		X			X
Even Blob-Instab.	EB	X	X			
Odd Blob-Instab.	OB	X	X			X
Oscillatory Instab.	OS	X	X		X	
Zig-Zag-Instability	ZZ		X		X	
Skewed Varicose Instab.	SV		X	X	X	
Osc. Skewed Var. Inst.		X	X	X	X	

condition

$$n + m > N_T \tag{58}$$

After the finite system of equations for the coefficients a_{mn} has been solved for given values of R and α, the solution can be compared with the solution obtained in the case when N_T is replaced by $N_T - 2$. If the change in typical physical properties described by the solution is sufficiently small, the solution can be regarded as a good approximation. Otherwise N_T must be increased.

4.3 Tertiary Solutions

In contrast to the rather universal form (57) of the secondary solutions, the tertiary solutions bifurcating from them reflect the specific physical conditions of the problem and thus exhibit a wide variety of shapes and styles. For this reason they are of special interest since much about characteristic dynamical mechanisms of the particular problem can be learned from the structure of the tertiary solutions. In order to analyze the instabilities of the solutions (57) we superimpose infinitesimal disturbances of the form

$$\tilde{\boldsymbol{X}} = \exp\{iby + idx + \sigma t\} \sum_{m,n} \tilde{a}_{mn} \exp\{im\alpha y\} \boldsymbol{G}_n(z) \tag{59}$$

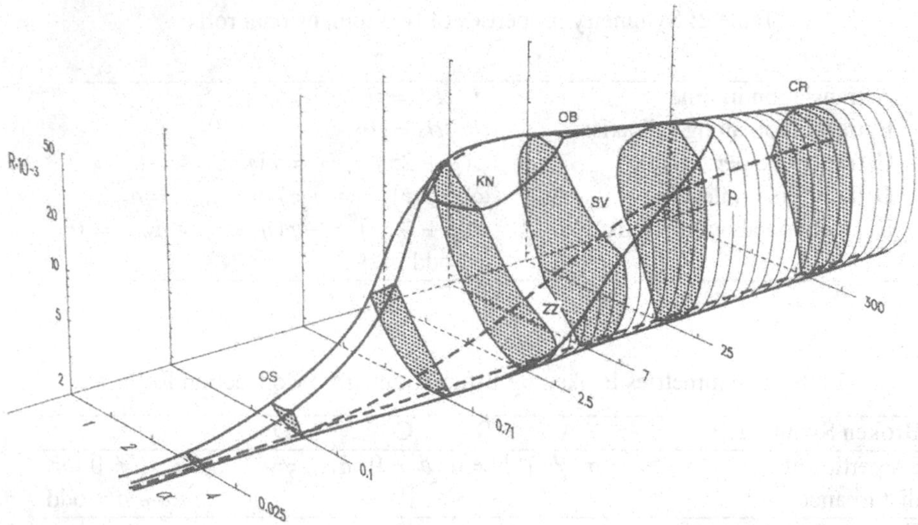

Figure 9: Region of stable convection rolls in the Ra-α-Pr parameter space. The region of stable rolls is bounded by surfaces corresponding to the onset of the instabilities listed in Table 2. Note that Pr corresponds to P in the figure increasing towards the right and R corresponds to Ra. The symbols CR, ... indicate the onset of the respective instabilities as listed in Table 3.

The y-dependence in this representation follows from Floquet's theory, while an exponential x- and t-dependence can be assumed because these variables do not appear explicitly in the equation for $\underline{\boldsymbol{X}}$. After introducing this ansatz into the equation

$$\underline{\boldsymbol{L}}\tilde{\boldsymbol{X}} - R\underline{\boldsymbol{K}}\tilde{\boldsymbol{X}} - \sigma\underline{\boldsymbol{M}}\tilde{\boldsymbol{X}} = \boldsymbol{N}(\boldsymbol{X},\tilde{\boldsymbol{X}}) + \boldsymbol{N}(\tilde{\boldsymbol{X}},\boldsymbol{X}) \tag{60}$$

and projecting it onto the space of expansion functions we obtain a linear algebraic eigenvalue problem for the coefficients $\tilde{a}_{mn}$ with the growth rate σ as eigenvalue. Of primary interest are the growth rates σ with largest real part σ_r as a function of the wavenumbers b and d. Whenever there exists a positive real part σ_r, the stationary solution $\boldsymbol{X}$ is unstable; otherwise it is regarded as stable. For the actual determination of the eigenvalues σ the same truncation condition (58) as for the steady solution must be employed. If there is a significant difference of relevant eigenvalues obtained for N_T and $N_T - 2$, then the stability of the stationary solution obtained for $N_T + 2$ should be analyzed. In the space spanned by the parameters R and α and other parameters of the problem, regions of stability can thus be determined which are bounded by hypersurfaces at which the largest value σ_r goes through zero. An example of such an enclosed stability region is shown in Figure 9. In Table 3 a number of typical instabilities that are encountered in the problem of Rayleigh-Bénard convection have been listed together with the symmetries of the roll solutions that are broken by them. As is evident, two or more symmetries are usually broken, but a number of symmetries are still preserved which tend to characterize the bifurcating three-dimensional tertiary solution. The two-dimensional Eckhaus-instability usually does not lead to a new solution, but instead tends to replace a roll solution in the unstable region with one in the

Table 4: Examples for Tertiary Solutions and Their Symmetries in Terms of the Complex Coefficients

Tertiary Solution	Reflection Symmetries	Inversion Symmetry
knot-/bimodal convection	$a_{-lmn} = a_{lmn} = a_{l-mn}$	$a_{lmn} = 0$ for $l+m+n =$ odd
undulating rolls	$a_{-lmn} = a_{lmn}, a_{l-mn} = (-1)^l a_{lmn}$	$a_{lmn} = 0$ for $m+n =$ odd
symmetric travelling wave convection or wavy rolls with Poiseuille flow	$a_{l-mn} = (-1)^l a_{lmn}$	$a_{lmn} = 0$ for $m+n =$ odd
Wavy rolls with Couette flow or wavy Taylor vortices in small gap limit	$a_{l-mn} = (-1)^l a_{lmn}$	$a_{lmn} = (-1)^{m+n} a_{-lmn}$
Travelling blob convection	$a_{l-mn} = a_{lmn}$	$a_{lmn} = 0$ for $l+m+n =$ odd

stable region. The actual evolution of this instability can be rather complex, see, for example, Kramer et al. (1988). An important role is played by the neutral disturbance mode

$$\tilde{\boldsymbol{X}} = \frac{\partial}{\partial y}\boldsymbol{X} \tag{61}$$

which satisfies (60) with $\sigma = 0$. This disturbance corresponds to an infinitesimal displacement of the stationary pattern in the y-direction. Some instabilities such as the zig-zag- or the oscillatory instability (in the case of stress-free boundaries) correspond to a small modification of the mode (61).

As the three-dimensional solutions bifurcate from the roll solution, they can be described by expressions of the form

$$\boldsymbol{X} = \sum_{l,m,n} a_{lmn} \exp\{il\alpha_x x + im\alpha_y y\}\boldsymbol{G}_n(z) \tag{62}$$

where we have assumed that the instability of interest has a vanishing imaginary part σ_i of the growth rate, i.e. the bifurcation is monotonic. When an instability with a finite value of σ_i occurs, it typically leads to travelling waves propagating along the x-axis which can also be described by the representation (62) if x is replaced by $\hat{x} = x - ct$. Numerous tertiary solutions can be described in this way; a partial list is given in Table 4. In addition to the usual conditions

$$a_{lmn} = a^{+}_{-l-mn} \qquad (^{+}\text{indicates complex conjugate})$$

which follow from the property that expression (62) describes a real quantity, we have assumed for the cases of Table 3 that the problem is symmetric with respect to the midplane of the layer. The functions $\boldsymbol{G}_n(z)$ can thus be separated into two symmetry classes with respect to the midplane $z = 0$, one corresponding to odd subscripts n, the other to even n. It is convenient to denote on this basis the symmetry properties of the tertiary solution (62), instead of introducing a cumbersome notation of the various symmetries.

The wavenumber α_y in expression (62) is not necessarily identical with the wavenumber α of the original roll representation (57). If a subharmonic instability occurs with $b = \alpha/2$, the representation (62) can be used as for example in the case of the staggered vortex solution that replaces transverse rolls in an inclined fluid layer (Nagata and Busse, 1983).

4.4 Quaternary Solutions and Higher Order Bifurcations

After the onset of three-dimensional tertiary solutions the continuous spatial symmetries such as the invariance with respect to the translation along the roll axis have been broken and have been replaced by reflection symmetries and inversion symmetries such as those shown in Table 3. The stability of the steady or travelling tertiary solutions can be investigated through the superposition of infinitesimal disturbances just as in the case (5) of the two-dimensional solution. Using the general Floquet ansatz

$$\tilde{\boldsymbol{X}} = \exp\{idx + iby + \sigma t\} \sum_{l,m,n} \tilde{a}_{lmn} \exp\{il\alpha_x x + im\alpha_y y\} \boldsymbol{G}_n(z) \tag{63}$$

we arrive at a linear homogeneous system of equations for the unknown coefficients $\tilde{a}_{lmn}$ with the growthrate σ as eigenvalue. When the maximum real part of σ as a function of d and b is less or equal to zero the tertiary solution is stable. Otherwise it is unstable. The same truncation parameter N_T should be used as for the representation (62) of the tertiary solution. But since no symmetry properties can be assumed for the complex coefficients $\tilde{a}_{lmn}$ in the general case, the stability analysis is much more demanding in terms of computer resources than the determination of the tertiary solution. Fortunately, many of the most interesting instabilities do not tend to change the periodicity interval in the x,y-plane and thus correspond to eigenvalues σ for which the real parts σ_r reach their maximum values at $d = b = 0$. In this case symmetry properties become indeed available which can be employed to reduce the computational effort. In the following section an example for this procedure will be discussed.

The most strongly growing disturbances of tertiary stationary solutions are often those with a non-vanishing imaginary part of σ_i. Since travelling wave type solutions are no longer possible after the translational invariance along the axis of the rolls has been broken, the time dependence must be taken into account explicitly. Time dependent three-dimensional solutions can be obtained through forward integration in time of the differential equations for the time dependent coefficients $a_{lmn}(t)$ in the representation for the quaternary solutions,

$$\boldsymbol{X} = \sum_{l,m,n} a_{lmn}(t) \exp\{il\alpha_x x + im\alpha_y y\} \boldsymbol{G}_n(z). \tag{64}$$

The system of differential equations is obtained, just as in the case of the algebraic equations of tertiary solutions (8) through projections of the equations of motion onto the space of the expansion functions. Alternatively, time periodic solutions of form (64) can be generated through a Fourier expansion

$$a_{lmn}(t) = \sum_{p=-M}^{M} a_{lmnp} \exp\{ip\omega t\} \tag{65}$$

where the complex coefficients a_{lmnp} obey nonlinear algebraic equations which can be solved through a Newton-Raphson iteration method. The truncation parameter M must be chosen sufficiently high such that the solution does not change significantly when M is replaced by $M - 1$. The frequency ω can be determined most easily through the choice of a particular phase, say $a_{1011} = a_{101-1}$. The equation corresponding to the latter coefficient thus becomes available as an equation for the unknown ω. Once the coefficients (65) including ω have been computed, the stability of solutions of the form (64) can be investigated, just as in the case of the tertiary solutions. The Fourier method offers advantages in this respect, since a stability analysis is not as easily possible when the coefficients $a_{lmn}(t)$ are computed through forward integration in time. The Fourier method has been applied by de la Torre and Busse (1995) in the case of thermal convection in a Hele-Shaw-channel.

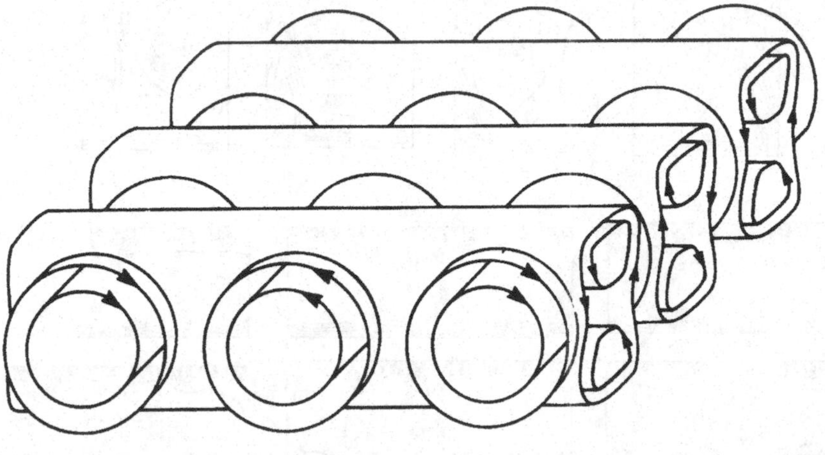

Figure 10: Schematic sketch of bimodal convection in a fluid layer heated from below.

4.5 Steady and Oscillatory Bimodal Convection

As an example of a tertiary solution we consider steady bimodal convection. Basically it corresponds to the superposition of a secondary roll pattern with smaller wavelength onto the given roll pattern as shown in the sketch of Figure 10. The smaller wavelength of the secondary rolls indicates that they take advantage of the thermal buoyancy stored in the thermal boundary layers close to the upper and lower rigid boundaries of the fluid layer. Through the onset of bimodal convection the convective heat transport becomes more efficient and the two roll patterns quickly reach comparable amplitudes as the Rayleigh number is increased beyond onset.

If the Prandtl number is in the range $10 \lesssim Pr \lesssim 10^2$ the bifurcation from rolls to bimodal cells is followed by a further bifurcation to oscillatory bimodal convection. The thermal boundary layers periodically thicken and blobs of fluid hotter or cooler than average circulate through the convection cells. These oscillations are characterized to some extent by a resonance between the time of circulation within the bimodal cell and the period of thickening and thinning of the

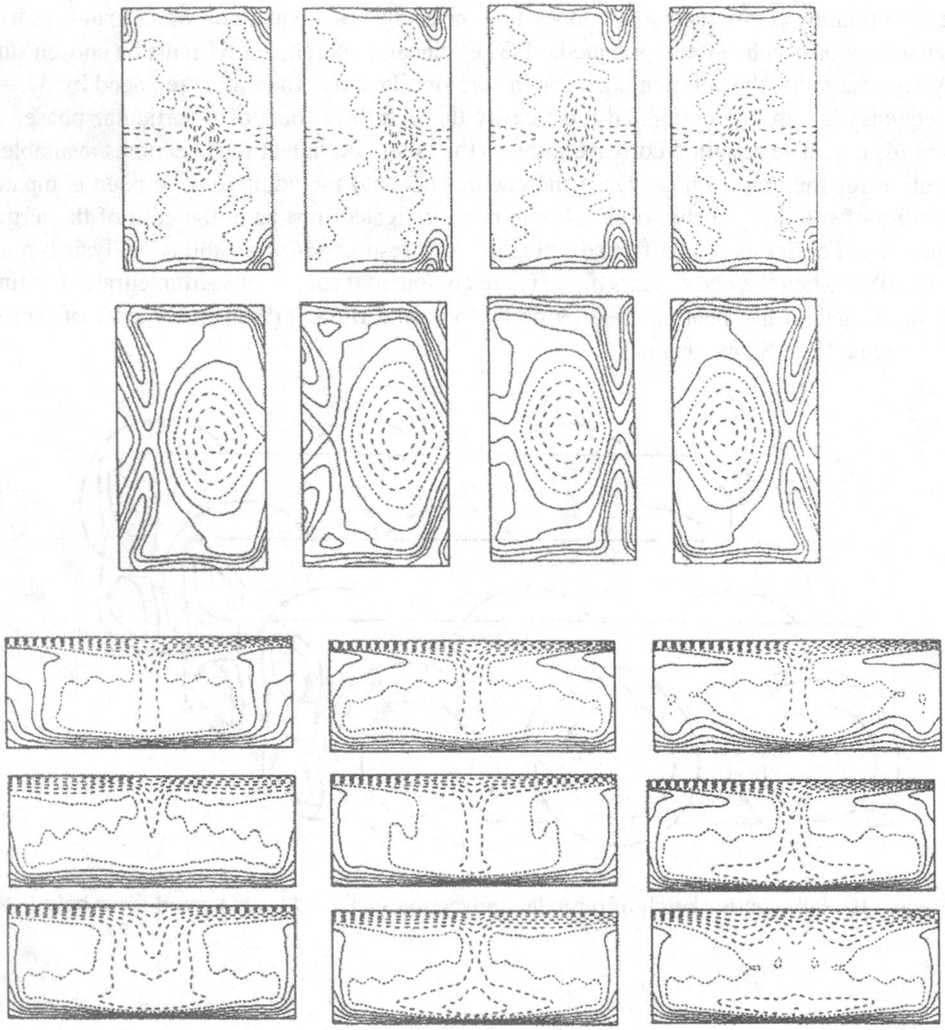

Figure 11: The upper part shows isotherms for wavy oscillatory bimodal convection in the middle plane $z = 0$ (uppermost row) and in the plane $z = -0.4$ close to the lower boundary at $z = -0.5$ (second row) at the times $t = n\pi/3\omega$ for $n = 0, 1, 2, 3$ (left to right) such that half a period of oscillation has passed between the first and the last picture. The lower part of the figure shows isotherms at the y, z-planes $x = 0, x = \pi/2\alpha_x$ and $x = \pi/\alpha_x$ (from top to bottom) at the times $t = n\pi/3\omega$ for $n = 0, 1, 2$. In the lower part of the figure y increases towards the right and z increases upwards while in the upper part x increases towards the right and y increases upwards. The parameter values $Pr = 30, Ra = 10^5, \alpha_x = 4.5, \alpha_y = 2.5$ have been used.

thermal boundary layers. The oscillations do thus not occur if the spatial periodicity interval of the bimodal cell is too small.

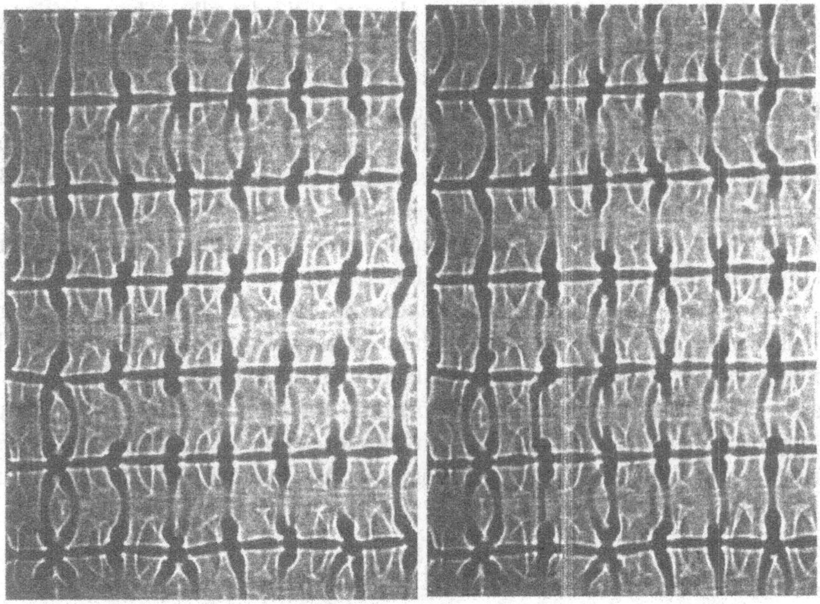

Figure 12: Shadowgraph observation of wavy oscillatory bimodal convection in a layer of silicon oil heated from below. The dark regions indicate hot rising fluid. The Prandtl and Rayleigh numbers are $Pr = 63$, $Ra = 1.5 \cdot 10^5$ and the wave numbers in the x-direction (towards the right) and y-direction are given by $\alpha_x = 4.08$, $\alpha_y = 2.04$. The right photograph has been taken 25 sec. after the left one which corresponds to nearly half a period. For details see Busse and Whitehead (1974).

There are actually two types of oscillatory bimodal convection, the symmetric one that does not change the spatial symmetry of steady bimodal convection and the other, called wavy oscillatory bimodal convection, which is characterized by the property that the set of coefficients $a_{lmn}(t)$ with

$$-a_{-lmn} = a_{lmn} = a_{l-mn} \quad \text{for} \quad l+m+n = \quad \text{odd and} \quad a_{lmn} = 0 \text{ otherwise} \tag{66}$$

are participating in the description of the solution in addition to those listed in Table 4 for bimodal convection. Figure 11 provides an impression of the time dependent structure of wavy oscillatory bimodal convection derived from the computations of Clever and Busse (1994) and in Figure 12 an experimental visualisation is shown. Clever and Busse (1994) also discuss a further transition introducing a large wavelength which is characteristic of knot convection (Clever and Busse, 1989). This large wavelength is typically given by the wavenumber $\alpha_x/3$ in the case of Figure 12 and is a characteristic property of high Rayleigh number convection at intermediate Prandtl numbers. It can clearly be discerned in turbulent convection at values $Ra \sim 10^6$ (Busse, 1994).

It should be kept in mind that the realisation of convection flows that are periodic in space and in time requires controlled initial conditions such that an approximately perfect roll pattern is realized after onset of convection. The transition to bimodal cells may occur in more inhomogeneous fashion since the steady bimodal attractor is sufficiently strong such that pattern imperfections can be eliminated in time except close to the sidewalls. The transition to oscillations usually occurs in a less homogeneous way and their phases tend to exhibit large scale variations. Without controlled initial conditions the convection flows at onset occur already in the form of patches of rolls with different horizontal orientations which tend to evolve in such a way that they ultimately reflect the geometrical configuration of the sidewalls of the layer. As the Rayleigh number increases the density of dislocations in the pattern increases rapidly and a chaotic structure of a kind of bimodal convection is realized when the Prandtl number is sufficiently high ($Pr \gtrsim 10$). The onset of oscillations in form of hot and cold blobs emerging from the thermal boundary layers occurs initially at a few spots where the convection pattern deviates most strongly from the ideal periodic form. Laboratory experiments thus exhibit in general a more turbulent scenario than that suggested by the sequence-of-bifurcation approach. The physical mechanisms, however, operate in the turbulent situation in qualitatively the same manner as in the case of the spatially and temporarily periodic solutions produced by the sequence of bifurcation approach. The latter method thus provides a sensible way towards an understanding of the processes occurring in turbulent convection as well as in other cases of fluid turbulence.

5 Upper Bounds on Turbulent Transports

5.1 Introductory Remarks

Scientists involved in the study of turbulent systems have often gained the impression that turbulent flows tend to maximize some transport quantity as for example the heat transport by convection in a fluid layer heated from below or the transport of angular momentum between differentially rotating coaxial cylinders. Malkus (1954a) has used the hypothesis of maximum heat transport in a formulation of a theory of turbulent convection and Howard (1963) was motivated by this attempt to formulate the problem of the determination of a rigorous upper bound on the heat transport by thermal convection. Busse (1969a) extended this approach through the consideration of bounds for transports by shear flows and introduced the multi-α-solutions for the corresponding variational problem (Busse, 1969b). In recent years interest in the theory of upper bounds has been revived by Doering and Constantin (1994) who introduced their "background" method. Kerswell (1998) has shown that both formulations lead to equivalent variational problems. We shall thus focus here on the older Howard-Busse-method.

In some respects the method of upper bounds for turbulent transports is a generalization of the energy method of stability which has been discussed in section 2. Through the upper bounds on transports the manifold of statistically stationary solutions is not only bounded from below with respect to the Reynolds number or Rayleigh number parameter but is also bounded from above with respect to their amplitude or, more exactly, their dissipation which is closely related to the turbulent transport as we shall see. More important than the bounds themselves are perhaps the extremalizing vector fields which deliver the upper bounds. As will be discussed in section 5.4 the vector fields have much in common with observed velocity fields and thus can provide some insights into properties of turbulence which are not available otherwise.

The bounds on properties of solutions of the Navier-Stokes equations are derived on the basis of certain energy integrals. The boundary conditions for the velocity field and other fields such as the temperature are usually taken into account accurately. An important constraint is offered by the equation of continuity. The quality of the bounds depends, of course, on the number of constraints that are taken into account. On the other hand, the mathematical work that is required for the derivation of bounds increases sharply with the number of constraints that are imposed.

Bounds are of particular interest in the case of asymptotically high values of the control parameters such as the Reynolds number in the case of turbulent shear flow or the Rayleigh number in the case of thermal convection. Upper bounds on the transports of momentum or heat in the asymptotic regime are of practical importance and also offer tests and comparisons with heuristically derived relationships. Fortunately the application of boundary layer methods permits the derivation of asymptotic solutions for the respective variational problems. These boundary layers represent another feature for which a comparison with observed fluid flows appears to be possible, especially in cases when the advection of a scalar field such as the temperature is the dominant nonlinear effect as for example in the case of thermal convection in a high Prandtl number fluid.

The optimum theory includes the energy stability theory as the special limit in which the turbulent transport vanishes. However, not all problems that can be treated by the energy stability method can also be treated by the theory of upper bounds. The latter theory always requires a spatial dimension with respect to which the problem on hand is homogeneous, such that an average with respect to this dimension is well defined. Actually, only problems for which an average with respect to two spatial dimensions can be defined have been treated so far by the Howard-Busse method or the Doering-Constantin approach because of the mathematical complexity of problems of lesser symmetry. The upper bound theory shares with the energy stability theory the property that the time dependent dynamics of the turbulent fluid flow is not well exhibited by extremalizing vector field. This property could be changed, however, when additional constraints are included in the theory.

In the following we first outline the optimum theory in the case of turbulent plane Couette flow and then discuss the multi-α-solutions in the case of convection in a porous layer.

5.2 Bounds on the Transport of Momentum

In order to outline the basic ideas of the bounding theory we consider the flow between two parallel plates which move in opposite directions with respect to each other which has already been introduced in section 2.1. Using the distance d between the plates as length scale and d^2/ν as time scale, we write the Navier-Stokes equations for the incompressible fluid in the form

$$\frac{\partial}{\partial t}\boldsymbol{v} + \boldsymbol{v}\cdot\nabla\boldsymbol{v} = -\nabla p + \nabla^2\boldsymbol{v} \tag{67}$$

$$\nabla\cdot\boldsymbol{v} = 0 \tag{68}$$

The boundary conditions are given by

$$\boldsymbol{v} = \mp\frac{1}{2}Re\,\boldsymbol{i} \text{ at } z = \pm\frac{1}{2} \tag{69}$$

where z is the coordinate perpendicular to the rigid plates.

It is convenient to separate the velocity field $\boldsymbol{v}$ into a mean and a fluctuating part

$$\boldsymbol{v} = \boldsymbol{U} + \boldsymbol{u} \quad \text{with} \quad \overline{\boldsymbol{v}} = \boldsymbol{U}.$$

where the average over planes $z =$ constant has been denoted by a bar. We also use the average over the entire fluid layer indicated by angular brackets, e.g.

$$\overline{uw} = \lim_{L\to\infty} \frac{1}{4L^2} \int_{-L}^{L}\int_{-L}^{L} uw \; dxdy, \quad \langle uw \rangle = \int_{-\frac{1}{2}}^{\frac{1}{2}} \overline{uw} dz \tag{70}$$

Because of the equation of continuity $\boldsymbol{U}$ does not have a z-component, and the average over planes $z =$ const. of equation (67) yields

$$\frac{\partial^2}{\partial z^2}\boldsymbol{U} - \frac{\partial}{\partial t}\boldsymbol{U} = \frac{\partial}{\partial z}\overline{w\boldsymbol{u}_2} \tag{71}$$

$$-\frac{\partial}{\partial z}\overline{p} = \frac{\partial}{\partial z}\overline{ww} \tag{72}$$

In these equations w and $\boldsymbol{u}_2$ denote the components of $\boldsymbol{u}$ perpendicular and parallel to the plates. After subtracting equations (71) and (72) from the corresponding components of equation (67) we obtain the following equation for the fluctuating velocity field $\boldsymbol{u}$,

$$\nabla^2\boldsymbol{u} - \frac{\partial}{\partial t}\boldsymbol{u} - \boldsymbol{U}\cdot\nabla\boldsymbol{u} - w\frac{\partial}{\partial z}\boldsymbol{U} - \boldsymbol{u}\cdot\nabla\boldsymbol{u} + \overline{\boldsymbol{u}\cdot\nabla\boldsymbol{u}} - \nabla(p - \overline{p}) = 0 \tag{73}$$

After multiplying this equation with $\boldsymbol{u}$, taking the average over the fluid layer and using the property that $\boldsymbol{u}$ vanishes at $z = \pm\frac{1}{2}$, we find

$$\langle |\nabla\boldsymbol{u}|^2\rangle - \frac{1}{2}\frac{d}{dt}\langle|\boldsymbol{u}|^2\rangle - \langle\overline{\boldsymbol{u}_2 w}\cdot\frac{\partial}{\partial z}\boldsymbol{U}\rangle = 0 \tag{74}$$

Since it must be required that $\boldsymbol{u}$ remains finite as x, y tend to infinity, the contributions from the surface integrals over surfaces besides those at $z = \pm\frac{1}{2}$ are negligible.

Relationship (74) can be simplified if we restrict the attention to fluid flow under stationary external conditions. We define this case by requiring that all mean quantities become time independent after the external conditions have been kept constant for a sufficiently long time such that transient processes have decayed. Alternatively the average could be defined as an ensemble average over many realizations of the same experiment. Accordingly the time derivative of $\boldsymbol{U}$ can be dropped and equation (71) yields

$$\frac{d}{dz}\boldsymbol{U} = \overline{w\boldsymbol{u}_2} - \langle w\boldsymbol{u}_2\rangle - Re\,\boldsymbol{i} \tag{75}$$

after the boundary condition (69) has been employed. Using equation (75) we obtain the final form of the energy balance (74)

$$\langle|\nabla\boldsymbol{u}|^2\rangle + \langle|\overline{\boldsymbol{u}_2 w} - \langle\boldsymbol{u}w\rangle|^2\rangle = Re\langle u_x w\rangle \tag{76}$$

where the identity

$$\langle|\overline{\boldsymbol{u}_2 w}|^2\rangle - \langle\boldsymbol{u}_2 w\rangle^2 = \langle|\overline{\boldsymbol{u}_2 w} - \langle\boldsymbol{u}_2 w\rangle|^2\rangle \tag{77}$$

has been used.

The momentum transport between the moving rigid plates is described by $-dU/dz\mid_{z=\pm\frac{1}{2}}$. The Reynolds number Re measures the momentum transport in the case of the laminar solution of the problem, $\boldsymbol{u} = 0$. Since $\langle u_x w \rangle$ is positive according to (76), the momentum transport by turbulent flow always exceeds the corresponding laminar value. Hence an upper bound on $\langle u_x w \rangle$ is of primary physical interest. We are thus led to the formulation of the variational problem: *Find the maximum $\mu(R)$ of $\langle \check{u}_x \check{w} \rangle$ among all solenoidal vector fields $\check{\boldsymbol{u}}$ that satisfy the boundary condition $\check{\boldsymbol{u}} = 0$ at $z = \pm\frac{1}{2}$ and the relationship*

$$\langle \mid \nabla \check{\boldsymbol{u}} \mid^2 \rangle + \langle \mid \overline{\check{\boldsymbol{u}}_2 \check{w}} - \langle \check{\boldsymbol{u}}_2 \check{w} \rangle \mid^2 \rangle = R \langle \check{u}_x \check{w} \rangle \quad \text{with} \quad \check{\boldsymbol{u}} = \check{\boldsymbol{u}}_2 + \boldsymbol{k}\check{w} \tag{78}$$

Here $\boldsymbol{k}$ denotes the unit vector in the z-direction. Obviously all statistically stationary solutions of (73), (75) are eligible vector fields of the variational problem. But the manifold of trial vector fields $\check{\boldsymbol{u}}$ is much larger and thereby allows a relatively simple formulation of the upper bound problem for $\langle \check{u}_x \check{w} \rangle$. The manifold of trial fields can be enlarged even further by dropping the constraint $\nabla \cdot \check{\boldsymbol{u}} = 0$. In this form the problem has been considered by Busse (1969a) and by Howard (1972).

Following Howard (1963) it can be shown that $\mu(R)$ is a monotonous function and thus the formulation of the variational problem can be inverted:

Find the minimum $R(\mu)$ of the functional

$$\mathcal{R}(\check{\boldsymbol{u}}, \mu) \equiv \frac{\langle \mid \nabla \check{\boldsymbol{u}} \mid^2 \rangle}{\langle \check{u}_x \check{w} \rangle} + \mu \frac{\langle \mid \overline{\check{\boldsymbol{u}}_2 \check{w}} - \langle \check{\boldsymbol{u}}_2 \check{w} \rangle^2 \mid \rangle}{\langle \check{u}_x \check{w} \rangle^2} \tag{79}$$

among all solenoidal vector fields $\check{\boldsymbol{u}}$ that satisfy the boundary condition $\check{\boldsymbol{u}} = 0$ at $z = \pm\frac{1}{2}$ and the condition $\langle \check{u}_x \check{w} \rangle > 0$.

Because the functional (79) is homogeneous, the extremalizing solution can be normalized to satisfy $\langle \check{u}_x \check{w} \rangle = \mu$. Accordingly the extremalizing solution obeys condition (78). On the other hand, any solution of the first variational problem is also a solution of the second problem. It thus follows that the two formulations are equivalent. In the limit $\mu \to 0$ the variational problem (79) becomes identical to the problem (13) which defines the energy stability limit, $Re_E \equiv R(0)$ for plane Couette flow. The optimum theory always includes the energy stability limit as a special case. On the other hand, since the upper bound theory depends on the separation of the turbulent field into a mean and a fluctuating part, its applications are more restricted than those of the energy stability theory.

In order to demonstrate the properties of the extremalizing vector fields $\check{\boldsymbol{u}}$ we turn to the simplest variational problem of the general form (79). We shall return to the particular case (79) in section 5.4.

5.3 Extremalizing Multi-α-Solutions

The most easily solvable variational problem of the general form (79) is obtained in the case of thermal convection in a porous medium (Busse and Joseph, 1972):

Given $\mu > 0$ find the minimum $P(\mu)$ of the functional

$$\mathcal{P}(\boldsymbol{u}, \theta, \mu) \equiv \frac{\langle \mid \boldsymbol{u} \mid^2 \rangle \langle \mid \nabla \theta \mid^2 \rangle + \mu \langle (w\theta - \overline{\langle w\theta \rangle})^2 \rangle}{\langle w\theta \rangle^2} \tag{80}$$

among all fields $\boldsymbol{u}, \theta$ *that satisfy* $\nabla \cdot \boldsymbol{u} = 0$ *and*

$$w = \theta = 0 \text{ at } z = \pm\frac{1}{2}. \tag{81}$$

Here w denotes $\boldsymbol{u} \cdot \boldsymbol{k}$ where $\boldsymbol{k}$ is the unit vector in the vertical z-direction. As before the bar indicates the average over the horizontal x-y-plane. The variational problem (80) and (81) is based on the Darcy-Boussinesq equations describing convection in a layer of a fluid saturated porous medium

$$Pr**(-1)B\left(\frac{\partial \boldsymbol{v}}{\partial t} + \boldsymbol{v} \cdot \nabla \boldsymbol{v}\right) = -\nabla \pi + \Theta \boldsymbol{k} - \boldsymbol{v} \tag{82}$$

$$0 = \nabla \cdot \boldsymbol{v} \tag{83}$$

$$\frac{\partial}{\partial t}\Theta + \boldsymbol{v} \cdot \nabla \Theta = Ra^{(p)} \boldsymbol{k} \cdot \boldsymbol{v} + \nabla^2 \Theta, \tag{84}$$

where B is the Darcy permeability coefficient divided by d^2 and $Ra^{(p)} = RaB$. Equations (82), (83) and (84) differ from equations (36), (37) and (38) only in that the viscous time scale, d^2/ν has been replaced by the thermal time scale, d^2/κ, and that $\nabla^2 \boldsymbol{v}$ is replaced by $-\boldsymbol{v}/B$. The equations (82), (83) and (84) give rise to the balances

$$\langle | \boldsymbol{v} |^2 \rangle = \langle \Theta \boldsymbol{v} \cdot \boldsymbol{k} \rangle \tag{85}$$

$$\langle | \nabla \Theta |^2 \rangle = Ra^{(p)} \langle \Theta \boldsymbol{v} \cdot \boldsymbol{k} \rangle - \langle (\overline{\boldsymbol{v} \cdot \boldsymbol{k} \Theta} - \langle \boldsymbol{v} \cdot \boldsymbol{k} \Theta \rangle)^2 \rangle \tag{86}$$

on which the formulation (80) of the variational problem is based. The trial fields $\boldsymbol{u}, \theta$ of the variational problem (80) correspond to the physical fields $\boldsymbol{v}$, Θ.

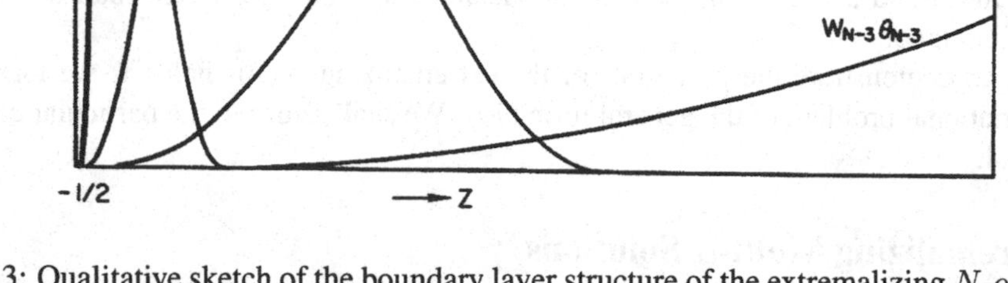

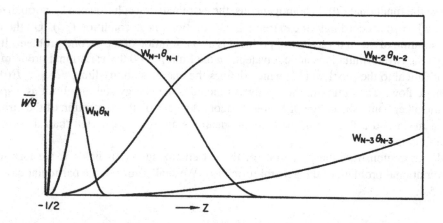

Figure 13: Qualitative sketch of the boundary layer structure of the extremalizing N-α-solution.

In order to eliminate the side constraint $\nabla \cdot \boldsymbol{u} = 0$ it is convenient to introduce the general representation for a solenoidal vector field,

$$\boldsymbol{u} = \nabla \times (\nabla \times \boldsymbol{k}\varphi) + \nabla \times \boldsymbol{k}\psi \tag{87}$$

From the general form of the dissipation integral

$$\langle| \boldsymbol{u} |^2\rangle = \langle\nabla^2\varphi\Delta_2\varphi\rangle + \langle| \nabla \times \boldsymbol{k}\psi |\rangle^2 \tag{88}$$

and the property $w = -\Delta_2\varphi \equiv -(\partial^2/\partial x^2 + \partial^2/\partial y^2)\varphi$ it is evident that the minimum of the functional (80) is obtained for $\nabla \times \boldsymbol{k}\psi = 0$. The variational problem thus depends only on the scalar variables φ, θ. Using w and $-\Delta_2\varphi$ exchangeably the Euler-Lagrange equations for a stationary value of (80) can be written in the form

$$\langle| \nabla\theta |^2\rangle\nabla^2 w - \{P\langle w\theta\rangle + \mu(\langle w\theta\rangle - \overline{w\theta})\}\Delta_2\theta = 0 \tag{89}$$

$$\langle\nabla^2\varphi\Delta_2\varphi\rangle\nabla^2\theta + \{P\langle w\theta\rangle + \mu(\langle w\theta\rangle - \overline{w\theta})\}w = 0 \tag{90}$$

The Euler-Lagrange equations are still nonlinear partial differential equations, but the nonlinear terms depend only on the z-coordinate, while the equations are linear with respect to the x, y-dependence. This property suggests solutions in the form of superpositions of waves as discussed below. Because of the homogeneity of the equations (89) and (90) in w and in θ we can impose the normalization conditions

$$\langle|\nabla\theta |^2\rangle = \langle\nabla^2\varphi\Delta_2\varphi\rangle = 1. \tag{91}$$

We then introduce the following general ansatz for the solutions of equations (89) and (90),

$$w = w^{(N)} \equiv \sum_{n=1}^{N} \alpha_n^{\frac{1}{2}} w_n(z)\phi_n(x, y) \tag{92}$$

$$\theta = \theta^{(N)} \equiv \sum_{n=1}^{N} \alpha_n^{-\frac{1}{2}} \theta_n(z)\phi_n(x, y) \tag{93}$$

where the functions $\phi_n(x, y)$ satisfy the equation

$$\Delta_2\phi_n = -\alpha_n^2\phi_n \tag{94}$$

Expressions (92) and (93) can thus be regarded as Fourier series expansions with respect to the x, y-dependence. Since only the overall wavenumber α_n enters the analysis, there is no need to distinguish between wave vectors in different directions in the Fourier series (92) and (93). The parameter N may tend to infinity. Without losing generality $\overline{\phi_n\phi_m} = \delta_{nm}$ can be assumed. After introducing the general ansatz (92) and (93) into the Euler-Lagrange equations (89) and (90) we obtain the system of equations

$$\left(\frac{\partial^2}{\partial z^2} - \alpha_n^2\right) w_n + \alpha_n\Psi\theta_n = 0 \tag{95}$$

$$\left(\frac{\partial^2}{\partial z^2} - \alpha_n^2\right) \theta_n + \alpha_n\Psi w_n = 0 \tag{96}$$

where the definition

$$\Psi \equiv P\sum_{m=1}^{N}\langle w_m\theta_m\rangle + \mu\sum_{m=1}^{N}(\langle w_m\theta_m\rangle - \overline{w_m\theta_m}) \tag{97}$$

has been used. The following properties can be proven (Busse and Joseph, 1972):

1. Extremalizing solutions of the system (95) and (96) exhibit the property

$$w_n = \theta_n \tag{98}$$

2. $\theta_n(z)$ is either symmetric or antisymmetric in z.

3. Since $\theta_n \equiv \theta_m$ follows from $\alpha_n = \alpha_m$, it can be assumed that all α_n are different.

4. The relationships

$$\langle \theta'_m \theta'_n \rangle = \alpha_m \alpha_n \langle \theta_m \theta_n \rangle \tag{99}$$

hold for all solutions for $m \neq n$. For the extremalizing solution this relationship also holds for $m = n$. Here θ'_m indicates the z-derivative of θ_m.

For a multi-α-solution with N non-vanishing functions $\theta_n(z)$ the functional (80) now assumes the form

$$\mathcal{P}^{(N)}(\theta_1, \ldots, \theta_N; \alpha_1, \ldots, \alpha_N; \mu) = \frac{I^2 + \mu\langle \left(\sum_{\nu=1}^{N} \theta_\nu^2 - \sum_{\nu=1}^{N} \langle \theta_\nu^2 \rangle \right)^2 \rangle}{\left(\sum_{\nu=1}^{N} \langle \theta_\nu^2 \rangle \right)^2} \tag{100}$$

where the definition

$$I \equiv \sum_{\nu=1}^{N} \{ \alpha_\nu^{-1} \langle \theta_\nu'^2 \rangle + \alpha_\nu \langle \theta_\nu^2 \rangle \} \tag{101}$$

has been introduced. For the discussion of the extremalizing solution of the variational problem (100) it is convenient to change the normalization condition and to use

$$\sum_{\nu=1}^{N} \langle \theta_\nu^2 \rangle = 1 \tag{102}$$

instead of (91). We also assume that the wave numbers α_n are ordered, $\alpha_N > \alpha_{N-1} > \ldots \alpha_1 > 0$. In the asymptotic case of large μ it is obvious that for the minimizing solution $\sum_\nu \theta_\nu^2$ must approach unity as closely as possible throughout the interval $-\frac{1}{2} < z < \frac{1}{2}$. Only near the boundaries $z = \pm\frac{1}{2}$ the boundary condition (81) prevents a close approach. Since derivatives of $\theta_n(z)$ will tend to increase near $z = \pm\frac{1}{2}$, the growth of the term I^2 will prevent a close approach to zero of the second term in the numerator of expression (100). Thus a balance between the two terms must be expected for the minimizing solution. The growth of the term I^2 can be moderated by assigning the sharpest increase at the boundaries to $\theta_N(z)$, since $\langle \theta_N^2(z) \rangle$ is divided by the largest wavenumber α_N. Of course the region over which $\theta_N^2(z)$ is finite must not become too large because of the contribution from the second term in the sum (101). It thus appears that the minimizing solution exhibits a boundary layer structure as shown in Figure 13 where the functions $\theta_n(z)$ take turns in making the main contribution to $\sum_\nu \theta_\nu^2$ as a function of the distance from the boundary $z = -\frac{1}{2}$. First $\theta_N(z)$ grows over a distance of the order μ^{-r_N} from zero at

$z = -\frac{1}{2}$ to one at the point $z + \frac{1}{2} = \mu^{-r_N}\hat{\zeta}_N$ where $\hat{\zeta}_N$ is of the order unity. $\theta_{N-1}(z)$ grows on the same scale of the order $\mu^{-r_{N-1}}$ over which $\theta_N(z)$ decays towards the interior such that $\theta_{N-1}^2 + \theta_N^2 = 1$ is satisfied for $\mu^{-r_N}\hat{\zeta}_N \leq z + \frac{1}{2} \leq \mu^{-r_{N-1}}\hat{\zeta}_{N-1}$ and so on. The same hierarchy of boundary layers develops at $z = \frac{1}{2}$. The functional (100) thus assumes the form

$$\mathcal{P}^{(N)}(\theta,\mu) = I^2 + 2\mu^{1-r_N}\int_0^{\hat{\zeta}_N}(1-\hat{\theta}_N^2)^2 d\zeta_N \tag{103}$$

with

$$I = \mu^{\frac{1}{2}r_1}b_1 + 2\sum_{n=1}^{N}\mu^{\frac{1}{2}(r_n - r_{n-1})}\left(\frac{1}{b_n}\int_0^{\hat{\zeta}_n}\hat{\theta}_n'^2 d\zeta_n + b_n\int_0^{\hat{\zeta}_n}(1-\hat{\theta}_{n-1}^2)d\zeta_{n-1}\right) \tag{104}$$

where the definitions

$$\theta_n(z) = \hat{\theta}_n(\zeta_n) \text{ for } \mid z \pm \frac{1}{2} \mid = \mu^{-r_n}\zeta_n \tag{105}$$

$$\alpha_n^2 = \mu^{r_n + r_{n-1}} b_n^2 \tag{106}$$

have been used. The scaling (106) is a consequence of relationship (99) for $m = n$. The second term in the sum of (104) must be replaced by zero for $n = 1$. Since the μ-dependence is now separated from the remaining dependencies of the functional (100), the exponents r_n can be determined by minimization of the functional as a function of μ. The result

$$r_n = n(N+1)^{-1} \text{ for } n = 1,\ldots,N$$

leads to the reduced form of the functional (103),

$$\mathcal{P}^{(N)}(\theta,\mu) = \mu^{1/(N+1)}\left(\hat{I}^2 + 2\int_0^{\hat{\zeta}_N}(1-\hat{\theta}_N^2)^2 d\zeta_n\right) \tag{107}$$

with

$$\hat{I} = b_1 + 2\sum_{n=1}^{N}\left(\frac{1}{b_n}\int_0^{\hat{\zeta}_n}\hat{\theta}_n'^2 d\zeta_n + b_n\int_0^{\hat{\zeta}_n}(1-\hat{\theta}_{n-1}^2)d\zeta_{n-1}\right). \tag{108}$$

The functions $\hat{\theta}_n(\zeta_n)$ can be determined as solutions of the boundary layer version of the Euler-Lagrange equations

$$\hat{\theta}_n'' + b_n b_{n+1}\hat{\theta}_n = 0 \text{ for } n = 1,\ldots,N-1 \tag{109}$$

$$\hat{I}\hat{\theta}_N'' + b_N(1-\hat{\theta}_N^2)\hat{\theta}_N = 0 \tag{110}$$

which can be easily solved,

$$\hat{\theta}_n = \pm\sin\sqrt{b_n b_{n+1}}\zeta_n \text{ for } 0 \leq \zeta_n \leq \hat{\zeta}_n \equiv \pi/2\sqrt{b_n b_{n+1}} \tag{111}$$

$$\hat{\theta}_N = \pm \tanh\left(\frac{b_N}{2\hat{I}}\right)^{\frac{1}{2}} \zeta_N \text{ for } 0 \le \zeta_N \le \infty. \tag{112}$$

The minimum $P^{(N)}(\mu)$ of the functional (105) can thus be evaluated,

$$P^{(N)}(\mu) = N(N+1)\pi^2 \left(\frac{64\mu}{9\pi^4 N}\right)^{1/(N+1)} \tag{113}$$

Numerical computations have shown that this asymptotic result provides a good approximation

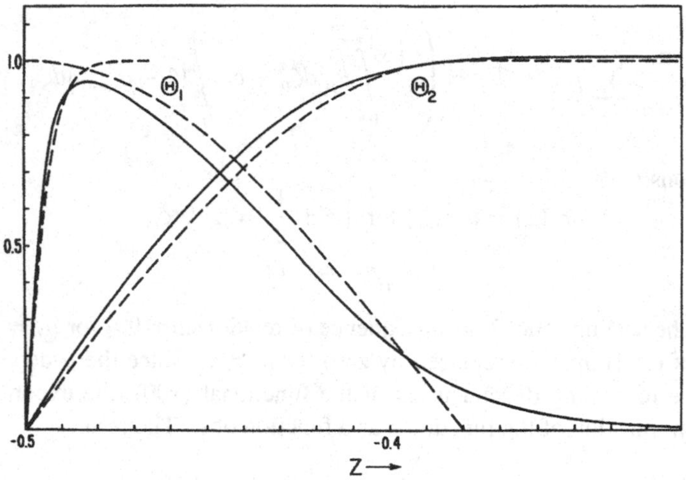

Figure 14: The two-α-solution at $R = 50\pi^2$. Numerical computations (solid lines) are compared with the results (111) and (112) for θ_1, θ_2 from the boundary layer theory. This figure differs from Figure 1 of Busse and Joseph (1972) in that an improved numerical approximation has been used.

even at finite values of μ. In particular, if the characteristic property of expression (113) is preserved so that the absolute minimum

$$P(\mu) = \min_N P^{(N)}(\mu) \tag{114}$$

corresponds to $N = 1$ for low values of μ and to increasing integers N as μ increases. The details of the transition from the N-α-solution to the $(N+1)$-α-solution can only be investigated numerically. To demonstrate the quality of boundary solutions (111) and (112) a comparison with the direct numerical integration of the Euler-Lagrange equations (95) and (96) (with $w_n = \theta_n$) is given in Figure 14. The results of Busse and Joseph (1972) indicate that the transition occurs in the form of a bifurcation in which the $(N+1)^{\text{th}}$ component of the solution first appears as a small perturbation in the Nth boundary layer of the N-α-solution. As a result the bound $P(\mu)$ appears to be a smooth curve without the kinks which would be suggested by a straightforward application of the asymptotic expressions (113).

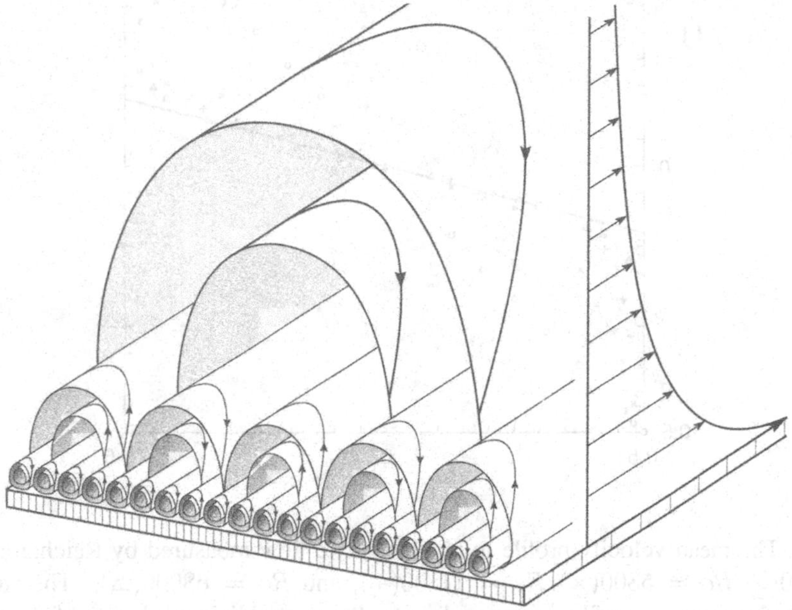

Figure 15: Qualitative sketch of the nested boundary layers which characterize the vector field of maximum transport. The profile of the mean shear is shown on the right side.

5.4 Similarities Between Extremalizing Vector Fields and Observed Turbulence

The extremalizing vector fields of the upper bound problems have in common with the observed turbulence that the wavenumber spectrum broadens as the N-α-solution is replaced by the $(N+1)$-α-solution with increasing control parameter. But the wavenumber spectrum of the extremalizing vector fields always remains discrete in contrast to the continuous spectra of most cases of physically realized turbulence. Before entering a more detailed discussion of similarities between observed turbulence and corresponding extremalizing vector fields, we must comment on the use of the asymptotic expressions provided by the boundary layer theory. For the cases of momentum transport in a plane Couette layer as outlined in section 5.1 or of the heat transport in an ordinary convection layer expressions similar to those derived in the preceding section 5.2 have been obtained (Busse, 1969, 1970). Since the N-α-solution for the extremalizing vector field provides the upper bound only in a finite interval of the control parameter, the assumption of an infinite ratio between the scales of subsequent boundary layers is not well satisfied. It turns out that this ratio assumes the value e^2 in the case of solutions (111), (112) and (113) for large n, N and the value 4 in all other cases of upper bound problems that have been studied. The

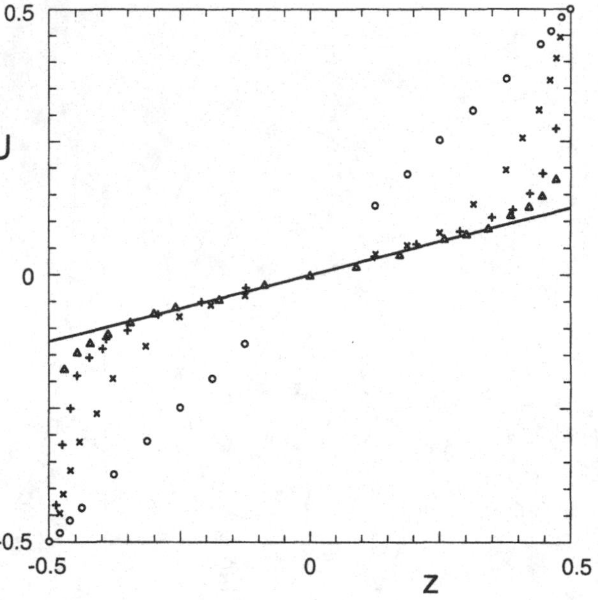

Figure 16: The mean velocity profile in plane Couette flow measured by Reichardt (1959) at $Re = 2400(\circ), Re = 5800(\times), Re = 11800(+)$, and $Re = 68000(\Delta)$. The straight line describes the asymptotic profile corresponding to the extremalizing solution of the variational problem (79) [from Busse, 1970].

extremalizing vector fields in the case of turbulent shear flows thus assume the form sketched in Figure 15. In contrast to the case of convection in either the porous medium layer or in the ordinary fluid layer where no horizontal orientation can be specified, the extremalizing vector fields in the case of shear flows must not depend on the coordinate in the direction of the shear flow. They share this property, of course, with the extremalizing solutions of the energy functional (13) as we have discussed in section 2.2. We now discuss the correspondence between extremalizing solutions and observations in more detail.

i) The bifurcation structure of bounds on transports

The bifurcation structure of the extremalizing vector fields is a consequence of the property that eddies with an increasing number of length scales are needed to accomplish an optimal transport as the control parameter increases. The same property can be observed in the transition from convection rolls to bimodal convection in a high Prandtl number fluid layer heated from below. Indeed, the Rayleigh numbers of the order of $2 \cdot 10^4$ at which the transitions to bimodal convection (Busse, 1967b; Krishnamurti, 1970; Busse and Whitehead, 1971) and from the 1-α-solution to the 2-α-solution (Straus, 1976) occur are about the same. A similar close correspondence between the experimentally observed transition (Combarnous and Le Fur, 1969) and the transition from the 1- to the 2-α-solution can be found in the problem of convection in a porous layer. At the higher transitions this correspondence is no longer as close although numerous higher transitions have been observed

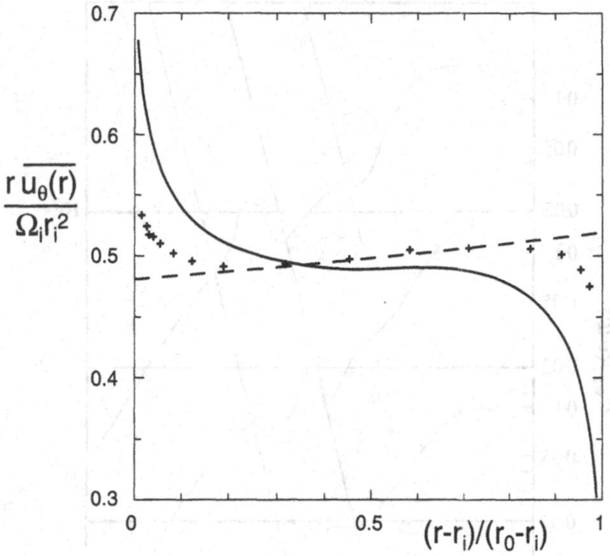

Figure 17: Measurements (+) by Smith and Townsend (1982) of the angular momentum density, $rU(r)$, normalized by the angular momentum of the inner cylinder, $\Omega_i r_i^2$, in comparison with the profile of a logarithmic layer model (solid line) (Lathrop et al., 1992) and with the profile of the extremalizing vector field in the limit of high Reynolds numbers (dashed line). A stationary outer cylinder with a radius ratio $\eta = 0.667$ has been used (After Busse, 1996).

in the dependence of the heat transport by turbulent convection on the Rayleigh number Ra. The results from different experiments (Malkus, 1954b; Willis and Deardorff, 1967; Chu and Goldstein, 1973) do not agree and appear to depend on the Prandtl number Pr. This is not surprising since most of the higher transitions involve time dependent processes which are not represented in the theory of upper bounds so far.

ii) Profiles of mean velocities and mean temperatures

Among the properties of turbulent flows that can easily be measured, the profiles of the time averaged velocity and temperature fields have received special attention. It is thus of interest to compare the measured profiles with the profiles corresponding to the extremalizing vector fields. In first approximation one may assume that any mean shear or mean temperature gradient will be wiped out by turbulent mixing except near the boundaries where steep gradients are generated. Indeed, in experiments on turbulent convection in fluid layers heated from below as well as in the case of the vector field extremalizing the heat transport an isothermal interior is found when the averages over planes $z =$ const. are taken. Surprisingly this property does not hold in the case of a shear layer as the comparison between measurements and prediction of the upper bound theory given in Figure 16 demonstrates. The extremalizing vector field does not need the drop of half of the velocity difference between the plates across the boundary layers in order to accomplish a maximum transport of momentum. Only $3/8$ are required. Although the experimentally

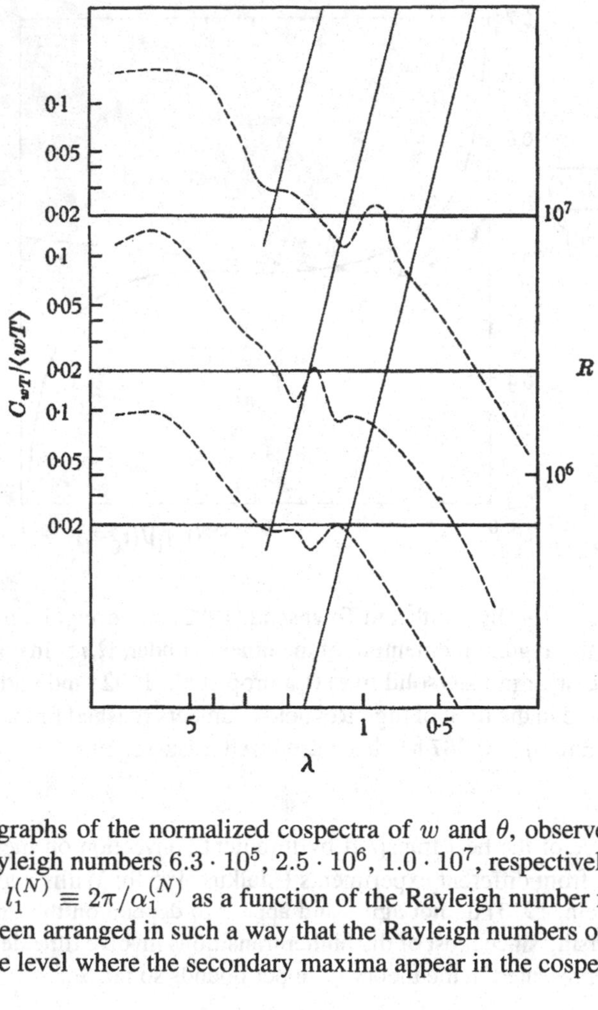

Figure 18: Three graphs of the normalized cospectra of w and θ, observed by Deardorff and Willis (1967) at Rayleigh numbers $6.3 \cdot 10^5$, $2.5 \cdot 10^6$, $1.0 \cdot 10^7$, respectively, are plotted on top of a figure showing $l_1^{(N)} \equiv 2\pi/\alpha_1^{(N)}$ as a function of the Rayleigh number for $N = 2, 3, 4$. The three graphs have been arranged in such a way that the Rayleigh numbers of both plots coincide approximately at the level where the secondary maxima appear in the cospectra.

observed shear in the middle of the Couette layer is somewhat less than $1/4$ of the applied shear for high values of Re, it does not seem to tend to zero for $Re \to \infty$ as might be expected on the basis of mixing length arguments.

In the case of the angular momentum transport by turbulent flow between differentially rotating coaxial cylinders the radius ratio enters as an additional parameter and more detailed comparisons between observed mean zonal flows and the prediction of the upper bound theory become possible. Unfortunately only a few measurements are available. One of them is displayed in Figure 17, where again the approach towards the "$1/4$-law" of the upper bound theory appears to be realized. The general expression for the mean zonal flow as a function of the dimensionless radius is given by (Busse, 1972)

$$U(r) = \frac{\Omega_i - \Omega_o}{4r}\frac{\eta^2}{1-\eta^2} + \frac{\Omega_i\eta^2(1-2\eta^2) + \Omega_o(2-\eta^2)}{2(1-\eta^4)} r \text{ for } \eta < r < 1 \qquad (115)$$

where $\Omega_i(\Omega_o)$ is the dimensionless angular velocity of the inner (outer) cylinder and η is the radius ratio. It will be of interest to see, whether an as good agreement without an adjustable parameter as shown in Figure 17 can be obtained in other cases of turbulent circular Couette flow.

iii) Discrete scales

The set of discrete wavenumbers characterizing the extremalizing vector field appears to be their most artificial feature when compared with the broad continuous wavenumber spectrum observed in turbulent flows. It must be remembered, however, that patterns of coherent structures in fully developed turbulence are notoriously difficult to measure. The fluctuations measured at a single point as a function of time which are interpreted as fluctuations in space via the Taylor hypothesis will usually yield a continuous spectrum even if, for instance, a perfect pattern of hexagons is advected by a mean flow. Few experiments can provide an instantaneous view of the two-dimensional structure of turbulence. Through the shadowgraph visualization technique (see, for instance, Busse and Whitehead, 1971, 1974) such a view can be obtained in the case of turbulent convection in a fluid layer heated from below. It is thus not surprising that a nearly stationary network of convection cells can be discerned in turbulent convection at Rayleigh numbers of several 10^6 (Busse, 1994). Regular cloud patterns as manifestations of convection in the atmosphere and the hierarchy of convection cells (granules, supergranules, giant cells) observed on the sun remind us that discrete scales may be the norm rather than the exception in highly turbulent, statistically stationary systems.

Measurements of spectral peaks that can be compared with the discrete scales of the extremalizing vector fields have been obtained by Deardorff and Willis (1967) as shown in Figure 18. While the correspondence between spatial peaks and extremalizing scales is tentative at best, more detailed comparisons appear to be possible when numerical simulations of convection with sufficiently large horizontal periodicity intervals are carried out. In the case of turbulent shear layers the longitudinal vortices in the laminar sublayer appear to be characterized by a distinct wavenumber.

(iv) Universal properties

The extremalizing vector fields of the variational problems exhibit universal properties, i.e. the solutions are similar for quite different situations of turbulent fluid flow. For example, the structures of shear flow boundary layers and of thermal boundary layers in a convection layer are identical for the extremalizing solutions when scaled appropriately. This similarity property can also be found in actual measurement as shown in Figure 19.

(v) Improved bounds

Finally, additional constraints will restrict the manifold of admissible vector fields in the variational problems and will lead to improved bounds. Vitanov and Busse (2001) have shown that a lower upper bound for the heat transport can be obtained by considering separate energy balances for the poloidal and toroidal parts of the velocity field in the case of convection in a layer heated from below and rotating about a vertical axis. The Euler-Lagrange equations for the extremalizing vector fields can be solved only numerically in this case, however.

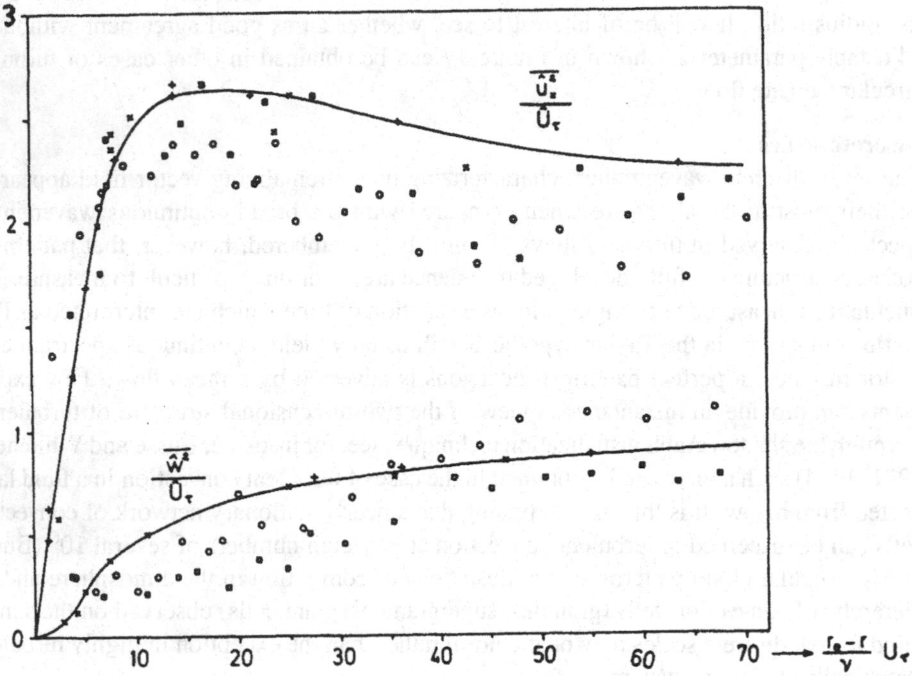

Figure 19: Root mean square (r.m.s.) values of the fluctuating components of the velocities in the streamwise direction, $\hat{u}_x/U_\tau$, and normal to the wall, $\hat{w}/U_\tau$, measured by Laufer (1954) at $Re = 2.5 \cdot 10^4 (+)$ and $Re = 2.5 \cdot 10^5 (\times x)$ are compared with the r.m.s. values of the temperature fluctuations $\hat{\theta}$ and of the vertical velocity $\hat{w}$ measured in turbulent thermal convection by Deardorff and Willis (1967). The latter values have been obtained for $Ra = 2.5 \cdot 10^6 (\circ)$ and $Ra = 2.5 \cdot 10^7 (\square)$ are plotted in units resulting from the correspondence of the variational problems (after Busse, 1970).

6 Concluding Remarks

The three approaches towards understanding turbulent fluid flow that have been outlined in sections 3, 4 and 5 of this review share the property that they do not involve a statistical analysis. Instead of statistical properties the approaches have focussed on mechanisms operating in turbulent systems. The approaches are limited in that they capture only parts of the properties of realized turbulence, but each approach yields unique insights which may not be accessible through other methods. The potentials of the approaches are far from being fully utilized and it is hoped that the results reported in this review will stimulate further studies.

References

Busse, F.H. (1967a). The stability of finite amplitude cellular convection and its relation to an extremum principle. *J. Fluid Mech.* 30:625–649.

Busse, F.H. (1967b). On the stability of two-dimensional convection in a layer heated from below. *J. Math. Phys.* 46:140–150.

Busse, F.H. (1969a). Bounds on the transport of mass and momentum by turbulent flow between parallel plates. *J. Applied Math. Phys. (ZAMP)* 20:1–14.

Busse, F.H. (1969b). On Howard's upper bound for heat transport by turbulent convection. *J. Fluid Mech.* 37:457–477.

Busse, F.H. (1970). Bounds for turbulent shear flow. *J. Fluid Mech.* 41:219–240.

Busse, F.H. (1972). A property of the energy stability limit for plane parallel shear flow. *Arch. Rat. Mech. Anals.* 47:28–35.

Busse, F.H. (1978). The optimum theory of turbulence. *Advances in Appl. Mech.* 18:77-121.

Busse, F.H. (1986). Phase-turbulence in convection near threshold. *Contemporary Mathematics* 56:1–8.

Busse, F.H. (1994). Spoke Pattern Convection. *ACTA MECHANICA (Suppl.)* 4:11–17.

Busse, F.H. (1996). Bounds for Properties of Complex Systems, pp. 1-9 in "Nonlinear Physics of Complex Systems", J. Parisi, S.C. Müller, W. Zimmermann (eds.), *Lecture Notes in Physics* 476

Busse, F.H., and Bolton, E.W. (1984). Instabilities of convection rolls with stress-free boundaries near threshold. *J. Fluid Mech.* 146:115–125.

Busse, F.H., and Heikes, K.E. (1980). Convection in a rotating layer: A simple case of turbulence. *SCIENCE* 208:173–175.

Busse, F.H., and Joseph, D.D. (1972). Bounds for heat transport in a porous layer. *J. Fluid Mech.* 54:521–543.

Busse, F.H., Kropp, M., and Zaks. M. (1992). Spatio-temporal structures in phase-turbulent convection. *Physica D* 61:94–105.

Busse, F.H., and Whitehead, J.A. (1971). Instabilities of convection rolls in a high Prandtl number fluid. *J. Fluid Mech.* 47:305–320.

Busse, F.H., and Whitehead, J.A. (1974). Oscillatory and collective instabilities in large Prandtl number convection. *J. Fluid Mech.* 66:67–79.

Chu, T.Y., and Goldstein, R.J. (1973). Turbulent convection in a horizontal layer of water. *J. Fluid Mech.* 60:141–159.

Clever, R.M., and Busse, F.H. (1979). Nonlinear properties of convection rolls in a horizontal layer rotating about a vertical axis. *J. Fluid Mech.* 94:609–627.

Clever, R.M., and Busse, F.H. (1989). Three-dimensional knot convection in a layer heated from below. *J. Fluid Mech.* 198:345–363.

Clever, R.M., and Busse, F.H. (1992). Three-dimensional convection in a horizontal fluid layer subjected to a constant shear. *J. Fluid Mech.* 234:511–527.

Clever, R.M., and Busse, F.H. (1994). Steady and Oscillatory Bimodal Convection. *J. Fluid Mech.* 271:103–118.

Combarnous, M., and Le Fur, B. (1969). Transfert de chaleur par convection naturelle dans une couche poreuse horizontale. *C.R. Acad. Sc. Paris* 269:1005–1012.

Deardorff, J.W., and Willis, G.E. (1967). Investigation of turbulent thermal convection between horizontal plates. *J. Fluid Mech.* 28:675–704.

Doering, C.R., and Constantin, P. (1994). Variational bounds on energy dissipation in incompressible flows: shear flow. *Phys. Rev. E* 49:4087–4099.

Goldstein, R.J., and Graham, D.J. (1969). Stability of a horizontal fluid layer with zero shear boundaries. *Phys. Fluids* 12:1133–1137.

Golubitsky, M., Swift, J.W., and Knobloch, E. (1984). Symmetries and Pattern Selection in Rayleigh-Bénard Convection. *Physica D* 100:249–276.

Grossmann, S. (2000). The onset of shear flow turbulence. *Rev. Mod. Phys.* 72:603–618.

Heikes, K.E., and Busse, F.H. (1980). Weakly nonlinear turbulence in a rotating convection layer. *Annals N.Y. Academy of Sciences* 357:28–36.

Howard, L.N. (1963). Heat transport by turbulent convection. *J. Fluid Mech.* 17:405–432.

Joseph, D.D. (1976). *Stability of fluid motions, vol. 1.* Springer, Berlin Heidelberg, New York.

Kerswell, R.R. (1998). Unification of variational principles for turbulent shear flows: The background method of Doering-Constantin and Howard-Busse's mean-fluctuation formulation. *Physica D* 121:175–192.

Kramer, L., Schober, H.R., and Zimmermann, W. (1988). Pattern competition and the decay of unstable patterns in quasi-one-dimensional systems. *Physica D* 31:212–226.

Krishnamurti, R. (1970). On the transition to turbulent convection. Part 1. The transition from two to three-dimensional flow. *J. Fluid Mech.* 42:295–307.

Küppers, G., and Lortz, D. (1969). Transition from laminar convection to thermal turbulence in a rotating fluid layer. *J. Fluid Mech.* 35:609–620.

Lathrop, D.P., Fineberg, J., and Swinney, H.L. (1992). Transition to shear-driven turbulence in Couette-Taylor flow. *Phys. Rev. A* 46:6390–6405.

Laufer, J. (1954). The Structure of Turbulence in Fully Developed Pipe Flow. *NACA Rep.* 1174.

Malkus, W.V.R. (1954a). The heat transport and spectrum of thermal turbulence. *Proc. Roy. Soc. London* A225:196–212.

Malkus, W.V.R. (1954b). Discrete transitions in turbulent convection. *Proc. Roy. Soc. London* A225:185–195.

Nagata, M. (1990). Three-dimensional finite-amplitude solutions in plane Couette flow: bifurcation from infinity. *J. Fluid Mech.* 217:519–527.

Nagata, M., and Busse, F.H. (1983). Three-dimensional tertiary motions in a plane shear layer. *J. Fluid Mech.* 135:1–26.

Nicodemus, R., Grossmann, S., and Holthaus, M. (1997). Variational bound on energy dissipation in plane Couette flow. *Phys. Rev. E* 56:6774–6786.

Reichardt, H. (1959). Gesetzmäßigkeiten der geradlinigen turbulenten Couetteströmung. Mitt. Max-Planck-Institut für Strömungsforschung, Göttingen, Nr. 22

Schmiegel, A., and Eckhardt, B. (1997). Fractal Stability Border in Plane Couette Flow. *Phys. Rev. Lett.* 79:5250–5253.

Schmiegel, A., and Eckhardt, B. (2000). Persistent turbulence in annealed plane Couette flow. *Europhysics. Letts.* 51:395–400.

Schmitt, B.J., and Wahl, W. von (1992). Decomposition of Solenoidal Fields into Poloidal Fields, Toroidal Fields and the Mean Flow. Applications to the Boussinesq-Equations. in "The Navier-Stokes Equations II - Theory and Numerical Methods". J.G. Heywood, K. Masuda, R. Rautmann, S.A. Solonnikov, eds. Springer Lecture Notes in Mathematics 1530:291–305.

Serrin, J. (1959). On the stability of viscous fluid motions. *Arch. Rat. Mech. Anal.* 3:1–13.

Smith, G.P., and Townsend, A.A. (1982). Turbulent Couette flow between concentric cylinders at large Taylor numbers. *J. Fluid Mech.* 123:187–217.

Straus, J.M. (1976). On the Upper Bounding Approach to Thermal Convection at Moderate Rayleigh Numbers. II. Rigid Boundaries. *Dyn. Atmos. Oceans* 1:77–90.

de la Torre-Juarez, M., and Busse, F.H. (1995). Stability of two-dimensional convection in a porous medium. *J. Fluid Mech.* 292:305–323.

Vitanov, N.K., and Busse, F.H. (2001). Bounds on the convective heat transport in a rotating layer. *Phys. Rev. E* 63:16303–16310.

Walden, R.W., and Ahlers, G. (1981). Non-Boussinesq and penetrative convection in a cylindrical cell. *J. Fluid Mech.* 109:89–114.

Willis, G.E., and Deardorff, J.W. (1967). Confirmation and renumbering of the discrete heat flux transitions of Malkus. *Phys. Fluids* 10:1861–1866.

Xi, H.-W., Li, X.-J., and Gunton, J.D. (1997). Direct Transition to Spatio-temporal Chaos in Low Prandtl Number Fluids. *Phys. Rev. Lett.* 78:1046–1049.

Zippelius, A., and Siggia, E.D. (1982). Disappearance of stable convection between free-slip boundaries. *Phys. Rev. A.* 26:1788–1790.

Renormalization Methods Applied to Turbulence Theory

David McComb

Department of Physics and Astronomy, The University of Edinburgh, Mayfield Road, Edinburgh EH9 3JZ, Scotland

Abstract

In these notes the turbulence problem is interpreted as one in many-body physics. We consider the many-body problem, along with the concept of renormalization, and show how the methods of renormalized perturbation theory have been applied to turbulence. Then a review and assessment of the two-point, two-time closures which arise from renormalized perturbation theory is given. After that, renormalization group is introduced as a more limited but potentially more rigorous way of applying perturbation methods to turbulence with particular relevance to the sub-grid modelling problem. We conclude with a brief discussion of the possible application of these methods to more realistic problems in shear flows.

1 Turbulence as a Problem in Many-Body Physics

In its Fourier wavenumber space representation turbulence is a problem of strong inter-mode coupling. The 'strength' of the nonlinear term is proportional to the Reynolds number and can be varied in the laboratory between zero and (in effect) infinity, while the structure of the Navier-Stokes equation (NSE) in k-space resembles that of the Schroedinger equation in quantum theory.

The problem of strong coupling is characteristic of 'many-body' physics. In a statistical theory it leads to the moment closure problem. In a numerical simulation, it makes it difficult to reduce the number of degrees of freedom to be explicitly simulated.

1.1 Renormalization

In order to explain what is meant by renormalization we temporarily leave fluid turbulence and consider the well known statistical problem of 'N particles in a box'. First we consider the easy problem of the perfect gas - where the particles only interact briefly through hard-sphere collisions. Then we consider the true many-body problem, where the interaction between particles has to be taken into account.

Renormalization is a term from quantum field theory ($\sim$ 1938). Nowadays it is widely used in statistical and many-body physics. Originally it was a method of removing divergences in quantum field theory. In many-body physics it is a method for handling *interactions*. Typical examples of interactions in physics include: electrostatic force between electrons, intermolecular forces in liquids and gases and spin-spin coupling in a magnetic lattice.

1.1.1 Non-interacting large-N systems

Start with an easy example: kinetic theory of a perfect gas. (This is NOT a many-body problem!). Consider N particles, each of mass m and velocity u in a box. This system has energy eigenstates E_i where i is an integer. At equilibrium, the probability of the system being in a particular eigenstate $|i\rangle$ is

$$P(E_i) \sim e^{-E_i/kT}, \tag{1}$$

where k is the Boltzmann constant and T is the absolute temperature. We can write this as

$$P(E_i) = \frac{e^{\,-E_i/kT}}{Z}, \tag{2}$$

where $Z \equiv$ partition function $\equiv$ sum over states, or

$$Z = \sum_i e^{-E_i/kT}. \tag{3}$$

As the individual particles are non-interacting, we can write $Z = (Z_1)^N$, where Z_1 is the partition function for a single-particle. The 'bridge equation' for free energy is:

$$F = -kT \ln Z = -NkT \ln Z_1, \tag{4}$$

that is, Z factorizes.

From the free energy, we can get many physical quantities using standard thermodynamic relations. The essential feature of the non-interacting (perfect) gas is that one can do the sum over states. We can factorize Z in terms of single-particle partition functions.

It is usual to work with the Hamiltonian formulation. We introduce the Hamiltonian $H \equiv H(p,q)$ where p is momentum and q is generalised (canonical) position. For a single particle

$$H = \frac{p^2}{2m} + V(q) \equiv \frac{mu^2}{2} + V(q), \tag{5}$$

where $V(q)$ is an external potential, and u is the speed of a particle. The Hamiltonian for an ideal system is

$$H = \sum_{i=1}^{N} H_i = \sum_{i=1}^{N} p_i^2/2m. \tag{6}$$

That is, there is no potential in the perfect gas case.

1.1.2 Interacting large-N systems

Now, switch on a potential between *pairs* of particles, thus:

$$H_i = \frac{p_i^2}{2m} + \sum_j V_{ij}(q); \qquad j \neq i. \tag{7}$$

This could be an electrostatic potential, if the particles are charged, or perhaps a molecular potential. Note $j \neq i$ as a particle cannot interact with itself. Due to the coupling term, the partition sum no longer factorizes, therefore it is difficult to calculate the partition function Z. This can be put in general form

$$H = \sum_i H_i + \sum_{i,j} H_{ij}. \tag{8}$$

Hence

$$Z = \sum_{states} e^{-H/kT} = \sum_{states} e^{-(\sum_i H_i + \sum_{i,j} H_{ij})/kT}. \tag{9}$$

1.1.3 Quasi-particles

We now introduce the concept of the quasi-particle. Replace the interaction term in the Hamiltonian by the *average* effect of all the other particles on one particle. That is, we approximate the Hamiltonian by

$$H \simeq \sum_i H_i^{'}, \tag{10}$$

where $H_i^{'}$ is the *effective* Hamiltonian of the ith quasi-particle. Each of these has a portion of the interaction energy added to their single-particle form. We can treat the interacting system as a perfect gas of *quasi-particles*. Thus we can use elementary statistical mechanics to calculate Z.

1.1.4 Example of a quasi-particle: an electron in a plasma or electrolyte

Referring to Fig. 1, the potential at a distance r from one electron in isolation is

$$V(r) = e/r, \tag{11}$$

where e is the electron charge. In a plasma, a cloud of charge around any one electron *screens* the Coulomb potential. According to the self-consistent theory of Debye and Hückel (McComb, 1990), the actual potential goes as:

$$V(r) = \frac{e \exp\{-r/\lambda\}}{r}, \tag{12}$$

where λ is the Debye-Hückel length and depends on the number density of electrons and the absolute temperature. Note that the Debye-Hückel potential is just the Coulomb potential with the replacement

$$e \to e \times \exp\{-r/\lambda\}, \tag{13}$$

That is, 'bare charge' → 'renormalized charge'.

1.1.5 Example of an interacting system: a ferromagnet

Picture a magnetic material as a simple cubic lattice with a tiny magnet at each node. At high temperatures, each little magnet points in a different direction but they can be lined up by an external field. As the temperature is reduced, spins can line up permanently (even in zero external field) due to interactions (cooperative effect). The highest temperature at which ferromagnetism can exist is called the critical temperature, or T_C.

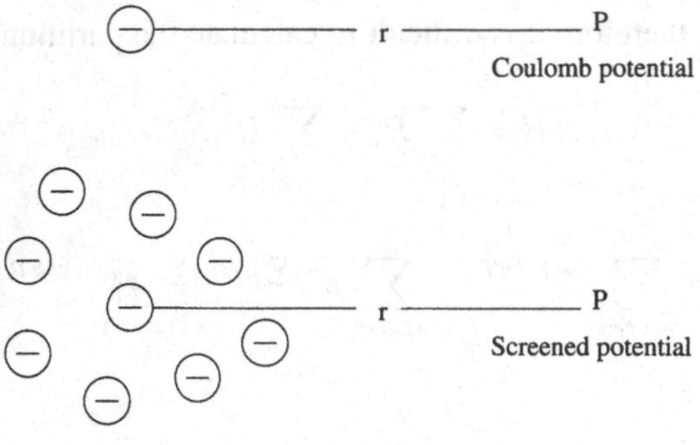

Figure 1: A cloud of electrons screens the potential due to any one electron.

1.1.6 Example of a model for an interacting system: spin-$\frac{1}{2}$ Ising model

In a general model of a magnetic lattice, the spin vector (individual magnetic moment) can point in any direction. In the Ising model, spin can only point up or down. If σ_i is the spin at lattice site i, then we have either $\sigma_i = \frac{1}{2}$ (spin up) or $\sigma_i = -\frac{1}{2}$ (spin down). The total system Hamiltonian is

$$H = -J \sum_{\langle i,j \rangle} \sigma_i \sigma_j - B \sum_i \sigma_i, \tag{14}$$

where B is an external magnetic field. Here J is the coupling constant and $\langle i, j \rangle$ restricts the sum to nearest neighbours on the lattice.

A permanent magnet can exist when $B = 0$ because there is an internal field due to the 'coupling term'. In this case, the probability distribution is

$$P \sim e^{-H/kT} \sim e^{\frac{-J}{kT} \sum_{\langle i,j \rangle} \sigma_i \sigma_j}. \tag{15}$$

Note the technical problem: the coupling term does NOT allow us to factorize Z into (Z_1^N) where Z_1 is a single-site partition function. Note also that if we increase the temperature T, then $(-J/kT)$ becomes small $\Longrightarrow$ reduced *effective* coupling $\Longrightarrow$ demagnetisation.

1.2 Renormalized Perturbation Theory

There is essentially only one general theoretical method of tackling problems involving coupling or nonlinearity and that is perturbation theory. Assume that interactions between particles will lead to perturbations of the 'perfect gas' solution. Begin by considering the problem from the macroscopic point of view. The perfect gas law is consistent with the neglect of interactions between the molecules. It also fails to allow for the fraction of the available volume which the molecules occupy. We expect the perfect gas law to be a good approximation for a gas which is not too dense and not too cold. For a system of N molecules, occupying a fixed volume V, the perfect gas law, usually written as

$$PV = NkT, \tag{16}$$

tells us the pressure P. However, if we rewrite this in terms of the number density $n = N/V$, then we can assume that this must be the limiting form (at low densities) of some more complicated law thus,

$$P = nkT + O(n^2). \tag{17}$$

Formally, it is usual to anticipate that the exact form of the law may be written as the expansion

$$PV = NkT[B_1(T) + B_2(T)n + B_3(T)n^2 + \dots]. \tag{18}$$

This is known as the virial expansion and the coefficients are referred to as:

$B_1(T)$: The first virial coefficient;

$B_2(T)$: The second virial coefficient;

$B_3(T)$: The third virial coefficient;

and so on, to any order. The coefficients depend on temperature because, for a given density, the effective strength of the particle interactions will depend on the temperature. It should also be noted that the status of equation (3), is that of a plausible guess. Next, we begin the process of seeing to what extent such a guess is supported by microscopic considerations.

1.2.1 The configurational partition function

Consider N interacting particles in phase space. The partition function can be written as:

$$\mathcal{Z} = \frac{1}{N!h^{3N}} \int\int\int\int e^{-E(\mathbf{q},\mathbf{p})/kT} \times d\mathbf{p}_1 \dots d\mathbf{p}_N d\mathbf{q}_1 \dots d\mathbf{q}_N. \tag{19}$$

We can factor out the integration with respect to $\mathbf{p}$, by writing the exponential as

$$e^{-E(\mathbf{p},\mathbf{q})/kT} = e^{-\sum_{i=1}^{N} p_i^2/2mkT} \times e^{-\Phi(\mathbf{q})/kT}, \tag{20}$$

and so write the total partition function for the system as

$$\mathcal{Z} = \mathcal{Z}_0 Q, \tag{21}$$

where $\mathcal{Z}_0$ is the perfect gas partition function. The **configurational integral** Q is given by

$$Q = \frac{1}{V^N} \int\int e^{-\Phi(\mathbf{q})/kT} d\mathbf{q}_1 \dots d\mathbf{q}_N. \tag{22}$$

We shall restrict our attention to two-body potentials where

$$\Phi(\mathbf{q}) = \sum_{i<j=1}^{N} \phi(|\mathbf{q}_i - \mathbf{q}_j|) \equiv \sum_{i<j=1}^{N} \phi_{ij}, \tag{23}$$

and hence the function $\Phi(\mathbf{q})$ will be written as the double sum over ϕ_{ij} from now on.

1.2.2 Perturbation expansion of the configuration integral

Now we consider the use of the perturbation expansion in terms of a book-keeping parameter. This is introduced as an arbitrary factor, just as if it were the usual 'small quantity' in perturbation theory. It is used to keep track of the various orders of terms during an iterative calculation. Unlike the conventional perturbation parameter, it is not small and in fact is put equal to unity at the end of the calculation.

If the potential is, in some sense, weak we can expand out the exponential as a power series and truncate the resulting expansion at low order. In general, for any exponential we have

$$e^x = 1 + x + \frac{x^2}{2!} + \frac{x^3}{3!} \dots \frac{x^s}{s!} + \dots = \sum_{s=0}^{\infty} \frac{x^s}{s!}. \tag{24}$$

Expanding the exponential in (6) in this way gives us

$$Q = V^{-N} \int\int \sum_{s=o}^{\infty} \left(\frac{-\lambda}{kT}\right)^s \frac{1}{s!} \left(\sum_{i<j=1}^{N} \phi_{ij}\right)^s \mathbf{dq}_1 \dots \mathbf{dq}_N. \tag{25}$$

Here we have introduced the factor λ as a 'book-keeping' parameter ($\lambda = 1$). Any possibility of low-order truncation depends on integrals being well-behaved and this in turn depends very much on the nature of ϕ. Also, combinatorial effects increase with order λ^s, as follows:
$\mathbf{s = 0}$:

$$Q_0 = V^{-N} \int\int \mathbf{dq}_1 \dots \mathbf{dq}_N = 1, \tag{26}$$

where, of course,

$$\int \mathbf{dq}_1 = V, \dots, \int \mathbf{dq}_N = V. \tag{27}$$

$\mathbf{s = 1}$:

$$\begin{aligned} (-kT)Q_1 &= V^{-N} \int\int \left(\sum_{i<j=1}^{N} \phi_{ij}\right) \mathbf{dq}_1 \dots \mathbf{dq}_N \\ &= V^{-N} \int\int (\phi_{12} + \phi_{13} + \phi_{23} + \phi_{14} + \dots) \mathbf{dq}_1 \dots \mathbf{dq}_N \\ &= V^{-2} \left[\int\int \phi_{12} \mathbf{dq}_1 \mathbf{dq}_2 + \int\int \phi_{13} \mathbf{dq}_1 \mathbf{dq}_3 + \dots \right]; \end{aligned} \tag{28}$$

and so on.

Noting that Q_1 is made up of many identical integrals, it follows that we need evaluate only one of these integrals, and may then multiply the result by the number of pairs which can be chosen from N particles. Hence

$$Q_1 = -\frac{1}{2} N(N-1) V^{-1} \int \frac{\phi_{12}}{kT} \mathbf{dr}_{12}. \tag{29}$$

We have made the change of variables $\mathbf{r}_{12} = |\mathbf{q}_1 - \mathbf{q}_2|$, and the integration with respect to the centroid coordinate $\mathbf{R} = (\mathbf{q}_1 + \mathbf{q}_2)/2$ cancels one of the factors $1/V$. Higher orders get more complicated and, as we shall see, diagram methods can be helpful. But the real problem is unsatisfactory behaviour when we attempt to take the thermodynamic limit:

$$Lt\ N/V \to n, \text{ as } N, V \to \infty.$$

The expansion fails this test as, at any order s, there are various dependencies on n, so that it does not take the expected form as in equation (18). (In mathematical terms, the expansion is inhomogeneous.)

This problem is not as serious as it might first appear. Our trick is to work with $\ln Q$ rather than Q (in other words, with the free energy due to interactions) to get a new series. In practice this amounts to a rearrangement of the perturbation expansion such that one finds an infinite series of terms associated with n, n^2, n^3, and so on. Each of these infinite series must be summed to give a coefficient in our new expansion in powers of n. We shall discuss some aspects of doing this in the next section, along with the helpful introduction of Mayer functions.

However we close here by reconsidering what we mean by saying that the potential is 'weak'. Intuitively, we can see that if the density is low, on average the particles will be far apart and hence the contribution of the interaction potential to the overall potential energy will be small. Also, we note that the potential energy (just like the kinetic energy) always appears divided by the factor kT, and so for large temperatures the arguments of the exponentials will be small. Thus, for either low densities or high temperatures the exponentials can be expanded out and truncated at low order.

1.2.3 The density expansion and the virial coefficients

We shall use statistical mechanics to explore the general method of calculating the virial coefficients. From equation (25), we may write the configurational integral as

$$Q = \frac{1}{V^N} \int\int e^{-\sum_{i<j} \phi_{ij}/kT} \mathrm{d}\mathbf{q}_1 \ldots \mathrm{d}\mathbf{q}_N = \frac{1}{V^N} \int\int \prod_{i<j} e^{-\phi_{ij}/kT} \mathrm{d}\mathbf{q}_1 \ldots \mathrm{d}\mathbf{q}_N. \tag{30}$$

Now we introduce the Mayer functions f_{ij}, which are defined such that

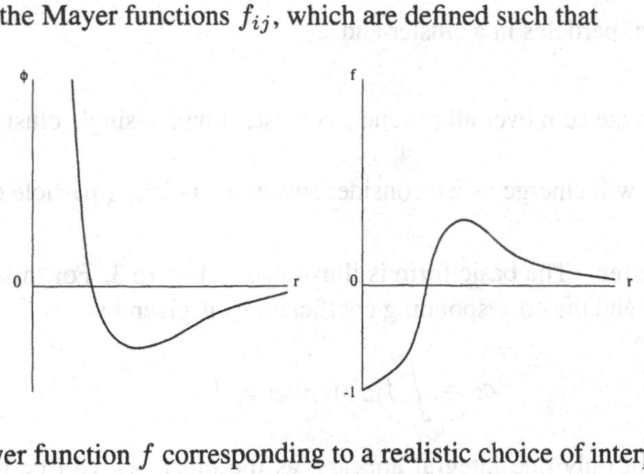

Figure 2: The Mayer function f corresponding to a realistic choice of interparticle potential ϕ.

$$f_{ij} = e^{-\phi_{ij}/kT} - 1. \tag{31}$$

These possess the useful property (see Figure 2) that as $r \to 0$, $f_{ij} \to -1$ for $\phi_{ij} \to \infty$, and they change the product of (30) into a sum.

Upon substitution of (31), equation (30) for the configurational integral becomes:

$$Q = \frac{1}{V^N}\int\int\prod_{i<j}(1+f_{ij})\mathbf{dq}_1\ldots\mathbf{dq}_N$$

$$= \frac{1}{V^N}\int\int\left[1+\sum_{i<j}f_{ij}+\sum_{i<j}\sum_{k<l}f_{ij}f_{kl}+\ldots\right]\mathbf{dq}_1\ldots\int\mathbf{dq}_N. \tag{32}$$

We should note three points about this:

1. f_{ij} is negligibly small in value unless the molecules making up the pair labelled by i and j are close together.
2. The terms in equation (32) involve molecular clusters. For this reason the multiple integrals in (32) are known as cluster integrals.
3. The expansion given in equation (32) is known as the virial cluster expansion.

The graphical representation of the types of integral which occur in the expansion in (32) is shown in Figure 3 for two-particle and three-particle clusters. Each circle corresponds to a particle, along with the corresponding integration over its position coordinate. The line joining any pair of circles labelled by i and j stands for the interaction f_{ij}. We shall find it helpful to characterise the terms of (32) by means of coefficients c_l, such that

$$c_l = \int\int\sum^{*}\left[\prod_{i<j}f_{ij}\right]\mathbf{dq}_1\ldots\mathbf{dq}_{l-1}. \tag{33}$$

Here l is the number of particles in a cluster and

$$\sum^{*} \equiv \text{the sum over all products consistent with a single cluster.} \tag{34}$$

The significance of c_l will emerge as we consider successively larger particle clusters.

The two-particle cluster The basic form is illustrated in Figure 3. For this case, the index in equation (15) is $l = 2$ and the corresponding coefficient c_2 is given by

$$c_2 = \int f_{12}(\mathbf{q}_1)\mathbf{dq}_1 \equiv I_2. \tag{35}$$

It should be noted that only one integral appears, as the other one can be done directly by a change of variables to cancel a factor $1/V$. The overall contribution to the configuration integral is easily worked out in this case, as we just multiply by the number of pairs of particles, which is $N(n-1)/2$. This is what one does in order to calculate the second virial coefficient.

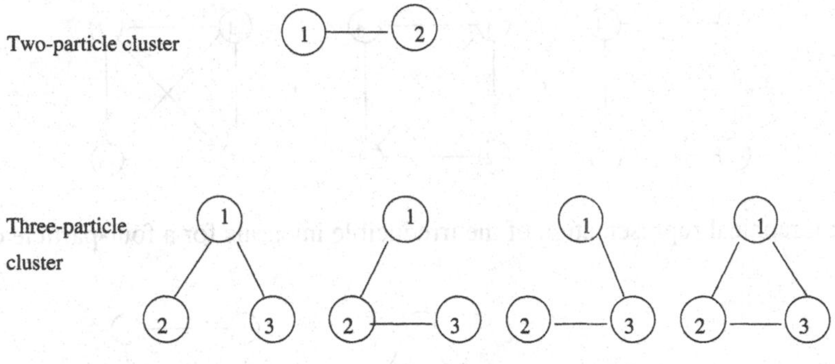

Figure 3: Graphical representation of the integrals for two-particle and three-particle clusters.

The three-particle cluster This is the case $l = 3$ and as we may see from Figure 3, there are four possible arrangements. The first three give the same contribution, so we shall just consider the first of these as typical and multiply our answer by a factor three. Considering the first of these terms, on the extreme left hand side of the figure, we may integrate over either $\mathbf{q}_1$ or $\mathbf{q}_2$ and obtain I_2, just as in the two-particle case. Integrating again over the volume of one of the remaining two particles also gives I_2 and hence the contribution of this type of arrangement is I_2^2, and with three such arrangements in all, we have

$$c_3 = 3I_2^2 + \int\int f_{12}f_{23}f_{13}d\mathbf{q}_1 d\mathbf{q}_2, \tag{36}$$

where the last term corresponds to the closed triangle in Figure 3. We note that the linkages between the interactions are such that it is not possible to factor this into simpler forms and hence we give it its own name and write it as follows:

$$I_3 = \int\int f_{12}f_{23}f_{13}d\mathbf{q}_1 d\mathbf{q}_2, \tag{37}$$

Thus

$$c_3 = 3I_2^2 + I_3. \tag{38}$$

We refer to I_3 and I_2 as irreducible integrals and in general we may expect that the numerical value of any c_l will be determined by sums and products of such integrals.

The four-particle cluster We shall not provide a full analysis of the four-particle case, but will merely concentrate on the irreducible integrals. In this instance, we have $l = 4$ and the three possibilities for irreducible integrals at this order are shown in Figure 4. The corresponding mathematical forms may be written down as follows:

$$I_{40} = \int\int\int f_{12}f_{23}f_{34}f_{14}d\mathbf{q}_1 d\mathbf{q}_2 d\mathbf{q}_3; \tag{39}$$

$$I_{41} = \int\int\int f_{12}f_{23}f_{34}f_{14}f_{13}d\mathbf{q}_1 d\mathbf{q}_2 d\mathbf{q}_3; \tag{40}$$

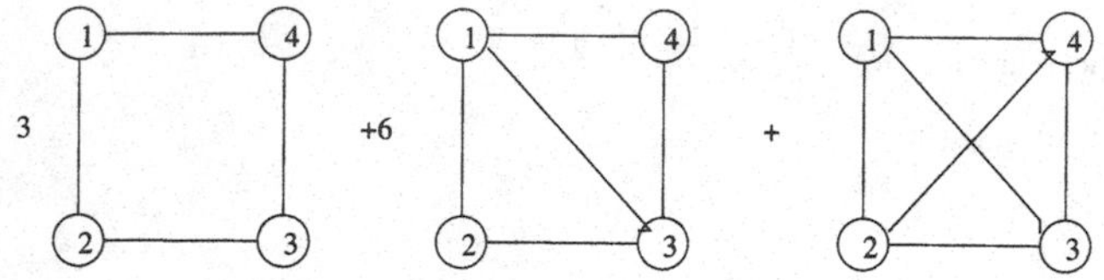

Figure 4: Graphical representation of the irreducible integrals for a four-particle cluster.

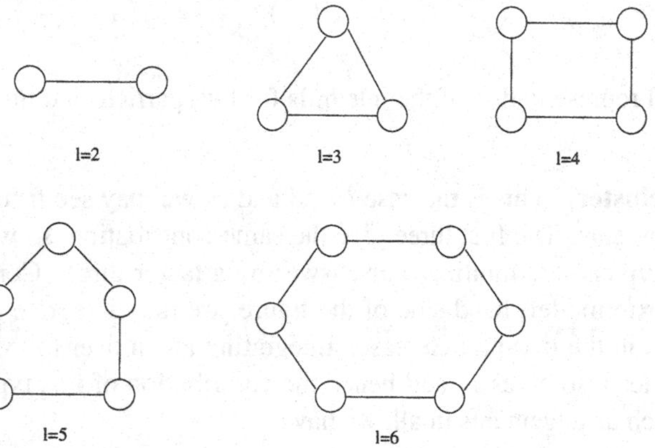

Figure 5: Irreducible cluster diagrams contributing to Debye-Hückel theory up to sixth order.

$$I_{42} = \int\int\int f_{12} f_{23} f_{34} f_{14} f_{13} f_{24} \mathbf{dq}_1 \mathbf{dq}_2 \mathbf{dq}_3. \tag{41}$$

It can be shown by repeating the analysis of the preceding sections that

$$c_4 = 16 I_2^2 + 12 I_2 I_3 + I_4, \tag{42}$$

where

$$I_4 = 3 I_{40} + 6 I_{41} + I_{42}. \tag{43}$$

The full analysis goes on to obtain the general term and hence the general form of the configuration integral to any order in particle clusters. The procedure is both complicated and highly technical and ultimately depends on the model taken for the interaction potential. Lastly, in Figure 5, we illustrate the perturbation theory of an electrolyte, where a sum over irreducible diagrams leads to the Debye-Hückel potential.

1.3 Review of Perturbation Theory

The use of series methods to solve differential equations should be familiar to all. Perturbation methods have a successful history in fluid mechanics, especially when there is a parameter (such as the Mach number) which can be taken to be small.

In many-body problems, the perturbation series is unlikely to be convergent even if there is a small parameter. Expansion in the 'coupling constant' nearly always runs into technical problems. Often these can be solved by a second expansion in (for instance) the temperature, the particle density or even the dimensionality. Such techniques are usually referred to as regularization. It may be possible to sum the series exactly or, more likely, to sum a specific class of terms. This is known as partial summation. All these techniques lead to renormalization, usually with the introduction of quasi-particles.

Lastly, a good 'rule of thumb' for perturbation theory is that the 'closer' the soluble zero-order model is to the exact problem the better. Further reading on the subject of perturbation theory in many-body physics will be found in the book by Reichl (1980) and in fluid dynamics in the book by van Dyke (1975).

2 Perturbation Methods in Turbulence Theory

The background to this section can be found in the book by McComb (1990) or the review article (McComb, 1995).

2.1 Turbulence: What is the Problem?

To begin, let us express it symbolically:

$$NSE: \qquad L_0 u = M uu,$$

where M stands for derivatives and the effects of the pressure and NSE is the Navier Stokes equation. Average term by term to get an evolution equation for $\bar{u} = \langle u \rangle$:

$$L_0 \langle u \rangle = M \langle uu \rangle. \tag{44}$$

We only know the relation between moments if we know the distribution $P[u]$. To obtain an equation for $\langle uu \rangle$ multiply NSE through by u and $\langle \cdot \rangle$:

$$L_0 \langle uu \rangle = M \langle uuu \rangle, \tag{45}$$

which introduces a new unknown $\langle uuu \rangle$. Form an equation for this new unknown: multiply through NSE by uu and then $\langle \cdot \rangle$:

$$L_0 \langle uuu \rangle = M \langle uuuu \rangle, \tag{46}$$

and so on. To solve (1), we need the solution of (2); to solve (2), we need the solution of (3); to solve (3), we need solution of (4); and so on ... *ad infinitum.*

2.2 The Idealized Turbulence Problem and the Basic Equations

In fluid dynamics, researchers are interested in a wide range of turbulent flows which are almost invariably inhomogeneous and anisotropic. The theoretical physicist likes to study situations which are the exact opposite! Fortunately the concept of isotropic turbulence in a box - once regarded with great suspicion - is now very much more acceptable due to its implementation in the direct numerical simulation of turbulence. In this section, we explain how the problem is formulated theoretically and give the basic equations in both real (**x**) and wavenumber (**k**) space.

2.2.1 Turbulence: the actual problem

'Real' turbulence occurs in pipes, jets, wakes, boundary-layers and other more complicated configurations. It is driven by a forcing which is deterministic and often constant in space and time (e.g. pressure gradient, mean rate of shear). The resulting turbulence is due to *instability*. The chaotic phenomenon which we observe is accompanied by energy and momentum flows in both **x**-space and **k**-space.

2.2.2 Turbulence in a box: the idealized problem

To make the turbulence problem as much as possible like other problems in physics, we consider an idealized 'turbulence in a box'. The situation is artificial and we have to apply a random forcing to the NSE in order to achieve stationarity. If we apply the forcing only to low-**k** modes (large eddies) we can hope that behaviour at high-**k** (small eddies) is like that in turbulent shear flows (i.e universal). The forcing we choose is multivariate normal (or Gaussian) and we have to specify its strength.

We consider an infinitely repeating box of incompressible turbulence of side L with periodic boundary conditions. Our turbulence is isotropic turbulence (statistically invariant under rotation and reflection of axes) homogeneous turbulence (statistically invariant under translation of axes) and stationary turbulence (statistically invariant over time). To ensure stationarity, the turbulence is stirred at low wavenumbers and this energy is transferred through the modes, to be dissipated at a rate ε.

2.2.3 Turbulence in a box: formulation of the problem

The statistical problem is formulated in terms of the Eulerian velocity field $u_\alpha(\mathbf{x}, t)$. Most of the time, we work in Fourier, or *wavenumber* space. Given a velocity field in real space, $\mathbf{u}(\mathbf{x}, t)$, we define the velocity field in Fourier space $\mathbf{u}(\mathbf{k}, t)$ by

$$\mathbf{u}(\mathbf{x}, t) = \sum_{\mathbf{k}} \mathbf{u}(\mathbf{k}, t) e^{i\mathbf{k}\cdot\mathbf{x}}. \tag{47}$$

The wavevectors, **k**, are defined by

$$k_\alpha = \frac{2\pi}{L} n_\alpha, \quad n_\alpha = 0, \pm 1, \pm 2, \ldots, \pm\infty. \tag{48}$$

The Fourier coefficients are given by

$$u_\alpha(\mathbf{k}, t) = \int x \, u_\alpha(\mathbf{x}, t) e^{-i\mathbf{k}\cdot\mathbf{x}} dx^3. \tag{49}$$

Often it is convenient to take the limit $L \to \infty$ and use the Fourier transform

$$u_\alpha(\mathbf{x}, t) = \left(\frac{1}{2\pi}\right)^3 \int k u_\alpha(\mathbf{k}, t) e^{i\mathbf{k}\cdot\mathbf{x}} dk^3. \tag{50}$$

The ensemble average is given by:

$$Q(k) = \langle |u^2(k, t)| \rangle = \lim_{N\to\infty} \sum_{n=1}^{N} \{u^2(k, t)\}_n. \tag{51}$$

2.2.4 The pair-correlation tensor

By definition the general correlation tensor is:

$$Q_{\alpha\beta}(\mathbf{x},\mathbf{x}';t,t') = \langle u_\alpha(\mathbf{x},t)u_\beta(\mathbf{x}',t')\rangle . \tag{52}$$

Homogeneous turbulence:

$$\begin{aligned} Q_{\alpha\beta}(\mathbf{x},\mathbf{x}';t,t') &= Q_{\alpha\beta}(\mathbf{x}-\mathbf{x}';t,t') \\ &= Q_{\alpha\beta}(\mathbf{r};t,t'). \end{aligned} \tag{53}$$

Taking the Fourier transform:

$$\langle u_\alpha(\mathbf{k},t)u_\beta(\mathbf{k}',t')\rangle = \delta(\mathbf{k}+\mathbf{k}')Q_{\alpha\beta}(\mathbf{k};t,t'). \tag{54}$$

Stationary turbulence:

$$\begin{aligned} Q_{\alpha\beta}(\mathbf{k};t,t') &= Q_{\alpha\beta}(\mathbf{k};t-t') \\ &= Q_{\alpha\beta}(\mathbf{k};\tau). \end{aligned} \tag{55}$$

Isotropic turbulence:

$$\begin{aligned} Q_{\alpha\beta}(\mathbf{k};t,t') &= \delta_{\alpha\beta}Q(\mathbf{k};t,t') - \frac{k_\alpha k_\beta}{k^2}Q(\mathbf{k};t,t') \\ &= D_{\alpha\beta}(\mathbf{k})Q(\mathbf{k};t,t'). \end{aligned} \tag{56}$$

Projection operator:

$$D_{\alpha\beta}(\mathbf{k}) = \delta_{\alpha\beta} - \frac{k_\alpha k_\beta}{k^2}. \tag{57}$$

Note that this form satisfies the continuity condition in both tensor indices.

2.2.5 Navier-Stokes equation (NSE) in x-space

Conservation of momentum takes the form:

$$\frac{\partial u_\alpha}{\partial t} + u_\beta\frac{\partial u_\alpha}{\partial x_\beta} = -\frac{1}{\rho}\frac{\partial p}{\partial x_\alpha} + \nu\nabla^2 u_\alpha, \tag{58}$$

where

$$\nabla^2 = \partial^2/\partial x_\beta \partial x_\beta. \tag{59}$$

Continuity equation (Conservation of Mass) gives:

$$\frac{\partial u_\beta(\mathbf{x},t)}{\partial x_\beta} = 0. \tag{60}$$

2.2.6 NSE in k-space

Conservation of momentum:

$$\left(\frac{\partial}{\partial t} + \nu k^2\right) u_\alpha(\mathbf{k},t) = -ik_\alpha p(\mathbf{k},t)$$
$$-ik_\beta \int u_\alpha(\mathbf{k}-\mathbf{j},t) u_\beta(\mathbf{j},t) \mathrm{d}^3 j. \tag{61}$$

Continuity equation:

$$k_\alpha u_\alpha(\mathbf{k},t) = 0. \tag{62}$$

2.2.7 Solenoidal NSE in k-space

The Fourier transformed Navier-Stokes equation reads

$$\frac{\partial}{\partial t} u_\alpha(\mathbf{k},t) + \int ik_\beta u_\beta(\mathbf{j},t) u_\alpha(\mathbf{k}-\mathbf{j},t) \mathrm{d}^3 j =$$
$$-ik_\alpha p(\mathbf{k},t) - \nu k^2 u_\alpha(\mathbf{k},t). \tag{63}$$

By using continuity one can get an expression for $p(\mathbf{k},t)$:

$$p(\mathbf{k},t) = -\frac{1}{k^2} \int k_\beta k_\gamma u_\beta(\mathbf{j},t) u_\gamma(\mathbf{k}-\mathbf{j},t) \mathrm{d}^3 j. \tag{64}$$

Thus leading to the solenoidal NSE:

$$\left(\frac{\partial}{\partial t} + \nu k^2\right) u_\alpha(\mathbf{k},t) =$$
$$M_{\alpha\beta\gamma}(\mathbf{k}) \int u_\beta(\mathbf{j},t) u_\gamma(\mathbf{k}-\mathbf{j},t) \mathrm{d}^3 j, \tag{65}$$

where

$$M_{\alpha\beta\gamma}(\mathbf{k}) = (2i)^{-1} \left[k_\beta D_{\alpha\gamma}(\mathbf{k}) + k_\gamma D_{\alpha\beta}(\mathbf{k})\right]. \tag{66}$$

2.2.8 The Navier-Stokes equation with stirring forces

We add an arbitrary stirring force to the RHS:

$$\left(\frac{\partial}{\partial t} + \nu_0 k^2\right) u_\alpha(\mathbf{k},t)$$
$$= f_\alpha(\mathbf{k},t)$$
$$+ M_{\alpha\beta\gamma}(\mathbf{k}) \int u_\beta(\mathbf{j},t) u_\gamma(\mathbf{k}-\mathbf{j},t) \mathrm{d}^3 j. \tag{67}$$

The stirring force f has multivariate normal distribution, with autocorrelation chosen to be of the form:

$$\langle f_\alpha(\mathbf{k},t) f_\beta(\mathbf{k}',t') \rangle =$$
$$2W(k)(2\pi)^d D_{\alpha\beta}(\mathbf{k}) \delta(\mathbf{k}+\mathbf{k}') \delta(t+t'). \tag{68}$$

$W(k)$is a measure of the rate at which the stirring force does work on the fluid. The rate at which the force does work on the fluid must equal the dissipation rate:

$$\int W(k)dk = \varepsilon. \tag{69}$$

2.3 The Physics of Turbulent Energy Transport and Dissipation

In order to have a physical basis for renormalization in turbulence, we need to consider the physics of the energy cascade. In this section, we introduce the energy spectrum, the dissipation rate (along with the Kolmogorov dissipation wavenumber) and a very important quantity known as the transport power. This latter quantity is a measure of the rate at which energy is transferred by nonlinear (or inertial effects) through wavenumber space.

2.3.1 Energy balance for fluid motion

We can obtain a general result for *all* fluid motion where the fluid is acted upon by a force $\mathbf{f}(\mathbf{x}, t)$. Consider a fluid of density ρ and kinematic viscosity ν occupying a volume V. Total energy of the fluid motion is given by:

$$E_T = \frac{1}{2}\sum_{\alpha}\int_V \rho U_\alpha^2 dv. \tag{70}$$

We can obtain an equation for E_T directly from the NSE. This is usually done in x-space. The balance equation is found to take the form:

$$\frac{dE_T}{dt} = \int_u \rho U_\alpha f_\alpha dV - \int_V \rho\varepsilon dV, \tag{71}$$

where ε is the energy dissipation per unit mass of fluid per unit time. A steady state exists when $u_\alpha f_\alpha = \varepsilon$. i.e. rate at which work is done by the force equals the dissipation rate. When the external forces are zero, the kinetic energy of the fluid flow dies away at a rate given by ε. The non-linear and pressure terms do no net work on the system.

2.3.2 The energy spectrum

For the single-time case, $t = t'$, equation (13) reduces to:

$$Q_{\alpha\beta}(\mathbf{k}; t, t') = D_{\alpha\beta}(\mathbf{k})Q(k; t - t') = D_{\alpha\beta}(\mathbf{k})Q(k), \tag{72}$$

where

$$Q(k) \equiv Q(k; 0), \tag{73}$$

and is called the *spectral density*. For total turbulent energy (per unit mass of fluid), we set $\alpha = \beta$ and sum over $\alpha = 1, 2$ and 3. The energy spectrum $E(k, t)$ is related to the spectral density by

$$E(k, t) = 4\pi k^2 Q(k, t) \equiv 4\pi k^2 Q(k, t, t). \tag{74}$$

It is also related to the total turbulent energy E per unit mass of fluid by:

$$E = \int_0^\infty E(k)dk. \tag{75}$$

In order to ensure stationarity, energy is added to the system at low wavenumbers by a source term $W(k)$ which satisfies

$$\int_0^\kappa \mathrm{d}kW(k) = \varepsilon, \tag{76}$$

for some $\kappa \ll k_d$, where $k_d = (\varepsilon/\nu_0^3)^{1/4}$ is the Kolmogorov dissipation wavenumber, and ε is the rate of energy input. Note that this is more specific than in equation (71).

We form the energy-balance equation from the NSE, thus:

$$\left(\frac{\partial}{\partial t} + 2\nu_0 k^2\right) E(k,t) = W(k) + T(k,t). \tag{77}$$

The non-linear term is defined in terms of the third-order moment $Q_{\beta\gamma\alpha}$, thus:

$$Q_{\beta\gamma\alpha}(\mathbf{j}, \mathbf{k}-\mathbf{j}, -\mathbf{k}; t) = \langle u_\beta(\mathbf{j},t) u_\gamma(\mathbf{k}-\mathbf{j},t) u_\alpha(-\mathbf{k},t)\rangle. \tag{78}$$

2.3.3 The energy transfer spectrum

We choose $\mathbf{k}$ as the polar axis and introduce

$$\mu \equiv \cos\theta_{kj} \tag{79}$$

where θ_{kj} is the angle between the vectors $\mathbf{k}$ and $\mathbf{j}$. The non-linear term takes the form

$$T(k,t) = \int_0^\infty \int_{-1}^1 T(k,j,\mu)\mathrm{d}j\mathrm{d}\mu \tag{80}$$

with

$$\begin{aligned} T(k,j,\mu) &= -8i\pi^2 k^2 j^2 \int \{j_\gamma Q_{\beta\gamma\beta}(\mathbf{j}, \mathbf{k}-\mathbf{j}, -\mathbf{k}) \\ &- k_\gamma Q_{\beta\gamma\beta}(-\mathbf{j}, \mathbf{j}-\mathbf{k}, \mathbf{k})\}\, \mathrm{d}^3 j. \end{aligned} \tag{81}$$

For simplicity

$$T(k,t) = \int_0^\infty T(k,j)\,\mathrm{d}j, \tag{82}$$

with

$$T(k,j) = \int_{-1}^1 T(k,j,\mu)\,\mathrm{d}\mu. \tag{83}$$

We have the antisymmetry

$$T(k,j,\mu) = -T(j,k,\mu). \tag{84}$$

Or, alternatively,

$$T(k,j) = -T(j,k) \tag{85}$$

In this formulation conservation of energy follows as:

$$\int_0^\infty \int_0^\infty T(k,j)\,\mathrm{d}k\mathrm{d}j = 0, \tag{86}$$

Or, when extended to numerical simulation,

$$\int_{k_{min}}^{k_{max}} \int_{k_{min}}^{k_{max}} T(k,j)\,\mathrm{d}k\mathrm{d}j = 0. \tag{87}$$

2.3.4 Conservation of energy

A criterion for the existence of an inertial range can be obtained by integrating both sides of equation (77) with respect to k. We find that for any K_I in the inertial range, we have the condition

$$\int_{K_I}^\infty T(k)\,\mathrm{d}k = -\int_0^{K_I} T(k)\,\mathrm{d}k = \varepsilon. \tag{88}$$

The inertial-range spectrum takes the form:

$$E(k) = \alpha\varepsilon^{2/3}k^{-5/3}. \tag{89}$$

The prefactor α is generally known as the Kolmogorov constant. This result for the energy spectrum has received ample experimental confirmation and the recent survey by Sreenivasan (1995) would seem to place the matter beyond reasonable doubt, with the value of the prefactor given by $\alpha = 1.620 \pm 0.168$. Recently, Yeung and Zhou (1997) made a critical examination of the way in which the data resulting from numerical simulations is interpreted and came to the conclusion that the value of the Kolmogorov prefactor is $\alpha = 1.62$, although no error bounds were given. This is supported by the numerical simulation of Young who adopted the same criteria and found $\alpha = 1.624 \pm 0.122$.

2.3.5 Dissipation rate in wavenumber space

By definition $\varepsilon = -dE/dt$ for freely decaying turbulence. The energy balance can be rewritten as:

$$\frac{dE}{dt} = -\varepsilon = -\int_0^\infty 2\nu k^2 E(k,t)dk. \tag{90}$$

Hence the dissipation rate is also given by:

$$\varepsilon = \int_0^\infty 2\nu k^2 E(k,t)dk. \tag{91}$$

The factor of k^2 ensures that dissipation is a high$-k$ (or small eddy) effect. The region in $k-$space where dissipation occurs is characterized by the Kolmogorov dissipation wavenumber: $k_d = (\varepsilon/\nu^3)^{\frac{1}{4}}$.

2.3.6 Transport power $\Pi(\kappa, t)$

For some given wavenumber $k = \kappa$, the transport power is the rate at which energy is transferred from modes with $k \leq \kappa$ to modes with $k \geq \kappa$. It is given by:

$$\Pi(\kappa, t) = \int_{\kappa}^{\infty} T(k, t) dk. \tag{92}$$

By the antisymmetry of $T(k, j)$:

$$\Pi(\kappa, t) = \int_{\kappa}^{\infty} \int_{0}^{\kappa} T(k, j) \mathrm{d}k \mathrm{d}j. \tag{93}$$

Also by antisymmetry of $T(k, j)$:

$$\Pi(\kappa, t) = -\int_{0}^{\kappa} T(k, t) dk. \tag{94}$$

2.3.7 Detailed energy balance in wavenumber

For true stationarity of isotropic turbulence we must add an input spectrum $W(k)$, which satisfies (76). Then $dE(k, t)/dt = 0$ and the energy balance becomes:

$$T(k) + W(k) - 2\nu k^2 E(k) = 0; \tag{95}$$

or,

$$\int_{0}^{\infty} T(k, j) dj = W(k) - 2\nu k^2 E(k) = 0. \tag{96}$$

At sufficiently high Reynolds numbers, we assume there is a wavenumber κ such that the input effects are below it and dissipation effects above it. That is, for a well-posed problem:

$$\int_{0}^{\kappa} W(k) dk \simeq \varepsilon \simeq -\int_{\kappa}^{\infty} 2\nu k^2 E(k) dk. \tag{97}$$

We can obtain two detailed energy balance equations by first integrating each term with respect to k from zero up to κ and then from infinity down to κ.

First,

$$\int_{0}^{\kappa} \int_{\kappa}^{\infty} T(k, j) \mathrm{d}k \mathrm{d}j + \int_{0}^{\kappa} W(k) dk = 0, \tag{98}$$

i.e. energy supplied directly by input term to modes with $k \leq \kappa$ is transferred by the nonlinearity to modes with $j \geq \kappa$. Thus $T(k)$ behaves like a dissipation and absorbs energy.

Second,

$$\int_{\kappa}^{\infty} \int_{0}^{\kappa} T(k, j) \mathrm{d}k \mathrm{d}j - \int_{\kappa}^{\infty} 2\nu k^2 E(k) dk = 0, \tag{99}$$

i.e. nonlinearity transfers energy from modes with $j \leq \kappa$ to modes with $k \geq \kappa$, where it is dissipated into heat. Thus in this range of wavenumbers $T(k)$ behaves like a source and emits energy which is then dissipated by viscosity.

2.4 Renormalized Perturbation Theory and the Turbulence Closure Problem

This part deals with those fundamental turbulent closures - usually for two-point, two-time correlations of the fluctuating velocity - which have been obtained by some general operations which do not involve specific assumptions about the nature of turbulent flows, nor do they have disposable constants which can be fitted by comparison with experiment. They should be distinguished in this respect from turbulence models, and in the context of physics they are properly described as theories. In all cases, they are obtained by some method of summing terms in perturbation series to all orders and they differ one from another by the way in which they do this.

2.4.1 Perturbation treatment of the Navier-Stokes equation

We begin by discussing the perturbation expansion of the NSE about a simple model problem which corresponds to a fluid being stirred by some external agency, without nonlinear effects being present. The perturbation expansion represents the effect of nonlinear coupling but is wildly divergent, so that we cannot truncate it at low order. We give a simplified explanation of how truncation at low order can be made if we renormalize the perturbation series.

2.4.2 Perturbation expansion of NSE

Insert a book-keeping parameter λ in front of the nonlinear term in the stirred NSE.

$$\left[\frac{\partial}{\partial t}+\nu k^2\right] u_\alpha(\mathbf{k},t) = f_\alpha(\mathbf{k},t) + \lambda M_{\alpha\beta\gamma}(\mathbf{k}) \int \mathbf{u}_\beta(\mathbf{j},\mathbf{t})\mathbf{u}_\gamma(\mathbf{k}-\mathbf{j},\mathbf{t})\mathrm{d}^3\mathbf{j}. \tag{100}$$

Here $\lambda = 0$ (linear system) or $\lambda = 1$ (nonlinear system) i.e. λ is also a control parameter. If we scaled variables in a suitable way, we could replace λ by a Reynolds number. Hence the perturbation expansion in λ is *effectively* in powers of a Reynolds number.

The zero-order 'model' system We base our perturbation approach on a model. Take $\lambda = 0$; NSE becomes:

$$\left[\frac{\partial}{\partial t}+\nu k^2\right] u_\alpha(\mathbf{k},t) = f_\alpha(\mathbf{k},t). \tag{101}$$

Choose $\mathbf{f}(\mathbf{k},t)$ to have Gaussian statistics (*note* turbulence itself is NOT Gaussian). The zero-order ($\lambda = 0$) velocity field is solution of above equation:

$$\mathbf{u}^{(0)}(\mathbf{k},t) = \int e^{-\nu k^2(t-t')}\mathbf{f}(\mathbf{k},t')dt' \equiv \int G^{(0)}(\mathbf{k};t-t')f(\mathbf{k},t')dt'. \tag{102}$$

This is *stirred* fluid motion, valid only in the limit of zero Reynolds number. Because $\mathbf{f}$ is Gaussian, so also is $\mathbf{u}^{(0)}$.

The 'actual' system Set $\lambda = 1$: the full NSE is restored. The nonlinear mixing effect of the nonlinear term is to couple together modes with different wavenumbers. The physical effect is to induce an *exact*, non-Gaussian velocity field $\mathbf{u}(\mathbf{k}, t)$. This may be written as a perturbation series in powers of λ:

$$\mathbf{u}(\mathbf{k}, t) = \mathbf{u}^{(0)}(\mathbf{k}, t) + \lambda \mathbf{u}^{(1)}(\mathbf{k}, t) + \lambda^2 \mathbf{u}^{(2)}(\mathbf{k}, t) + \ldots \tag{103}$$

The coefficients $\mathbf{u}^{(1)}, \mathbf{u}^{(2)} \ldots$ are calculated iteratively in terms of $\mathbf{u}^{(0)}$.

2.4.3 The perturbation expansion

Good news! The Gaussian model is soluble because we can factor moments of the $\mathbf{u}^{(0)}$ to all orders.

1. All odd-order moments $\langle u^{(0)}u^{(0)}u^{(0)}\rangle$ etc., are zero.
2. All even-order moments $\langle u^0 u^{(0)}u^{(0)}u^{(0)}\rangle$ etc., can be expressed as products of the second-order moments $\langle u^{(0)}u^{(0)}\rangle$.

Bad news! The resulting perturbation expansion for the exact second-order moment $\langle uu\rangle$ is wildly divergent. Nevertheless, we go ahead anyway! Introduce a simplified notation: NSE becomes

$$L_{0k}u_k = \lambda M_{kjl}u_j u_l + f_k, \tag{104}$$

where

$$L_{0k} \equiv \frac{\partial}{\partial t} + \nu k^2, \tag{105}$$

and subscripts stand for all the variables (wavevector, tensor index and time). The zero-order solution becomes

$$u_k^{(0)} = G_k^{(0)} f_k, \tag{106}$$

and the perturbation expansion becomes

$$u_k = u_k^{(0)} + \lambda u_k^{(1)} + \lambda^2 u_k^{(2)} + \ldots. \tag{107}$$

2.4.4 The iterative calculation of coefficients

For convenience, we invert the operator on the LHS of NSE to write:

$$u_k = u_k^{(0)} + \lambda G_k^{(0)} M_{kjl} u_j u_l. \tag{108}$$

Now substitute the perturbation expansion for $\mathbf{u}$ into the above equation and multiply out:

$$\begin{aligned} u_k^{(0)} + \lambda u_k^{(1)} + \lambda^2 u_k^{(2)} + \cdots = u_k^{(0)} + \lambda G_k^{(0)} M_{kjl} \times \\ \times [u_j^{(0)} + \lambda u_j^{(1)} + \lambda^2 u_j^{(2)} + \ldots] \\ \times [u_l^{(0)} + \lambda u_l^{(1)} + \lambda^2 u_l^{(2)} + \ldots]. \end{aligned} \tag{109}$$

Equate terms at each order in λ, thus:

$$\text{order } \lambda^0 : u_k^{(0)} = G_k^0 f_k. \tag{110}$$

$$\text{order } \lambda^1 : u_k^{(1)} = G_k^{(0)} M_{kjl} u_j^{(0)} u_l^{(0)}. \tag{111}$$

$$\text{order } \lambda^2 : u_k^{(2)} = 2G_k^0 M_{kjl} u_j^{(0)} u_l^{(1)}. \tag{112}$$

$$\text{order } \lambda^3 : u_k^{(3)} = 2G_k^0 M_{kjl} u_j^{(0)} u_l^{(2)} + G_k^{(0)} M_{kjl} u_j^{(1)} u_l^{(1)}. \tag{113}$$

And so on, to any order. Then we can substitute successively for $u^{(1)}, u^{(2)}$ etc., in terms of $u^{(0)}$.

2.4.5 Explicit form of the coefficients

First order is already in correct form:

$$u_k^{(1)} = G_k^{(0)} M_{kjl} u_j^{(0)} u_l^{(0)}. \tag{114}$$

At second order, substitute for $u^{(1)}$:

$$u_k^{(2)} = 2G_k^{(0)} M_{kjl} u_j^{(0)} G_l^{(0)} M_{lpq} u_p^{(0)} u_q^{(0)}. \tag{115}$$

At third order, substitute for $u^{(1)}$ and $u^{(2)}$ as appropriate:

$$\begin{aligned} u_k^{(3)} &= 4G_k^{(0)} M_{kjl} u_j^{(0)} G_l^{(0)} M_{lpq} u_p^{(0)} G_q^{(0)} M_{qrs} u_r^{(0)} u_s^{(0)} \\ &+ G_k^{(0)} M_{kjl} G_j^{(0)} M_{jpq} u_p^{(0)} u_q^{(0)} G_l^{(0)} M_{lrs} u_r^{(0)} u_s^{(0)}; \end{aligned} \tag{116}$$

and so on.

2.4.6 Equation for the exact second moment

The exact second moment is then given by

$$\begin{aligned} Q_k &= \langle u_k u_{-k} \rangle = \langle u_k^{(0)} u_{-k}^{(0)} \rangle + \langle u_k^{(0)} u_{-k}^{(2)} \rangle \\ &+ \langle u_k^{(1)} u_{-k}^{(1)} \rangle + \langle u_k^{(2)} u_{-k}^{(0)} \rangle + O(\lambda^4). \end{aligned} \tag{117}$$

Substituting in for the coefficients $\mathbf{u}^{(1)}, \mathbf{u}^{(2)}$ etc., yields.

$$\begin{aligned} Q_k &= Q_k^{(0)} + 2G_k^{(0)} M_{kjl} M_{lpq} G_l^{(0)} \langle u_k^{(0)} u_j^{(0)} u_p^{(0)} u_q^{(0)} \rangle \\ &+ G_k^{(0)} M_{kjl} M_{-kpq} G_k^{(0)} \langle u_j^{(0)} u_l^{(0)} u_p^{(0)} u_q^{(0)} \rangle \\ &+ 2G_k^{(0)} M_{kjl} M_{lpq} G_l^{(0)} \langle u_k^{(0)} u_j^{(0)} u_p^{(0)} u_q^{(0)} \rangle \\ &+ O(\lambda^4). \end{aligned} \tag{118}$$

2.4.7 Factorizing the zero-order moments

We won't go higher than fourth-order moments: we take the first such term as an example. We use a property of Gaussian statistics:

$$\begin{aligned}\langle u_k^{(0)} u_j^{(0)} u_p^{(0)} u_q^{(0)} \rangle = &\langle u_k^{(0)} u_j^{(0)} \rangle \langle u_p^{(0)} u_q^{(0)} \rangle \\ &+\langle u_k^{(0)} u_p^{(0)} \rangle \langle u_j^{(0)} u_q^{(0)} \rangle \\ &+\langle u_k^{(0)} u_q^{(0)} \rangle \langle u_j^{(0)} u_p^{(0)} \rangle . \end{aligned} \tag{119}$$

For isotropic, homogeneous fields we recall (for example) that

$$\langle u_k^{(0)} u_{k'}^{(0)} \rangle = \delta_{kk'} D_k Q_k^{(0)} . \tag{120}$$

Hence

$$\begin{aligned}\langle u_k^{(0)} u_j^{(0)} u_p^{(0)} u_q^{(0)} \rangle = &\delta_{kj}\delta_{pq} D_k D_p Q_k^{(0)} Q_p^{(0)} \\ &+\delta_{kp}\delta_{pj} D_k D_j Q_k^{(0)} Q_j^{(0)} \\ &+\delta_{kq}\delta_{jp} D_k D_j Q_k^{(0)} Q_j^{(0)} . \end{aligned} \tag{121}$$

Recall $l = |\mathbf{k} - \mathbf{j}|$, thus $\delta_{kj} M_{lqp} = 0$ as $k = j \implies l = 0$, and so the first term gives zero. Second and third terms are identical (with dummy variables swapped),

$$\langle u_k^{(0)} u_j^{(0)} u_p^{(0)} u_q^{(0)} \rangle = 2\delta_{kq}\delta_{jp} D_k D_j Q_j^{(0)} Q_k^{(0)} . \tag{122}$$

With these results, the equation for $Q(k)$ becomes

$$\begin{aligned} Q_k = Q_k^0 &+ G_k^0 M_{kjl} M_{lkj} G_l^0 D_k D_j Q_j^0 Q_k^0 \\ &+ 2G_k^0 M_{kjl} M_{-klj} G_k^0 D_j D_l Q_j^0 Q_l^0 \\ &+ 4G_k^0 M_{kjl} M_{jlk} G_j^0 D_l D_k Q_l^0 Q_k^0 + O(\lambda^4). \end{aligned} \tag{123}$$

We can combine all the M's and D's into a *simple* coefficient: $L(k, j, l)$. Hence:

$$\begin{aligned} Q_k &= Q_k^0 + G_k^0 L(k,j,l) G_l^0 Q_j^0 Q_k^0 \\ &+ G_k^0 L(k,j,l) G_k^0 Q_j^0 Q_l^0 + G_k^0 L(k,j,l) G_j^0 Q_l^0 Q_k^0 \\ &+ O(\lambda^4), \end{aligned} \tag{124}$$

where the coefficient takes the form:

$$L(k,j,l) = 4M_{kjl} M_{lkj} D_k D_j = 2M_{kjl} M_{-klj} D_j D_l . \tag{125}$$

2.5 Deriving a Transport Equation for the Second Moment

Let's remind ourselves what we are trying to do. In reduced notation, we have the NSE:

$$L_{0k} u_k = f_k + M_{kjl} u_j u_l . \tag{126}$$

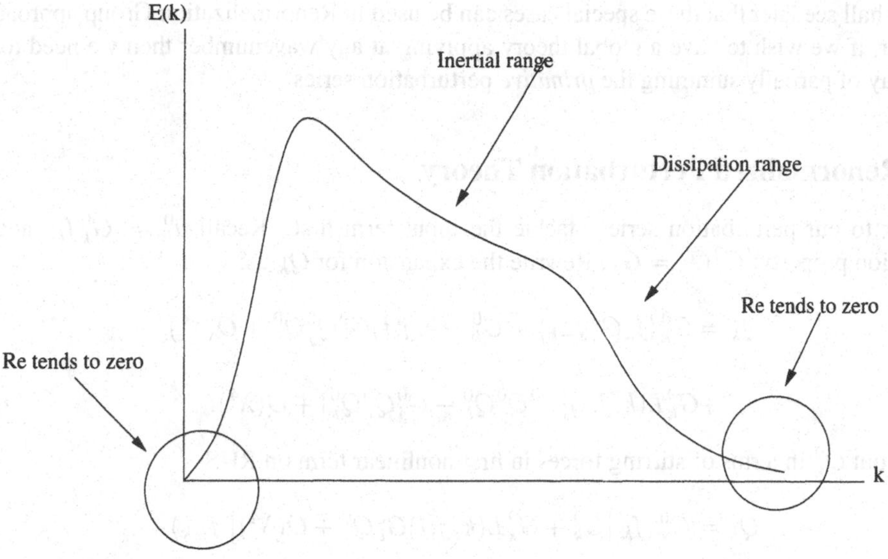

Figure 6: Regions of wavenumber space where the local Reynolds number is small.

Multiply through by u_{-k} and average:

$$L_{0k}Q_k = \langle f_k u_{-k}\rangle + M_{kjl}\langle u_j u_l u_{-k}\rangle. \tag{127}$$

The first term on RHS is the *input term*, due to stirring forces. This can be worked out exactly by perturbation theory. The second term on the RHS is the energy transfer due to nonlinearity. This can only be treated approximately.

2.6 Overview of Perturbation Theory for Turbulence

Our perturbation series is divergent as the number of terms rises with the 'order'. This is typical of many-body problems. Some form of regularization or renormalization will be required. The expansion parameter (or coupling constant) for the Navier-Stokes equation is always a Reynolds number. Most definitions of Reynolds number have a value greater than unity when turbulence is present.

We can define a local, spectral Reynolds number at wavenumber k as

$$R(k) = \frac{k^2 V(k)}{\nu} = \frac{[E(k)]^{1/2}}{\nu k^{1/2}} \tag{128}$$

Here ν is the kinematic viscosity and $V(k)$ is the rms Fourier coefficient. From an examination of the turbulence energy spectrum, we see that there are two circumstances under which $R(k)$ is small, viz;

1. $R(k) \to 0$ as $k \to 0$.
2. $R(k) \to 0$ as $k \to \infty$.

We shall see later that these special cases can be used in Renormalization Group approaches. However, if we wish to have a global theory applying at any wavenumber then we need to find some way of partially summing the *primitive* perturbation series.

2.7 Renormalized Perturbation Theory

Go back to our perturbation series: tackle the input term first. Recall $u_k^0 = G_k^0 f_k$; and the contraction property: $G_k^0 G_k^0 = G_k^0$. Rewrite the expansion for Q_k as:

$$Q_k = G_k^0 \langle f_k G_k^0 f_{-k} \rangle + G_k^0 L(k,j,l) G_l^0 Q_j^0 Q_k^0 + O(\lambda^4)$$

$$+ G_k^0 L(k,j,l) [G_k^0 Q_j^0 Q_l^0 - G_j^0 Q_e^0 Q_k^0] + O(\lambda^4). \tag{129}$$

Expand out Q_k^0 in terms of stirring forces in first nonlinear term on RHS.

$$Q_k = G_k^0 \langle f_k \left[G_k^0 + G_k^0 L(k,j,l) G_l^0 Q_j^0 + O(\lambda^4) \right] f_{-k} \rangle$$

$$+ G_k^0 L(k,j,l) \left[G_k^0 Q_j^0 Q_l^0 - G_j^0 Q_l^0 Q_k^0 \right] + O(\lambda^4). \tag{130}$$

Identify the quantity in square brackets as the *exact* response function G_k. Hence: recalling that $\langle f_k f_{-k} \rangle = W_k$:

$$Q_k = G_k^0 G_k W_k + G_k^0 L(k,j,l) [G_k^0 Q_j^0 Q_l^0 - G_j^0 Q_l^0 Q_k^0] + O(\lambda^4). \tag{131}$$

This equivalence leads to

$$G_k = G_k^0 + G_k^0 L(k,j,l) G_l^0 Q_j^0 + O(\lambda^4). \tag{132}$$

Operate with L_{0k} from left, using $L_{0k} G_k^0 = \delta(t - t')$:

$$L_{0k} G_k = \delta(t - t') + L(k,j,l) G_l^0 Q_j^0 + O(\lambda^4). \tag{133}$$

Similarly, operate on the equation for Q_k with L_{0k} from the left.

$$L_{0k} Q_k = W_k + L(k,j,l) \left[G_k^0 Q_j^0 Q_l^0 - G_j^0 Q_l^0 Q_k^0 \right] + O(\lambda^4). \tag{134}$$

In this way, we obtain two transport equations as (in principle) expansions in λ to all orders. In our case, λ is not small ($= 1$) and the nonlinear term is large, hence this is wildly divergent. To have any chance of succeeding we must be able to sum certain classes of terms to *all* orders. For example, in some perturbation expansions, we can recognise a geometric series, which can be summed exactly. One way of partial summing, is to replace Q^0, G^0 by Q and G on RHS of both equations. This can be justified (to some extent) on topological grounds using diagrams. This means replacing the *mythical* Q_k^0 by the *actual* Q_k. It also means replacing the *actual* G_k^0 by the *mythical* G_k. In this respect we differ totally from quantum field theory where "bare" Green's functions are replaced by renormalized *observable* Green's functions.

2.8 RPT Transport Equations

We make the replacements $Q^0 \longrightarrow Q, G^0 \longrightarrow G$: the important thing is that we do *both*. That is what makes it a renormalization process.

For the covariance we get:

$$L_{0k}Q_k = W_k + L(k,j,l)[G_kQ_jQ_l - G_jQ_lQ_k] + O(\lambda^4). \qquad (135)$$

For the response function we get:

$$L_{0k}G_k = \delta(t-t') + L(k,j,l)G_lQ_j + O(\lambda^2). \qquad (136)$$

3 Two-Point Closures and their Assessment

We give an overview of the various RPTs of turbulence, without going into details of the differences between them. We then concentrate on two theories, viz., the Direct-Interaction Approximation (or DIA); and the Local Energy Transfer theory (or LET). We give detailed equations for both these theories and then present graphs giving the results of their numerical integration for: (a) the free decay of turbulence; (b) forced isotropic turbulence.

3.1 Renormalized Perturbation Theories

The essential feature of renormalized perturbation theories is how you choose the *mythical* $G_{\mathbf{k}}$ in order to generate a good value for the *observable* Q_k. This was realised when the first RPT (Kraichnan's Direct Interaction Approximation) was found to be incompatible with the Kolmogorov spectrum. Essentially DIA is the set of equations we derived in the previous lecture, although Kraichnan obtained them in a different way. Virtually all theories give the same equation for $Q(k;t,t')$ but differ in the way their response function is chosen.

3.1.1 The covariance equations: time-dependent forms

The pioneering theories were due to Kraichnan (1959), Edwards (1964) and Herring (1965). Kraichnan's two-time covariance equation (DIA) may be written as:

$$\left[\frac{\partial}{\partial t} + \nu k^2\right] Q(k;t,t')$$
$$= \int L(\mathbf{k},\mathbf{j}) \left[\int_0^{t'} G(k;t',t'')Q(j;t,t'')Q(|\mathbf{k}-\mathbf{j}|;t,t'')\mathrm{d}t''\right.$$
$$\left. - \int_0^{t} G(j;t,t'')Q(k;t'',t')Q(|\mathbf{k}-\mathbf{j}|;t,t'')\mathrm{d}t''\right] \mathrm{d}^3 j. \qquad (137)$$

The coefficient $L(\mathbf{k},\mathbf{j})$ is given by:

$$L(\mathbf{k},\mathbf{j}) = \frac{[\mu(k^2+j^2) - kj(1+2\mu^2)]\,(1-\mu^2)kj}{k^2+j^2-2kj\mu}. \qquad (138)$$

The closure is completed by an equation for the response-function $G(k,t,t')$.

The single-time covariance equation due to Edwards (EFP) may be written as:

$$\left[\frac{d}{dt}+2\nu k^2\right]Q(k,t)=2\int L(\mathbf{k},\mathbf{j})\frac{Q(|\mathbf{k}-\mathbf{j}|,t)\,[Q(j,t)-Q(k,t)]}{\omega(k)+\omega(j)+\omega(|\mathbf{k}-\mathbf{j}|)}\mathrm{d}^3 j. \tag{139}$$

An energy input $W(k)$ can be added to the right-hand side to ensure stationarity. The function $\omega(k)$ is the renormalized eddy damping rate and an equation may be found for this quantity by considering conservation of energy. Herring's self-consistent field (SCF) approach is a more general version of the EFP approach and for the time-dependent case is closely related to the DIA equations. There are various closure models, such as EDQNM and TFM, which have the same covariance equation as EFP. The distinction here is: models have one or more adjustable parameters, whereas theories do not.

3.1.2 Covariance equations: stationary case

The EFP covariance equation can be obtained from the DIA equation as follows: Substitute into (1) the assumed time dependencies:

$$Q(k,t-t')=Q(k)\exp\{-\omega(k)|t-t'|\}; \tag{140}$$

and

$$\begin{aligned} G(k,t-t') &= \exp\{-\omega(k)(t-t')\} \quad \text{for} \quad t>t'; \\ &= 0 \quad \text{for} \quad t<t'. \end{aligned} \tag{141}$$

Then, integrate the DIA forms over intermediate times. The result (with some re-arrangement) is:

$$W(k)-2\nu k^2 Q(k)=\int L(k,j)\frac{Q(|\mathbf{k}-\mathbf{j}|)\,[Q(k)-Q(j)]}{\omega(k)+\omega(j)+\omega(\mathbf{k}-\mathbf{j}|)}\,\mathrm{d}^3 j. \tag{142}$$

This simple form is helpful in understanding certain properties, e.g. conservation of energy by the nonlinear term and behaviour of the system in the limit of infinite Reynolds number.

3.1.3 Summary: RPTs of the first kind

None of these theories gives the Kolmogorov $-5/3$ spectrum as its solution at large Reynolds numbers.

Kraichnan (1959): DIA
Edwards (1964): EFP
Herring (1965): SCF
Phythian (1969): another type of SCF
Balescu and Senatorski (1970): yet another SCF

3.1.4 Summary: RPTs of the second kind

These theories all claim to give the Kolmogorov $-5/3$ spectrum as solution at infinite Reynolds numbers.

Lagrangian or mixed coordinate system

Kraichnan (1965): ALHDI
Kraichnan and Herring (1978): SBALHDI
Horner and Lipowsky (1979): a Lagrangian theory
Kaneda (1981): LRA
Kida and Gotoh (1997): LRA

Eulerian coordinate system

Edwards and McComb (1969): maximal entropy method
McComb (1978): LET
Qian (1983): a variational method
Nakano (1988): another derivation of LET

3.2 Numerical Assessment of RPTs

Some RPTs have been computed for freely decaying isotropic turbulence with generally good results. The main investigations are: Kraichnan (1964a), Herring and Kraichnan (1972) and McComb and Shanmugasundaram (1984). A key quantity is the Taylor microscale λ. For decaying isotropic turbulence this is

$$\lambda(t) = \left[5E(t)/\int_0^\infty k^2 E(k,t)\mathrm{d}k\right]^{1/2}. \tag{143}$$

Note that in decaying turbulence most quantities depend on the time. We then characterise the turbulence by the Taylor-Reynolds number R_λ:

$$R_\lambda = \lambda u/\nu, \tag{144}$$

where u is the r.m.s. velocity and ν is the viscosity. More recently, Quinn (2000) has computed DIA and LET for forced isotropic turbulence and compared the results with those obtained from a direct numerical simulation. The procedure is to fix the initial conditions and then integrate the equations forward in time for both DIA and LET. This implies that we choose an arbitrary initial spectrum. This is done numerically on a computer and requires discretization in both wavenumber and time. Some representative results are shown in Figures 7 - 14.

3.3 The DIA and LET Closures

We shall concentrate on two RPTs: DIA (Kraichnan 1959) and LET (McComb 1978). Other theories: use mixed Lagrangian-Eulerian coordinate systems and are very complicated; or are not time dependent; or reduce to either DIA or LET.

The basic ansatz of DIA is that there is a response function such that

$$\delta u_\alpha(\mathbf{k},t) = \int_{-\infty}^{t} \hat{G}_{\alpha\beta}(\mathbf{k},t,t')\delta f_\beta(\mathbf{k},t')\,dt', \tag{145}$$

and that this *infinitesimal response* function can be renormalised. The resulting response equation is

$$\begin{aligned}&\left[\frac{\partial}{\partial t} + \nu k^2\right] G(k;t,t') \\ &+ \int L(\mathbf{k},\mathbf{j}) \int_{t'}^{t} G(k;t'',t')G(j;t,t'')Q(|\mathbf{k}-\mathbf{j}|;t,t'')\,\mathrm{d}^3\mathbf{j}\mathrm{d}t'' \\ &= 0.\end{aligned} \tag{146}$$

LET introduces a *renormalised propagator* H such that :

$$Q_{\alpha\beta}(\mathbf{k},t,t') = H_{\alpha\gamma}(\mathbf{k};t,t')Q_{\gamma\beta}(\mathbf{k};t,t). \tag{147}$$

The resulting propagator equation is

$$\begin{aligned}&\left[\frac{\partial}{\partial t} + \nu k^2\right] H(k;t,t') \\ &+ \int L(\mathbf{k},\mathbf{j}) \int_{t'}^{t} H(k;t'',t')H(j;t,t'')Q(|\mathbf{k}-\mathbf{j}|;t,t'')\,\mathrm{d}^3\mathbf{j}\mathrm{d}t'' \\ &= \frac{1}{Q(k;t',t')} \int L(\mathbf{k},\mathbf{j}) \left[\int_0^{t'} \mathrm{d}t'' Q(|\mathbf{k}-\mathbf{j}|;t,t'')\right. \\ &\left.\{H(k;t',t'')Q(j;t,t'') - Q(k;t',t'')H(j;t,t'')\}\right] \mathrm{d}^3\mathbf{j}.\end{aligned} \tag{148}$$

The governing equations for DIA can be obtained by setting the right hand side of the LET equation for the exact response function equal to zero.

3.3.1 DIA and LET as mean field theories

In order to derive the DIA equations one must take the step

$$\begin{aligned}\langle \hat{G}(t-t')u(t)u(t')\rangle &= \langle \hat{G}(t-t')\rangle\langle u(t)u(t')\rangle \quad (149)\\ &= G(t-t')Q(t-t'). \quad (150)\end{aligned}$$

This is a mean-field approximation. LET begins with the postulate

$$u_\alpha(\mathbf{k},t) = \hat{H}_{\alpha\beta}(\mathbf{k},t,t')u_\beta(\mathbf{k},t'). \tag{151}$$

Multiplying through by $u_\alpha(-\mathbf{k},t')$ and averaging

$$\langle u_\alpha(\mathbf{k},t)u_\gamma(-\mathbf{k},t')\rangle = \langle \hat{H}_{\alpha\beta}(\mathbf{k};t,t')u_\beta(\mathbf{k},t')u_\gamma(-\mathbf{k},t')\rangle. \tag{152}$$

If, as in DIA, we take the propagator and velocity field to be uncorrelated, we can write

$$Q_{\alpha\gamma}(\mathbf{k};t,t') = H_{\alpha\beta}(k;t,t')Q_{\beta\gamma}(\mathbf{k};t',t'), \tag{153}$$

where

$$H_{\alpha\beta}(k;t,t') = \langle \hat{H}_{\alpha\beta}(\mathbf{k};t,t') \rangle. \tag{154}$$

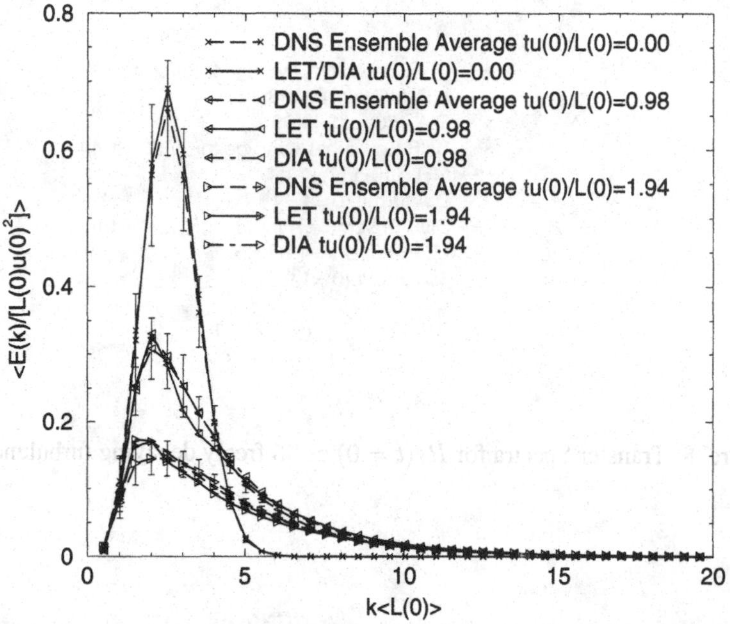

Figure 7: Energy Spectra for $R_\lambda(t=0) \simeq 95$ freely decaying turbulence

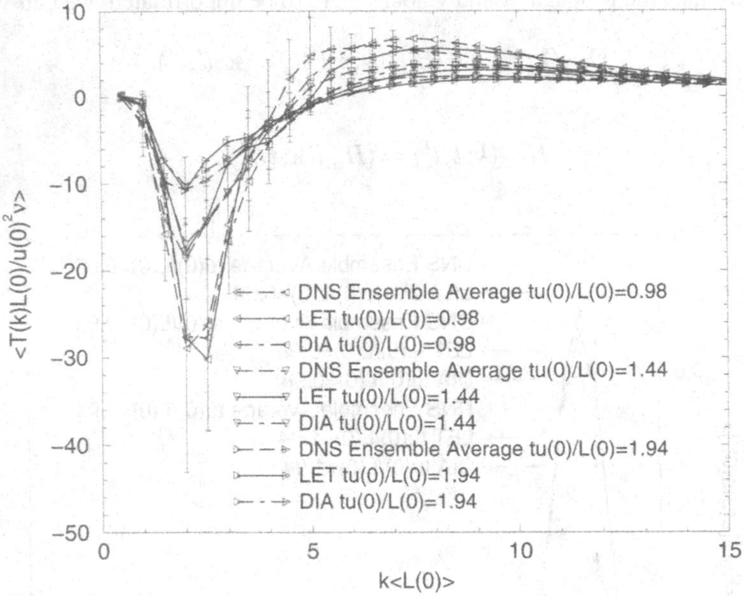

Figure 8: Transfer Spectra for $R_\lambda(t=0) \simeq 95$ freely decaying turbulence

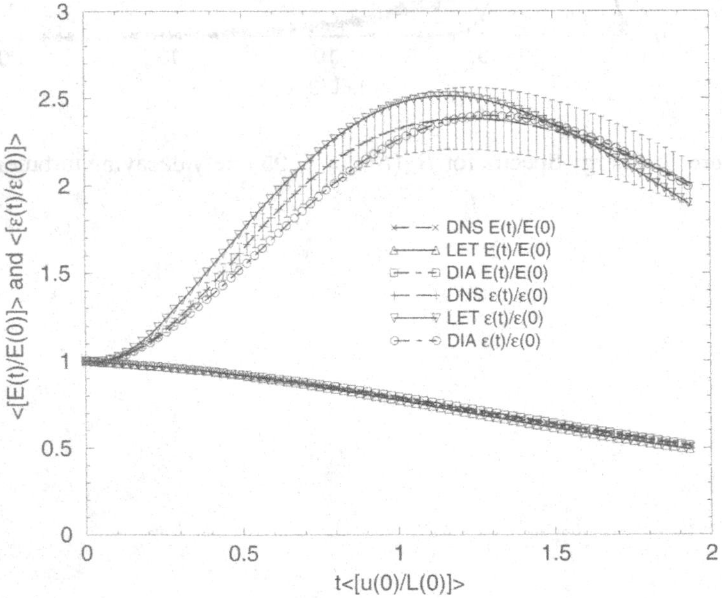

Figure 9: Total Energy per unit mass and Dissipation Rate per unit mass for $R_\lambda(t=0) \simeq 95$ freely decaying turbulence

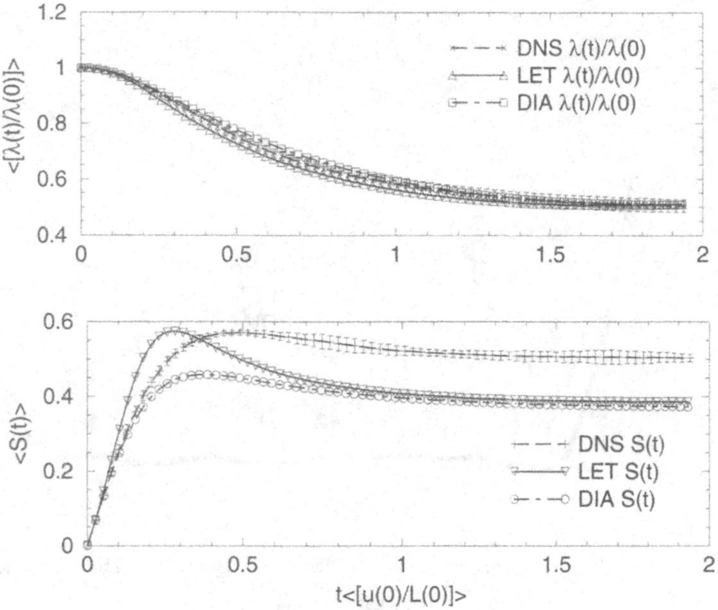

Figure 10: Microscale Length (upper group of curves) and Velocity Derivative Skewness (lower group of curves) for $R_\lambda(t = 0) \simeq 95$ freely decaying turbulence

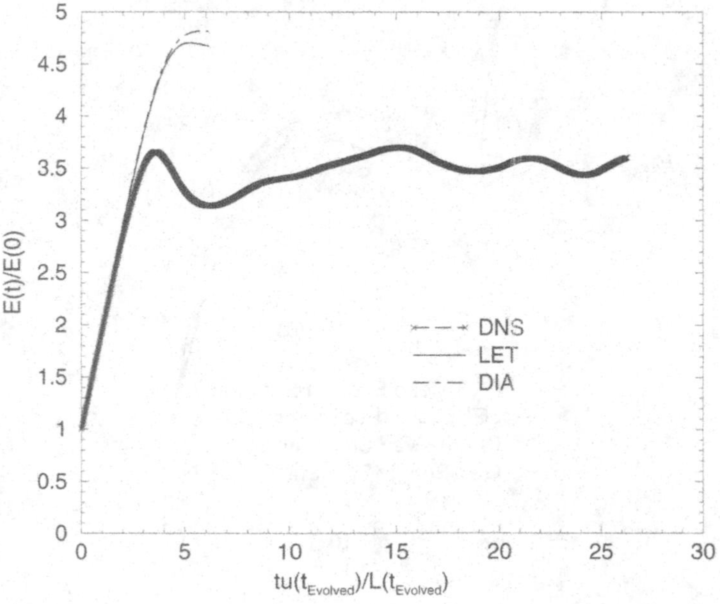

Figure 11: Total Energy per unit mass for $R_\lambda(t_{Evolved}) \simeq 232$ in forced turbulence (N.B. LET and DIA calculations ceased at $tu/L \simeq 6.14$)

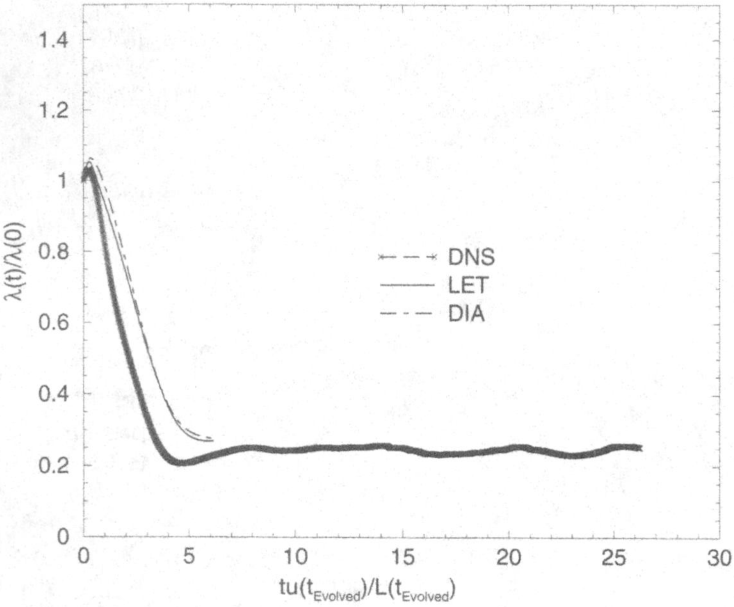

Figure 12: Microscale Length for $R_\lambda(t_{Evolved}) \simeq 232$ in forced turbulence

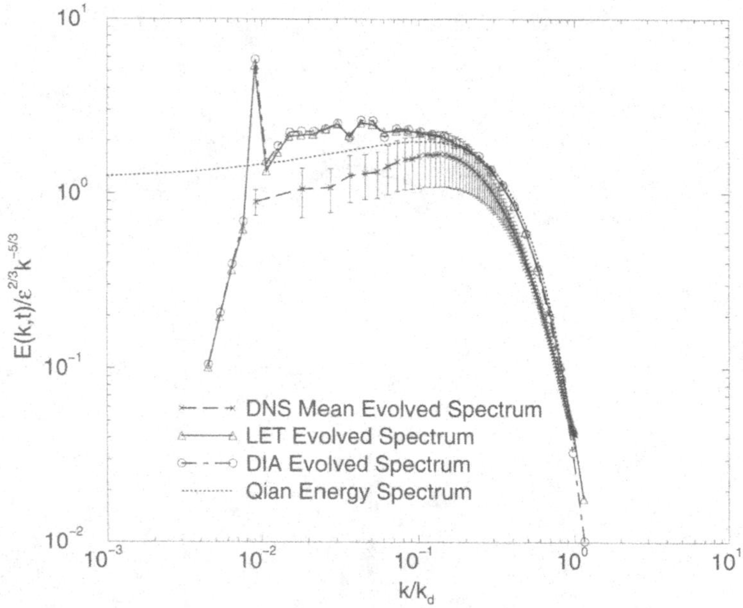

Figure 13: Comparison of the compensated energy spectrum at $R_\lambda(t_{Evolved}) \simeq 232$ in forced turbulence with the *ad hoc* energy spectrum due to Qian (1984)

4 Renormalization Group (RG) Methods

In many-body physics, RG is a technique for reducing the number of degrees of freedom in a problem. It has had its greatest successes in critical phenomena - notably the solution of the three-dimensional Ising model of magnetism. Probably its application in turbulence will be to formulate a reduced numerical problem, perhaps in the manner of LES, but with a rigorous solution to some aspects of the problem of subgrid modelling.

4.1 Renormalization Group in Microscopic Physics

For a proper introduction to this beautiful technique, we really have to look at the para-ferro magnetic transition, and in particular at the Ising model. In this section we give a general background discussion of RG and then concentrate on specific calculations in one and two dimensions. An important part of this discussion is the role played by concepts of *geometrical similarity*, *self-similarity* and *scale invariance*.

Renormalisation Group (RG) is a successful technique in *microscopic* physics which comes in a variety of forms. It can be: a method for eliminating divergences; a way of reducing the number of degrees of freedom, a framework which helps one to understand the physics of a complicated system, and a way of controlling the approximations made when one does not really know what one is doing! These are all 'flavours' and are not mutually exclusive.

4.1.1 RG: a general definition

RG is a method for reducing the number of degrees of freedom in a problem with many degrees of freedom. In the context of fluid turbulence, this means eliminating Fourier modes. There are always two steps involved in one iteration.

Step 1: eliminate some modes (e.g. by averaging or by summing).

Step 2: rescale the equations so that it looks as if no modes were eliminated.

These two steps are iterated until one reaches a *fixed point*, where the 'picture' does not change any more.

4.2 Discrete Dynamical Systems as Toy Models

In physics we are accustomed to think of dynamics as being the subject dealing with the motion of bodies under the influence of forces. We expect such motion to take the form of trajectories, in which positions vary continuously with time, and which are in accord with Newton's laws (or their relativistic generalization). However, nowadays there is also a wide class of problems coming under the heading of discrete dynamical systems, in which system properties vary in a series of discrete steps. In mathematics, it is said that such systems are described by the iteration of a simple map. We shall introduce the subject by means of two simple examples, each of which illustrates the idea of a fixed point. Then we shall discuss the idea of a fixed point a little more formally, and conclude by discussing the behaviour of a specific system with two fixed points.

4.2.1 Example: repaying a loan

Suppose you borrow £$A(0)$, at 1% interest per month, and repay the loan at £20 per month. After one month, you owe:

$$A(1) = A(0) + 0.01A(0) - 20 = 1.01A(0) - 20;$$

and after $n+1$ months:

$$A(n+1) = 1.01A(n) - 20. \tag{155}$$

The amount you can borrow is limited by both the interest rate and the repayment rate. For example, if we arbitrarily choose three different initial amounts £$A(0)$ =£1000, £3000 and £2000, in that order, then equation (1) gives us:

$$\begin{array}{llll} \text{(a)} & A(0) = £1000, & A(1) = £990, & A(2) = £979.90\ldots \\ \text{(b)} & A(0) = £3000, & A(1) = £3010, & A(2) = £3020.10\ldots \\ \text{(c)} & A(0) = £2000, & A(1) = £2000, & A(3) = £2000\ldots. \end{array}$$

In case (a), the amount you owe decreases with time, whereas for case (b) it increases with time. However, case (c) is a constant solution and the value £2000 is a fixed point of the system. The fixed point is also known as the critical point or, the equilibrium value or, the constant solution.

4.2.2 Definition of a fixed point

Consider a first-order dynamical system

$$A(n+1) = f(A(n)). \tag{156}$$

A number a is a fixed point of the system if

$$A(n) = a \quad \text{for all} \quad n,$$

when $A(0) = a$. That is, when

$$A(n) = a,$$

is a constant solution, the value a is a fixed point. We can put this in the form of a theorem, viz., The number a is a fixed point of

$$A(n+1) = f(A(n)) \qquad \text{if and only if} \qquad a = f(a). \tag{157}$$

For completeness, we should note that a dynamical system may have many fixed points. For example if there is an $A^2(n)$ term, then there are two fixed points; an $A^3(n)$ term, then three fixed points; and so on. In general, the more nonlinear a system is, the more fixed points it will have.

4.2.3 Example: A dynamical system with two fixed points.

Consider the dynamical system

$$A(n+1) = [A(n)+4]A(n)+2. \tag{158}$$

It can be shown that this has two fixed points, $a = -1$ and -2. That is, for an initial value $A(0) = -2$, we find that $A(n) = -2$ for all n, and for $A(0) = -1$, we find that $A(n) = -1$ for all n. Now take different initial values $A(0) = -1.01, -0.99, -2.4$. We sketch the result in Figure 15, from which it may be concluded that:

$a = -1$ is a repelling fixed point (unstable equilibrium);

$a = -2$ is an attracting fixed point (stable equilibrium).

It should be noted that $A(0) = -0.99$ is not within the 'basin of attraction' of $a = -2$.

4.2.4 General considerations RG: background in quantum field theory

In turbulence, perturbation theory led to coupled expansions for the exact covariance Q_k and the exact Green's function G_k. In quantum field theory, there is only the Green's function. The "bare" Green's function G_{0k} corresponds to an isolated particle propagating through free space.

In a 'real' situation, such as scattering (or other interactions) the *exact* Green's function is obtained as a perturbation series in G_{0k}, thus:

$$G_k = G_{0k} + \int G_{0j} dj G_{0k} + \ldots.$$

The integral over the bare Green's function diverges at $j = 0$. This can be fixed up by various kinds of renormalization. The mass of the particle appears in the Green's function. The bare mass m_0 can be renormalized to m. In fact m is the observable and so counter-terms to cancel the divergence can be absorbed into m_0 which is never observed! 'Tricks' of this kind *cannot* be applied to turbulence. One must be careful about what one can carry over from quantum mechanics to a classical mechanics problem.

4.2.5 RG: background in statistical physics

RG was introduced as an *ad hoc* technique in quantum field theory to cure divergences ($\sim$ 1955). In 1970, Wilson (1983) united the technique with Kadanoff's idea of 'block spins' in magnetism to give a physical meaning to the process. A *fixed point* of the RG transformation corresponded to *scale invariance* in this new picture. So also did the existence of a critical point: e.g. the onset of permanent magnetism.

4.2.6 RG in critical phenomena

Consider an Ising model of a ferromagnet: 'spins' on a lattice. The smallest scale is a the distance between lattice sites. The largest scale is the physical size of the magnet: normally this is taken to be infinite. At very high temperatures, spins are aligned at random. No order exists at any

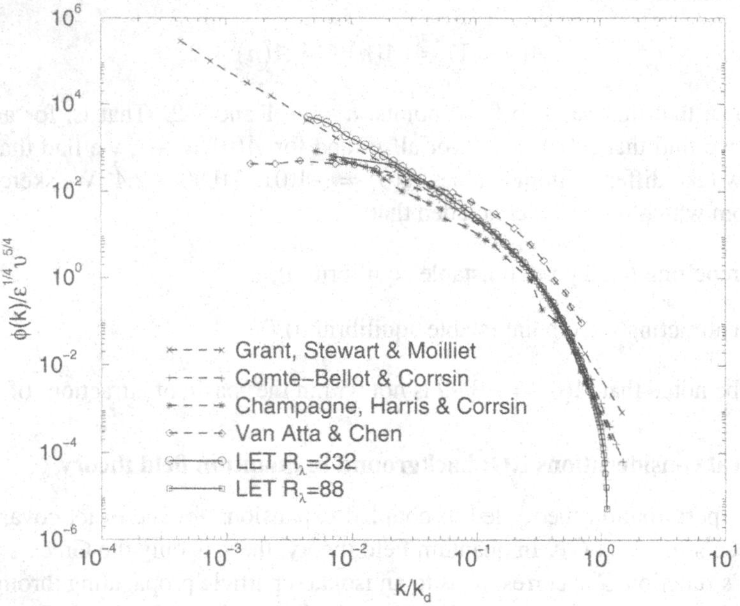

Figure 14: Comparison of the LET one-dimensional energy spectrum for $R_\lambda(t_{Evolved}) \simeq 88$ and 232 in forced turbulence with some experimental results

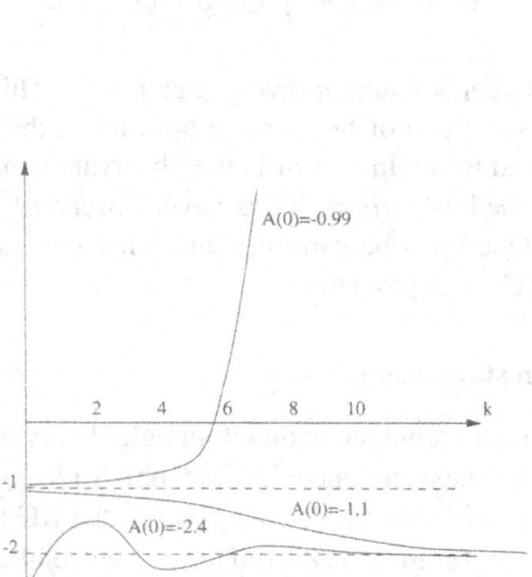

Figure 15: Illustration of recursion relations leading to fixed points.

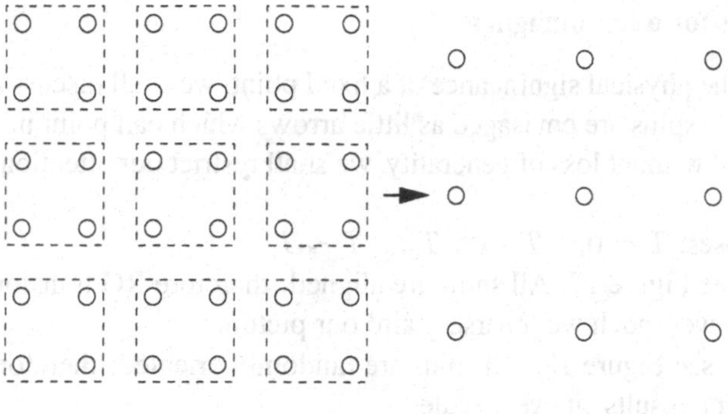

Figure 16: 'Blocking lattice spins'.

scale greater than a. The picture is the same on every scale. At the critical point, all spins are lined up in one direction and again the picture is the same on every scale. As the critical point is approached from above, a series of coarse-graining transformations will eliminate order. At temperatures close to the critical point, correlations (i.e. alignments of spins) can exist on all length scales from a to infinity.

4.2.7 Turbulence as a critical phenomenon

Turbulence also involves many length and time scales. Basically there is no way to say that some are more important than others. Compare this with other physical problems: e.g. laminar flow of a fluid involves scales much larger than those of molecular motions. Hence we can average out the molecular motions and replace their mean effect by a coefficient of viscosity. There are two possible critical points for turbulence. First, there is the transition from laminar to turbulent flow. Second, there is the onset of scaling behaviour (i.e. the Reynolds number becomes large enough for the development of a $-5/3$ inertial range).

4.2.8 An RG transformation: 2-d Ising Model

Represent spin up by $\oplus$, spin down by $\bigcirc$ and divide the 2-d lattice into blocks of 4 spins. Take the spin of each block to be the same as the *majority*, and located at the centre of the block. If a block has 2 spins up and two spins down, toss a coin to determine the spin of the block. This process reduces the number of spins on the lattice by a factor of 4 and increases the lattice spacing by a factor of 2. This is illustrated in Figure 16.

4.2.9 An RG Transformation: 2-D Heisenberg Model

This differs from the Ising model in that spin vectors can point in any direction (i.e. not just up or down). Divide the 2-D lattice into blocks of 4 spins. Take the spin of each block to be the *average* of the four spins and located at the centre of the block. Like the *majority rule* transformation, this RG transformation reduces the number of spins by a factor of 4 and increases the lattice spacing by a factor of 2. Note that the picture seen after each application of this transformation will not change if either all the spins are aligned in one direction or all the spins have random orientation.

4.2.10 Fixed points for a ferromagnet

In order to illustrate the physical significance of a fixed point, we shall discuss a Heisenberg-type model, in which lattice spins are envisaged as little arrows which can point in any direction. For sake of simplicity, and without loss of generality, we shall restrict our attention to a lattice in two dimensions.

Consider three cases: $T = 0, \quad T \to\gg T_c, \quad T \sim T_c^>$

Case 1: $T = 0$: see Figure 17. All spins are aligned: therefore RG transformation must give the same results, however much we 'coarse grain' our picture.

Case 2: $T \to \infty$: see Figure 18. All spins are randomly oriented: therefore RG transformation must give the same results, at every scale.

Case 3: At $T \geq T_c$: see Figure 19. The spins are a random sea, with islands of 'correlation', each of which only exists for a brief time as a fluctuation. For a finite correlation length ξ, a finite number of RG coarse-graining transformations will 'hide' ordering effects. As $\xi(T) \to \infty$ (for $T \to T_c^>$), then no finite number of RG scale changes will hide the correlations. As a result we can identify:

$$\text{fixed point} \equiv \text{critical point.}$$

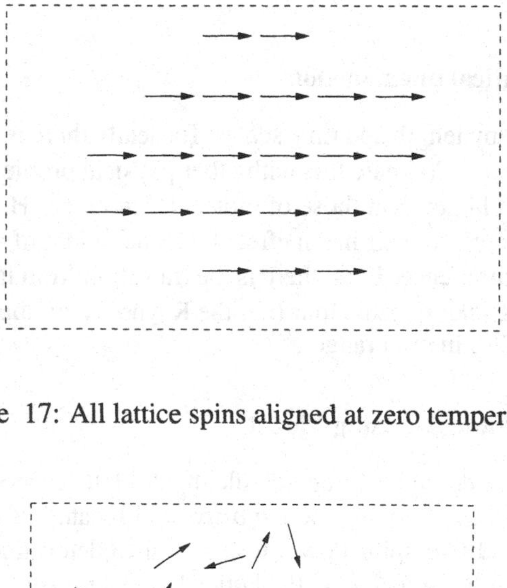

Figure 17: All lattice spins aligned at zero temperature.

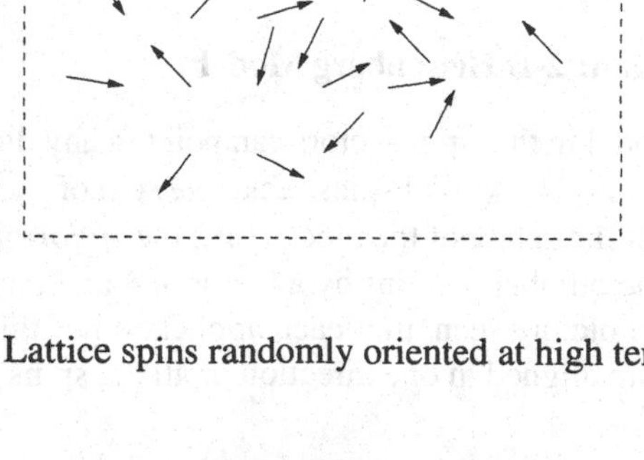

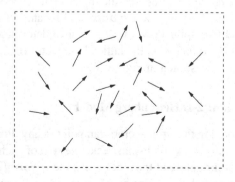

Figure 18: Lattice spins randomly oriented at high temperatures.

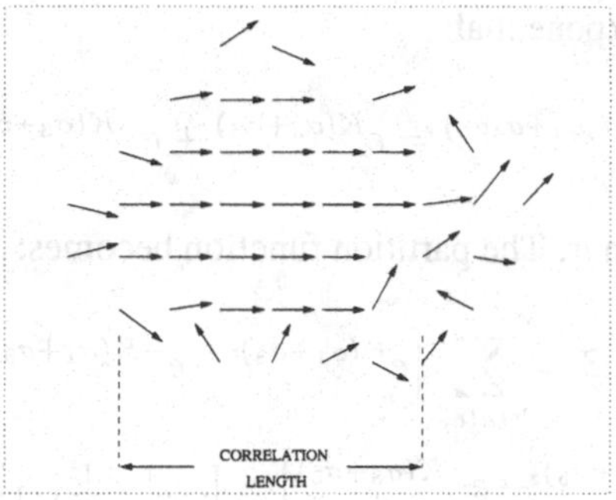

Figure 19: Correlated groups of spins as the critical temperature is approached from above.

To sum this up, for a system of this kind we can see that the iteration of RG transformations could reveal the presence of three different kinds of fixed point, viz:

1. Low-temperature, perfect order.
2. High-temperature, complete disorder.
3. The critical fixed point.

From the point of view of describing critical phenomena, the low-temperature and high-temperature fixed points are trivial.

4.2.11 Example: Calculation of the partition function for the $1d$ Ising model

The partition function is given by:

$$\mathcal{Z} = \sum_{states} e^{J\sum_{<i,j>} \sigma_i \sigma_j / kT},$$

where $< i,j >\equiv$ sum over nearest neighbours. Introduce the coupling constant K such that $K \equiv J/kT$, hence

$$\mathcal{Z} = \sum_{states} e^{K\sum_{<i,j>} \sigma_i \sigma_j}.$$

The procedure is then as follows. Sum over all $N/2$ even-numbered spins:

$$\mathcal{Z} = \sum_{states} e^{K[\sigma_1\sigma_2 + \sigma_2\sigma_3 + \sigma_3\sigma_4 + \sigma_4\sigma_5 + \dots]}$$

$$= \sum_{states} e^{K(\sigma_1\sigma_2 + \sigma_2\sigma_3)} \times e^{K(\sigma_3\sigma_4 + \sigma_4\sigma_5)} \times e^{K(\dots)}$$

Now, sum over states $\sigma_2 = \pm 1$ in 1st exponential:

$$\sum e^{K(\sigma_1\sigma_2 + \sigma_2\sigma_3)} = e^{K(\sigma_1 + \sigma_3)} + e^{-K(\sigma_1 + \sigma_3)}.$$

Similarly for $\sigma_4 = \pm 1$ in 2nd exponential:

$$\sum e^{K(\sigma_3\sigma_4+\sigma_4\sigma_5)} = e^{K(\sigma_3+\sigma_5)} + e^{-K(\sigma_3+\sigma_5)};$$

and so on, for σ_6, σ_8 etc., $\forall$ even σ. The partition function becomes:

$$\mathcal{Z} = \sum_{states} \left[e^{K(\sigma_1+\sigma_3)} + e^{-K(\sigma_1+\sigma_3)}\right] \\ \times \left[e^{K(\sigma_3+\sigma_5)} + e^{-K(\sigma_3+\sigma_5)}\right] \times [\ldots] \times [\ldots] \ldots$$

Next step is find a value K' of the coupling *and* a function $f(K)$ such that

$$e^{K(\sigma_1+\sigma_3)} + e^{-K(\sigma_1+\sigma_3)} = f(K)e^{K'\sigma_1\sigma_3},$$

which of course will apply for σ_3, σ_5 etc. *Note* $f(K)$ must NOT be a function of σ_1, σ_3. For instance, $\sigma_1 = 1$ and $\sigma_3 = 1$ (or $\sigma_1 = -1, \sigma_3 = -1$) yields

$$e^{2K} + e^{-2K} = fe^{K'},$$

whereas for $\sigma_1 = 1, \sigma_3 = -1$ (or *vice versa*) we get

$$2 = fe^{<}K'.$$

Solving these two simultaneous equations gives

$$K' = \frac{1}{2}\ln\cosh(2K),$$

and

$$f(K) = 2\cosh^{1/2}(2K).$$

So we can write $\mathcal{Z}$ as

$$\mathcal{Z} = \sum_{states} \left[f(K)e^{K'\sigma_1\sigma_3}\right]\left[f(K)e^{K'\sigma_3\sigma_5}\right] \times [\ldots] \times \ldots,$$

and hence

$$\mathcal{Z} = f(K)^{N/2} \sum_{states} e^{K'[\sigma_1\sigma_3+\sigma_3\sigma_5+\ldots]}$$

The remaining sum is the partition function for $N/2$ spins with new coupling constant K'. So we can write our partially summed partition function as

$$\mathcal{Z}(N, K) = f^{N/2}(K)\mathcal{Z}(N/2, K'). \tag{159}$$

4.2.12 The Free Energy F

F is extensive therefore we must have

$$F \sim \ln \mathcal{Z} = N\zeta \text{ (say)},$$

where ζ doesn't depend on N, but will depend on K. Take the logarithm of both sides of equation (159) and put

$$\ln \mathcal{Z}(N,K) = N\zeta(K),$$

to get

$$N\zeta(K) = \frac{N}{2}\ln f(K) + \frac{N}{2}\zeta(K').$$

Now divide across by N/2, rearrange and substitute for $f(K)$ to get

$$\zeta(K') = 2\zeta(K) - \ln[2\cosh^{1/2}(2K)].$$

Taken with the solution for K', this gives us a *recursion relation* to calculate ζ (i.e.$\mathcal{Z}$) for other values from any given value. This recursion is always such that $K' < K$, hence it goes in the direction of reducing K.

We can invert these relations to go the other way:

$$K = \frac{1}{2}\cosh^{-1}(e^{2K'}), \tag{160}$$

and

$$\zeta(K) = \frac{1}{2}\ln 2 + \frac{1}{2}K' + \frac{1}{2}\zeta(K'). \tag{161}$$

Let us remind ourselves that the coupling constant is

$$K = \frac{J}{kT},$$

where $J \equiv$ exchange energy of a pair of spins For fixed J and high T, K is small.

4.2.13 Numerical calculation of the partition function

Step 1 Begin at high temperatures: coupling const K' is small: spin-spin interaction negligible. Therefore for free spins, partition function is just the number of arrangements is $\mathcal{Z} \sim 2^N$. Say $K' \sim 0.01$. Therefore:

$$\zeta(K' = 0.01) = \ln 2.$$

Step 2 Substitute for K' in (6), putting:

$$K' = 0.01.$$

Get

$$K = 0.100334.$$

Step 3 Calculate $\zeta(K)$ from (7) using

$$\zeta(K' = 0.01) = \ln 2.$$

Hence

$$\zeta(K) = \frac{1}{2}\ln 2 + \frac{1}{2} \times 0.01 + \frac{1}{2}\ln 2.$$

Carrying on this way, we get the same result (to six significant figures) as the $1d$ Ising exact result:

$$\mathcal{Z} = \left[2\cosh\left(\frac{J}{kT}\right)\right]^N.$$

but we only get trivial fixed points at $K = 0$ or $K = \infty$. In higher dimensions, it is not so simple as this procedure generates further interaction terms. In $2d$, one gets a nontrivial fixed point $K = K_c$.

4.3 Calculation of the Partition Function for the $2d$ Ising Model

We can carry out the same procedure for a $2d$ lattice and sum over alternate spins. A typical term resulting from this is:

$$\mathcal{Z} = \sum \ldots [e^{K(\sigma_1+\sigma_2+\sigma_3+\sigma_4)}] + [e^{-K(\sigma_1+\sigma_2+\sigma_3+\sigma_4)}] \ldots$$

We continue to follow the procedure as for $1d$, but in $2d$ we find that new couplings are generated (i.e. not just nearest neighbours):

$$\mathcal{Z} = f(K)^{N/2}\left[K_1 \sum_{nn} \sigma_i\sigma_j + K_2 \sum_{nnn} \sigma_i\sigma_j \right.$$
$$\left. K_3 \sum\sum \sigma_i\sigma_j\sigma_r\sigma_s\right],$$

where nn denotes 'sum over nearest neighbours'; nnn is 'sum over next nearest neighbours'; and the double sum denotes a sum over sets of four spins around a square. The new coupling constants are given by:

$$K_1 = \frac{1}{4}\ln\cosh(4K);$$
$$K_2 = \frac{1}{8}\ln\cosh(4K);$$
$$K_3 = \frac{1}{8}\ln\cosh(4K) - \frac{1}{2}\ln\cosh(2K).$$

Approximations are needed if we are to make this work: we really need the form:

$$\mathcal{Z}(N, K) = f^{N/2}\mathcal{Z}(N/2, K'),$$

if this is to work out like the $1d$ case. One way is to set $K_2 = K_3 = 0$. Then we get a result like the $1d$ problem, but unfortunately with no nontrivial fixed point. A better approximation is to set $K = K_1 + K_2$, $K_3 = 0$. Now one gets a nontrivial fixed point, with a value of the critical temperature within 20% of the exact result. It is not easy to improve on this. For instance, if K_3 is included, then a poorer result is obtained.

4.4 The application of RG to Turbulence

We again consider the relevance of self-similarity, but this time in the context of turbulence. If we represent the kinematic viscosity of the fluid by ν_0, then the corresponding Kolmogorov dissipation wavenumber may be denoted by $k_d^{(0)}$. Now, if we have a procedure in which modes in a small band of wavenumbers are averaged out and their mean effect replaced by an increment to the viscosity, then we make the replacement $\nu_0 \to \nu_1 = \nu_0 + \delta\nu_0$. The NSE is now on a reduced set of wavenumbers, but with an increased viscosity and a reduced Kolmogorov dissipation wavenumber $k_d^{(1)}$. Accordingly, solutions which scaled as $f(k/k_d^{(0)})$ may now scale as $f(k/k_d^{)1)})$, so that rescaling appropriately may lead to scale-invariance. In fact it usually takes about 5 or 6 iteration cycles before scale invariance is demonstrated by the recursion reaching a fixed point.

4.4.1 RG for turbulence

Introduce the RG approach to turbulence by dividing up the velocity field at $k = k_1$ as follows:

$$u_\alpha(\mathbf{k}, t) = \begin{cases} u_\alpha^<(\mathbf{k}, t) & \text{for} \quad 0 \leq k \leq k_1 \\ u_\alpha^>(\mathbf{k}, t) & \text{for} \quad k_1 \leq k \leq k_0, \end{cases}$$

where k_1 is defined by

$$k_1 = (1 - \eta)k_0,$$

and

$$0 \leq \eta \leq 1.$$

In principle, the RG approach involves 2 stages:

1. Solve NSE on $k_1 \leq k \leq k_0$. Substitute that solution for the mean effect of the high-k modes into the NSE on $0 \leq k \leq k_1$. This results in an increment to the viscosity $\nu_0 \to \nu_1 = \nu_0 + \delta\nu_0$.
2. Rescale the basic variables, so that the NSE on $0 \leq k \leq k_1$ looks like the original Navier-Stokes equation on $0 \leq k \leq k_0$.

The algorithm is then applied to successively lower wavenumbers:

$$\begin{aligned} k_1 &= (1 - \eta)k_0 \\ k_2 &= (1 - \eta)k_1 \\ &\vdots \end{aligned}$$

where $0 < \eta < 1$.

The filtered equations of motion are:

$$\begin{aligned} \left(\frac{\partial}{\partial t} + \nu_0 k^2\right) u_\alpha^<(\mathbf{k}, t) = M_{\alpha\beta\gamma}^< \int \Big\{ & u_\beta^<(\mathbf{j}, t) u_\gamma^<(\mathbf{k} - \mathbf{j}, t) \\ & + 2u_\beta^<(\mathbf{j}, t) u_\gamma^>(\mathbf{k} - \mathbf{j}, t) \\ & + u_\beta^>(\mathbf{j}, t) u_\gamma^>(\mathbf{k} - \mathbf{j}, t) \Big\} \, d^3 j + f_\alpha(\mathbf{k}, t). \end{aligned} \tag{162}$$

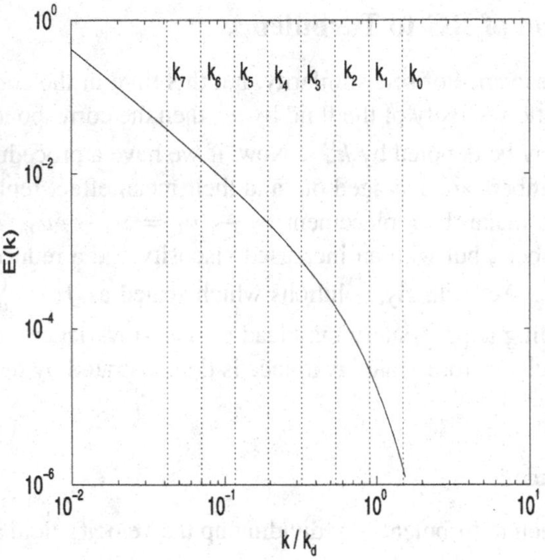

Figure 20: Wavenumber bands for iterative mode elimination.

$$\left(\frac{\partial}{\partial t} + \nu_0 k^2\right) u_\alpha^>(\mathbf{k},t) = M_{\alpha\beta\gamma}^> \int \Big\{ u_\beta^<(\mathbf{j},t) u_\gamma^<(\mathbf{k}-\mathbf{j},t) + 2u_\beta^<(\mathbf{j},t) u_\gamma^>(\mathbf{k}-\mathbf{j},t) + u_\beta^>(\mathbf{j},t) u_\gamma^>(\mathbf{k}-\mathbf{j},t) \Big\} \, d^3 j. \qquad (163)$$

The band-filtered spectral density is:

$$Q^>(k) = \langle u^>(\mathbf{k},t) u^>(-\mathbf{k},t) \rangle, \qquad (164)$$

where

$$u^>(\mathbf{k},t) = 0 \text{ for } 0 \le k \le k_1,$$

and

$$u^>(\mathbf{k},t) = u(\mathbf{k},t) \text{ for } k_1 \le k \le k_0.$$

4.4.2 Technical problems: 1 Need for a conditional average

The normal ensemble average is illustrated by Figure 21. In practice, RG is often applied using a filtered average, as in Figure 22. Denote filtered ensemble average by $\langle \ldots \rangle^>$. Then

$$\langle u_\alpha^<(\mathbf{k},t) \rangle^> = u_\alpha^<(\mathbf{k},t);$$

$$\langle u_\alpha^>(\mathbf{k},t) \rangle^> = 0.$$

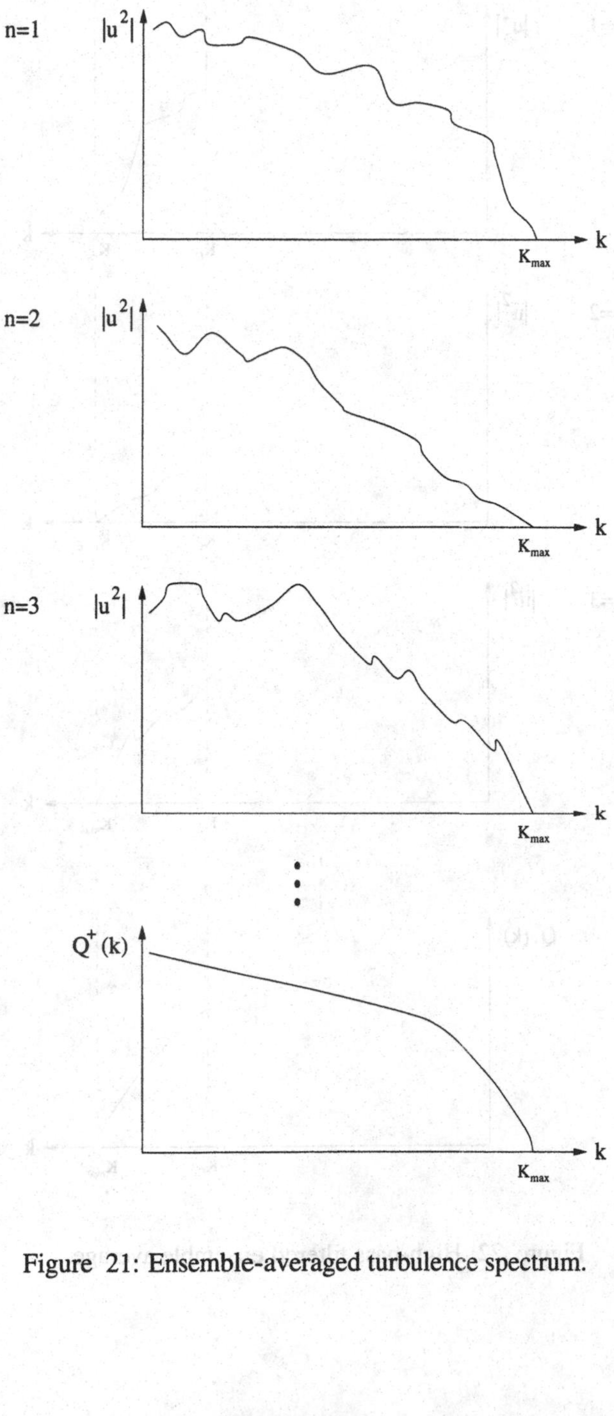

Figure 21: Ensemble-averaged turbulence spectrum.

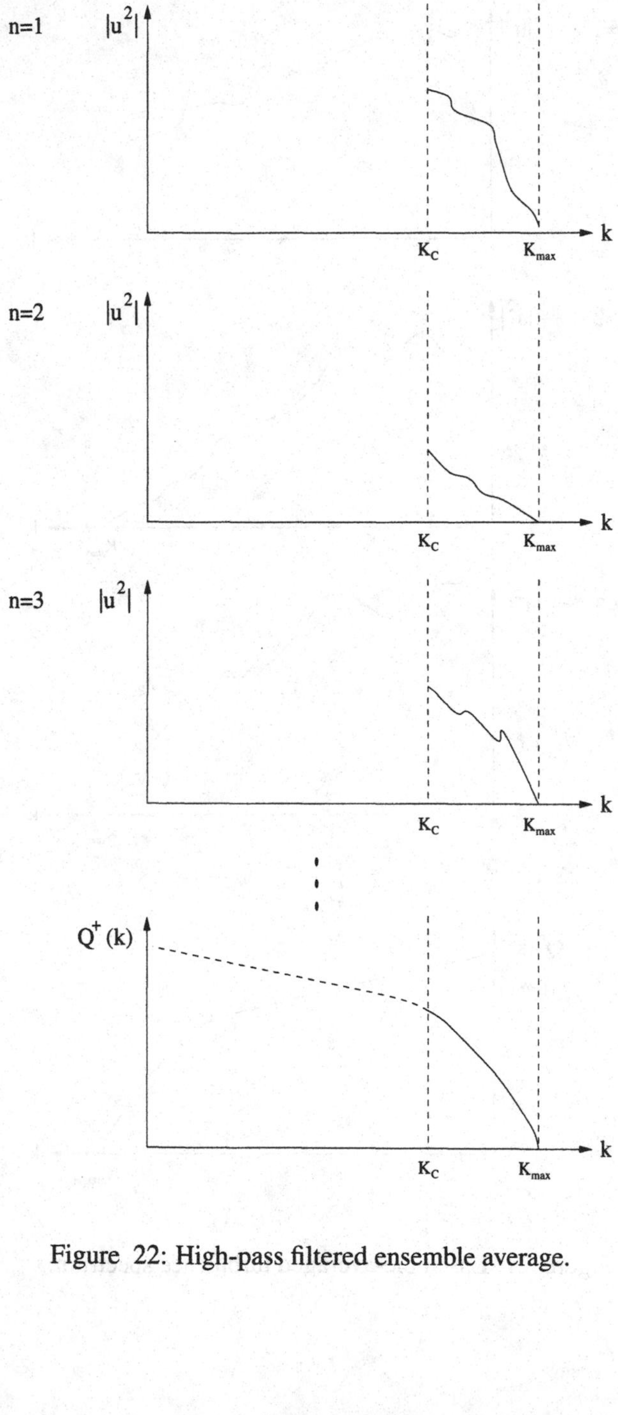

Figure 22: High-pass filtered ensemble average.

Average the low-k equation of motion

$$\left[\frac{\partial}{\partial t}+\nu_0 k^2\right] u_\alpha^<(\mathbf{k},t) = M_{\alpha\beta\gamma}^<(k)\int\Big\{u_\beta^<(j)u_\gamma^<(\mathbf{k}-\mathbf{j}) + 2u_\beta^<(j)\langle u_\gamma^>(\mathbf{k}-\mathbf{j})\rangle^> + \langle u_\beta^>(\mathbf{j})u_\gamma^>(\mathbf{k}-\mathbf{j})\rangle^>\Big\}\,\mathrm{d}^3 j+\mathbf{f}. \quad (165)$$

Second term on RHS = 0, as $\langle u^>\rangle^> = 0$. Third term on RHS = 0 from homogeneity, as $M(\mathbf{k})Q^>(\mathbf{j})\delta(\mathbf{j}+\mathbf{k}-\mathbf{j}) \sim M(0) = 0$. Hence a filtered ensemble average gives the strange result that there is no turbulence problem! Terms involving $u^>$ are just averaged away.

4.4.3 Technical problems: 2 Higher-order nonlinearities

We wish to eliminate the $u^>$ from the equation for the $u^<$. If we 'solve' the high-k equation: by inverting the linear operator $(\partial/\partial t+\nu_0 k^2)\equiv L_0$:

$$u_{\mathbf{k}}^> = L_0^{-1}M_{\mathbf{k}}^>\int[u_j^<u_{k-j}^< + 2u_j^<u_{k-j}^> + u_j^>u_{k-j}^>]\,\mathrm{d}^3 j.$$

And substitute back into the low-k equation:

$$L_0 u_{\mathbf{k}}^< = M_k^<[u_j^<u_{k-j}^< + 2u_j^<L_0^{-1}M_{k-j}u_{k-j}^<u_{k-j-\gamma}^< + \ldots].$$

Note the occurrence of the triple nonlinearity $u^<u^<u^<$ which breaks the form-invariance of the NSE.

4.5 Pioneering Applications of RG to Fluid Motion.

Forster et al. (1976) adapted the RG theory of dynamical critical phenomena due to Ma and Mazenko to the case of stirred fluid motion. In order to apply the theory, they chose the wavenumber cut-off to exclude cascade effects. They did not claim that this was a theory of turbulence. Despite this, numerous further investigations have been made along these lines and are claimed to be theories of turbulence. Rose (1977) applied the theory to a more realistic situation but only considered passive scalar convection.

McComb (1982) developed the theory of iterative averaging which was applicable to turbulence and was the first to recognize the need for a new kind of average. Yakhot and Orszag (1986) have based an approach to turbulence modeling on the work of Forster *et al.*

4.5.1 Forster, Nelson and Stephen

Start with NSE driven by stirring forces. Choose a wavenumber cutoff at $k=\Lambda$, which is lower in k-space than the start of the inertial range. The stirring force has a Gaussian distribution over the ensemble, zero mean, and white noise correlation as in equation (68). This makes the velocity field take certain statistical properties, viz.,

$$\langle u_0^>(\mathbf{k},t)\rangle = 0;$$
$$\langle u_0^> u_0^> u_0^>\rangle = 0;$$
$$M^<(\mathbf{k})\langle u_0^>(\mathbf{k},t)u_0^>(\mathbf{k}-\mathbf{j},t)\rangle = 0.$$

Split the velocity field into high- and low-k components. Do a perturbation expansion of the high-k field, forming equations for the evolution of those terms. Substitute these in low-k equation and average using the above rules. This gets round the problem of the conditional average by averaging over the $f^>$ at constant $f^<$.

We are then left with a term linear in $u^<$: this leads to an effective viscosity. The triple term vanishes in $k \longrightarrow 0$ limit. Rescale the variables, leading to renormalization of parameters. Using infinitesimal wavenumber bands of elimination leads to differential equations for parameters.

FNS show that the renormalized coupling parameter $\overline{\lambda}$ reaches a fixed point if $\epsilon = 4+y-d > 0$, and as the bandwidth parameter $l \rightarrow \infty$. Resulting energy spectrum is of the form:

$$E(k) \simeq k^{-5/3+2(d-y)/3}.$$

Criticism of the theory, made by Eyink (1994), is that it is not technically possible to average over the $f^>$ assuming independence from $u^<$ to give the effective dynamics of $u^<$. Such an approximation will be uncontrolled since $u^<$ get dependence on $f^>$ through coupling to $u^>$. An average over $f^>$ forces with $u^<$ fixed will change distribution of $f^>$ in an unknown way.

5 Relevance of RG to the Large-Eddy Simulation of Turbulence

We restrict our attention to the subgrid modelling problem for homogeneous, isotropic, stationary turbulence, with zero mean velocity. Our main emphasis is on recognizing the importance of conditionally averaging the subgrid stresses. We also distinguish the relative importance of different categories of subgrid stress according to their effect on the following characteristics of the resolved scales: (a) the evolution of the velocity field; (b) the evolution of the energy spectrum; and (c) the inertial transfer of energy and the dissipation rate. The turbulence which we consider only occurs an approximation in nature but can be represented experimentally by numerical simulation. This section is divided into three parts, as follows:

1. We discuss the formulation of LES and introduce an *ad hoc* subgrid viscosity, along with a correction term. We note that the form of the correction term may change from one conservation relation (viz., momentum, energy transfer and dissipation) to another.

2. We perform a conditional average of the resolved-scales equation of motion and derive the corresponding equations for the energy spectrum and dissipation rate.

   - We show that a correction term vanishes when one unconditionally averages (and so does not contribute to the evolution of the energy spectrum).
   - A further correction term vanishes when integrated over wavenumber space to form the dissipation relation.

3. We calculate an effective viscosity, by eliminating modes in successive bands, to represent the effect of the conditionally–averaged 'subgrid–subgrid' stress and show that it can only contribute to a renormalized dissipation rate.

5.1 LES: Statement of the Problem

The statistical problem is formulated in terms of the Eulerian velocity field $u_\alpha(\mathbf{x}, t)$. We work in wavenumber $(\mathbf{k})$ space, thus:

$$u_\alpha(\mathbf{k}, t) = \int u_\alpha(\mathbf{x}, t) e^{-i\mathbf{k}\cdot\mathbf{x}}\, \mathrm{d}^3 x. \tag{166}$$

We wish to simulate the energy spectrum on the interval $0 \le k \le k_C$, interactions with the *subgrid* modes in the interval $k_C \le k \le k_{max}$ being represented analytically by a subgrid model. Note that we are only concerned with high-wavenumber modes in the inertial and dissipation ranges. To have a well-posed problem, we specify ε, the rate of energy transfer through the inertial range of wavenumbers. The specification of the numerical problem is completed by fixing the number of degrees of freedom in the system. We do this by introducing a maximum wavenumber k_{max} through a modification of the dissipation integral as

$$\varepsilon = \int_0^\infty 2\nu_0 k^2 E(k)\, \mathrm{d}k \simeq \int_0^{k_{max}} 2\nu_0 k^2 E(k)\, \mathrm{d}k, \tag{167}$$

where $E(k)$ is the energy spectrum and ν_0 is the kinematic viscosity of the fluid.

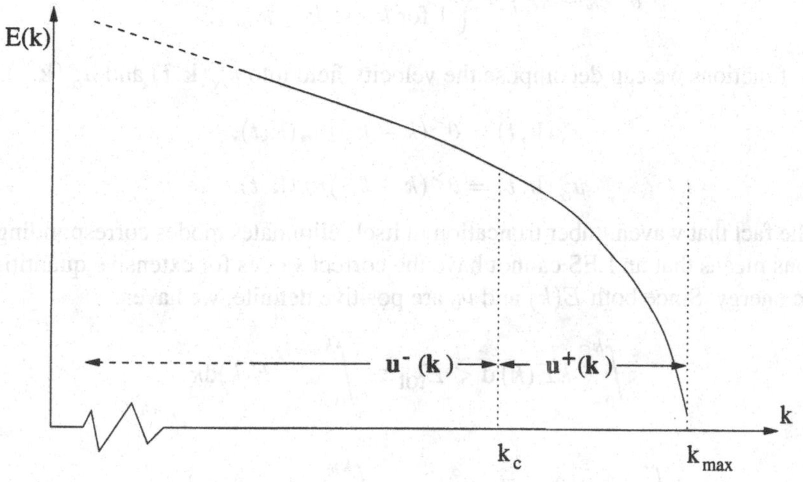

Figure 23: Filtered DNS energy spectrum.

5.1.1 Filtered variables

In general, the number of degrees of freedom may be reduced by filtering. We start by introducing a sharp cutoff filter at wavenumber $k = k_C$ where $k_C \ll k_{max}$. This filter is defined using the two unit step functions, viz.,

$$\theta^<(k - k_C) = \begin{cases} 1 \text{ for } 0 \le k \le k_C; \\ 0 \text{ for } k_C < k \le k_{max}, \end{cases} \tag{168}$$

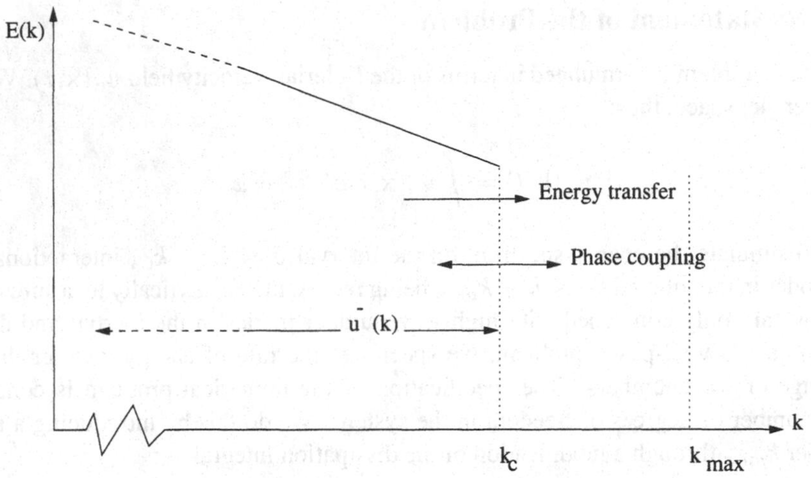

Figure 24: Truncated or LES energy spectrum.

and

$$\theta^>(k - k_C) = \begin{cases} 0 \text{ for } 1 \le k \le k_C; \\ 1 \text{ for } k_C < k \le k_{max}. \end{cases} \tag{169}$$

Using these functions we can decompose the velocity field into $u_\alpha^<(\mathbf{k}, t)$ and $u_\alpha^>(\mathbf{k}, t)$, where

$$u_\alpha^<(\mathbf{k}, t) = \theta^<(k - k_C)u_\alpha(\mathbf{k}, t); \tag{170}$$

$$u_\alpha^>(\mathbf{k}, t) = \theta^>(k - k_C)u_\alpha(\mathbf{k}, t). \tag{171}$$

However, the fact that wavenumber truncation in itself eliminates modes corresponding to small-scale motions means that an LES cannot have the correct values for extensive quantities like the total kinetic energy. Since both $E(k)$ and ν_0 are positive definite, we have:

$$\int_0^{k_C} kE(k)\,\mathrm{d} < E_{\text{tot}} \simeq \int_0^{k_{max}} E(k)\mathrm{d}k \tag{172}$$

and

$$\int_0^{k_C} 2\nu_0 k^2 E(k)\,\mathrm{d}k < \varepsilon \simeq \int_0^{k_{max}} 2\nu_0 k^2 E(k)\,\mathrm{d}k \tag{173}$$

In other words, truncation reduces both the amount of energy in the system and the rate at which energy is dissipated.

5.1.2 The conditional average and its impossibility

The need for a conditional average when eliminating modes has been widely, if not universally, recognised. However, we have

$$u_\alpha^<(\mathbf{k}, t) = \int \theta^<(k - k_C)u_\alpha(\mathbf{x}, t)e^{-i\mathbf{k}\cdot\mathbf{x}}\,\mathrm{d}^3x; \tag{174}$$

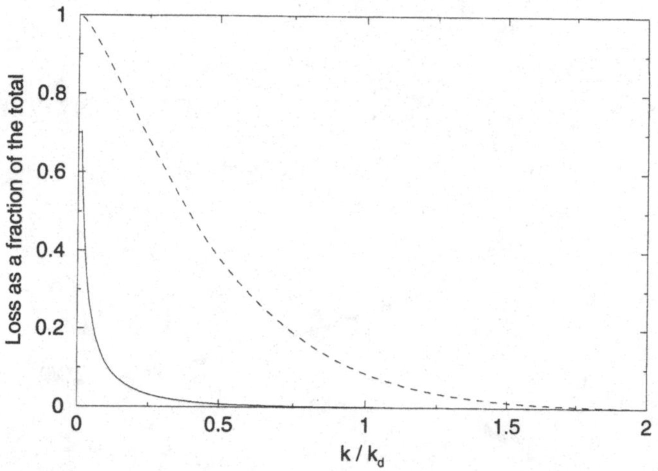

Figure 25: Loss of energy and dissipation rate due to spectral truncation: The continuous line represents turbulent kinetic energy while the dashed line is the dissipation rate.

and

$$u_\alpha^>(\mathbf{k},t) = \int \theta^>(k - k_C) u_\alpha(\mathbf{x},t) e^{-i\mathbf{k}\cdot\mathbf{x}}\, \mathrm{d}^3x. \tag{175}$$

Evidently, if we average either of $u_\alpha^\pm(\mathbf{k},t)$, the only thing which is actually averaged is the field $u_\alpha(\mathbf{x},t)$. Therefore, if in any kind of averaging process $u_\alpha^<(\mathbf{k},t)$ is left unaveraged, it follows immediately that $u_\alpha^>(\mathbf{k},t)$ is also left unaveraged. This leaves us with the question: how does one average out the effect of high-k modes while leaving the low-k modes unaffected? The answer may lie in the concept of unpredictability and, particularly, in formulating and evaluating the conditional average as an *approximation*.

This involves specifying the condition for the average with some degree of imprecision in the low-k modes and then relying on the chaotic nature of turbulence to amplify this uncertainty in the high-k modes. We have referred to this property of turbulence as *local chaos*. This hypothesis has been examined in high-resolution numerical experiments by Machiels (1997), McComb et al. (1997) with encouraging results.

5.2 Conservation Equations for the Explicit Scales

We begin by decomposing the NSE, in terms of the filtered variables. We obtain an equation of motion for the low-k modes, thus:

$$\begin{aligned}\left(\frac{\partial}{\partial t} + \nu_0 k^2\right) u_\alpha^<(\mathbf{k},t) = M_{\alpha\beta\gamma}^<(\mathbf{k}) \int \Big\{ & u_\beta^<(\mathbf{j},t) u_\gamma^<(\mathbf{k}-\mathbf{j},t) \\ & + 2u_\beta^<(\mathbf{j},t) u_\gamma^>(\mathbf{k}-\mathbf{j},t) \\ & + u_\beta^>(\mathbf{j},t) u_\gamma^>(\mathbf{k}-\mathbf{j},t) \Big\}\, \mathrm{d}^3 j \end{aligned} \tag{176}$$

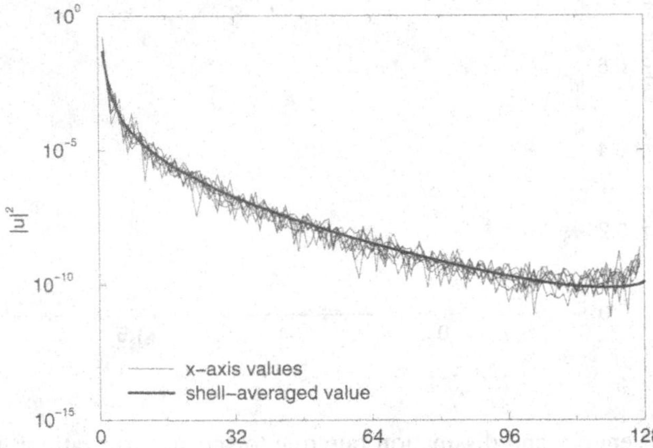

Figure 26: Chaotic behaviour of instantaneous energy spectra.

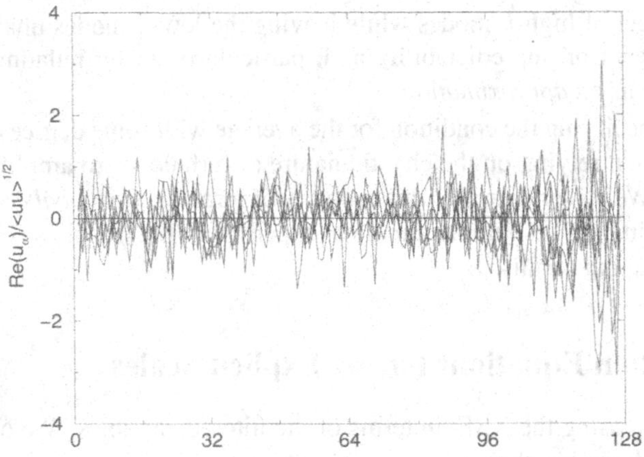

Figure 27: Chaotic behaviour of Fourier components of velocity.

Each term in this equation is defined on the wavenumber interval $0 \leq k \leq k_C$. For example,

$$M^{<}_{\alpha\beta\gamma}(\mathbf{k}) = \theta^{<}(k - k_C)M_{\alpha\beta\gamma}(\mathbf{k}). \tag{177}$$

We are still left with a version of the turbulence closure problem in which the dynamical equation for the retained modes depends directly on modes in the range which we wish to filter out. This equation expresses the principle of conservation of momentum for the explicit scales. We may use it, as we used the full NSE, to derive conservation relations for kinetic energy and energy dissipation for the explicit scales and this is our next step.

5.2.1 Ensemble-averaged conservation equations and the *ad hoc* effective viscosity

The effects of coupling between the retained and eliminated modes are often represented by an enhanced, wavenumber-dependent viscosity $\nu(k|k_C)$ acting upon the explicitly simulated modes. Then, the general dynamical equation for use in a low-wavenumber simulation might be expected to take the form

$$\left(\frac{\partial}{\partial t} + \nu(k|k_C)k^2\right) u^{<}_{\alpha}(\mathbf{k}, t) = M^{<}_{\alpha\beta\gamma}(\mathbf{k}) \int u^{<}_{\beta}(\mathbf{j}, t)u^{<}_{\gamma}(\mathbf{k} - \mathbf{j}, t)\mathrm{d}^3 j + S^{<}_{\alpha}(\mathbf{k}|k_C; t), \tag{178}$$

$S^{<}_{\alpha}(\mathbf{k}|k_C; t)$ has been added as a correction term. This term reflects the inadequacy of the effective viscosity concept, and incorporates additional effects of the kind which are often referred to as 'eddy noise' or 'backscatter'. This equation is exact and as yet no approximation has been made. We may then obtain a modified energy balance equation, for the low-wavenumber simulation.

$$2\nu(k|k_C)k^2 E^{<}(k) = W(k) + T^{<}(k) + 8\pi k^2 \langle \hat{S}^{<}_{\alpha}(\mathbf{k}|k_C; t)u^{<}_{\alpha}(-\mathbf{k}, t)\rangle \tag{179}$$

We have assumed that $\kappa \ll k_C$, where κ is the upper endpoint of the forcing spectrum. Integrating both sides of this equation with respect to k over the interval 0 to k_C, we then obtain a modified form of the energy dissipation relation as:

$$\int_0^{k_C} 2\nu(k|k_C)k^2 E(k)\,\mathrm{d}k = \varepsilon + \int_0^{k_C} 8\pi k^2 \langle \tilde{S}^{<}_{\alpha}(\mathbf{k}|k_C; t)u^{<}_{\alpha}(-\mathbf{k}, t)\rangle\,\mathrm{d}k \tag{180}$$

Note that in these equations, we write the correction terms as $S^{<}$, $\hat{S}^{<}$ and $\tilde{S}^{<}$, respectively. This is because we should allow for the possibility that $S^{<}$ contains terms which vanish under the average or under the subsequent wavenumber integration. As we shall see shortly, this is in fact the case. The effective viscosity, as it appears in the momentum equation is not (to employ the language of quantum physics) an observable. It may, however, be an observable in the equation for the dissipation rate, provided there exist circumstances under which the term in $\tilde{S}^{<}$ vanishes.

5.2.2 Local energy transfer?

Another limitation on the concept of a 'subgrid' eddy viscosity arises in the equation for the energy spectrum. Consider the case of local transfers near the cut-off wavenumber k_C. With wavenumber triads subject to the restrictions:

$$0 \leq k \leq k_C \qquad k_C \leq j, \quad \text{and} \quad |\mathbf{k} - \mathbf{j}| \leq k_{max}.$$

It follows from an exact symmetry that energy transfer is zero for the case:

$$k \approx j \approx |\mathbf{k}-\mathbf{j}| \approx k_C.$$

Evidentally the term local must be interpreted with some caution in this context.

5.3 Conditionally-Averaged Conservation Equations

Previously we have defined the conditional average in terms of a conditionally sampled ensemble, but we can adapt the usual definition of conditional average and write it in terms of a limit. Let us consider a fluctuation in the low-k modes, thus:

$$u_\alpha^<(\mathbf{k},t) \rightarrow u_\alpha^<(\mathbf{k},t) + \delta u_\alpha^<(\mathbf{k},t). \tag{181}$$

We may define the conditional average of any functional $F[u_\alpha(\mathbf{k},t)]$ in the limit of $\delta u_\alpha^<(\mathbf{k},t) \rightarrow 0$, as

$$\langle F[u_\alpha(\mathbf{k},t)]\rangle_c = \langle F[u_\alpha(\mathbf{k},t)] | u_\alpha^<(\mathbf{k},t)\rangle, \tag{182}$$

such that

$$\langle u_\alpha^<(\mathbf{k},t)\rangle_c = u_\alpha^<(\mathbf{k},t), \tag{183}$$

where the subscript 'c' on the angle brackets stands for 'conditional average'. Note that we employ the standard notation for a conditional average on the right hand side of equation, where the vertical bar separates the argument from the 'condition'.

Our next step is to take the conditional average of the low-pass filtered NSE,

$$\begin{aligned}\left(\frac{\partial}{\partial t} + \nu_0 k^2\right) u_\alpha^<(\mathbf{k},t) = M_{\alpha\beta\gamma}^<(\mathbf{k}) \int \Big\{ &\langle u_\beta^<(\mathbf{j},t) u_\gamma^<(\mathbf{k}-\mathbf{j},t)\rangle_c \\ &+ 2\langle u_\beta^<(\mathbf{j},t) u_\gamma^>(\mathbf{k}-\mathbf{j},t)\rangle_c \\ &+ \langle u_\beta^>(\mathbf{j},t) u_\gamma^>(\mathbf{k}-\mathbf{j},t)\rangle_c \Big\} \mathrm{d}^3 j\end{aligned} \tag{184}$$

The conditional averages are themselves stochastic variables, fluctuating on the interval $0 \le k \le k_C$. For the moment we shall treat them as completely unknown variables and lump them into a correction term $H_\alpha^<$, say. To do this, we add and subtract quantities, in such a way as to leave it unaffected, thus:

$$\begin{aligned}\left(\frac{\partial}{\partial t} + \nu_0 k^2\right) u_\alpha^<(\mathbf{k},t) = {} & M_{\alpha\beta\gamma}^<(\mathbf{k}) \int u_\beta^<(\mathbf{j},t) u_\gamma^<(\mathbf{k}-\mathbf{j},t)\mathrm{d}^3 j \\ & + H_\alpha^<(\mathbf{k}|k_C),\end{aligned} \tag{185}$$

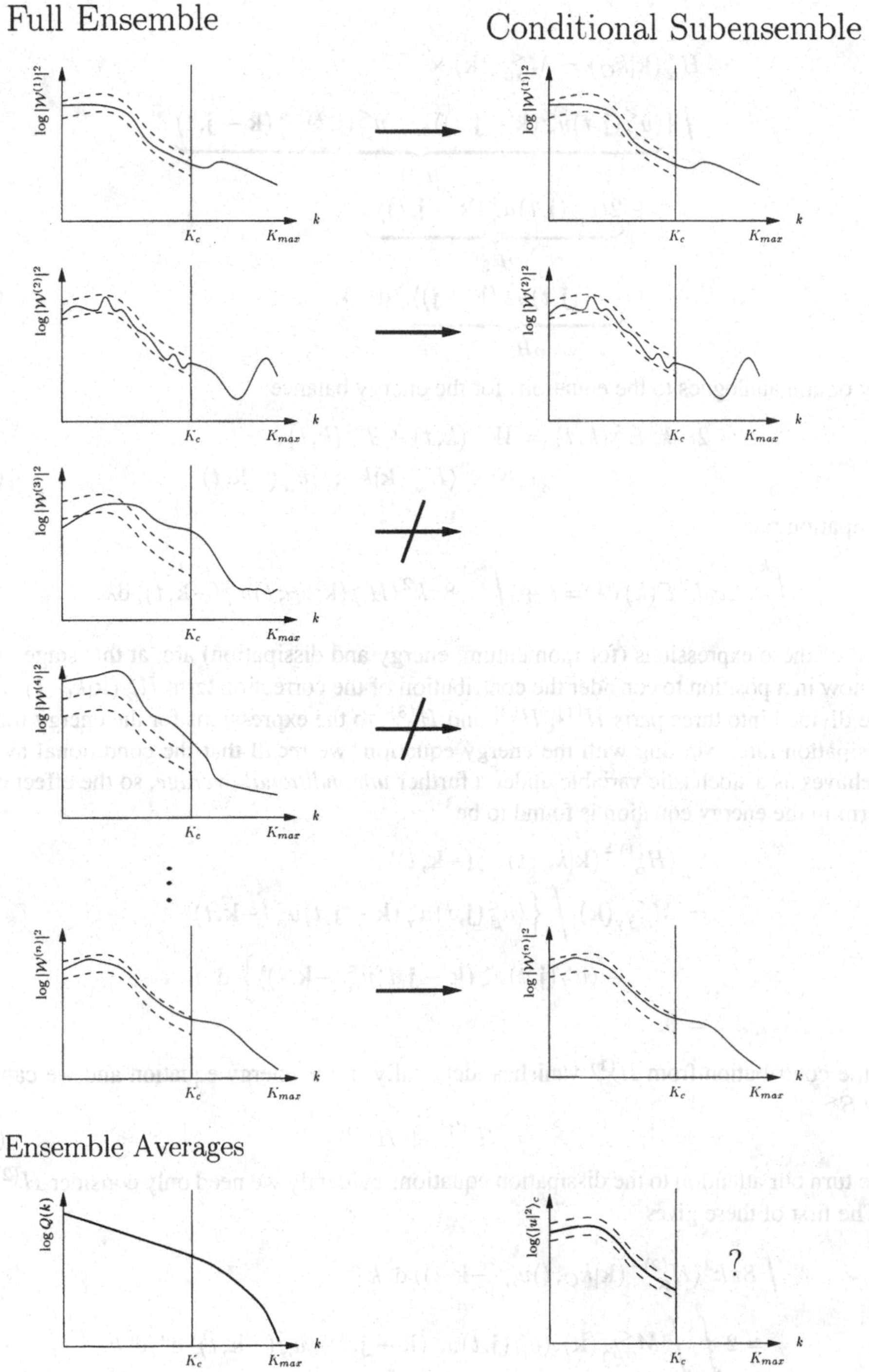

Figure 28: A conditionally-sampled ensemble.

where

$$H_\alpha^<(\mathbf{k}|k_C) = M_{\alpha\beta\gamma}^<(\mathbf{k}) \times$$
$$\int \{ \underbrace{\langle u_\beta^<(\mathbf{j},t)u_\gamma^<(\mathbf{k}-\mathbf{j},t)\rangle_c - u_\beta^<(\mathbf{j},t)u_\gamma^<(\mathbf{k}-\mathbf{j},t)}_{H^{(1)}}$$
$$+ \underbrace{2\langle u_\beta^<(\mathbf{j},t)u_\gamma^>(\mathbf{k}-\mathbf{j},t)\rangle_c}_{H^{(2)}}$$
$$+ \underbrace{\langle u_\beta^>(\mathbf{j},t)u_\gamma^>(\mathbf{k}-\mathbf{j})\rangle_c}_{H^{(3)}} \, \mathrm{d}^3 j\}. \tag{186}$$

We may obtain analogues to the equations for the energy balance:

$$2\nu_0 k^2 E^<(k,t) = W^<(k,t) + T^<(k,t) + 8\pi k^2 \langle H_\alpha^<(\mathbf{k}|k_C;t) u_\alpha^<(-\mathbf{k},t)\rangle, \tag{187}$$

and dissipation rate:

$$\int_0^{k_C} 2\nu_0 k^2 E(k)\,\mathrm{d}k = \varepsilon + \int_0^{k_C} 8\pi k^2 \langle H_\alpha^<(\mathbf{k}|k_C;t) u_\alpha^<(-\mathbf{k},t)\rangle\,\mathrm{d}k. \tag{188}$$

All three of these expressions (for momentum, energy and dissipation) are, at this stage, exact. We are now in a position to consider the contribution of the correction term $H_\alpha^<(\mathbf{k}|k_C;t)$, which we have divided into three parts $H^{(1)}$, $H^{(2)}$ and $H^{(3)}$, to the expressions for the energy transfer and dissipation rate. Starting with the energy equation, we recall that the conditional average itself behaves as a stochastic variable under a further *unconditional average*, so the effect of the $H^{(1)}$ term in the energy equation is found to be

$$\begin{aligned}&\langle H_\alpha^{(1)-}(\mathbf{k}|k_C;t) u_\alpha^<(-\mathbf{k},t)\rangle \\ &= M_{\alpha\beta\gamma}^<(\mathbf{k}) \int \Big\{ \langle u_\beta^<(\mathbf{j},t) u_\gamma^<(\mathbf{k}-\mathbf{j},t) u_\alpha^<(-\mathbf{k},t)\rangle \\ &\qquad - \langle u_\beta^<(\mathbf{j},t) u_\gamma^<(\mathbf{k}-\mathbf{j},t) u_\alpha^<(-\mathbf{k},t)\rangle \Big\}\,\mathrm{d}^3 j \\ &= 0.\end{aligned} \tag{189}$$

Hence the contribution from $H^{(1)}$ vanishes identically in the energy equation and we can now identify $\hat{S}^<$

$$\hat{S}^< = H^{(2)-} + H^{(3)-} \tag{190}$$

Next we turn our attention to the dissipation equation: evidently we need only consider $H^{(2)}$ and $H^{(3)}$. The first of these gives

$$\begin{aligned}&\int 8\pi k^2 \langle H_\alpha^{(2)-}(\mathbf{k}|k_C;t) u_\alpha^<(-\mathbf{k},t)\rangle\,\mathrm{d}^3 k \\ &= 2\int\int M_{\alpha\beta\gamma}^<(\mathbf{k}) \langle\langle u_\beta^<(\mathbf{j},t) u_\gamma^>(\mathbf{k}-\mathbf{j},t)\rangle_c u_\alpha^<(-\mathbf{k},t)\rangle\,\mathrm{d}^3 j\mathrm{d}^3 k \\ &= \int_0^{k_C}\int_0^{k_C} \theta^>(|\mathbf{k}-\mathbf{j}| - k_C) T(k,j,\mu)\mathrm{d}j\mathrm{d}k = 0,\end{aligned} \tag{191}$$

due to the antisymmetry of $T(k, j, \mu)$, under interchange of $\mathbf{k}$ and $\mathbf{j}$. Accordingly, we now identify $\tilde{S}^<$ as

$$\tilde{S}^< = H^{(3)-}. \tag{192}$$

It should however be noted that this result holds only because both wavenumbers are in the same interval. This is not the case regarding the contribution from $H^{(3)}$, which is

$$\begin{aligned}\int 8\pi k^2 \langle H_\alpha^{(3)-}(\mathbf{k}|k_C;t) u_\alpha^<(-\mathbf{k},t)\rangle \, \mathrm{d}^3 k &= \\ = 2\int\int M_{\alpha\beta\gamma}^<(\mathbf{k}) &\left\{\langle\langle u_\beta^>(\mathbf{j},t) u_\gamma^>(\mathbf{k}-\mathbf{j},t)\rangle_c u_\alpha^<(-\mathbf{k},t)\rangle\right\} \mathrm{d}^3 j \mathrm{d}^3 k \\ = 2\int\int M_{\alpha\beta\gamma}^<(\mathbf{k}) &\left\{\langle u_\beta^>(\mathbf{j},t) u_\gamma^>(\mathbf{k}-\mathbf{j},t) u_\alpha^<(-\mathbf{k},t)\rangle\right\} \mathrm{d}^3 j \mathrm{d}^3 k\end{aligned} \tag{193}$$

Again, this is just a special case of the spectral transfer term $T(k, j, \mu)$, for wavenumbers in the ranges $j, |\mathbf{k}-\mathbf{j}| \geq k_C$ and $k \leq k_C$. Hence we may write the analogue of the dissipation integral as

$$\begin{aligned}\int_0^{k_C} 2\nu_0 k^2 E(k)\, \mathrm{d}k = \varepsilon + \\ \int_0^{k_C}\int_{k_C}^{k_{max}} \theta^>(|\mathbf{k}-\mathbf{j}| - k_C) T(k, j, \mu)\, \mathrm{d}j \mathrm{d}k,\end{aligned} \tag{194}$$

for the truncated range of wavenumbers $0 \leq k \leq k_C$.

5.4 Quasi-Stochastic Estimate of the Renormalized Dissipation Rate

It is clear that the correction to the dissipation rate due to truncation of the wavenumber range is controlled by the subgrid stress involving $u^> u^>$, and in particular by its conditional average which is evaluated with the explicit-scales field $u^<$ held constant. Recently McComb and Johnston (1999, 2000) and Johnston and McComb (2000) have given a method for the approximate calculation of this conditional average, which we shall refer to as the *quasi-stochastic estimate* (or QSE) of the conditional average and denote by $\langle\cdot\rangle_{QSE}$. We introduce this by making the replacement:

$$\nu(k|k_C) k^2 u_\alpha^<(\mathbf{k},t) = M_{\alpha\beta\gamma}^<(\mathbf{k}) \int \langle u_\beta^>(\mathbf{j},t) u_\gamma^>(\mathbf{k}-\mathbf{j},t)\rangle_{QSE}\, \mathrm{d}^3 j. \tag{195}$$

We then rewrite the filtered NSE in terms of the QSE, by adding and subtracting quantities, in such a way as to leave it unaffected, to find

$$\begin{aligned}\left(\frac{\partial}{\partial t} + \nu(k|k_C)k^2\right) u_\alpha^<(\mathbf{k},t) \;=\; & M_{\alpha\beta\gamma}^<(\mathbf{k}) \int u_\beta^<(\mathbf{j},t) u_\gamma^<(\mathbf{k}-\mathbf{j},t)\, \mathrm{d}^3 j \\ & + S_\alpha^<(\mathbf{k}|k_C),\end{aligned} \tag{196}$$

where

$$S_\alpha^<(\mathbf{k}|k_C) = M_{\alpha\beta\gamma}^<(\mathbf{k}) \int \{ \underbrace{\langle u_\beta^<(\mathbf{j},t) u_\gamma^<(\mathbf{k}-\mathbf{j},t)\rangle_c - u_\beta^<(\mathbf{j},t) u_\gamma^<(\mathbf{k}-\mathbf{j},t)}_{S^{(1)}}$$

$$+ \underbrace{2\langle u_\beta^<(\mathbf{j},t) u_\gamma^>(\mathbf{k}-\mathbf{j},t)\rangle_c}_{S^{(2)}}$$

$$+ \underbrace{\langle u_\beta^>(\mathbf{j},t) u_\gamma^>(\mathbf{k}-\mathbf{j})\rangle_c - \langle u_\beta^>(\mathbf{j},t) u_\gamma^>(\mathbf{k}-\mathbf{j},t)\rangle_{QSE}}_{S^{(3)}} \} \, \mathrm{d}^3 j. \quad (197)$$

Note that we have labelled the correction term S in order to use the same notation as in the earlier analogous equation. Also note that $H^{(1)-} = S^{(1)-}$ and $H^{(2)-} = S^{(2)-}$. However, $S^{(3)-}$ contains the correction term involving the quasi-stochastic estimate:

$$\int 8\pi k^2 \langle S_\alpha^{(3)-}(\mathbf{k}|k_C;t) u_\alpha^<(-\mathbf{k},t)\rangle \, \mathrm{d}^3 k =$$

$$= 2 \int \int M_{\alpha\beta\gamma}^<(\mathbf{k}) \Big\{ \langle \langle u_\beta^>(\mathbf{j},t) u_\gamma^>(\mathbf{k}-\mathbf{j},t)\rangle_c u_\alpha^<(-\mathbf{k},t)\rangle -$$

$$- \langle \langle u_\beta^>(\mathbf{j},t) u_\gamma^>(\mathbf{k}-\mathbf{j},t)\rangle_{QSE} \, u_\alpha^<(-\mathbf{k},t)\rangle \Big\} \, \mathrm{d}^3 j \mathrm{d}^3 k$$

$$= 2 \int \int M_{\alpha\beta\gamma}^<(\mathbf{k}) \Big\{ \langle u_\beta^>(\mathbf{j},t) u_\gamma^>(\mathbf{k}-\mathbf{j},t) u_\alpha^<(-\mathbf{k},t)\rangle -$$

$$- \langle \langle u_\beta^>(\mathbf{j},t) u_\gamma^>(\mathbf{k}-\mathbf{j},t)\rangle_{QSE} \, u_\alpha^<(-\mathbf{k},t)\rangle \Big\} \, \mathrm{d}^3 j \mathrm{d}^3 k \quad (198)$$

as the correction to the dissipation rate. Note that in this case the two terms on the right hand side will cancel under any circumstances in which the QSE is a good approximation for the exact conditional average. For the moment we note that the renormalized dissipation rate now takes the form:

$$\int_0^{k_C} 2\nu(k|k_C) k^2 E^<(k) \, \mathrm{d}k = \varepsilon +$$

$$\int_0^{k_C} \int_{k_C}^{k_{max}} \theta^>(|\mathbf{k}-\mathbf{j}| - k_C) T(k,j,\mu) \mathrm{d}j \mathrm{d}k -$$

$$2 \int \int M_{\alpha\beta\gamma}^<(\mathbf{k}) \times$$

$$\langle \langle u_\beta^>(\mathbf{j},t) u_\gamma^>(\mathbf{k}-\mathbf{j},t)\rangle_{QSE} \, u_\alpha^<(-\mathbf{k},t)\rangle \, \mathrm{d}^3 j \mathrm{d}^3 k. \quad (199)$$

The appearance of the renormalized viscosity on the left hand side is balanced by the appearance of the term involving the QSE on the right hand side.

6 Application of Renormalization Methods to Real Flows.

The full calculation of the contribution by the conditionally-averaged stresses to the subgrid stress has been given by McComb and Johnston (2001). Here we give an outline of the method.

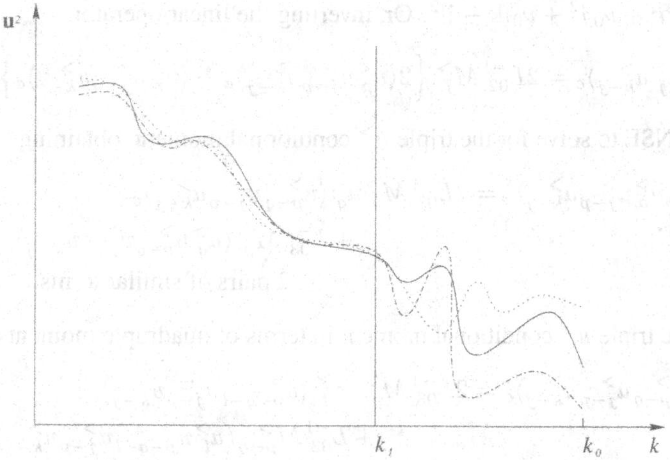

Figure 29: Fuzzy conditional average.

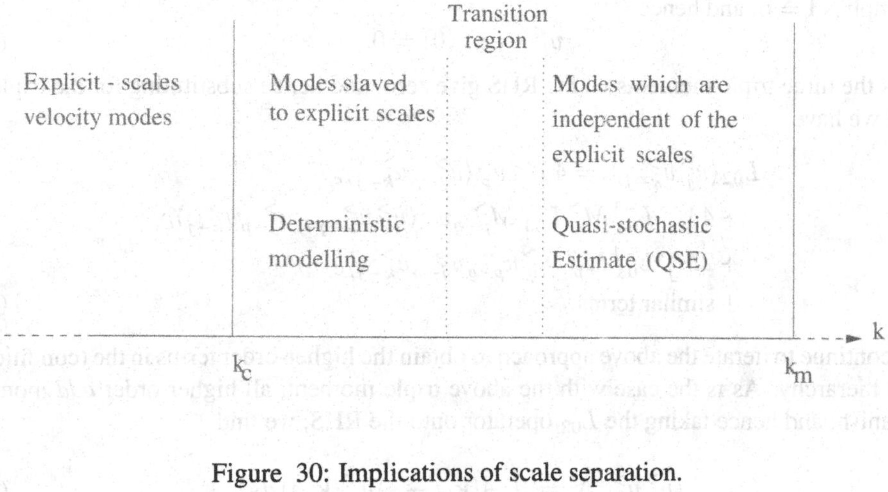

Figure 30: Implications of scale separation.

6.1 Iterative Evaluation of the Subgrid Stress

From now on we use an abbreviated notation with all indices and variables contracted into a single subscript, e.g. $u_\alpha(\mathbf{k},t) \rightarrow u_k$. We conditionally average the low-pass filtered NSE,

$$L_{01}u_k^< = M_k^< u_j^< u_{k-j}^< + M_k^< \langle u_j^> u_{k-j}^> \rangle_c, \tag{200}$$

for $j, k-j \sim k_m$ (asymptotically), where $L_{01} \equiv \partial/\partial t + \nu_0 k^2$. In practice, we evaluate the conditional average by using the high-pass filtered NSE to solve iteratively for $\langle u_j^> u_{k-j}^> \rangle_c$. We first obtain

$$L_{02}\langle u_j^> u_{k-j}^> \rangle_c = 2M_j^> \left\{ 2\langle u_p^< u_{j-p}^> u_{k-j}^> \rangle_c + \langle u_p^> u_{j-p}^> u_{k-j}^> \rangle_c \right\}, \tag{201}$$

where $L_{02} \equiv \partial/\partial t + \nu_0 j^2 + \nu_0|\mathbf{k}-\mathbf{j}|^2$. Or, inverting the linear operator,

$$\langle u_j^> u_{k-j}^> \rangle_c = 2L_{02}^{-1} M_j^> \left\{ 2\langle u_p^< u_{j-p}^> u_{k-j}^> \rangle_c + \langle u_p^> u_{j-p}^> u_{k-j}^> \rangle_c \right\}. \tag{202}$$

Next we use the NSE to solve for the triple $u^>$ conditional moment, obtaining

$$\begin{aligned} \langle u_p^> u_{j-p}^> u_{k-j}^> \rangle_c = {} & 2L_{03}^{-1} M_p^> u_q^< \langle u_{p-q}^> u_{j-p}^> u_{k-j}^> \rangle_c \\ & + L_{03}^{-1} M_p^> \langle u_q^> u_{p-q}^> u_{j-p}^> u_{k-j}^> \rangle_c \\ & + 2 \text{ pairs of similar terms.} \end{aligned} \tag{203}$$

Now solve for the triple $u^>$ conditional moment in terms of quadruple moments

$$\begin{aligned} \langle u_{p-q}^> u_{j-p}^> u_{k-j}^> \rangle_c = {} & 2L_{03}^{-1} M_{p-q}^> u_l^< \langle u_{p-q-l}^> u_{j-p}^> u_{k-j}^> \rangle_c \\ & + L_{03}^{-1} M_{p-q}^> \langle u_l^> u_{p-q-l}^> u_{j-p}^> u_{k-j}^> \rangle_c \\ & + 2 \text{ pairs of similar terms.} \end{aligned} \tag{204}$$

We can see that the moment on the LHS will be zero unless $\mathbf{k} = \mathbf{q}$, as will the quadruple moment on the RHS. However, the triple moment on the RHS will only be non zero if $\mathbf{k} - \mathbf{q} - \mathbf{l} = \mathbf{0}$, which implies $\mathbf{l} = \mathbf{0}$, and hence

$$u_l^< \to u^<(0) = 0. \tag{205}$$

Thus the three triple moments on the RHS give zero, and hence substituting for the triple $u^>$ moment we have

$$\begin{aligned} L_{02} \langle u_j^> u_{k-j}^> \rangle_c = {} & 4M_j^> u_p^< \langle u_{j-p}^> u_{k-j}^> \rangle_c \\ & + 4M_j^> L_{03}^{-1} M_p^> L_{03}^{-1} M_{p-q}^> u_q^< \langle u_l^> u_{p-q-l}^> u_{j-p}^> u_{k-j}^> \rangle_c \\ & + 2M_j^> L_{03}^{-1} M_p^> \langle u_q^> u_{p-q}^> u_{j-p}^> u_{k-j}^> \rangle_c \\ & + \text{similar terms.} \end{aligned} \tag{206}$$

We can continue to iterate the above approach to obtain the higher-order terms in the (conditional) moment hierarchy. As is the case with the above triple moment, all higher-order *odd* moments of $u^>$ vanish, and hence taking the L_{02} operator onto the RHS, we find

$$\langle u_j^> u_{k-j}^> \rangle_c = \int A(\mathbf{k}, t-s) u^<(\mathbf{k}, s)\, ds, \tag{207}$$

where $A(\mathbf{k}, t-s)$ is an expansion in even-order moments of $u^>$, the form of which may be inferred (for the first two orders) to be:

$$\begin{aligned} A(\mathbf{k}, t-s) = {} & 4L_{02}^{-1} M_j^> \langle u_{j-p}^> u_{k-j}^> \rangle_c \\ & + 4L_{02}^{-1} M_j^> L_{03}^{-1} M_p^> L_{03}^{-1} M_{p-q}^> \langle u_l^> u_{p-q-l}^> u_{j-p}^> u_{k-j}^> \rangle_c \\ & + \text{similar terms.} \end{aligned} \tag{208}$$

Substituting back, the filtered NSE may thus be re-written as

$$L_0 u_k^< = M_k^< u_j^< u_{k-j}^< + M_k^< \int A(\mathbf{k}, t-s) u^<(\mathbf{k}, s)\, ds. \tag{209}$$

6.1.1 Perturbative calculation of the conditional average

We introduce a perturbation expansion based upon the local Reynolds number at $k = k_c$. To do this, we first introduce the dimensionless variables ψ_α, $\mathbf{k}'$ and t' according to

$$u_\alpha(\mathbf{k}, t) = V(k_c)\psi_\alpha(\mathbf{k}', t'); \tag{210}$$

$$\mathbf{k}' = \mathbf{k}/k_c; \tag{211}$$

$$t' = t/\tau(k_c), \tag{212}$$

where $V(k_c)$ is the r.m.s. velocity at k_c and $\tau(k_c)$ is a representative timescale. Having defined these variables, we can then introduce the *local* Reynolds number

$$R(k_c) = \frac{k_c^2 V(k_c)}{\nu}, \tag{213}$$

a result which we can further show satisfies the condition:

$$R(k_c) \leq \left(\frac{\alpha}{2\pi}\right)^{1/2} \left(\frac{k_c}{k_d}\right)^{-4/3}, \tag{214}$$

where α is the Kolmogorov constant. For any realistic value of α this implies that $R(k_c)$ is less than unity for any k_c greater than $0.6k_d$

6.1.2 Truncation of the moment expansion

Formally we introduce a perturbation series in $\psi^>$

$$\psi^> = \psi_0^> + R(k_c)\psi_1^> + R(k_c)^2\psi_2^> + \ldots \tag{215}$$

We find, to $\mathcal{O}(R^4)$,

$$\begin{aligned} L_0'\psi_{k'}^< &= R(k_c)\psi_{j'}^<\psi_{k'-j'}^< \\ &+ 4R(k_c)^2 M_{k'}^< L_{02}'^{-1} M_{j'}^> \psi_{p'}^< \langle\psi_{j'-p'}^> \psi_{k'-j'}^>\rangle_c \\ &+ 4R(k_c)^4 M_{k'}^< L_{02}'^{-1} M_{j'}^> L_{03}'^{-1} M_{p'}^> L_{03}'^{-1} M_{p'-q'}^> \\ &\qquad \times \psi_{q'}^< \langle\psi_{l'}^> \psi_{p'-q'-l'}^> \psi_{j'-p'}^> \psi_{k'-j'}^>\rangle_c \\ &+ \text{similar terms.} \end{aligned} \tag{216}$$

To a first approximation, assuming that k_c is in the region where $R(k_c) < 1$, it is reasonable to neglect terms of order R^4 and greater as being negligible w.r.t. the order R^2 term. Accordingly, in order to simplify our calculation, we truncate at second order, thus:

$$\begin{aligned} L_0'\psi_{k'}^< &= R(k_c)\psi_{j'}^<\psi_{k'-j'}^< \\ &+ 4R(k_c)^2 M_{k'}^< L_{02}'^{-1} M_{j'}^> \psi_{p'}^< \langle\psi_{j'-p'}^> \psi_{k'-j'}^>\rangle_c \\ &+ \text{similar terms.} \end{aligned} \tag{217}$$

6.2 The RG Calculation of the Effective Viscosity

We rewrite the low-pass filtered NSE as

$$\left(\frac{\partial}{\partial t}+\nu_0 k^2\right)u_k^< = M_k^< u_j^< u_{k-j}^< + \frac{4M_k^< M_j^> \langle u_{j-p}^> u_{k-j}^> \rangle_c}{\nu_0 j^2 + +\nu_0|\mathbf{k}-\mathbf{j}|^2} u_p^< . \tag{218}$$

The conditional average may be performed, as a stochastic estimate using our hypothesis,

$$\lim_{\zeta\to 0}\langle u_{j-p}^> u_{k-j}^> \rangle_c = \lim_{|\mathbf{j}-\mathbf{p}|,|\mathbf{k}-\mathbf{j}|\to k_m} \langle u_{j-p}^> u_{k-j}^> \rangle + \text{higher order terms.} \tag{219}$$

Since the velocity field is homogeneous, isotropic and stationary, we have the result that

$$\langle u_{j-p}^> u_{k-j}^> \rangle = Q^>(|\mathbf{k}-\mathbf{j}|)D_{k-j}\delta(\mathbf{k}-\mathbf{p}). \tag{220}$$

In order to evaluate the limit involved in the conditional average we write $\langle u_{j-p}^> u_{k-j}^> \rangle_c$ in terms of a Taylor series expansion of $Q^>(|\mathbf{k}-\mathbf{j}|)$ about k_m

Introducing this step, the mode coupling term may be rewritten, also invoking the isotropy of the system, as

$$\frac{4M_k^< M_j^> \langle u_{j-p}^> u_{k-j}^> \rangle_c}{\nu_0 j^2 + \nu_0|\mathbf{k}-\mathbf{j}|^2} u_p^<$$

$$= \frac{L(\mathbf{k},\mathbf{j})\left(Q^>(l)\big|_{l=k_m} + (l-k_m)\left.\frac{\partial Q^>(l)}{\partial l}\right|_{l=k_m}\right)}{\nu_0 j^2 + \nu_0|\mathbf{k}-\mathbf{j}|^2} u_k^< , \tag{221}$$

where $l=|\mathbf{k}-\mathbf{j}|$, and $L(\mathbf{k},\mathbf{j}) = -2M_k^< M_j^> D_{k-j}$, as given by equation (138).

Thus we may rewrite the explicit scales-equation as

$$\left(\frac{\partial}{\partial t}+\nu_1(k) k^2\right)u_k^< = M_k^< u_j^< u_{k-j}^< , \tag{222}$$

where

$$\nu_1(k) = \nu_0 + \delta\nu_0(k), \tag{223}$$

and

$$\delta\nu_0(k) = \frac{1}{k^2}\int \frac{L(\mathbf{k},\mathbf{j})\left(Q^>(l)\big|_{l=k_m} + (l-k_m)\left.\frac{\partial Q^>(l)}{\partial l}\right|_{l=k_m}\right)}{\nu_0 j^2 + \nu_0|\mathbf{k}-\mathbf{j}|^2} \mathrm{d}^3 j. \tag{224}$$

6.2.1 Recursion relations for the effective viscosity

The above procedure may be extended to further shells as follows:

1. Set $u_k^< = u_k$ in the filtered equation, so that we have a new NSE with effective viscosity $\nu_1(k)$ defined on the interval $0 < k < k_1$

2. Decompose into $u^<$ and $u^>$ modes at wavenumber k_2, such that $u^>$ is now defined in the band $k_2 \leq k \leq k_1$

   Repeat the above procedure to eliminate modes in the band $k_2 \leq k \leq k_1$

Define the nth shell in this procedure by

$$k_n = (1-\eta)^n k_m, \qquad 0 \leq \eta \leq 1, \tag{225}$$

then by induction:

$$\nu_{n+1}(k) = \nu_n(k) + \delta\nu_n(k); \tag{226}$$

and, similarly,

$$\delta\nu_n(k) = \frac{1}{k^2}\int \frac{L(\mathbf{k},\mathbf{j})\left(Q^>(l)\big|_{l=k_n} + (l-k_n)\left.\frac{\partial Q^>(l)}{\partial l}\right|_{l=k_n}\right)}{\nu_n(j)j^2 + \nu_n(|\mathbf{k}-\mathbf{j}|)|\mathbf{k}-\mathbf{j}|^2}\,\mathrm{d}^3 j. \tag{227}$$

The fixed point of the RG calculation is indicated by the scaled forms of the above equations becoming invariant under successive iterations. To obtain the scaled forms of the equations, assume that the energy spectrum in the band is of the form $E(k) = \alpha\varepsilon^r k^s$ and introduce the scaling transformation $k = k_n k'$. If we then impose the consistency requirement that $\nu_n(k)$ and $\delta\nu_n(k)$ scale in the same manner, we obtain the scaled recursion relation

$$\tilde{\nu}_{n+1}(k') = h^{4/3}\tilde{\nu}_n(hk') + h^{-4/3}\delta\tilde{\nu}_n(k'), \tag{228}$$

where

$$\delta\tilde{\nu}_n(k') = \frac{1}{4\pi k'^2}\int \frac{L(\mathbf{k}',\mathbf{j}')Q'}{\tilde{\nu}_n(hj')j'^2 + \tilde{\nu}_n(hl')l'^2}\,\mathrm{d}^3 j', \tag{229}$$

for the wavenumber bands $0 \leq k' \leq 1$ and $1 \leq j', l' \leq h^{-1}$ and

$$Q' = h^{11/3} - \frac{11}{3}h^{14/3}(l' - h^{-1}) + \text{higher order terms}, \tag{230}$$

where h is defined by the relation $k_{n+1} = hk_n$, that is $h = 1-\eta$.

Results from iterating these equations are shown in Figures 31-35.

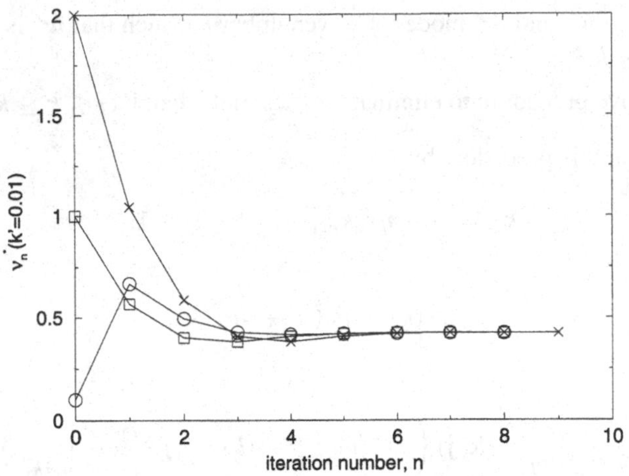

Figure 31: Variation of subgrid viscosity with iteration cycle.

6.3 Some results from large-eddy simulations based on the use of RG to calculate the subgrid viscosity

Results for LES of isotropic turbulence based on the effective viscosity calculated above are shown in Figures 36-39. For a full account see the paper by McComb et al. (2001).

6.4 Inhomogeneous and anisotropic flows

The key quantity in our work is the two-point (or spectral) correlation. In general, this will be required for turbulence fields which are both inhomogeneous and isotropic. That is,

$$\langle u_\alpha(\mathbf{x},t)u_\beta(\mathbf{x}',t')\rangle = Q_{\alpha\beta}(\mathbf{x},\mathbf{x}';t,t'), \tag{231}$$

where α and β take the values $1, 2$ and 3.

This correlation tensor is a formidable object: it is a 3×3 matrix of functions and each of these functions depends on six scalar spatial variables and two times. Compare this with the isotropic, homogeneous stationary case: there the correlation matrix is given by one function. The one correlation function depends on one scalar spatial variable and one time.

How does the general correlation tensor relate to the single-point correlation tensor? The isotropic, homogeneous correlation tensor?

The answers to these questions can be found by considering how we would measure the correlation tensor. Assume that we have two anemometers, situated respectively at $\mathbf{x}$ and $\mathbf{x}'$, and each capable of measuring velocity components in the directions x_1, x_2 and x_3.

Instead of varying $\mathbf{x}$ and $\mathbf{x}'$ independently, we can construct a traversing apparatus to do the following: Vary the distance between the anemometers: $\mathbf{r} = \mathbf{x} - \mathbf{x}'$ and vary their absolute position in space: $\mathbf{R} = (\mathbf{x} + \mathbf{x}')/2$.

In these relative ($\mathbf{r}$) and centroid ($\mathbf{R}$) coordinates, the correlation tensor may be written as

$$\langle u_\alpha(\mathbf{x},t)u_\beta(\mathbf{x}',t')\rangle = Q_{\alpha\beta}(\mathbf{r},\mathbf{R};t,t')\rangle. \tag{232}$$

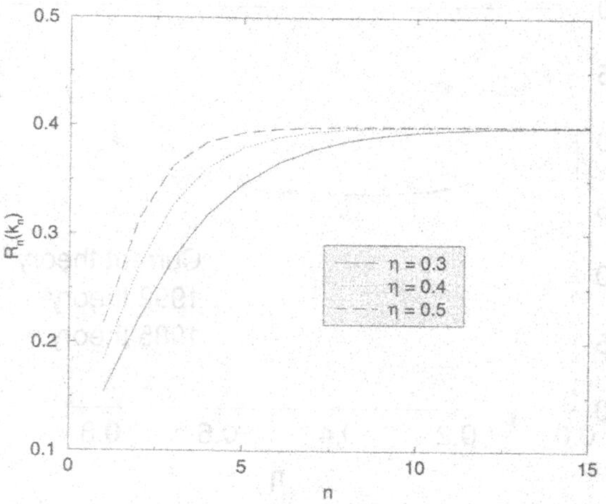

Figure 32: Evolution of local Reynolds number with iteration cycle.

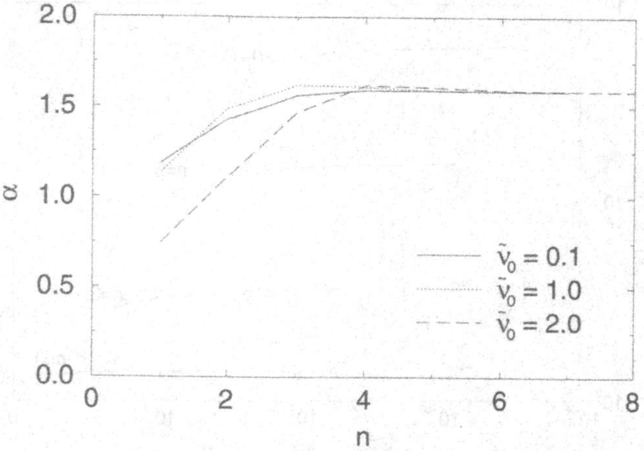

Figure 33: Kolmogorov prefactor reaching a fixed point for various initial viscosities.

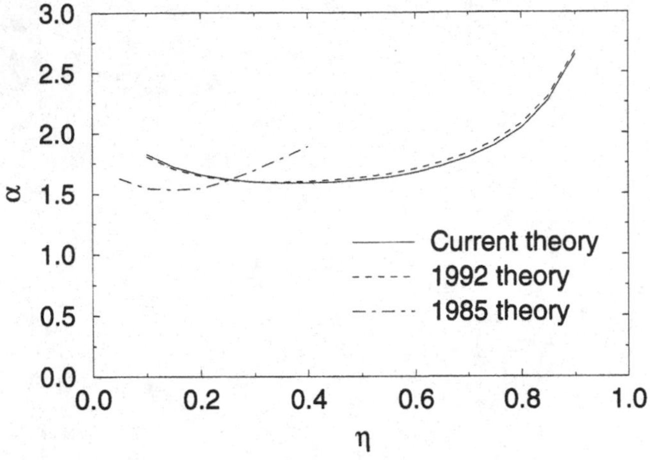

Figure 34: Variation of the Kolmogorov prefactor with bandwidth.

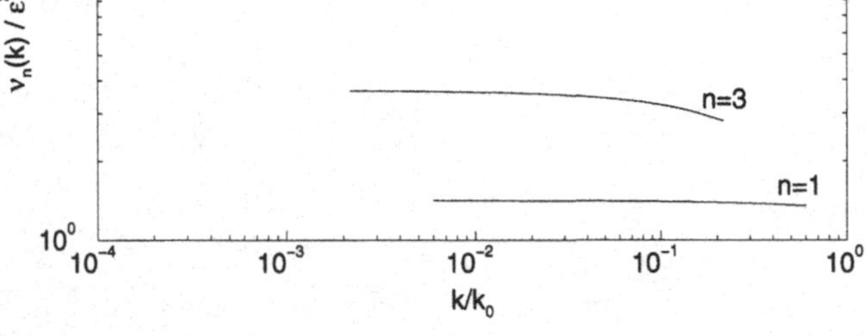

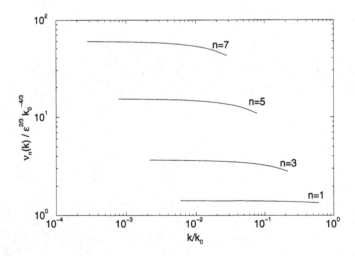

Figure 35: Evolution of unscaled viscosity with iteration cycle.

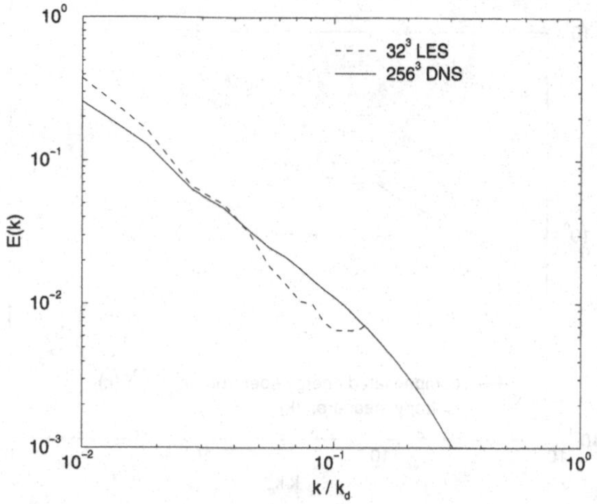

Figure 36: The time-averaged energy spectrum for a 32^3 LES with the two-field theory as subgrid model compared to that of a 256^3 DNS with the same input parameters

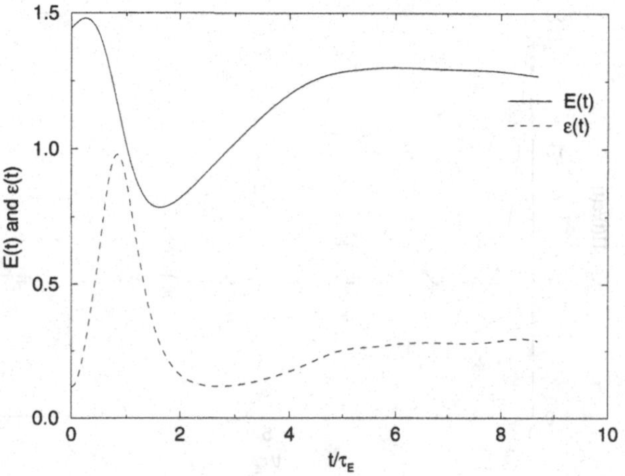

Figure 37: The evolution of total energy E and dissipation rate ε as a function of time (scaled on eddy turnover time) in a 128^3 LES using the two-field theory as a subgrid model

We could do the same transformation on the time variables but will not pursue that. To recover the single-point form, set $\mathbf{x} = \mathbf{x}'$ and

$$Q_{\alpha\beta}(\mathbf{r}, \mathbf{R}; t, t')|_{\mathbf{x}=\mathbf{x}'} = Q_{\alpha\beta}(\mathbf{x}; t, t'). \tag{233}$$

To relate to the spectral treatment, Fourier transform with respect to $\mathbf{r}$, thus

$$Q_{\alpha\beta}(\mathbf{r}, \mathbf{R}; t, t') \rightarrow Q_{\alpha\beta}(\mathbf{k}, \mathbf{R}; t, t'). \tag{234}$$

If the field is not too inhomogeneous, one could expand in powers of R about $R = 0$. If field is

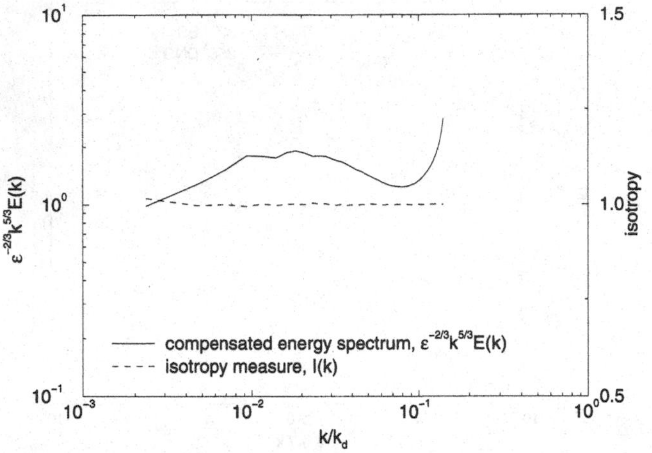

Figure 38: The time-averaged compensated energy spectrum plotted with a measure of the isotropy $I(k)$ from our 128^3 LES using the two-field subgrid viscosity

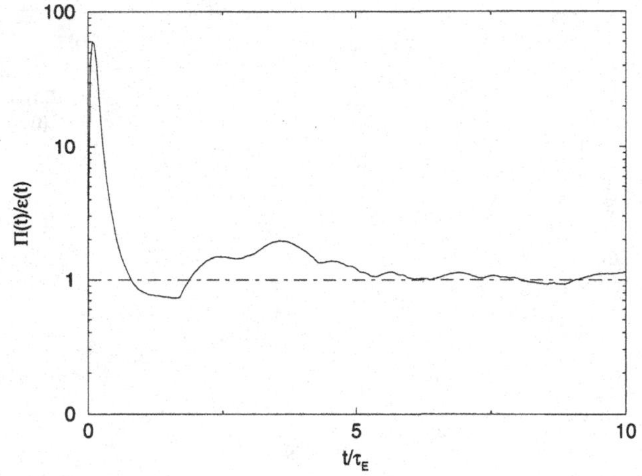

Figure 39: The maximum transport power $\Pi(t)$ scaled on the dissipation rate $\varepsilon(t)$ as a function of time scaled on eddy-turnover time. Results are from our 128^3 LES based on the two-field eddy viscosity

not too isotropic, one could expand in spherical harmonics about the isotropic case

$$Q_{\alpha\beta}(\mathbf{k}) = D_{\alpha\beta}(\mathbf{k})Q(k). \tag{235}$$

In real flows, there will be mean velocities and mean pressure gradients. The instantaneous velocity $U_\alpha(\mathbf{x},t)$ can be decomposed as

$$U_\alpha(\mathbf{x},t) = \overline{U}_\alpha(\mathbf{x},t) + u_\alpha(\mathbf{x},t). \tag{236}$$

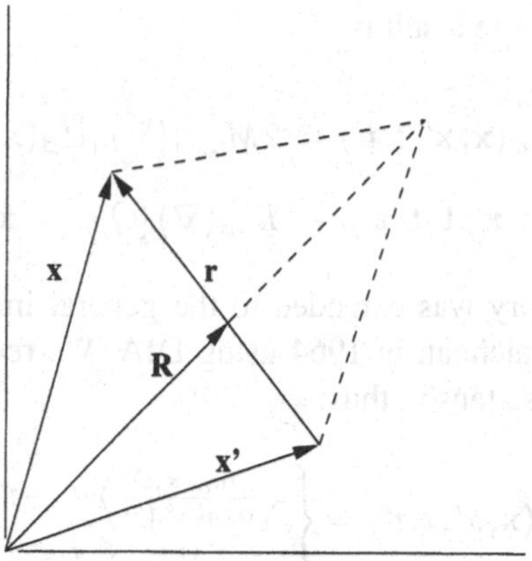

Figure 40: Relative and absolute coordinates.

Here $\overline{U}_\alpha(\mathbf{x},t) \equiv$ mean velocity and $u_\alpha(\mathbf{x},t) \equiv$ fluctuation in the velocity about the mean. Formally, pressure can be obtained in terms of the velocity field by solving the Poisson equation, which is obtained by taking the divergence of all terms in NSE and invoking continuity. Accordingly, we carry on the solenoidal formulation that we used in the spectral case. The NSE can be written as:

$$\begin{aligned}\frac{\partial U_\alpha}{\partial t} - \nu\nabla^2 U_\alpha &= M_{\alpha\beta\gamma}(\nabla)\left[U_\beta(\mathbf{x},t)U_\gamma(\mathbf{x},t)\right] \\ &\quad - L_{\alpha\beta}(\nabla)\left[U_\beta(\mathbf{x},t)\right] \\ &\quad - \frac{\partial P_{ext}}{\partial x_\alpha}. \qquad (237)\end{aligned}$$

Here $M_{\alpha\beta\gamma}(\nabla)$ is the Fourier transform of $M_{\alpha\beta\gamma}(\mathbf{k})$; $L_{\alpha\beta}(\nabla)$ is a surface term; and P_{ext} is an externally imposed pressure gradient.

Reynolds decomposition and averaging the NSE leads to the Reynolds equation for the mean velocity:

$$\begin{aligned}\left(\frac{\partial}{\partial t} - \nu\nabla^2\right)\overline{U}_\alpha(\mathbf{x},t) &= -\frac{1}{\rho}\frac{\partial P_{ext}}{\partial x_\alpha} + M_{\alpha\beta\gamma}(\nabla)\left[\overline{U}_\beta(\mathbf{x},t)\overline{U}_\gamma(\mathbf{x},t)\right. \\ &\quad \left.+ Q_{\beta\gamma}(\mathbf{x},\mathbf{x}';t,t)\right] - L_{\alpha\beta}(\nabla)\overline{U}_\beta(\mathbf{x},t). \qquad (238)\end{aligned}$$

Subtracting Reynolds equation from the NSE yields an equation for the fluctuating velocity:

$$\begin{aligned}\left(\frac{\partial}{\partial t} - \nu_0\nabla^2\right)u_\alpha(\mathbf{x},t) &= 2M_{\alpha\beta\gamma}(\nabla)\left[\overline{U}_\beta(\mathbf{x},t)u_\gamma(\mathbf{x},t)\right] \\ &\quad + M_{\alpha\beta\gamma}(\nabla)\left[u_\beta(\mathbf{x},t)u_\gamma(\mathbf{x},t) - Q_{\beta\gamma}(\mathbf{x},\mathbf{x},t,t)\right] - L_{\alpha\beta}(\nabla)\left[u_\beta(\mathbf{x},t)\right]. \qquad (239)\end{aligned}$$

To form an equation for the pair-correlation, multiply each term of the above equation by

$u_\sigma(\mathbf{x}', t')$ Average throughout. The result is:

$$\left(\frac{\partial}{\partial t} - \nu\nabla^2\right) Q_{\alpha\sigma}(\mathbf{x}, \mathbf{x}'; t, t') = 2M_{\alpha\beta\gamma}(\nabla)\left[\overline{U}_\beta(\mathbf{x}, t) Q_{\gamma\sigma}(\mathbf{x}, \mathbf{x}'; t, t')\right]$$
$$+M_{\alpha\beta\gamma}(\nabla)\left[Q_{\beta\gamma\sigma}(\mathbf{x}, \mathbf{x}, \mathbf{x}'; \mathbf{t}, \mathbf{t}, \mathbf{t}')\right] - L_{\alpha\beta}(\nabla)\left[Q_{\beta\sigma}(\mathbf{x}, \mathbf{x}'; t, t')\right]. \tag{240}$$

Renormalized perturbation theory was extended to the general inhomogeneous case by Allen in 1963 who used EFP and Kraichnan in 1964 using DIA. We require a generalization of the response function to the response tensor, thus:

$$G_{\alpha\beta}(\mathbf{x}, x'; t, t') = \begin{cases} \left\langle \frac{\partial u_\alpha(\mathbf{x},t)}{\partial f_\beta(\mathbf{x}',t')} \right\rangle & t > t' \\ 0 & t < t'. \end{cases} \tag{241}$$

Here $f_\beta(\mathbf{x}', t')$ is a force applied to the fluid at position $\mathbf{x}'$ and time t'. The DIA expression for the triple moment is given by

$$Q_{\beta\gamma\sigma}(\mathbf{x}, \mathbf{x}', \mathbf{x}''; t, t't'') = 2\int\int_0^t \mathrm{d}s G_{\beta\rho}(\mathbf{x}, \mathbf{y}; t, s) M_{\rho\delta\epsilon}(\nabla)$$
$$\times Q_{\gamma\sigma}(\mathbf{x}', \mathbf{y}; t', s) Q_{\sigma\epsilon}(\mathbf{x}'', \mathbf{y}; t's)\mathrm{d}^3y + \text{two similar terms}. \tag{242}$$

These equations pose a very large computational problem for even the simplest flows, but one can think of possible strategies:

1. Consider flows with weak dependence on the spatial coordinate and use a low-order truncation in Taylor series.
2. Seek scaling transformations to reduce the dimensionality (e.g. Leslie, 1973, Oberlack et al., 2001).
3. Apply to only part of a flow and use some other model or simulation technique for the rest of the flow.
4. Use as a starting point for the derivation of single-point model equations.

These approaches, although by no means new ideas, have scarcely been tested. We shall return to this point presently, but for the moment we note that even the simpler less comprehensive RG method still poses a formidable challenge when we seek to apply it to real flows.

Generalizing conditional mode elimination (RG) to the inhomogeneous case, the increment to viscosity becomes

$$\delta\nu_n(k) = \frac{1}{k^2}\int \frac{L(\mathbf{k}, \mathbf{j}, \nabla)\left[Q^>(l, \mathbf{R})|_{l=k_n} + (l - k_n)\partial Q^>(l, \mathbf{R})/\partial l|_{l=k_n}\right.}{\nu_n(j, \mathbf{R}) j^2 + \nu_n(|\mathbf{k} - \mathbf{j}|, \mathbf{R})|k - j|^2} \mathrm{d}^3 j \tag{243}$$

There would have to be a different $\delta\nu_n$ for each coordinate direction (that is, the effective viscosity becomes a tensor), and after that one would adopt exactly the same strategies as for the two-point closures, viz; weak-dependence, mixed-modelling, derivation of single-point models.

6.5 Discussion

We conclude by noting that renormalized perturbation theories of turbulence are approaching their half-century but, despite their obvious successes, and the enormous growth in the volume of turbulence research, they receive little attention. It is difficult to see where any improvement in their fundamental status can come, as the general operation of truncating a renormalized perturbation series has already been very successful in producing theories which perform so well on the limited tests which have been carried out so far. We are inclined to argue that what is needed is much more attention to be given to applying them to problems. Turbulence research is becoming more eclectic. For instance, the study of shell models is an attempt to emulate the study of the Ising model which dominates much of condensed matter and particle theory nowadays. With this thought in mind, we would like to make the radical proposal that the renormalized perturbation theories provide a natural playground for the turbulence theorist and that what is needed is an increasing realization that hard fundamental problems will only be solved if they are studied for their own sake.

Obviously there are some existing lines of research in this area. For instance, we note in particular the application of models of the EDQNM-type to inhomogeneous turbulence (e.g. Bertoglio and Jeandel (1987), Burden (1991), Cambon and Scott (1999) and Godeferd et al. (2001) for reviews). But, when compared to other activities in turbulence research this is very much a minor activity, and the situation for RPTs is worse as they seem to have been almost totally ignored. The only sustained investigation known to us is Frederiksen's application of DIA, LET and SCF to two-dimensional turbulence (Frederiksen et al., 1994, 2000). Of course it is natural that the quasi-Markovian models should appeal to researchers as they possess some of the essential features of the RPTs with relative computational simplicity and the additional benefit that their errors can be controlled (at least to some extent) by adjusting a constant. Nevertheless, if the subject is to make progress on the fundamental issues of closure, it will be necessary to study the RPTs in more detail and in a wider range of applications than hitherto.

References

Allen, J. (1963). *Formulation of the theory of turbulent shear flow*. Ph.D. Dissertation, Victoria University of Manchester.

Balescu, R., and Senatorski, A. (1970). A new approach to the theory of fully developed turbulence. *Ann.Phys(NY)* 58:587.

Bertoglio, J.-P., and Jeandel, D. (1987). A simplified spectral closure for inhomogeneous turbulence applicable to the boundary layer. In Durst, F., ed., *Turbulent Shear Flows 5*, 323. Berlin: Springer-Verlag.

Burden, A. D. (1991). Towards an EDQNM-Closure for inhomogeneous turbulence. In Johansson, A. V., and Alfredsson, P. H., eds., *Advances in Turbulence 3*, 387. Berlin: Springer-Verlag.

Cambon, C., and Scott, J. F. (1999). Linear and nonlinear models of anisotropic turbulence. *Ann. Rev. Fluid Mech.* 31:1.

Edwards, S., and McComb, W. (1969). Statistical mechanics far from equilibrium. *J.Phys.A* 2:157.

Edwards, S. (1964). The statistical dynamics of homogeneous turbulence. *J. Fluid Mech.* 18:239.

Eyink, G. (1994). The renormalization group method in statistical hydrodynamics. *Phys. Fluids* 6(9):3063–3078.

Forster, D., Nelson, D., and Stephen, M. (1976). Long-time tails and the large-eddy behaviour of a randomly stirred fluid. *Phys. Rev. Lett.* 36(15):867–869.

Frederiksen, J. S., and Davies, A. G. (2000). Dynamics and spectra of cumulant update closures for two-dimensional turbulence. *Geophys. Astrophys. Fluid Dynamics* 92:197.

Frederiksen, J., Davies, A., and Bell, R. (1994). Closure theories with non-gaussian restarts for truncated two dimensional turbulence. *Phys. Fluids* 6(9):3153.

Godeferd, F. S., Cambon, C., and Scott, J. F. (2001). Two-point closures and their applications: report on workshop. *J. Fluid Mech.* 436:393.

Herring, J., and Kraichnan, R. (1972). *Comparison of some approximations for isotropic turbulence Lecture Notes in Physics*, volume 12. Springer, Berlin. chapter Statistical Models and Turbulence, 148.

Herring, J. (1965). Self-consistent field approach to turbulence theory. *Phys. Fluids* 8:2219.

Horner, H., and Lipowsky, R. (1979). On the Theory of Turbulence: A non Eulerian Renormalized Expansion. *Z.Phys.B* 33:223.

Johnston, C., and McComb, W. (2000). Renormalized expression for the turbulent energy dissipation rate. *Phys. Rev E* 63:015304.

Kaneda, Y. (1981). Renormalized expansions in the theory of turbulence with the use of the Lagrangian position function. *J. Fluid Mech.* 107:131–145.

Kida, S., and Gotoh. (1997). A Lagrangian direct-interaction approximation for homogeneous isotropic turbulence. *J. Fluid Mech.* 345:307–345.

Kraichnan, R., and Herring, J. (1978). A strain-based Lagrangian-history turbulence theory. *J. Fluid Mech.* 88:355.

Kraichnan, R. H. (1959). The structure of isotropic turbulence at very high Reynolds numbers. *J. Fluid Mech.* 5:497–543.

Kraichnan, R. H. (1964a). Decay of isotropic turbulence in the Direct-Interaction Approximation. *Phys. Fluids* 7(7):1030–1048.

Kraichnan, R. (1964b). Direct-interaction approximation for shear and thermally driven turbulence. *Phys. Fluids* 7(7):1048.

Kraichnan, R. H. (1965). Lagrangian-history closure approximation for turbulence. *Phys. Fluids* 8(4):575–598.

Leslie, D. (1973). *Developments in the theory of modern turbulence*. Clarendon Press, Oxford.

Machiels, L. (1997). Predictability of small-scale motion in isotropic fluid turbulence. *Phys. Rev. Lett.* 79(18):3411–3414.

McComb, W. D., and Johnston, C. (1999). Conditional mode elimination with asymptotic freedom for isotropic turbulence at large Reynolds numbers. In Vassilicos, J., ed., *Turbulence Structure and Vortex Dynamics, Proceedings of the Symposium at the Isaac Newton Institute Cambridge*. CUP.

McComb, W. D., and Johnston, C. (2000). Elimination of turbulence modes using a conditional average with asymptotic freedom. *J. Phys. A:Math. Gen.* 33:L15–L20.

McComb, W. D., and Johnston, C. (2001). Conditional mode elimination and scale-invariant dissipation in isotropic turbulence. *Physica A* 292:346.

McComb, W. D., and Shanmugasundaram, V. (1984). Numerical calculations of decaying isotropic turbulence using the LET theory. *J. Fluid Mech.* 143:95–123.

McComb, W., Yang, T.-J., Young, A., and Machiels, L. (1997). Investigation of renormalization group methods for the numerical simulation of isotropic turbulence. In *Proc. Eleventh Symposium on Turbulent Shear Flows*.

McComb, W. D., Hunter, A., and Johnston, C. (2001). Conditional mode-elimination and the subgrid-modelling problem for isotropic turbulence. *Physics of Fluids* 13:2030.

McComb, W. (1978). A theory of time dependent, isotropic turbulence. *J.Phys.A:Math.Gen.* 11(3):613.

McComb, W. D. (1982). Reformulation of the statistical equations for turbulent shear flow. *Phys. Rev. A* 26(2):1078–1094.

McComb, W. (1990). *The Physics of Fluid Turbulence*. Oxford University Press.

McComb, W. (1995). Theory of turbulence. *Rep. Prog. Phys.* 58:1117–1206.

Nakano, T. (1988). Direct interaction approximation of turbulence in the wave packet representation. *Phys. Fluids* 31:1420.

Oberlack, M., McComb, W., and Quinn, A. (2001). Solution of functional equations and reduction of dimension in the local energy transfer theory of incompressible, three-dimensional turbulence. *Phys. Rev. E* 63:026308–1.

Phythian, R. (1969). Self-consistent perturbation series for stationary homogeneous turbulence. *J.Phys.A* 2:181.

Qian, J. (1983). Variational approach to the closure problem of turbulence theory. *Phys. Fluids* 26:2098.

Quinn, A. P. (2000). *Local Energy Transfer theory in forced and decaying isotropic turbulence*. Ph.D. Dissertation, University of Edinburgh.

Reichl, L. (1980). *A modern course in statistical physics*. Edward Arnold, London.

Rose, H. (1977). Eddy diffusivity, eddy noise and subgrid-scale modelling. *J. Fluid Mech.* 81(4):719–734.

Sreenivasan, K. R. (1995). On the universality of the Kolmogorov constant. *Phys. Fluids* 7:2778.

van Dyke, M. (1975). *Perturbation Methods in Fluid Mechanics*. Stanford, CA: Parabolic Press.

Wilson, K. G. (1983). The renormalization group and critical phenomena. *Rev. Mod. Phys.* 55(3):583–600.

Yakhot, V., and Orszag, S. (1986). Renormalization Group analysis of turbulence. I. Basic theory. *J. Sci. Comp.* 1(1):3–51.

Yeung, P. K., and Zhou, Y. (1997). Universality of the Kolmogorov constant in numerical simulations of turbulence. *Phys. Rev. E* 56:1746.

From Rapid Distortion Theory to Statistical Closure Theories of Anisotropic Turbulence

Claude Cambon

Laboratoire de Mécanique des Fluides et d'Acoustique U.M.R.
CNRS 5509, Ecole Centrale de Lyon, 69131 Ecully Cedex, France

Abstract

An overview of non-local theories and models is given, ranging from linear to nonlinear. The background principles are presented and illustrated mainly for incompressible, homogeneous, anisotropic turbulence. In that case, *which includes effects of mean gradients and body forces and related structuring effects*, the complete rapid distortion theory (RDT) solution is shown to be a building block for constructing a full nonlinear closure theory. Firstly, a general overview of the closure problem is presented, which accounts for both the nonlinear problem and the non-local problem. A classical spectral description is introduced for the fluctuating flow and its multi-point correlations. Applications to stably-stratified and rotating turbulence are discussed. In this particular context, homogeneous turbulence is revisited in the presence of dispersive waves, taking advantage of the close relationship between recent theories of weakly nonlinear interactions, or 'wave-turbulence' , and classical two-point closure theories. Applications to weak turbulence in compressible flows are touched upon. Extensions of the frontiers of rapid distortion theory (RDT) and multi-point closures are discussed, especially developments leading towards inhomogeneous turbulence. Recent works related to zonal RDT and stability analyses for wavepacket disturbances to non-parallel rotational base flows are presented.

1 Introduction

Two-point statistical closures (Direct Interaction Approximation, DIA , Eddy Damped Quasi Normal Markovian, EDQNM , and the Test Field Model, TFM) were initially mainly developed for the special case of homogeneous, isotropic turbulence during the ground-breaking studies of the 60's and 70's (see *e.g.* Kraichnan, 1959, Orszag, 1970, Monin and Yaglom, 1975, among many others), but have since then been extended to some *anisotropic* and even *inhomogeneous* flows, areas in which work continues today.

Although such models are aimed at strongly nonlinear turbulence, their mathematical structure is closely related to that of weakly nonlinear theories (see Cambon and Scott, 1999, and references therein). For example, the theory of weak turbulence (see Benney and Saffman, 1966), which has recently seen considerable interest in the geophysical context, presents strong similarities with two-point closures, even if very few studies illustrating the connections between the two approaches have appeared to date. Thus, the case of anisotropic, incompressible, homogeneous turbulence subject to different anisotropizing influences, such as rotation or stratification, and of weakly compressible turbulence, even in the isotropic case, present challenges which are currently being addressed using both two-point techniques and asymptotic theories of weak turbulence. It is important to extend the domain of applicability of two-point closures by incorporating results from linear theory (RDT, using methods from stability theory) and weakly nonlinear analyses, results which include at least some aspects of the *real dynamics* of the flow. Other methods, such as renormalisation or homogenisation, may also help in developing two-point closures. Two-point models are intrinsically more realistic than one-point models, describing more of the physics of turbulence, such as the continuum of different scales, and providing a correct treatment of pressure fluctuations (*via* the formalism of projection onto solenoidal modes in the incompressible case). The fact that two-point closures can be used to describe different turbulence scales has proved, and will no doubt continue to prove, useful in the construction of subgrid models in LES, but two-point modelling is by no means limited to this single application, important though it may be.

Throughout, our aim is to illustrate the importance of linear mechanisms, mostly in the form of mean gradients and body forces, which render the turbulence anisotropic. For this reason, we restrict attention to models and theories capable of handling anisotropy. This chapter is organised as follows. The general problem of closure is introduced below, with emphasis on both non-local and nonlinear aspects. Section 2 gives the background for linear theory, with Rapid Distortion Theory and a recent zonal stability analysis. The statistical formalism for homogeneous anisotropic turbulence is given in section 3, and generic nonlinear, two-point, closures, are introduced in section 4. Typical applications, such as rotation and stable stratification are presented in section 5 and 6, respectively. Effects of compressibility are touched upon in section 7, and inhomogeneous flows are addressed in section 8. Concluding comments are given in section 9.

1.1 Background Equations

The velocity and pressure fields are first split into mean and fluctuating components and equations for their time evolution are derived from the basic equations of motion of the fluid. Assuming incompressibility, as we shall do in this chapter unless explicitly stated otherwise, this gives the mean flow equations

$$\frac{\partial \overline{u}_i}{\partial t} + \overline{u}_j \frac{\partial \overline{u}_i}{\partial x_j} = -\frac{\partial \overline{p}}{\partial x_i} + \nu \frac{\partial^2 \overline{u}_i}{\partial x_j \partial x_j} - \underbrace{\frac{\partial \overline{u'_i u'_j}}{\partial x_j}}_{\text{Reynolds stress term}} \tag{1}$$

$$\frac{\partial \overline{u}_i}{\partial x_i} = 0 \tag{2}$$

and the equations for the fluctuating component

$$\frac{\partial u'_i}{\partial t} + \overline{u}_j \frac{\partial u'_i}{\partial x_j} + u'_j \frac{\partial \overline{u}_i}{\partial x_j} + \underbrace{\frac{\partial}{\partial x_j}(u'_i u'_j - \overline{u'_i u'_j})}_{\text{Nonlinear term}} = \underbrace{-\frac{\partial p'}{\partial x_i}}_{\text{Pressure term}} + \underbrace{\nu \frac{\partial^2 u'_i}{\partial x_j \partial x_j}}_{\text{Viscous term}} \tag{3}$$

and

$$\frac{\partial u'_i}{\partial x_i} = 0 \tag{4}$$

Here, $\overline{u}_i$ and $\overline{p}$ are the mean velocity and 'pressure' (pressure divided by density), while u_i' and p' are the corresponding fluctuating quantities, usually interpreted as representing turbulence.

At various points, we will describe related work in the area of hydrodynamic stability. In so doing, it is recognised that equations (3) and (4) for the fluctuating flow are essentially the same as those for a perturbation u_i', about a basic flow, $\overline{u}_i$, with an additional forcing term, $\partial\overline{u_i'u_j'}/\partial x_j$, in the inhomogeneous case. Although the aims of stability theory (to characterise growth of the perturbation) and of the theory of turbulence (to determine the statistics of u_i') are different, we believe it is nonetheless valuable to draw parallels between the two fields of study. It is our hope that in so doing we will encourage specialists in both areas to become more conversant with each others work.

Equation (3) is now used to derive equations for the time evolution of velocity moments, i.e. averages of products of u_i' with itself at one or more points in space. Setting up the equations for the n'th order velocity moments at n points, one discovers that there are two main difficulties. Firstly, the term in (3) which is nonlinear in the fluctuations leads to the appearance of $(n+1)$'th order moments in the evolution equation at n'th order. Secondly, the pressure term introduces pressure-velocity moments.

The pressure field is intimately connected with the incompressibility condition. Indeed, taking the divergence of (3) leads to a Poisson equation

$$\nabla^2 p' = -\frac{\partial^2}{\partial x_i \partial x_j}(u_i'\overline{u}_j + \overline{u}_i u_j' + u_i'u_j' - \overline{u_i'u_j'}) \tag{5}$$

for the 'pressure' fluctuations. Solution of this equation by Green's functions expresses p' at any point in space in terms of an integral of the velocity field over the entire volume of the flow, together with integrals over the boundaries, the details of whose expression in terms of velocity do not concern us here. Thus, the pressure at a given point is non-locally determined by the velocity field at all points of the flow, resulting in the equations for the velocity moments being integro-differential when the pressure-velocity moments are expressed in terms of velocity alone. It should be observed that non-locality is not specific to the use of statistical methods, but is intrinsic to the physics of incompressible fluids, for which the pressure field responds instantaneously and non-locally to changes in the flow to maintain incompressibility. The source term in the Poisson equation (5) consists of parts which are linear and nonlinear in the velocity fluctuation, feeding through into corresponding components of p and hence of the pressure-velocity terms in the evolution equation for the n-point velocity moments.

1.2 The Closure Problem: Non-Linearity and Non-Locality

Both the nonlinear pressure component and the nonlinear term appearing directly in (3) contribute to the closure problem, namely that the equation for the n'th order velocity moments involves $(n+1)$'th order moments. In consequence, no finite subset of the infinite hierarchy of integro-differential equations describing the velocity moments at all orders is complete, reflecting the fundamental difficulty of the turbulence problem, viewed through the classical statistical description in terms of moments. The origin of the closure problem is the non-linearity of the Navier-Stokes equations, which feeds through into the moment equations, both directly and via the nonlinear part of the pressure fluctuations. Non-locality , of itself, does not lead

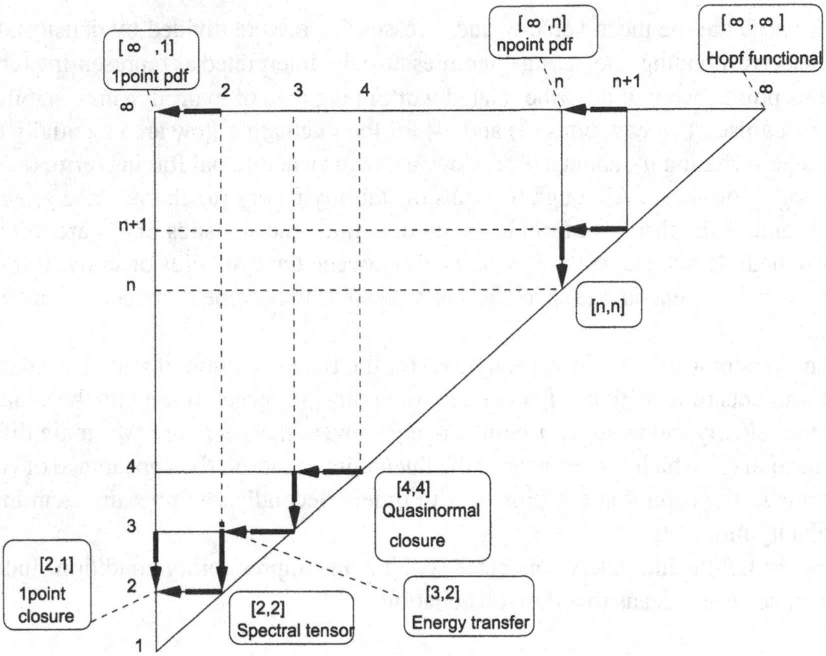

Figure 1: Synthetic scheme for statistical closures

to problems, although the technical difficulties associated with integro-differential, rather than differential equations, are nontrivial.[1]

The non-local problem of closure is removed from consideration only in models for *multi-point* statistical correlations, e.g. double correlations at two points or triple correlations at three points, so that in such models the problem of closure is determined by non-linearity alone.

On the other hand, the knowledge of a probability density function (pdf) for the velocity is equivalent to the knowledge of all the statistical moments up to any order. Hence the problem of the open hierarchy of the moment-equations, mentioned above, is avoided in a pdf approach. Accordingly, the problem of closure induced by the non-linearity is removed from consideration using a pdf approach, but the non-local problem of closure remains, so that the equations for a local velocity pdf involve a two-point velocity pdf, and equations for a n-point velocity pdf involve a $(n+1)$-point velocity pdf (Lundgren, 1967). Thus, an open hierarchy of equations is recovered using a pdf approach but with respect to a multi-point spatial description!

In order to present all the consequences of the above discussion, a synthetic scheme using a triangle is shown in Figure 1 and discussed as follows. The vertical axis bears the ordering of the statistical moments, from 1 (the mean velocity), 2 (second order moments), until an arbitrary high order. For convenience, the moments of order up to 2 are centered, so that they only involve the fluctuating velocity field. Each vertical order n corresponds to a number of different points for a possible multi-point description of the n-order moment under consideration along the horizontal

[1]Non-locality ought to be only understood in *physical space* here. Of course, the operators linked to pressure and dissipation will appear as local in Fourier space, but this is only mathematical convenience, as we will see in the next sections. Discussing the possible degree of locality of nonlinear interactions in Fourier space is outside our scope, too.

axis, from 1 (single-point), 2 (two-point), until n. In other words, the vertical axis can display the open hierarchy due to non-linearity, whereas the horizontal one deals with the non-locality. Each point of the triangle can characterize a level of description, for instance the point $[3,2]$ represents triple correlations at two points (those that drive the spectral energy transfer and the energy cascade). In addition, the problem of closure can be stated by looking at the adjacent points (if any) just above and just to the left. For instance, The main problem which concerns engineering, when solving Reynolds-averaged Navier Stokes equations, is expressing the flux of the Reynolds stress tensor: this can be expressed by an arrow from $[2,1]$ to $[1,1]$. Then, the equation that governs the Reynolds-stress tensor $[2,1]$ needs extra information (not given by $[2,1]$ itself, hence the closure problem) on second-order two-point terms $[2,2]$ (involved in the 'rapid' pressure-strain rate term and the dissipation term), triple-order one- $[3,1]$ and two-point $[3,2]$ terms (involved in the 'slow' pressure-strain rate and diffusion terms). Of course, the Reynolds stress tensor $[2,1]$ is directly derived from second-order correlations at two points $[2,2]$, illustrating a simple rule of 'concentration of the information' from right to left. The non-local problem of closure, due to pressure and dissipation terms, is removed from consideration, leaving only the hierarchy due to non-linearity, when looking only at $[n,n]$ correlations (located on the hypotenuse of the triangle in Figure 1): the equation that governs $[2,2]$ needs only extra information on $[3,2]$; the equation that governs $[3,3]$ needs only extra-information on $[4,3]$; the latter two examples, which are directly involved in classical two-point closures, will be discussed in section 4. The arrow from $[n+1,n]$ to $[n,n]$ gives an obvious generalization of the optimal way to use multipoint closures, and illustrates the open hierarchy of equations *due to the non-linearity only*. Often the closure relationship holds at the level $[n+1,n+1]$, from which is readily derived the level $[n+1,n]$. For instance, the quasi-normal assumption, which is involved in all 'multi-point' closures, as well as in wave-turbulence theories, calls into play the $[4,4]$ level.

Regarding the pdf approach, we are concerned with the upper horizontal side of the triangle. It seems to be consistent to relate to the point $[\infty,1]$ a description in terms of a local velocity pdf —knowledge of which is equivalent to the knowledge of all the one-point moments (the complete vertical line below). Accordingly, the arrow from $[\infty,2]$ to $[\infty,1]$ shows the need for extra-information on the two-point pdf in the equations that govern local pdf. In the same way, the arrow from $[\infty,n+1]$ to $[\infty,n]$ shows the link of n- to $(n+1)$-point pdf (Lundgren, 1967), and illustrates the open hierarchy of equations due to non-locality only.

The last limit concerns the ultimate point $[\infty,\infty]$. It is consistent to consider that the limit of a joint-pdf of velocity values at an infinite number of points is equivalent to the functional pdf description of Hopf (1952). In this case we reach the top right point of the triangle and there is no need for any extra-information; accordingly the Hopf-equation is closed, and it is possible to derive from it any multi-point pdf or statistical moment. It is interesting to point out that the bottom left point $[1,1]$ gives the most crude information about the velocity field – its mean value – whereas the opposite point $[\infty,\infty]$ gives the most sophisticated.

As a last general comment, our synoptic scheme clearly shows that the problem of closure, which reflects a loss of information at a given level of statistical description, can be removed from consideration, at least partially, if additional degrees of freedom are introduced in order to enlarge the configuration-space. For instance, to introduce as a new dependent variable the vector which joins the two points in a two-point second -order description allows removal of the problem of closure due to non-locality, which is present using a single-point second-order description. The introduction, as a new dependent variable, of the test-value Υ_i of the random velocity field u_i' in a pdf approach

$$P(\Upsilon_i, \mathbf{x}, t) = \overline{\delta(u_i'(\mathbf{x}, t) - \Upsilon_i)}$$

allows removal of the problem of closure due to nonlinearity, which is present in any description in terms of statistical moments. Finally, any problem of closure is removed using the Hopf-equation but the price to pay is an incredibly complicated configuration-space! The probabilistic description, which is of practical interest regarding a *concentration scalar* field rather than a velocity field, is extensively addressed in the context of combustion modelling, and will no longer be considered in this chapter.

2 Rapid Distortion Theory and Related Linear Stability Analysis

2.1 RDT for Homogeneous Anisotropic Turbulence

The simplest multi-point closure consists of the drastic measure of dropping all nonlinear terms in (3) before averaging. If one also drops the viscous term, in keeping with the high Reynolds number associated with the large scales of turbulence, the result is known as rapid distortion theory (RDT), introduced by Batchelor and Proudman (1954) (see Townsend, 1956,1976, Hunt and Carruthers, 1990; and especially Cambon and Scott, 1999, sections 2 and 5, for recent reviews). In neglecting nonlinearity entirely, the effects of the interaction of turbulence with itself are supposed to be small compared with those resulting from mean-flow distortion of turbulence. One often has in mind flows such as weak turbulence encountering a sudden contraction in a channel or flows around an aerofoil. Implicit is the idea that the time required for significant distortion by the mean flow is short compared with that for turbulent evolution in the absence of distortion. Linear theory can also be envisaged, at least over short enough intervals of time, whenever physical influences leading to linear terms in the fluctuation equations dominate turbulent flows, such as strongly stratified or rotating fluid or a conducting fluid in a strong magnetic field. For such cases, the term 'rapid distortion theory' is probably a little misleading.

Thanks to linearity, time evolution of u_i' may be formally written as

$$u_i'(\mathbf{x}, t) = \int \mathcal{G}_{ij}(\mathbf{x}, \mathbf{x}', t, t') u_j'(\mathbf{x}', t') d^3\mathbf{x}' \qquad (6)$$

where $\mathcal{G}_{ij}(\mathbf{x}, \mathbf{x}', t, t')$ is a Green's function matrix expressing evolution from time t' to time t. Whereas u_i' is a random quantity, varying from realisation to realisation of the flow, $\mathcal{G}_{ij}$ is deterministic and can, in principle, be calculated for a given $\overline{u}_i(\mathbf{x}, t)$. From $\mathcal{G}_{ij}$ and the initial turbulence, (6) may be used to determine later time behaviour.

Another simplifying assumption which is often made is that the size of turbulent eddies, L, is small compared with the overall length scales of the flow, ℓ, which might be the size of a body encountering fine-scale free-stream turbulence (see *e.g.* Hunt and Carruthers, 1990). In that case, one uses a local frame of reference convected with the mean velocity and approximates the mean velocity gradients as uniform, but time-varying. Thus, the mean velocity is approximated by

$$\overline{u}_i = A_{ij}(t) x_j \qquad (7)$$

in the moving frame of reference. In the example of fine-scale turbulence encountering a body, one may imagine following a particle convected by the mean velocity, which sees a varying mean velocity gradient, $A_{ij}(t)$, even when the mean flow is steady.

If we consider that the mean flow with space-uniform velocity gradients (7) is extensional, filling all the space, its presence can be consistent with statistical homogeneity for the fluctuating flow. This is a common background for homogeneous RDT and recent linear stability analysis (Bayly, 1986, among others). [2] Equations (1) and (3) can be simplified by dropping the Reynolds stress term in both, so that (1) reduces to a particular Euler equation with solution of type (7). As a consequence, the trace-free matrix $\mathbf{A}$ is subjected to the conditions that $d\mathbf{A}/dt + \mathbf{A}^2$ be symmetric, or

$$\epsilon_{ijk}\left(\frac{dA_{jk}}{dt} + A_{jn}A_{nk}\right) = 0 \quad A_{ii} = 0 \tag{8}$$

In the linear limit, the fluctuating fields (u'_i, p') satisfy the modified equation (3) with the advection-distortion parts written in terms of $A_{ij}(t)$, or

$$\underbrace{\frac{\partial u'_i}{\partial t} + A_{jk}x_k\frac{\partial u'_i}{\partial x_j}}_{\text{advection}} + A_{ij}u'_j + \frac{\partial p'}{x_i} = 0 \tag{9}$$

Its solution is most easily obtained via Fourier analysis, with elementary components of the form

$$u'_i(\boldsymbol{x}, t) = a_i(t)\exp\left(\imath\boldsymbol{k}(t)\cdot\boldsymbol{x}\right) \tag{10}$$

$$p'(\boldsymbol{x}, t) = b(t)\exp\left(\imath\boldsymbol{k}(t)\cdot\boldsymbol{x}\right) \tag{11}$$

with $\imath^2 = -1$. In term of the amplitudes above, simplified equation (9) gives

$$\frac{da_i}{dt} + \imath a_i x_j\left(\frac{dk_j}{dt} + A_{nj}k_n\right) + A_{ij}a_j + \imath k_i b = 0$$

Time-dependency of the wavevector allows to simplify the advection term by setting

$$\frac{dk_i}{dt} + A_{ji}k_j = 0 \tag{12}$$

More classically, the pressure contribution b is removed from consideration, Using the incompressibility constraint, which amounts to

$$k_i a_i = 0, \tag{13}$$

Applying the projection operator

$$P_{in} = \delta_{in} - \frac{k_i k_n}{k^2} \tag{14}$$

[2]It is important to stress that the feedback of the Reynolds stress tensor in (1) vanishes due to statistical homogeneity (zero gradient of any averaged quantity), so that the mean flow (7) has to be a particular solution of the Euler equations and can be considered as a base flow for stability analysis. In turn, the form (7) is consistent with maintaining homogeneity of the fluctuating flow governed by (3) and (4), provided homogeneity holds for the initial data. This explains why homogeneous RDT can have the same starting point as a rigorous and complete linear stability analysis in this case, before the random initialisation of the fluctuating velocity field is considered in (19).

to the last $\boldsymbol{a}$-equation, one finds

$$\frac{da_i}{dt} - \frac{k_i}{k^2} k_n \frac{da_n}{dt} + P_{in} A_{nj} a_j = 0$$

with $k_i \frac{da_i}{dt} = -\frac{dk_i}{dt} a_i = A_{ni} k_n a_i$, using (12), so that $\boldsymbol{a}$ is found to satisfy the following ODE

$$\frac{da_i}{dt} = -\underbrace{\left(\delta_{in} - 2\frac{k_i k_n}{k^2}\right) A_{nj}\, a_j}_{M_{ij}} \tag{15}$$

The linear system of simple ordinary differential equations (12) and (15), are referred to as Townsend equations . In the matrix $\mathbf{M}$, the factor $\frac{k_i k_n}{k^2}$ reflects the contribution from the fluctuating pressure term, with a prefactor 2 which takes into account advection in wave-space. As usual, spectral analysis allows for a straightforward treatment of the non-local dependence of pressure upon velocity. On the other hand, time dependency of the wavevector represents the convection of the plane wave $\exp[\imath \boldsymbol{k}(t) \cdot \boldsymbol{x}]$ by the base flow. Both the direction and magnitude of $\boldsymbol{k}$ change as wavecrests rotate and approach, or separate, from each other due to mean velocity gradients.

General solutions, which are valid for any initial data, are expressed in terms of linear transfer matrices, as follows

$$k_i(t) = B_{ij}(t, t_0) k_j(t_0) \tag{16}$$

$$a_i(t) = G_{ij}(t, t_0) a_j(t_0) \tag{17}$$

with universal values for $\mathbf{B}$ and $\mathbf{G}$ at $t = t_0$ recalled below.

In the equations above, it is perhaps clearer to specify the wavevector dependency in $\boldsymbol{a}$ and $\mathbf{G}$, especially if we combine elementary solutions of the form (10) via Fourier synthesis

$$u_i'(\boldsymbol{x}, t) = \int \hat{u}_i(\boldsymbol{k}, t) \exp(\imath \boldsymbol{k} \cdot \boldsymbol{x})\, d^3\boldsymbol{k}. \tag{18}$$

Accordingly, the RDT solution writes

$$\widehat{u}_i[\boldsymbol{k}(t), t] = G_{ij}(\boldsymbol{k}, t, t_0) \widehat{u}_j[\boldsymbol{k}(t_0), t_0] \tag{19}$$

in which the Green's function is eventually determined by the initial conditions [3]

$$G_{ij}(\boldsymbol{k}, t_0, t_0) = \delta_{ij} - \frac{K_i K_j}{K^2}, \quad K_i = k_i(t_0) \tag{20}$$

As for the time-dependency of the wavevector, $\mathbf{B}$ can be directly linked to the Cauchy matrix $\mathbf{F}$ through

$$B_{ij}(t, t_0) = F_{ji}^{-1}(t, t_0) \tag{21}$$

The general definition of the Cauchy matrix for an arbitrary (mean) flow is recalled in the next subsection, but the following definition in term of matrix exponentiation is enough to close the complete relationship for RDT.

$$\mathbf{F} = e^{\int_{t_0}^{t} \mathbf{A}(t) dt}, \quad \mathbf{F}(t_0, t_0) = \delta_{ij} \tag{22}$$

[3] A different initialization $G_{ij} = \delta_{ij}$ was prescribed by Townsend (1956,1976). The equation (20) presents some advantages, since $k_i G_{ij} = 0$ can be satisfied at any time, and the RDT Green's function can be more easily related to the Kraichnan's response function.

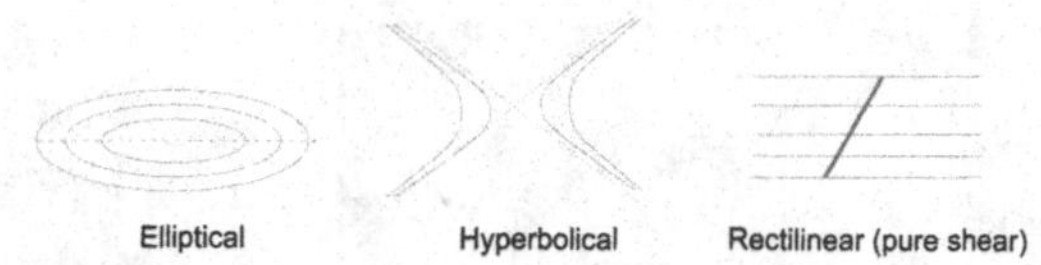

Figure 2: Sketch of isovalues of the streamfunction for the steady mean flow in homogeneous RDT: the three canonical cases

At this stage, it may be noticed that homogeneous RDT gathers enough features for solving two problems:

- A deterministic problem, which consists in solving the initial value linear system of equations for a_i, in the most general way. This is done by determining the spectral Green function, which is also the key quantity requested in linear stability analysis.
- A statistical problem which is useful for the prognostic of statistical moments of $\boldsymbol{u}'$ and p'. Interpreting the initial amplitude $\hat{\boldsymbol{u}}(\boldsymbol{k}(t_0), t_0)$ as a random variable with a given dense $\boldsymbol{k}(t_0)$-spectrum, equation (19) yields prediction of statistical moments though products of the basic Green's function.

Typical RDT (or linear stability) results can be summarized now, only regarding the deterministic problem, for base flows of the type (7) with (8). Irrotational mean flows, with $A_{ij} = A_{ji}$ yield simple RDT solutions, in which both **F** and **G** display dominant exponential growth, reflecting pure stretching of vorticity disturbances, in accordance with hyperbolical instability. Rotational mean flows yield more complicated linear solutions, and only the steady case has received much attention (Craik and coworkers, Bayly and coworkers performed recent developments in unsteady cases, see *e.g.* Bayly *et al.* , 1996). Conditions (8) imply that **A** writes as

$$\mathbf{A} = \begin{pmatrix} 0 & S - \Omega_0 & 0 \\ S + \Omega_0 & 0 & 0 \\ 0 & 0 & 0 \end{pmatrix} \tag{23}$$

in the steady, rotational case, when axes are chosen appropriately, where $S, \Omega_0 \geq 0$. This corresponds to steady plane flows, combining vorticity $2\Omega_0$ and irrotational straining S. The related streamfunction (sketched on Figure 2) is

$$\psi = -\frac{S}{2}(x_1^2 - x_2^2) + \frac{\Omega_0}{2}(x_1^2 + x_2^2) \tag{24}$$

The problem with arbitrary S and Ω_0 was analysed in my Ph. D. thesis (Cambon, 1982), in order to generalise classical RDT results, which were restricted to pure strain and pure shear. For $S > \Omega_0$, the mean flow streamlines are open and hyperbolic, and RDT results are qualitatively close to those of pure strain. For $S < \Omega_0$, the mean flow streamlines are closed and elliptic about the stagnation point at the origin. This case is the most surprising, with **F** time-periodic, but with **G** capable of generating exponential growth for $\boldsymbol{k}$-directions concentrated about special angles ($k_3/k \sim \pm 1/2$ if $S \ll \Omega_0$). RDT can therefore be relevant for explaining the mechanism for *elliptical flow instability* (Cambon, 1982, Bayly, 1986). The limiting case, $S = \Omega_0$, correspond to pure shearing of straight mean streamlines; RDT solutions by Townsend (1956,1976) or Moffat (1967) reflect algebraic growth in the language of stability analysis.

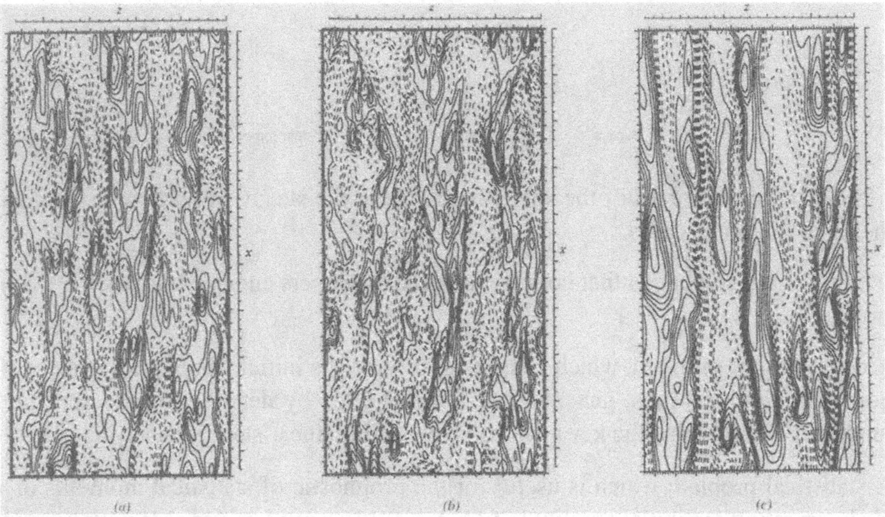

Figure 3: Contours of streamwise fluctuating velocity from (a) direct numerical simulation (DNS), and (b) rapid distortion theory (RDT) calculations for uniformly sheared homogeneous turbulence, and (c) direct numerical simulation of plane channel flow near a wall (horizontal plane $y^+ \sim 10$. The streamwise elongation of turbulent structures resulting from shear appears clearly, as does the strong similarity between RDT and DNS results. From Lee, Kim and Moin (1990).

2.1.1 Relevance of homogeneous RDT

RDT can predict qualitative trends, and even quantitative ones for single and two-point statistical quantities, which are often dimensionless and characterize anisotropy. Most usual quantities are Reynolds Stress components $\overline{u_i' u_j'}$, with nondimensional deviatoric tensor b_{ij}, and integral lengthscales $L_{ij}^{(n)}$ for different velocity components (subscripts i and j) and different directions of two-point separation (superscript n) for them. The anisotropy reflected in the latter lengthscales can be very different from the Reynolds Stress anisotropy, and therefore cannot be derived from the knowledge of b_{ij}. Qualitative relevance of RDT solutions can appear even for particular realizations (snapshots) of the fluctuating velocity field, when compared to full DNS. This is illustrated by Figure 3, which is borrowed from Lee *et al.* (1990), in the case of pure plane shear and plane channel flow near the wall. Accordingly, the tendency to create elongated 'streaklike' structures by a strong mean shear is inherent to this 'homogeneous RDT' operator, independently of the presence of a wall. In contrast, the tendency to create columnar structures in a rotating flow cannot be understood by means of RDT alone.

2.1.2 Poloidal-toroidal decomposition, and Craya-Herring frame of reference

The poloidal-toroidal decomposition (See *e.g.* Busse, 2001, in the present book) is used to represent a three-component divergence-free velocity field in terms of two independent scalar terms,

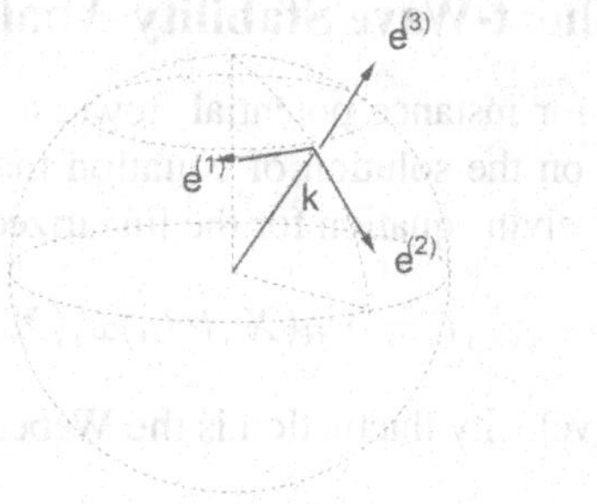

Figure 4: Polar-spherical system of coordinates for $\boldsymbol{k}$ and related 'Craya-Herring' frame of reference.

taking advantage of the presence of a privileged direction $\boldsymbol{n}$.

$$\boldsymbol{u}' = \nabla \times (s_{pol}\boldsymbol{n}) + \nabla \times [\nabla(s_{tor}\boldsymbol{n})] \tag{25}$$

the axial vector $\boldsymbol{n}$ being chosen along the vertical direction, without loss of generality. As a caveat, some care is needed to represent a Vertically Sheared Horizontal Flow (VSHF hereinafter, after Smith and Waleffe (2002)) or $\boldsymbol{u}'(\boldsymbol{x}\cdot\boldsymbol{n}, t)$, with $\boldsymbol{u}'\cdot\boldsymbol{n} = 0$, with this decomposition.

In Fourier space, the above decomposition yields a pure geometrical one, or

$$\hat{\boldsymbol{u}} = \boldsymbol{k} \times \boldsymbol{n}(\imath\hat{s}_{pol}) - \boldsymbol{k} \times (\boldsymbol{k} \times \boldsymbol{n})(\hat{s}_{tor}) \tag{26}$$

and it appears immediately that the Fourier mode related to vertical wavevector direction, or $\boldsymbol{k} \parallel \boldsymbol{n}$, has zero contribution; this 'hole' in the spectral description yields missing the VSHF mode in physical space. In order to complete the decomposition, one can define an orthonormal frame of reference, which is nothing else than the local reference frame of a polar-spherical system of coordinates for $\boldsymbol{k}$ (see Figure 4).

$$\boldsymbol{e}^{(1)} = \frac{\boldsymbol{k} \times \boldsymbol{n}}{|\boldsymbol{k} \times \boldsymbol{n}|} \quad \boldsymbol{e}^{(2)} = \boldsymbol{e}^{(3)} \times \boldsymbol{e}^{(1)} \quad \boldsymbol{e}^{(3)} = \frac{\boldsymbol{k}}{k} \tag{27}$$

for $\boldsymbol{k} \times \boldsymbol{n} \neq 0$, and $\boldsymbol{e}^{(1)}, \boldsymbol{e}^{(2)}, \boldsymbol{e}^{(3)}$ may coincide with the fixed frame of reference, with $\boldsymbol{e}^{(3)} = \boldsymbol{n}$ for $\boldsymbol{k} \parallel \boldsymbol{n}$. In the turbulence community, the local frame $(\boldsymbol{e}^{(1)}, \boldsymbol{e}^{(2)})$ of the plane normal to the wavevector is often referred to as Craya-Herring frame . Accordingly, the divergence-free velocity field in wave-space has only two components in the Craya-Herring frame, or

$$\hat{\boldsymbol{u}}(\boldsymbol{k}, t) = u^{(1)}\boldsymbol{e}^{(1)} + u^{(2)}\boldsymbol{e}^{(2)} \tag{28}$$

For $\boldsymbol{k} \times \boldsymbol{n} \neq 0$, $u^{(1)}$ and $u^{(2)}$ are directly linked to the toroidal mode and the poloidal mode, respectively. For $\boldsymbol{k} \times \boldsymbol{n} = 0$, they correspond to the VSHF mode. RDT equations can be written in the Craya-Herring frame, resulting in a reduced Green's function with only four independent components (Cambon, 1982). A similar decomposition is used in Bayly *et al.* (1996). Finally, the 'wave-vortex' decomposition introduced by Riley *et al.* (1981) in the particular context of stably stratified turbulence (see section 6), is also a particular case of (25).

2.2 Zonal RDT and Short-Wave Stability Analysis

For irrotational mean flows, for instance potential flows, a tractable form of inviscid RDT in physical space can be based on the solution of equation that governs the fluctuating vorticity ($\omega_i = \epsilon_{ijk} u'_{j,k}$), a particular Kelvin equation for the linearized case without mean vorticity:

$$\omega_i(\boldsymbol{x}, t) = F_{ji}(\boldsymbol{X}, t, t_0)\omega_j(\boldsymbol{X}, t_0) \tag{29}$$

A useful related equation for velocity fluctuation is the Weber equation:

$$u'_i(\boldsymbol{x}, t) = F_{ji}^{-1}(\boldsymbol{X}, t, t_0)u'_j(\boldsymbol{X}, t_0) + \frac{\partial \phi(\boldsymbol{x}, t)}{\partial x_i} \tag{30}$$

The mean flow may involve complex trajectories, which are defined by

$$x_i = \overline{x}_i(\boldsymbol{X}, t_0, t) \quad \text{with} \quad \dot{\overline{\boldsymbol{x}}} = u_i(\boldsymbol{x}, t) \tag{31}$$

in which Lagrangian coordinates $\boldsymbol{X}$ denote the initial position at time t_0 of a particle , which reaches the position $\boldsymbol{x}$ at time t, and the overdot holds for the related 'mean' material derivative. The Cauchy matrix, or mean displacement gradient matrix, is

$$F_{ij}(\boldsymbol{X}, t, t_0) = \frac{\partial \overline{x}_i}{\partial X_j} \tag{32}$$

Of course, the complete solution of type (6) requires that the potential term in the right-hand-side of (30) be expressed in terms of initial data. This can be done using incompressibility condition with relevant boundary conditions, and even applications to compressible flows are possible (Goldstein, 1978). Of course, in the general incompressible case, integral non-local dependency, as in (6) reappears through the solution for ϕ in (30).

As soon as the mean flow is rotational, equations such as (29) or (30) are no longer valid to tackle inhomogeneous RDT. Assuming weak inhomogeneity, considerably more progress can be made without the need for irrotationality of the mean flow, although simplifications occur in the latter case. As discussed earlier, turbulence which is fine-scale compared with the overall dimensions of the flow can be treated under RDT by following a notional particle moving with the mean velocity. Thus, the results obtained for strictly homogeneous turbulence can be extended to the weakly inhomogeneous case, but with a mean velocity gradient matrix $A_{ij}(t)$ which reflects the $\partial \bar{u}_i / \partial x_j$ seen by the moving particle.

Even if the Green's function related to the canonical base flow (7)-(8) can give interesting information for linear stability analysis and short-time development of turbulence, this problem is somewhat unphysical in the absence of typical lengthscales for variation of the base flow gradients and disturbances. For instance, the Green's function in (19) only depends on the orientation, not on the modulus, of the wavevector. Rather than to consider perturbations with arbitrary wavelength k^{-1} in the presence of the flow (7), it is more physical to consider a base flow whose velocity gradients vary over a typical lengthscale ℓ, and to restrict the validity of the zonal stability analysis to perturbations of much shorter wavelength $k^{-1} \ll \ell$. In so doing, the disturbance field should locally experience advection and distortion effects by the base flow, similarly to the effects of an extensional flow with space-uniform gradients. Given *a priori* a lengthscale separation between base and disturbance flows, one can imagine to look through a mathematical magnifying glass at the vicinity of real base trajectories. This idea has been formalised in the context of flow stability (see the short-wave 'geometric optics' of Lifschitz and Hameiri, 1991) using an asymptotic approach based on the WKB method, which is traditionally used to analyse the theoretical ray limit (i.e. short waves) in wave problems (see *e.g.* Lighthill, 1978). The perturbation solution is written as

$$u'_i(\mathbf{x},t) = a_i(\mathbf{x},t)\exp[\imath\Phi(\mathbf{x},t)/\epsilon] \tag{33}$$

with a similar expression for the fluctuating pressure, with $b(\boldsymbol{x},t)$ amplitude, where Φ is a real phase function and ϵ is a small parameter expressing the small scale of the "waves" represented by (33), while $a_i(\mathbf{x},t)$ for the velocity, and $b(\boldsymbol{x},t)$ for the pressure, are complex amplitudes which are expanded in powers of ϵ according to the WKB technique, e.g. $a_i = a_i^{(0)} + \epsilon a_i^{(1)} + \ldots$. Inserting (33) into linearized equation (3) and (4) yield: $\dot{\Phi} a_i^{(0)} + b^{(0)}\frac{\partial\Phi}{x_i} = 0$ and $k_i a_i^{(0)} = 0$ at the leading ϵ^{-1} order. Consequently, it is found that $b^{(0)} = 0$ and that

$$\dot{\Phi} = \frac{\partial\Phi}{\partial t} + \overline{u}_j\frac{\partial\Phi}{\partial x_j} = 0 \tag{34}$$

i.e. the wave crests of (33) are convected by the mean flow, with its trajectories given by (31). It is then apparent that (33) is locally a plane-wave Fourier component of wavenumber

$$k_i(\mathbf{x},t) = \epsilon^{-1}\frac{\partial\Phi}{\partial x_i} \tag{35}$$

The spatial derivatives of $\dot{\Phi} = 0$ yield an eikonal equation:

$$\dot{k}_i = -A_{ji}(t)k_j \tag{36}$$

where, as before, $A_{ij} = \partial\overline{u}_i/\partial x_j$ and the dot represents the mean-flow material derivative $\partial/\partial t + \overline{u}_i\partial/\partial x_i$. Finally, at the next ϵ^0 order, one obtains

$$\dot{a}_i^{(0)} = -M_{ij}(t)a_j^{(0)} \tag{37}$$

with M_{ij} as in (15), after elimination of the pressure using the leading-order incompressibility condition $k_i a_i^{(0)} = 0$.

Equations (36) and (37) have exactly the same form as the basic equations of homogeneous RDT (Townsend's equations) and therefore, together with (31), describe the weakly inhomogeneous case at leading order. The only difference is that, rather than being simple time derivatives, the dots represent mean-flow material derivatives, implying that one should follow mean flow trajectories which differ from one to another. In homogeneous RDT, the different classes of disturbances are only labelled by the direction of the initial wavevector $\boldsymbol{K} = \boldsymbol{k}(t_0)$, and all trajectories, such as $\psi = constant$ in (24) are equivalent. In the zonal RDT approach, it is necessary to add the Lagrangian coordinate vector $\boldsymbol{X}$ for labelling different trajectories. In agreement with classic continuum mechanics, one has

$$d\overline{x}_i = F_{ij}dX_j + \overline{u}_i dt \tag{38}$$

when diferentiating the mean trajectory equation $x = \overline{x}(X, t_0, t)$, so that $\dot{\Phi} = 0$ and (36) correspond to

$$k \cdot \delta x = K \cdot \delta X, \qquad k_i(X, t) = F_{ji}^{-1}(X, t_0, t) K_j$$

which generalizes (16)-(21). The latter equations themselves actually correspond to $k \cdot x = K \cdot X$.

The stability context is no longer discussed here for the sake of brevity (see at the end of section 5).

3 Background for HAT: Statistics and Dynamics

3.1 Second Order Statistics

Going back to homogeneous turbulence, we aim at taking into account the possible distorting effects of a mean flow of type (7) with (8), or effects of body forces, so that anisotropy is essential. Hence, attention is restricted to *Homogeneous Anisotropic Turbulence* (HAT). Fourier transform is an unvaluable tool to tackle equations for velocity and pressure fluctuations, considered as random variables, as well as their statistical multipoint correlations matrices. In the following scheme, we consider from left to right, the velocity field (basic random variable), the related two-point correlation tensor, and its single-point counterpart, the Reynolds Stress tensor. On the line just below, are quoted the quantities in Fourier space that carry out the same information. The arrow $\rightarrow$ denotes a contraction of the information.

$$\begin{array}{ccccc} u'_i(x, t) & \rightarrow & \overline{u'_i(x, t) u'_j(x + r, t)} & \rightarrow & \overline{u'_i u'_j} \\ \\ \hat{u}_i(k, t) & \rightarrow & \overline{\hat{u}_i^* \hat{u}_j} \sim \hat{R}_{ij}(k, t) & & \nearrow \end{array}$$

Inverse Fourier transform that connect u' to $\hat{u}$ is given in (18), and is applied to the two-point correlation tensor, so that

$$\overline{u'_i(x, t) u'_j(x + r, t)} = \int \hat{R}_{ij}(k, t) \exp(\imath k \cdot r) d^3 k \tag{39}$$

One may recall here that the direct Fourier transform writes

$$\hat{R}_{ij}(k, t) = \frac{1}{(2\pi)^3} \int R_{ij}(r, t) \exp(-\imath k \cdot r) d^3 r \tag{40}$$

(note the prefactor $\frac{1}{(2\pi)^3}$ in (40), and not in (39). According to (39), the Reynolds Stress tensor, which is obtained by setting $r = 0$ in R_{ij}, derives from its spectral counterpart $\hat{\mathbf{R}}$ through a 3D integral

$$\overline{u'_i(x, t) u'_j(x, t)} = \int \hat{R}_{ij}(k, t) d^3 k \tag{41}$$

the latter equation corresponding to the arrow $\nearrow$ in the scheme above. The last interesting equation, that is abridged by $\hat{u}_i \rightarrow \overline{\hat{u}_i^* \hat{u}_j}$, is

$$\overline{\hat{u}_i^*(p, t) \hat{u}_j(k, t)} = \hat{R}_{ij}(k, t) \delta(k - p) \tag{42}$$

Two alternative ways can be used to construct equations for statistical quantities in spectral space. On the one hand, an equation for $\overline{u'_i(\boldsymbol{x},t)u'_j(\boldsymbol{x}+\boldsymbol{r},t)}$ can be derived by combining equations (3) (see *e.g.* Oberlack, 2002); then the equation for $\hat{R}_{ij}$, is obtained by applying (40). On the other hand, an equation can be directly obtained in Fourier space for $\hat{u}_i(\boldsymbol{k},t)$, from which the equation for $\hat{R}_{ij}$ is derived using (42). At least in homogeneous turbulence, the second way is simpler, and will be used in what follows, since the pressure term can be solved in the simplest way in the equation for $\hat{u}_i$. The first way (Craya, 1958), even if more cumbersome, has the advantage of applying Fourier-transform only on statistical (smooth) quantities, without need for distribution theory. The reader is referred to Batchelor (1953) and to Mathieu and Scott (2000), section 6, for detailed analysis of Fourier series and their limit in an infinite box, if both random variables and their statistical moments are considered.

As an example of the first way, the RDT equation for second order statistics is readily derived from (19), using (42) as

$$\hat{R}_{ij}[\boldsymbol{k}(t),t] = G_{in}(\boldsymbol{k},t,t_0)G_{jm}(\boldsymbol{k},t,t_0)\hat{R}_{nm}[\boldsymbol{k}(t_0),t_0] \tag{43}$$

Given an initial $\hat{R}_{ij}$ at $t=t_0$, for instance isotropic, one can calculate it at later times using (43), provided the Green's function $G_{ij}(\boldsymbol{k},t,t')$ is known. The determination of G_{ij} is thus the main problem in applying homogeneous RDT in practice. [4]

The Green's function also will appear in models allowing for nonlinearity through formal solutions of the moment equations, in which the nonlinear terms are treated as forcing of the linear part.

3.2 Nonlinear Dynamics

In the presence of a mean flow consistent with statistical homogeneity of the fluctuating one, nonlinear equation (3) writes

$$\dot{\widehat{u}}_i + M_{ij}\widehat{u}_j = s_i - \nu k^2\widehat{u}_i \tag{44}$$

where

$$\dot{\widehat{u}}_i = \frac{\partial \hat{u}_i}{\partial t} + \frac{\partial \hat{u}_i}{\partial k_m}\frac{dk_m}{dt} = \frac{\partial \hat{u}_i}{\partial t} - A_{lm}k_l\frac{\partial \hat{u}_i}{\partial k_m} \tag{45}$$

corresponds to linear advection by the mean flow (7), and $M_{ij} = A_{mj}(\delta_{im} - 2k_ik_m/k^2)$ gathers linear distortion and pressure terms, as in (15). Once nonlinear and viscous terms are added (right-hand-side), (44) generalises the linear inviscid equation for which the RDT solution is (19).

The nonlinear term s_i is given by

$$s_i(\boldsymbol{k},t) = -\imath P_{ijk}(\boldsymbol{k})\int_{\mathbf{p}+\mathbf{q}=\mathbf{k}} \widehat{u}_j(\mathbf{p},t)\widehat{u}_k(\mathbf{q},t)d^3\mathbf{p} \tag{46}$$

[4] One-point moments contain rather limited information compared with $\hat{R}_{ij}(\boldsymbol{k},t)$, but they are nonetheless usually among the first quantities to be calculated following an anisotropic spectral calculation, along with correlations lengths in different directions. Given $\hat{R}_{ij}$, for instance obtained using RDT, evaluation of the integral over 3D Fourier space can be a nontrivial task. For example, the RDT Green's function can be determined analytically in the case of simple shear, but the integrals in (41) are not straightforward and must be evaluated numerically or asymptotically (Rogers, 1991, Beronov and Kaneda, 2001)

in terms of a convolution integral, the usual expression of a quadratic nonlinearity, and $P_{ijk} = \frac{1}{2}(P_{ij}k_k + P_{ik}k_j)$ which arises from the elimination of pressure using the incompressibility condition $k_i\widehat{u}_i(\boldsymbol{k},t) = 0$.

The equations (44) to (46) are completely generic, and hold for other cases, including body forces and additional random variables, only changing the matrix **M** of the linear operator, and/or the influence matrix P_{ijk} in the convolution product which reflects quadratic nonlinearity. The evolution equation for $\hat{R}_{ij}$ (Craya, 1958), derived from (42) and (44), is

$$\dot{\hat{R}}_{ij} + M_{ik}\hat{R}_{kj} + M_{jk}\hat{R}_{ik} = T_{ij} - 2\nu k^2\hat{R}_{ij} \tag{47}$$

where the left-hand side arises from the linear inviscid part of (44), consisting of the term $\dot{\hat{R}}_{ij}$, which, as in (44), is a convective time derivative in $\boldsymbol{k}$-space, together with distortion components.

Although purely linear theory closes the equations without further ado and simplifies mathematical analysis, it is rather limited in its domain of applicability, ignoring as it does all interactions of turbulence with itself, including the physically important cascade process. Multi-point turbulence models which account for nonlinearity via closure lead to moment equations with a well-defined linear operator and nonlinear source terms. The view taken in this chapter is that, even when nonlinearity is significant, the behaviour of the linear part of the model often still has a significant influence. Thus, it is important to first understand the properties of the linearised model, an undertaking which is, moreover, mathematically more tractable than attacking the full model directly. As a bonus, linearised analysis often allows a simplified formulation of the nonlinear model using more appropriate variables.

4 The Nonlinear Problem of Closure

Given the, in our view, importance of allowing for anisotropy and the associated effects of mean flow gradients, we will not discuss the many isotropic models of spectral evolution which have been proposed (see, for instance, Monin and Yaglom, 1975, section 17). Instead we concentrate on a small number of models capable of handling the anisotropic case and which illustrate the way in which linear theory combines with nonlinear closures .

By reintroducing the viscous term, the RDT Green's function immediately generates the zeroth-order Kraichnan response function , which is a key quantity in any Renormalized Perturbation Theory (hereinafter referred to as RPT , Mc Comb, 2002).

$$G^{(0)}_{ij}(\boldsymbol{k},t,t') = G_{ij}(\boldsymbol{k},t,t')\exp\left(-\nu\int_{t'}^{t}k^2(\tau)d\tau\right) \tag{48}$$

The most general *nonlinear* response function should be found in adding an arbitrary external stirring term $\boldsymbol{f}(\boldsymbol{k}',t')$ to (44) and to express the response in term of velocity disturbance $\delta\hat{\boldsymbol{u}}$ to an infinitesimal perturbation $\delta\boldsymbol{f}$. Because of the presence of the primitive nonlinear term, this response function has to take into account an arbitrary advection term of type $u'\delta u'$. Hence it is necessarily strongly non-local even in spectral space, so that:

$$\delta\hat{u}_i(\boldsymbol{k},t) = \int\int_{t'}^{t} G^{(\tau)}_{ij}(\boldsymbol{k},\boldsymbol{k}',t,t')\delta f_j(\boldsymbol{k}',t')d^3\boldsymbol{k}'dt' \tag{49}$$

Its zeroth order is found when neglecting the advection terms $u'\delta u'$ with respect to those linked to the mean gradients, so that

$$G_{ij}^{(r)}(\boldsymbol{k},\boldsymbol{k}',t,t') = G_{ij}^{(0)}(\boldsymbol{k},t',t)\delta[\boldsymbol{k}' - {}^t\mathsf{F}^{-1}(t,t')\boldsymbol{k}] \tag{50}$$

which only displays a very simple non-locality in $\boldsymbol{k}$-space, due to the very particular form (7) of the advecting field. The matrix ${}^t\mathsf{F}^{-1}$ holds for transposed inverse Cauchy matrix , in agreement with (16) and (21).

It is outside the scope of this lecture to deeply revisit existing RPT's, and we will just illustrate how to incorporate the zeroth-order Green function into the simplest versions of these theories. Considering the right-hand-side of (44) as a given source term, this equation rewrites

$$\hat{u}_i(\boldsymbol{k},t) = G_{ij}^{(0)}(\boldsymbol{k},t,t_0)\hat{u}_j({}^t\mathsf{F}(t,t_0)\,\boldsymbol{k},t_0) + \int_{t_0}^{t} G_{ij}^{(0)}(\boldsymbol{k},t,t')s_j\left({}^t\mathsf{F}(t,t')\boldsymbol{k},t'\right)dt' \tag{51}$$

Similar equations can be derived for statistical multi-point moments of any order, for instance for the spectral tensor of second and triple order. Only symbolic forms of equations are needed, as follows:

$$u(t) = G^0 u(t_0) + \int G^0 u.u(t')dt' \tag{52}$$

which represents (51), and

$$\overline{uu}(t) = G^0G^0\overline{uu}(t_0) + \int G^0G^0\overline{uuu}(t')dt' \tag{53}$$

for the second order, as an integral form of (47), and finaly

$$\overline{uuu}(t) = G^0G^0G^0\overline{uuu}(t_0) + \int G^0G^0G^0\overline{uuuu}(t')dt' \tag{54}$$

at the third order.

All multi-point closures, even those that are based on sophisticated RPT's, assume formally a weak departure from a *linear and Gaussian reference state*; hence quasi-normal relationship is pivotal in all these aproaches. The problem is tackled at the level of fourth order correlations at four points, that are involved in the right-hand-side of the latter equation. General quasi-normal hypothesis amounts to assuming a linear relationship between fourth and third order *cumulants*, or

$$\overline{uuuu} - \sum \overline{uu}.\overline{uu} = -\frac{1}{\theta}\overline{uuu}. \tag{55}$$

Regarding the structure of the equation given above, it is important to give some preliminary remarks as follows:

- The latter two equations are abridged and 'symbolic', in the sense that the Fourier velocity components $\widehat{u}_i$ at different wave vectors are actually involved in place of u, so that $\overline{uuu}$ would represent a third order spectral tensor and $\overline{uu}$ would represent the second-order spectral tensor $\hat{R}_{ij}$. Accordingly, the 'true' equation abridged by (55) allows one to close the equation (54), and then to close the equation (47) for the second order spectral tensor:

the infinite hierarchy of open equations ($[n,n] \leftarrow [n+1,n] \leftarrow [n+1,n+1]$ in Figure 1) is broken at the fourth order. The detailed form of the cumulant in the left-hand-side of (55), which corresponds to the $[4,4]$-level is

$$\overline{\hat{u}_r\hat{u}_s\hat{u}_p\hat{u}_q} - \overline{\hat{u}_r\hat{u}_s}.\overline{\hat{u}_p\hat{u}_q} - \overline{\hat{u}_r\hat{u}_p}.\overline{\hat{u}_s\hat{u}_q} - \overline{\hat{u}_r\hat{u}_q}.\overline{\hat{u}_s\hat{u}_p} \quad (56)$$

in which the subscripts denote different velocity components related to different wave-vectors (four wave-vectors related to four separation vectors in physical space).

- Equation (55) is formally consistent with 'nearly linear' and 'nearly Gaussian' assumptions: since linear operators conserve the gaussianity, pure linear dynamics (reflected for instance by the so-called RDT) conserve the Gaussian properties if present in the initial data. Accordingly, all cumulants remain zero in this situation and both right-hand-side and left-hand-side of (55) vanish identically. Hence a linear relationship such as (55) is consistent with considering both right-hand-side and left-hand-side as formal 'weak' departures from gaussianity, caused by formal 'weak' nonlinearity.

The impact of a closure relationship such as (55) on the dynamics of triple correlations, that carry the energy cascade, is conventionally seen as follows: since (54) can be rewritten as

$$\frac{\partial \overline{uuu}}{\partial t} + \text{exact linear}(\overline{uuu})\text{terms}$$

$$+\frac{1}{\theta}\overline{uuu} = \sum \overline{uu}.\overline{uu},$$

the term $\sum \overline{uu}.\overline{uu}$ in the left-hand-side acts as a source term for increasing triple correlations, so that the contribution to the right-hand-side term, which comes from the closure relationship (55), appears as a damping term which exhibits the characteristic time denoted θ. An 'ad hoc' eddy-damping term is chosen using the EDQNM-type model, but the structure of more sophisticated two-point closure theories (Direct Interaction Approximation, DIA, Test Field Model, TFM) is not fundamentally different. Introducing the eddy damping conserves the general structure of the equations that result from the pure quasi-normal theory, in just changing the viscous factor νk^2 into $\mu = \nu k^2 + \mu'$ in the zeroth-order response function (48). the new 'renormalized' Green's function is denoted $\mathbf{G}^{ED}$. Symbolic equation (54) becomes

$$\overline{uuu}(t) = G^{ED}G^{ED}G^{ED}\overline{uuu}(t_o)$$

$$+\sum\int_{t_0}^{t}\int_{k+p+q=0} \underbrace{G^{ED}P.}_{k}\underbrace{G^{ED}\overline{uu}.}_{p}\underbrace{G^{ED}\overline{uu}}_{q} \quad (57)$$

The corresponding expression for the right-hand side T_{ij} of (47), which reflects triple correlations at two point, consists of an integral over the third-order spectral tensor. The previous quasi-normal procedure gives the typical relationship between T_{ij} and $\hat{R}_{ij}$,

$$T_{ij}(\boldsymbol{k},t) = \tau_{ij}(\boldsymbol{k},t) + \tau^*_{ji}(\boldsymbol{k},t) \quad (58)$$

where

$$\tau_{ij}(\boldsymbol{k},t) = P_{jkl}(\boldsymbol{k}) \int_{-\infty}^{t} \int_{\mathbf{k}+\mathbf{p}+\mathbf{q}=0} G_{im}(\boldsymbol{k},t,t') G_{kp}(\boldsymbol{p},t,t') G_{lq}(\boldsymbol{q},t,t') \hat{R}_{qn}(\boldsymbol{q}',t') \left[\frac{1}{2} P_{mnr}(\boldsymbol{k}') \hat{R}_{pr}(\boldsymbol{p}',t') + P_{pnr}(\boldsymbol{p}') \hat{R}_{mr}(\boldsymbol{k}',t') \right] d^3\boldsymbol{p}\, dt' \quad (59)$$

in which the triple product of Green's functions arises from the Green's function solution for the third-order moments and the notation $\mathbf{k}+\mathbf{p}+\mathbf{q}=\mathbf{0}$ on the integral sign means that $\mathbf{q}$ should be replaced by $-\boldsymbol{k}-\mathbf{p}$ throughout the integrand, representing interacting triads of wavenumbers $\boldsymbol{k}$, $\mathbf{p}$, $\mathbf{q}$ which form triangles. The vectors $\boldsymbol{k}'$, $\boldsymbol{p}'$, $\boldsymbol{q}'$ are derived from $\boldsymbol{k}$, $\boldsymbol{p}$, $\boldsymbol{q}$ through the relationship as in (49).

It gives the generic anisotropic structure of most of generalised classical theories dealing with two-point closure, provided the Green function inherited from RDT be replaced by a modified version, for instance including viscous terms and eddy damping as in EDQNM theory.

The problem of 'Markovianization' (last letter of the acronym EDQNM) will be rediscussed at length on a typical example, in section 5.

We should perhaps say a few words about two-time RPT's, which are more complicated than EDQNM, since they are based on spectral tensors involving two times, rather than $\hat{R}_{ij}(\boldsymbol{k},t)$. In DIA (Kraichnan, 1959), an additional response tensor, which appears as a simplified averaged version of $\mathbf{G}^{(r)}$ in (49) is introduced for which, like the spectral tensor, an evolution equation is formulated and closed using heuristic approximations and/or renormalized expansions, around a zero-order Gaussian state. The final evolution equation for the two-time spectral tensor contains an integral whose structure is much the same as the quasi-normal expression (59), with terms such as $G_{lq}(\mathbf{q},t,t')\hat{R}_{qn}(\mathbf{q}',t')$ replaced by the two-time spectral tensor $\hat{R}_{ln}(\mathbf{q}',t,t')$, leaving one remaining Green's function from the three-fold product, which is replaced by the response tensor. Kraichnan himself demonstrated that primitive DIA equations were not consistent with the Kolmogorov law, and proposed more sophisticated theories (LHDIA, mixing Lagrangian and Eulerian correlations) or alternative Markovian models (TFM). Finally, it is always possible to derive a one-time version from a two-time two-point closure theory, using a 'fluctuation-dissipation' theorem. Among various versions, two RPT's give a reasonable degree of sophistication, and could be used as a first step to improve EDQNM: the LET (Mc Comb, 1974 and Mc Comb, 2002) and the LRA by Kaneda (1981).

4.1 Reminder: the Isotropic Case

In general, RDT operators break statistical isotropy at any scale, even if the initial data are strictly isotropic. It should be borne in mind that isotropy imposes a very special form

$$\hat{R}_{ij}(\boldsymbol{k},t) = \frac{E(k,t)}{4\pi k^2} \left(\delta_{ij} - \frac{k_i k_j}{k^2} \right) \quad (60)$$

on the spectral tensor, where $E(k,t)$, with $k = |\boldsymbol{k}|$, is the usual energy spectrum, representing the distribution of turbulent energy over different scales and the quantity in brackets will be recognised as the projection matrix, $P_{ij}(\boldsymbol{k})$, as in (14). Thus, $\hat{R}_{ij}$ is determined by a single real scalar quantity, E, which is a function of the magnitude of $\boldsymbol{k}$ alone. Both the form of $\hat{R}_{ij}$ at a single point and its distribution over $\boldsymbol{k}$-space are strongly constrained by isotropy.

The Craya equation (47) reduces to the scalar Lin equation

$$\frac{\partial E}{\partial t} + 2\nu k^2 E = T \tag{61}$$

in which the third-order correlations are involved in the scalar spectral transfer term $T(k,t)$. This equation can be seen as a spectral counterpart of the third-order Kolmogorov equation discussed at length by Benzi (2002)..

In the absence of mean flow or body forces, the zeroth-order response function only displays laminar viscous effects, so that:

$$G_{ij}^{(0)}(\boldsymbol{k},t,t') = \left(\delta_{ij} - \frac{k_i k_j}{k^2}\right) e^{-\nu k^2 (t-t')} \tag{62}$$

In contrast with the cases which are emphasized in this chapter, such a zeroth-order state is very far from developed turbulence. Isotropic turbulence, however, allows dramatic simplifications for all statistical theories or models, and therefore is one of the most interesting canonical flow of reference. For instance, all classical two-point closure theories have a similar structure, as a non-local expression of T in terms of E

$$T(k,t) = \int_{\Delta_k} \frac{1}{\mu(k,t) + \mu(p,t) + \mu(q,t)} C(k,p,q) e(q,t)[e(p,t) - e(k,t)] dp dq \tag{63}$$

in which Δ_k denotes the domain so that the moduli k, p and q are the length of the sides of a triangle, $e(k,t) = E(k,t)/(4\pi k^2)$, and $C(k,p,q)$ is a known geometric factor. Different versions of statistical theories only differ from the expression of the damping factor μ, which gathers pure viscous contribution and nonlinear readjustment of the response function. The reader is referred to Monin and Yaglom (1975) and to Mc Comb (2002) for more detail.

4.2 Anisotropic Description

Independently of closure, the spectral tensor $\hat{R}_{ij}$ is not a general complex matrix, but has a number of special properties, including the fact that it is Hermitian, positive-definite, as follows from (42), and satisfies $\hat{R}_{ij} k_j = 0$, obtained from (42) and the incompressibility condition $k_j \hat{u}_j = 0$. Taken together, these properties mean that, instead of the 18 real degrees of freedom of a general complex tensor, $\hat{R}_{ij}$ has only four. Indeed, using a spherical polar coordinate system in $\boldsymbol{k}$-space, or (27) and (28), the tensor takes the form (see Cambon *et al.* , 1997 for details).

$$\hat{R} = \begin{pmatrix} e + Z_r & Z_i - \imath\mathcal{H}/k & 0 \\ Z_i + \imath\mathcal{H}/k & e - Z_r & 0 \\ 0 & 0 & 0 \end{pmatrix} \tag{64}$$

where the scalars $e(\boldsymbol{k},t)$ and $\mathcal{H}(\boldsymbol{k},t)$ are real, and $Z(\boldsymbol{k},t) = Z_r + \imath Z_i$ is complex. The quantity $e(\boldsymbol{k},t) = \frac{1}{2}\hat{R}_{ii}$ is the energy density in $\boldsymbol{k}$-space, whereas $\mathcal{H}(\boldsymbol{k},t) = \imath k_l \epsilon_{lij} \hat{R}_{ij}$ is the helicity spectrum and, along with Z, is zero in the isotropic case. Recall that the local frame of reference $\boldsymbol{e}^{(1)}, \boldsymbol{e}^{(2)}, \boldsymbol{e}^{(3)}$, in (27) is chosen direct and orthonormal, with $\boldsymbol{e}^{(3)} = k_i/k$.

Anisotropy is expressed through variation of these scalars with the direction of $\boldsymbol{k}$, as well as departures of $\mathcal{H}$ and Z from zero at a given wavenumber. Whatever spectral closure is used, the number of real unknowns may be reduced to the above four when carrying out numerical calculations, and presentation of the results can be simplified using these variables, particularly when the turbulence is axisymmetric.

In order to display the different contributions from $e, Z, \mathcal{H}$ with their related tensorial operators, (64) is rewritten as follows

$$\hat{R}_{ij} = e(\boldsymbol{k},t)P_{ij}(\boldsymbol{k}) + Re[Z(\boldsymbol{k},t)N_i(\boldsymbol{k})N_j(\boldsymbol{k})] + \imath\mathcal{H}(\boldsymbol{k},t)\epsilon_{ijn}\frac{k_n}{k} \quad (65)$$

in which P_{ij} denotes the projection operator, and $\boldsymbol{N} = \boldsymbol{e}^{(2)} - \imath\boldsymbol{e}^{(1)}$. The anisotropic structure is then analyzed by isolating in (65) the pure isotropic contribution (60), so that

$$Re\left(\hat{R}_{ij}\right) = \underbrace{\frac{E(k)}{4\pi k^2}P_{ij}}_{\text{Isotropic part}} + \underbrace{\left(e(\mathbf{k}) - \frac{E(k)}{4\pi k^2}\right)P_{ij}}_{\text{Directional anisotropy}} + \underbrace{Re\left(Z(\mathbf{k},t)N_iN_j\right)}_{\text{Polarization anisotropy}} \quad (66)$$

A three-fold splitting follows for any single-point correlation, for instance

$$\overline{u_iu_j} = \int \hat{R}_{ij}(\boldsymbol{k},t)d^3\boldsymbol{k} = q^2\left(\frac{\delta_{ij}}{3} + \underbrace{b_{ij}^{(e)} + b_{ij}^{(z)}}_{b_{ij}}\right) \quad (67)$$

for the Reynolds Stress tensor,

$$D_{ij} = \int \frac{k_ik_j}{k^2}2e(\boldsymbol{k},t)d^3\boldsymbol{k} = q^2\left(\frac{\delta_{ij}}{3} + 2b_{ij}^{(e)} + 0\right)$$

for the 'Dimensionality structure tensor' (Kassinos *et al.* , 2000), and

$$\overline{\omega_i\omega_j} = \omega^2\left(\frac{\delta_{ij}}{3} + b_{ij}^{(k^2e)} - b_{ij}^{(k^2z)}\right)$$

for the vorticity correlations tensor.[5]

The above relationship shows that the deviatoric tensor b_{ij} is the sum of two very different contributions, b^e_{ij} from the directional anisotropy (or dimensionality) and b^z_{ij} from polarization anisotropy . Surprising RDT results in rotating flows are explained by this decomposition (see section 5), and the formalism of Kassinos *et al.* (2000) appears as a byproduct of (65) in homogeneous turbulence, with the decomposition in terms of directional and polarization anisotropy lending support to componental and dimensional anisotropy.

In conclusion, it is worthwhile to point out that a fully anisotropic spectral (or two-point) description carries a very large amount of information, even if restricted to second-order statistics.

[5]Rather than the vorticity correlations tensor, Kassinos *et al.* (2000) introduced a 'Circulicity tensor' C_{ij}, which involves larger scales. This tensor corresponds to $C_{ij} = q^2\left(\frac{\delta_{ij}}{3} + b_{ij}^{(e)} - b_{ij}^{(z)}\right)$, with our notations

In the inhomogeneous case, the POD (proper orthogonal decomposition, Lumley (1967)) has renewed interest in second-order two-point statistics, but this technique is applied to strongly inhomogeneous quasi-deterministic flows. It is only said that POD spatial modes are Fourier modes in the homogeneous turbulence, without considering that a spectral tensor such as $\hat{\mathsf{R}}$ ought to be diagonalized in order to exhibit its eigenmodes as POD modes in the *anisotropic* case. In fact, diagonalising the real part of the tensor $\hat{\mathsf{R}}$, is an easy task using the above $e - Z$ decomposition (64): the principal components (nonzero eigenvalues) are $e + |Z|$ and $e - |Z|$, and the angular position of the principal axes (eigenvectors) in the Craya-Herring frame of reference ($\boldsymbol{e}^{(1)}, \boldsymbol{e}^{(2)}$, see (27)), is given by the phase of Z, at each $\boldsymbol{k}$.

4.3 Flows Dominated by Production and Flows Dominated by Neutral Wave-Effects

Before examining the simplest and most interesting applications of linear and nonlinear theory, it is wortwhile to anticipate the difficulties for passing from (linear) RDT to generalized quasi-normal (nonlinear) closure for homogeneous anisotropic turbulence, in the presence of mean flows given by (7, 24).

In the 'hyperbolic' and 'elliptic' cases, with $0 \neq S \neq \Omega_0$ in (24), the RDT Green's function can display exponential growth, at least for particular angles of $\boldsymbol{k}$ ($k_3/k \sim 1/2$ in the case $S \ll \Omega_0$). If the bare zeroth-order response function is only modified by eddy damping, convergence is not ensured for the time integral of the three-fold product GGG in the generic closure relationship (59).

A less critical situation occurs when $S = \Omega_0$ (pure plane shear), since the RDT Green's function yields only algebraic growth, so that the viscous term ensures convergence of the time integral involved in the closure. Nevertheless, it is very cumbersome to develop, and especially to solve numerically with enough accuracy, a complete anisotropic EDQNM model in this case. Recall that even calculation of single-point correlations resulting from viscous RDT at high St is not easy (Beronov and Kaneda, 2001). Direct Numerical Simulations suggest that fully nonlinear effects yield exponential growth for the turbulent kinetic energy, but computations are very sensitive to cumulated errors (remeshing, low angular resolution at small k). Such a transition from algebraic growth (linear, small time) to exponential growth (nonlinear) is not really described and explained. Interesting scaling laws, however, for possible exponential growth, follow from the Oberlack's approach (Oberlack, 2002) as well as from self-similarity arguments (Julian Scott, private communications).

Only for pure rotation, or $S = 0, \Omega_0 \neq 0$, the most general EDQNM versions were carried out towards complete achievement. In this case, the zeroth order state consists of superimposed oscillating modes of motion, with no amplification and no interaction: they correspond to neutral dispersive inertial waves. Time integral of a three-fold product of Green's functions converges, provided an infinitesimal viscous (or eddy damping) term is added. In the limit of small interaction, two-point closures and theories of wave-turbulence share an important background. Even if the latter are developed in the inviscid case, a vanishing damping term is also added, as a mathematical convenience, in order to regularise the resonant operators. Similar effects can be found in stably-stratified flows with and without rotation, in MHD (Magneto-Hydro-Dynamics) flows, and use of similar two-point closure/ wave-turbulence theories is particularly relevant (see Galtier *et al.* , 2000 for flows dominated by Alfvén waves).

This preliminary discussion justifies, to a certain extent, to discriminate flows dominated by production from flows dominated by waves, a distinction which is revisited in the conclusion. The first class is illustrated by classical shear flows, in which a nonzero 'production term' is displayed in the equations governing the Reynolds Stress tensor. This production is often related to growth of instabilities, when stability analysis is addressed. The second class is illustrated in the following two sections, as the privileged area to apply spectral closures. Note that the dynamics can be dominated by dispersive waves, which are neutral but for a small part of the configuration space, in which exponential amplification occurs. In the latter case, e.g. for flows with weak ellipticity ($S \ll \Omega_0$ in (24)), production of energy is nonzero, but classic single-point closure models are of poor relevance, since only particular orientations in wave-space are subjected to parametric instability.

5 Turbulence and Vortex Dynamics in a Rotating Fluid

Rotation of the reference frame is an important factor in certain mechanisms of flow instability, and the study of rotating flows is interesting from the point of view of turbulence modelling in fields as diverse as engineering (*e.g.* turbomachinery and reciprocating engines with swirl and tumble), geophysics and astrophysics. Effects of mean curvature or of advection by a large eddy can be tackled using similar approaches.

This problem can be directly related to the context of turbulence in the presence of a mean flow with space-uniform gradients, provided a pure antisymmetric form be chosen, or $A_{ij} = \epsilon_{ikj}\Omega_k$, with $\boldsymbol{\Omega}$ the angular velocity, but it is simpler to work with coordinates system and velocity vectors seen in the rotating frame. In this *non-Galilean* frame, rotation of the frame only introduces inertial forces, centrifugal and Coriolis. Since the former can be incorporated in the pressure term, only the latter has to be taken into account in the following Navier-Stokes equations in the rotating frame:

$$\frac{\partial \boldsymbol{u}}{\partial t} + 2\boldsymbol{\Omega} \times \boldsymbol{u} - \nu \nabla^2 \boldsymbol{u} + \boldsymbol{\nabla} p = -\mathbf{u} \cdot \boldsymbol{\nabla} \mathbf{u} \tag{68}$$

From several experimental, theoretical and numerical studies, in which rotation is suddenly applied to homogeneous turbulence, some agreed statements are summarised as follows (Bardina *et al.*, 1985, Jacquin *et al.* , 1990, Cambon *et al.* , 1992, Cambon *et al.* , 1997, Cambon, 2001).

- Rotation inhibits the energy cascade, so that the dissipation rate is reduced.
- The initial three-dimensional (3D) isotropy is broken through nonlinear interactions modified by rotation, so that a moderate anisotropy, consistent with a transition from a 3D to a 2D state, can develop.
- Both previous effects involve non-linear or 'slow' dynamics, and the second is relevant only in an intermediate range of Rossby numbers as found by Jacquin et al. (1990). This intermediate range is delineated by $Ro_L = u_{\mathrm{rms}}/(2\Omega L) < 1$ and $Ro_\lambda = u_{\mathrm{rms}}/(2\Omega\lambda) > 1$, in which u_{rms} is an axial rms velocity fluctuation, whereas L and λ denotes a typical integral lengthscale (macroscale) and a typical Taylor microscale respectively.

- If the turbulence is initially anisotropic, the 'rapid' effects of rotation (linear dynamics tackled in a RDT fashion) conserve a part of the anisotropy (called directional, $b_{ij}^{(e)}$) and damp the other part (called polarization anisotropy $b_{ij}^{(z)}$), resulting in a spectacular change of the anisotropy b_{ij} of the Reynolds Stress tensor (see equations (66) and (67)).

These effects, which are not at all taken into account by current one-point second order closure models (from $k-\varepsilon$ to $\overline{u_i'u_j'}-\varepsilon$ models), have motivated new modelling approaches by Cambon *et al.* (1992), Cambon *et al.* (1997), and to a lesser extent by Kassinos *et al.* (2000) for linear (or 'rapid') effects only. It is worth noticing that the modification of the dynamics by the rotation ultimately comes from the presence of inertial waves (Greenspan, 1968), having an anisotropic dispersion law, which are capable of changing the initial anisotropy of the turbulent flow and also can affect the non-linear dynamics. Contrary to a well-known interpretation, the Proudman theorem shows only that the 'slow manifold' (the stationary modes unaffected by the inertial waves) is the two-dimensional manifold at small Rossby number, but cannot predict the transition from 3D to 2D turbulence, which is a non-linear mechanism of transfer from all possible modes towards the 2D ones. This can be discussed from the vorticity equation, which is derived from (68)

$$\frac{\partial \omega_i}{\partial t} - 2\Omega_l \frac{\partial u_i}{\partial x_l} = \omega_l \frac{\partial u_i}{\partial x_l} + \nu \nabla^2 \omega_i \tag{69}$$

Nonlinear and viscous terms are gathered on the right-hand-side. The linear inviscid limit is found by discarding the right-hand-side, assuming very low Rossby number and very high Ekman number. The two-dimensional limit $\partial u_i/\partial x_\parallel = 0$, however, is only found if the additional assumption of slow motion is done. In Fourier space, the slow — and 2D — manifold corresponds to the wave plane normal to the rotation axis, or $k_\parallel = 0$.

5.1 The Linear Wave-Regime

Linearized inviscid equation writes

$$\frac{\partial \boldsymbol{u}}{\partial t} + 2\boldsymbol{\Omega} \times \boldsymbol{u} + \boldsymbol{\nabla} p = 0, \qquad \boldsymbol{\nabla}\cdot\boldsymbol{u} = 0 \tag{70}$$

Since the Coriolis force is not divergence-free, the pressure term has a nontrivial contribution to maintain the incompressibility constraint. The velocity can be eliminated between the latter equation and the Poisson equation for the pressure, for which a closed equation is found

$$\partial_t^2 \left(\nabla^2 p\right) + 4\Omega^2 \nabla_\parallel^2 p = 0 \tag{71}$$

Even if the primitive Poisson equation $\nabla^2 p = f$ is of the parabolical type, the equation (71)admit solutions in terms of propagating waves. Very surprinsing properties of these inertial waves are illustrated by the St Andrew-cross shaped structures from the experiment by Mc Ewan (1970) (see Figure 5 and 6). If a local harmonic forcing takes place, with frequency σ_0, in the tank rotating at angular velocity Ω, simplified solutions can be sought under normal modes, or $p = e^{\imath\sigma_0 t}\mathcal{P}$, so that the spatial part is governed by

$$[\sigma_0^2 \nabla_\perp^2 + (\sigma_0^2 - 4\Omega^2)\nabla_\parallel^2]\mathcal{P} = 0$$

which shows the possible transition from elliptic to hyperbolic nature, when σ_0 crosses the threshold 2Ω by decreasing values. Hence the sudden appearance of the cross-shaped structure for $\sigma_0 < 2\Omega$. In spite of the rather complex geometry, one can assume, in addition, that the disturbances are plane waves, or $p \sim e^{\imath(\boldsymbol{k}\cdot\boldsymbol{x}-\sigma t)}$. Injecting in (71), the classical dispersion law of inertial waves is recovered as

$$\sigma_k = \pm 2\Omega \frac{k_\parallel}{k} = \pm 2\Omega \cos\theta. \tag{72}$$

If one interprets the rays emanating from the small forcing zone in the figure as trace of isophase surfaces, so that the wave vector is normal to them, equation (72) with $\sigma_k = \sigma_0$ gives the angle θ (angle of $\boldsymbol{k}$ with respect to the vertical axis) in excellent agreement with the directions of the rays.

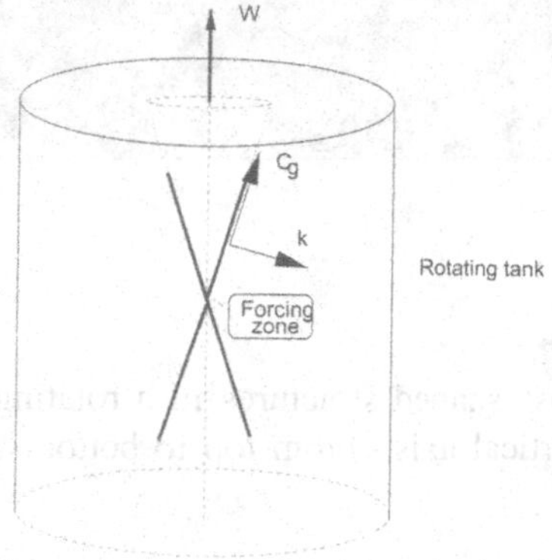

Figure 5: Sketch of the experiment by Mc Ewan.

Note that, without pressure, only the horizontal part of the flow is affected by circular periodic (constant frequency 2Ω) motion, but propagating waves cannot occur. Hence, fluctuating pressure (through its linkage to divergence-free condition) is responsible both for anisotropic dispersivity and for horizontal-vertical coupling.

Going back to velocity, an equation similar to (71) can be found for both poloidal and toroidal potentials in (25). Without forcing and boundary conditions, the specific 'RDT' problem writes

$$\frac{\partial \hat{u}_i}{\partial t} + 2\Omega P_{in}\epsilon_{n3j}\hat{u}_j = 0 \tag{73}$$

this equation is simpler than (15), since $\boldsymbol{x}$ and $\boldsymbol{u}$ in physical space are projected onto the rotating frame, so that there is no advection by the mean, and therefore no time-shift in the wave-vector. [6] Given the incompressibility constraint $\hat{\boldsymbol{u}}\cdot\boldsymbol{k} = 0$, it is easier to project the equation onto the local frame $(\boldsymbol{e}^{(1)}, \boldsymbol{e}^{(2)})$ normal to $\boldsymbol{k}$ in (27). The solution expresses that the initial Fourier component $\hat{\boldsymbol{u}}(\boldsymbol{k}, 0)$ is rotated about the axis $\boldsymbol{k}$ of an angle $2\Omega t k_3/k$ (also equal to $\sigma_k t$). A tractable diagonal

[6] of course, a strictly equivalent problem would exactly correspond to equations of subsection 2.1 in a Galilean frame of reference, for a pure antisymmetric gradient matrix, or $A_{ij} = \Omega\epsilon_{i3j}$, with $\boldsymbol{k}(\Omega t)$ following the solid-body rotating motion.

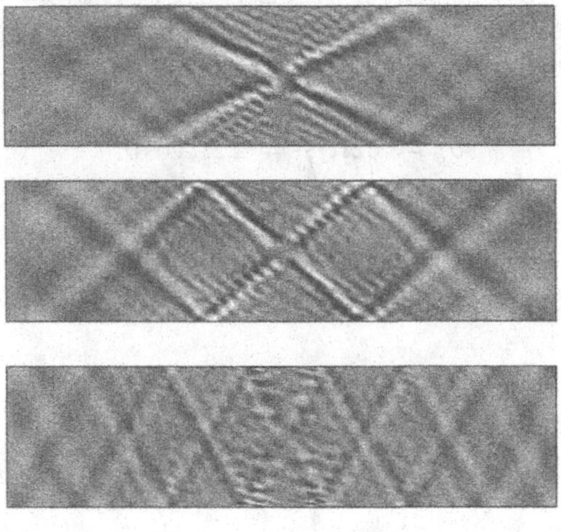

Figure 6: Saint Andrew's Cross shaped structures in a rotating flow, from DNS in a plane channel, rotating around the vertical axis. From top to bottom $2\Omega/\sigma_0 = 1.10, 1.33, 2.$. See Godeferd and Lollini (1998)

form of the RDT Green's function is found in terms of the two complex eigenvectors $\boldsymbol{N} = \boldsymbol{e}^{(2)} - \imath \boldsymbol{e}^{(1)}$ and $\boldsymbol{N}^* = \boldsymbol{N}(-\boldsymbol{k}) = \boldsymbol{e}^{(2)} + \imath \boldsymbol{e}^{(1)}$ in the plane normal to $\boldsymbol{k}$, or

$$G_{ij}^{RDT}(\boldsymbol{k}, t, t') = \sum_{s=\pm 1} N_i(s\boldsymbol{k}) N_j(-s\boldsymbol{k}) e^{\imath s \sigma_k (t-t')} \tag{74}$$

Diagonal decomposition is particularly useful in the context of pure rotation, since $\boldsymbol{N}$ and $\boldsymbol{N}^*$ more generally generate the eigenmodes of the Curl operator, and directly appear in the (e, Z, h) decomposition in (65). Accordingly, the RDT solution for the second-order spectral tensor writes

$$e(\boldsymbol{k}, t) = e(\boldsymbol{k}, t_0), \quad h(\boldsymbol{k}, t) = h(\boldsymbol{k}, t_0), \quad Z(\boldsymbol{k}, t) = e^{4\imath \sigma_k (t-t_0)} Z(\boldsymbol{k}, 0) \tag{75}$$

5.2 Linear-Nonlinear or Rapid-Slow ?

Simplified equations projected on the basis of eigenmodes $\boldsymbol{N}$ and $\boldsymbol{N}^*$ can be used for discussing both linear and nonlinear operators, as well as to develop closure theories for rotating turbulence. In terms of the amplitudes ξ_s, $s = \pm 1$, which are defined by

$$\hat{\boldsymbol{u}}(\boldsymbol{k}, t) = \xi_+(\boldsymbol{k}, t) \boldsymbol{N}(\boldsymbol{k}) + \xi_-(\boldsymbol{k}, t) \boldsymbol{N}(-\boldsymbol{k}) \tag{76}$$

the complete equation writes

$$[\frac{\partial}{\partial t} + \nu k^2 - \imath s \underbrace{\left(2\Omega \frac{k_\parallel}{k}\right)}_{\sigma_k}]\xi_s$$

$$= \sum_{s',s''=\pm 1} \int_{k+p+q=0} m_{ss's''}(\mathbf{k},\mathbf{p})\xi^*_{s'}(\mathbf{p},t)\xi^*_{s''}(\mathbf{q},t)d^3\mathbf{p} \tag{77}$$

with a diagonal form for the linear operator, and a modified quadratic form for the nonlinear term. The linear inviscid solution writes

$$\xi_s(\mathbf{k},t) = \exp\left(2\imath s\Omega t\frac{k_\parallel}{k}\right)\xi_s(\mathbf{k},0), \quad s = \pm 1$$

Replacing the initial condition by a new function $\boldsymbol{a}_s$, or

$$\xi_s(\mathbf{k},t) = \exp\left(2\imath s\Omega t\frac{k_\parallel}{k}\right)a_s(\mathbf{k},t), \quad s = \pm 1 \tag{78}$$

An equation for $\boldsymbol{a}_s$ is derived, in which linear operators are absorbed in the nonlinear one, as integrating factors.

$$\dot{a}_s = \sum_{s',s''=\pm 1}\int_{k+p+q=0} \exp\left[2\imath\Omega\left(s\frac{k_\parallel}{k} + s'\frac{p_\parallel}{p} + s''\frac{q_\parallel}{q}\right)t\right] m_{ss's''}(\boldsymbol{k},\boldsymbol{p})a^*_{s'}(\boldsymbol{p},t)a^*_{s''}(\boldsymbol{q},t)d^3\mathbf{p} \tag{79}$$

Note that the $\boldsymbol{a}_s$ play the role of slow amplitudes, and the problem could be analyzed by a technique of multiple (two) time-scale, setting $\boldsymbol{a}_s = \boldsymbol{a}_s(\boldsymbol{k},\epsilon t)$, with ϵ a (really) small parameter for asymptotic expansion, that can be related to a Rossby number. Such a refined analysis is not needed here, and we will just retain from equation (79) the importance of resonant triads, $\sigma_k \pm \sigma_p \pm \sigma_q = 0$, which correspond to zero value of the phase term in the right-hand-side of (79), or

$$\frac{k_\parallel}{k} \pm \frac{p_\parallel}{p} \pm \frac{q_\parallel}{q} = 0 \quad \text{with} \quad \boldsymbol{k} + \mathbf{p} + \mathbf{q} = 0. \tag{80}$$

These resonant, or quasi-resonant, triads are found to dominate nonlinear slow motion, since the effect of the phase term in the left-hand-side of (79), if far from zero, is to severely damp the nonlinearity by a 'scrambling effect'. The complexity of the resonant surfaces is illustrated by the Figure 7. Going back to second order statistics, equation (47) reduces to the simple system

$$\left(\frac{\partial}{\partial t} + 2\nu k^2\right)e = T^{(e)} \tag{81}$$

$$\left(\frac{\partial}{\partial t} + 2\nu k^2 + 2\imath\sigma_k\right)Z = T^{(Z)} \tag{82}$$

$$\left(\frac{\partial}{\partial t} + 2\nu k^2\right)\mathcal{H} = T^{(h)} \tag{83}$$

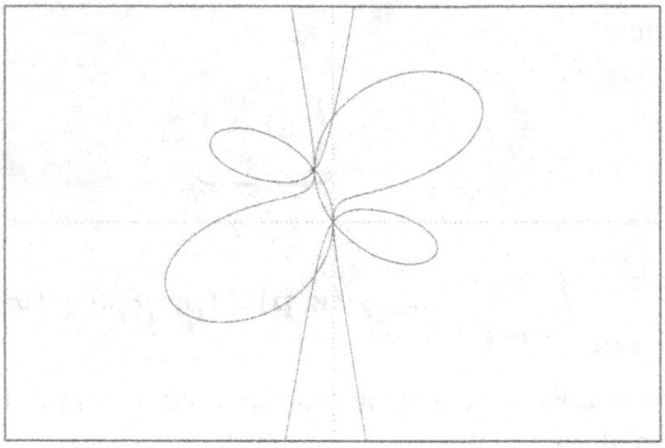

Figure 7: Visualization of resonant surfaces of inertial waves, given by (80), for a given orientation of $\boldsymbol{k}$. The locus of $\boldsymbol{p}$ is seen in the plane $p_2 = 0$, for $\theta_k = 1.1$. Complex loops appear for $\pi/3 < \theta_k < \pi/2$. Courtesy of F. S. Godeferd.

in terms of the set $(e, Z, \mathcal{H})$. Contribution from triple velocity correlations are gathered into the generalized spectral transfer terms $T^{(e,Z,h)}$, which derive from equations (58) and (59).If the above system of equation is started with 3D isotropic initial data, or $e(k,0) = E(k)/(4\pi k^2)$, $Z = \mathcal{H} = 0$, the anisotropy which should reflect the transition towards 2D structure can be created by the nonlinear spectral transfer terms only. This anisotropy consists of axisymmetry without mirror symmetry, leading to $e = e(k, \cos\theta = \frac{k_\parallel}{k}, t)$, $Z = Z(k, \cos\theta = \frac{k_\parallel}{k}, t)$, with $Z = 0$ if $\boldsymbol{k}$ is parallel to the vertical axis, in agreement with the symmetries of basic (rotating Navier-Stokes) equations , which ought to be satisfied by the closure theory.

It is now possible to discuss an optimal way to treat the 'Markovianisation' , or the time-dependency in the integrands that connect the transfer term to second order correlations. Closed equations (58) display three kinds of time-dependent terms:

1. Viscous, or viscous + damping, terms

$$\exp\left(\int_{t'}^{t} \mu dt''\right) \rightarrow V(t,t')$$

2. Terms from the RDT Green's function

$$\mathbf{G}(t,t') \rightarrow \exp[\pm \imath\sigma(t-t')]$$

3. Terms from the second-order spectral tensor (though quasi-normal assumption)

$$\hat{\mathbf{R}}(t') \rightarrow (e, Z, h)(t')$$

According to the markovianization procedure in classical EDQNM , we can assume that $V(t,t')$ is so rapidly decreasing in term of time-separation $\tau = t - t'$ that it is only concerned by the time integral in the closure equations, whereas the other terms take their instantaneous value, at $t' = t$, so that they are replaced by $\mathbf{G}(t,t)$ and $\hat{\mathbf{R}}(t)$, respectively. In other words, one considers only $V(t,t')$ as rapid, and the other terms as slow. This procedure, say EDQNM1, is not convenient for rotating turbulence, since the presence of $\mathbf{G}(t-t')$ in the closure relationship is responsible for breaking the initial isotropy. Using EDQNM1 started with isotropic initial data, isotropy is maintained, and no effect of system rotation can appear. A second step, say EDQNM2, consists of only 'freezing' (setting $t' = t$ in them) the (e, Z, h) terms, whereas the complete 'readjusted' response function, with both $V(t,t')$ and $\mathbf{G}(t,t')$ terms, is conserved in the time-integrand with its detailed time-dependency. An interesting result is that the time integral of the three-fold product of response functions yields a generic closure relationship as

$$T^{(e,Z,h)} = \sum_{s=\pm1, s'=\pm1, s''=\pm1} \int \frac{S^{ss's''}(e,Z,h)}{\mu_{kpq} + \imath(s\sigma_k + s'\sigma_p + s''\sigma_q)} d^3\mathbf{p} \tag{84}$$

Results from EDQNM2, regarding the creation of directional anisotropy, or description of the transition to 3D isotropy to 2D structure, are shown in Figure 8, and compared with high-resolution 528×128×128 LES. Recall that the development of angular dependency in $e(k, \cos\theta = \frac{k_\parallel}{k}, t)$, which amounts to a concentration towards the 2D slow manifold (sketched in Figure 12), results from nonlinear interactions mediated by $T^{(e)}$ in (81) and (84).

As pointed out by Julian Scott (private communication), the latter procedure can be questioned, in spite of its excellent numerical results, since it is not completely consistent with the basic rapid-slow decomposition suggested by (78). All the terms in the set (e, Z, h), which generates $\hat{\mathbf{R}}$, have not to be considered as 'slow', regarding RDT solution (75). Therefore, it is necessary to write

$$Z(t') = \exp(2\imath\sigma t')Z0(t'), \tag{85}$$

so that $Z0$ only appears as a slow variable, in complete agreement with (78). In so doing, this optimal procedure, say EDQNM3, yields freezing $e(t') = e(t)$, $h(t') = h(t)$, $Z0(t') = Z0(t)$ with keeping the (t') dependency under the integral for $Z(t') = \exp(2\imath\sigma t')Z0(t)$, and other terms $V(t,t')$, $\mathbf{G}(t,t')$, as before. This EDQNM3 version only slightly differs from EDQNM2, but presents decisive advantages. It is completely consistent with building EDQNM in terms of slow amplitudes using (78). As a bonus, an asymptotic development can be derived in the limit $\mu \ll 2\Omega$, which exactly coincides with Eulerian wave-turbulence theory (see Caillol and Zeitlin, 2000 in the similar case of stratified turbulence, and concluding remarks). Realizability can be demonstrated in this limit, and is not ensured in the EDQNM2 version.

5.3 Stability of Organised Eddies in a Rotating Frame

The Coriolis force alters the stability of 2D vortex flows through 3D disturbances. As an illustration, it is possible to consider 2D base flows more complex than those of homogeneous RDT in (24), as shown in Figures 9 and 10. For instance, the Taylor-Green flow in a rectangular cell (Figure 9) has an elliptic point in the core of each eddy, and an hyperbolic point in the corner of the four cells. The Stuart flow (Figure 10) is elliptic in the core region with hyperbolic points inserted between adjacent vortices (only a single vortex is shown in Figure 10, but one has to consider periodicity in the horizontal direction). The stability of these flows can be revisited in a rotating frame, using the short wave WKB theory by Lifschitz and Hameiri (1991), which

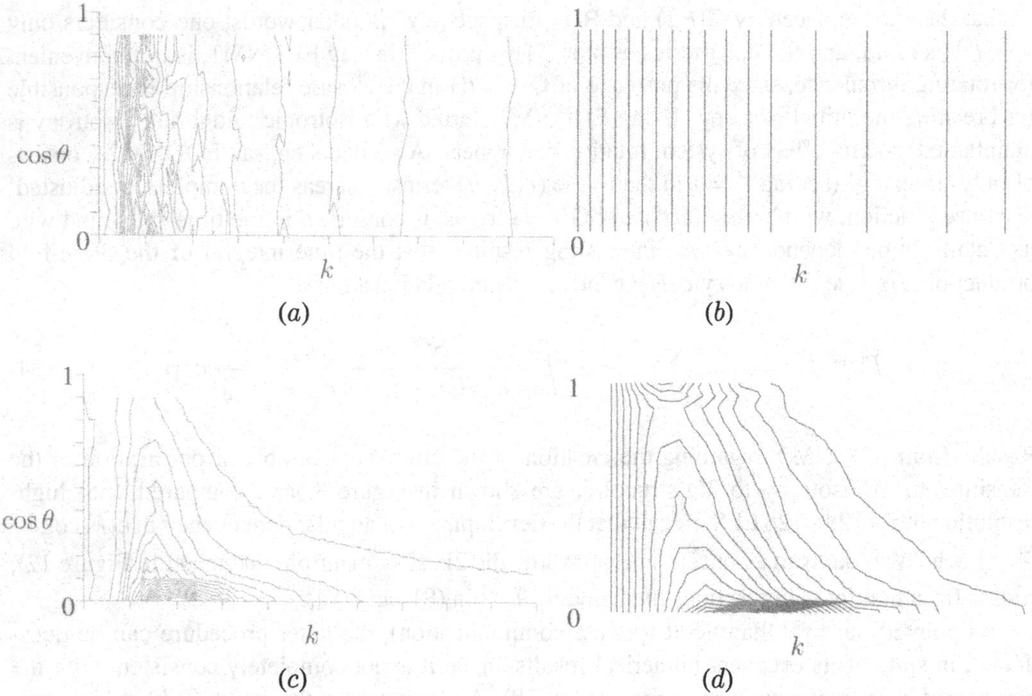

Figure 8: Isolines of kinetic energy $e(k, \cos\theta, t)$ for $512 \times 128 \times 128$ LES computations (*a*) at $\Omega = 0$ at time $t/\tau = 427$, (*b*) EDQNM2 with $\Omega = 0$; (*c*) LES with $\Omega = 1$ at $t/\tau = 575$; and (*d*) EDQNM2 calculation with $\Omega = 1$ at time $t/\tau = 148$. The vertical axis bears $\cos\theta_k$ (from 0 to 1 upwards) and the horizontal one the wave number k. (see Cambon *et al.* 1997, section 5.)

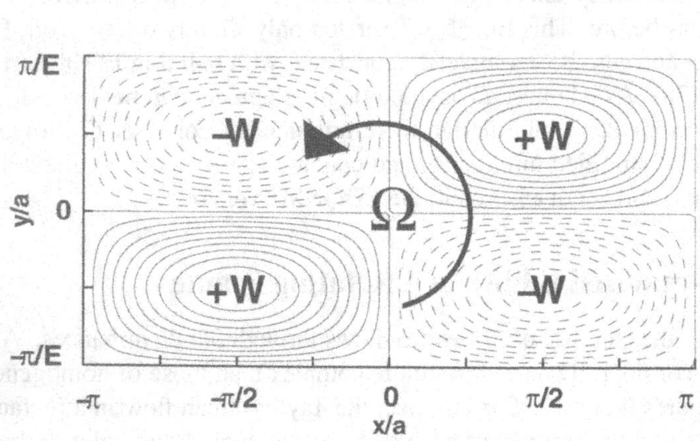

Figure 9: The Taylor-Green flow : iso-values of the vorticity. Case E = 2. (Courtesy Sipp *et al.* 1999)

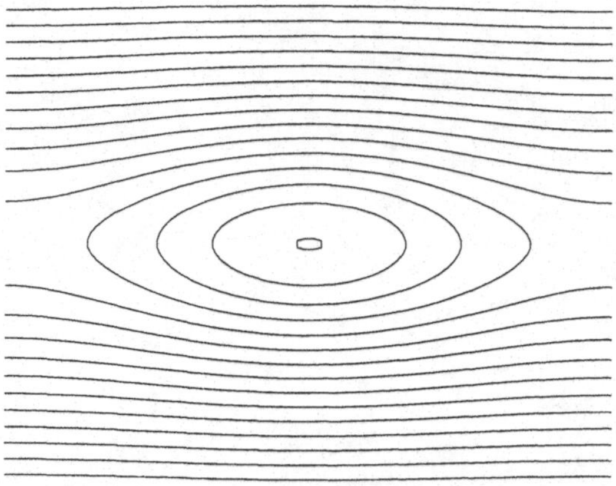

Figure 10: The Stuart flow. Isovalues of the streamfunction. Core ellipticity parameter $\rho = 1/3$.

amounts to a zonal RDT analysis (subsection 2.2). Such an analysis allows us to identify the role of elliptical and hyperbolic points in 3D instabilities altered by system rotation, but also to capture the centrifugal instability that may affect anticyclonic vortices (Sipp *et al.* , 1999 and Godeferd et al., 2001). The three kinds of instability and their possible competition were studied by solving the Townsend equations (36) and (37) along different trajectories. For each closed trajectory, a temporal Floquet parameter can be calculated from the zonal RDT Green's function; this parameter $\sigma(x_0, \theta)$ depends both on the space coordinate x_0, that labels the trajectory, and on the angle θ that gives the orientation of the wave-vector. A typical pattern of $\sigma(x_0, \theta)$ is shown in Figure 11, in the case of the rotating Stuart flow. The dominant instability is the centrifugal one in the particular anticyclonic case chosen in the figure. In addition, a typical elliptical instability branch emanates from the core (left part of the figure). One should point out that the 'geometric optics' for short wave disturbances can provide real insight into the nature of instabilities – e.g. elliptical, hyperbolic, and centrifugal – that occur in non-parallel flow, with and without system rotation. Classical massive eigenvalue problems provide little or no such insight.

5.4 Concluding Comments

Turbulence in a rotating frame illustrates the subtle interplay between linear and nonlinear processes and the significance of spectral anisotropy, especially the angular dependence of spectral energy which reflects the dimensionality. As evidenced by equations (81) to (83), the 3D isotropy, if initially stated, can only be broken by nonlinear interactions. In contast with isotropic free turbulence, RPT's can be based on an actual small parameter, the Rossby number, for exploring the 'slow' nonlinear tendencies. EDQNM2-3 models illustrate the role of resonant triads of inertial waves, as should do an Eulerian 'wave-turbulence' theory. In the limit of small Rossby number, all these approches give essentially the same equations, but a sophisticated version of the eddy damping would be desirable to describe a broader range of Rossby and Reynolds numbers. If the flow includes preexisting large-scale vortices, the stability analysis shows the importance of the Coriolis force, which alters, but even can create or inhibit classical elliptical or centrifugal instabilities. In spite of the evidence of the transition towards 2D structure in homogeneous decaying turbulence, the creation of large strong vortices from initially unstructured

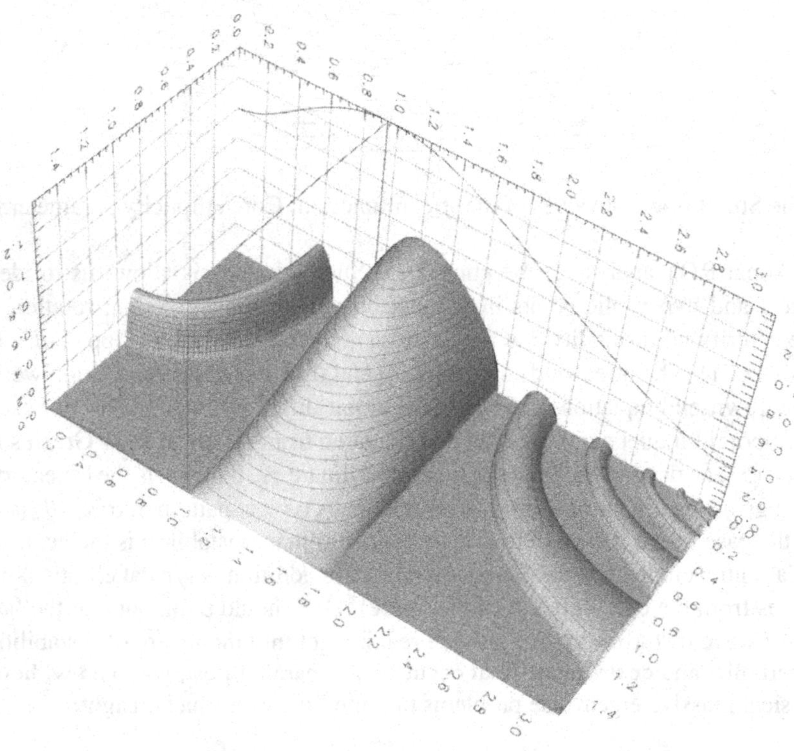

Figure 11: Floquet amplification parameter σ (plotted onto the vertical axis) as a function of the trajectory, indexed by the position x_0 (on the left), and the orientation θ (on the right) of the local wavevector to the spanwise (normal to the plane of the base flow) axis. x_0 varies from 0 (core) to π (periphery), and θ varies from 0 (pure spanwise modes) to $\pi/2$ (pure two-dimensional modes). Anticyclonic system rotation: the dimensionless vorticity at the core ($x_0 = 0$) of a Stuart cell is -7 and the related Rossby number is -5. See Godeferd *et al.* (2001)

turbulence by the Coriolis force, is difficult to observe in numerical simulations (free of numerical bias !) in the absence of some forcing effects (see Godeferd and Lollini, 1999; Cambon, 2001; Smith and Waleffe, 2002 and references therein).

6 Turbulence in a Stably-Stratified Fluid with and without Rotation

These cases continue to illustrate the 'flows dominated by wavy effects' with 'zero production' introduced at the end of section 3. For pure rotating turbulence, there is zero production of kinetic energy and for stratified rotating turbulence there is zero production of total (kinetic plus potential) energy. Linearised solutions of the Navier-stokes equations, with buoyancy force b within the Bousinesq assumption, are easily obtained in the presence of a uniform mean density gradient and in a rotating frame. For the sake of simplicity, the mean flow is restricted to a uniform vertical gradient of density and to a solid body rotation in the horizontal direction, with typical parameters N (the Brunt-Väisälä frequency) and Ω (the angular velocity).

$$(\partial_t + \boldsymbol{u}\cdot\boldsymbol{\nabla})\,\boldsymbol{u} + 2\Omega\boldsymbol{n}\times\boldsymbol{u} + \boldsymbol{\nabla}p - \nu\nabla^2\boldsymbol{u} = \bar{v} \tag{86}$$

$$(\partial_t + \boldsymbol{u}\cdot\boldsymbol{\nabla})b - \kappa\nabla^2 b = -N^2\boldsymbol{n}\cdot\boldsymbol{u} \tag{87}$$

$$\boldsymbol{\nabla}\cdot\boldsymbol{u} = 0 \tag{88}$$

for the fluctuating velocity $\boldsymbol{u}$, the pressure p divided by a mean reference density, and the buoyancy force $\bar{v}$. The vector $\boldsymbol{n}$ denotes the vertical unit upward direction aligned with both the gravitational acceleration $\boldsymbol{g} = -g\boldsymbol{n}$ and the angular velocity of the rotating frame $\boldsymbol{\Omega} = \Omega\boldsymbol{n}$. The buoyancy force is related to the fluctuating temperature field τ by $\bar{v} = -g\beta\tau = b\boldsymbol{n}$, through the coefficient of thermal expansivity β. With temperature stratification characterized by the vertical gradient γ, the Brunt-Väisälä frequency $N = \sqrt{\beta g\gamma}$ appears as the characteristic frequency of buoyancy-stratification. Hence the linear operators in equations (86) and (87) display the two frequencies N and 2Ω. Without loss of generality the fixed frame of reference is chosen such that $n_i = \delta_{i3}$. Therefore, u_3 is the vertical velocity component.

Analysis of the linear limit, mathematical treatment of equations in terms of eigenmodes, and closure methods for statistics in homogeneous anisotropic turbulence, can be developed as for the particular case of pure rotation, in section 5.

Without pressure fluctuation, the additional buoyancy and stratification terms yield oscillations for vertical velocity and buoyancy term, with frequency N. This simple motion reflects that the buoyancy force acts as a restoring force in the case of stable stratification. As for the case of pure rotation, the pressure term in (86) is needed to satisfy (88), and its role in the complete linear solution consists of coupling vertical and horizontal motion and of generating dispersive inertia-gravity waves. An important difference, however, is that a part of the horizontal motion remains steady in the linear limit, and therefore decoupled from 3D wave motion: this is related to the quasi-geostrophic mode, which is pivotal in dynamical meteorology.

In the unbounded case, or for periodic boundary conditions, the different modes, wavy and steady, are easily found in Fourier space, and a tractable RDT solution is found in terms of them. Pressure fluctuation is removed from consideration in the Fourier-transformed equations by using the local frame in the plane normal to the wave vector (27), taking advantage of (88), so that the problem in five components (u_1, u_2, u_3, p, b) in physical space is reduced to a problem in three components, two solenoidal velocity components $(u^{(1)}, u^{(2)})$ and a component for $\hat{b}$, in Fourier space. The three-component set $(u^{(1)}, u^{(2)}, \hat{b})$ is not a true vector, and this can complicate further mathematical developments in terms of its eigenmodes and statistical correlations. So it is more convenient to gather these three components into a new vector $\hat{v}$, whose inverse 3D Fourier transform, $\boldsymbol{v}$, is real. $\hat{\boldsymbol{v}}$, can be written as

$$\hat{\boldsymbol{v}} = \hat{\boldsymbol{u}} + \imath \frac{1}{N} \hat{b} \frac{\boldsymbol{k}}{k} \tag{89}$$

so that its three components are $u^{(1)}$, $u^{(2)}$, and $\imath \frac{1}{N} \hat{b}$ in the frame (27). The scaling of the contribution from the buoyancy force allows one to define twice the total energy spectral density as

$$\hat{v}_i^* \hat{v}_i = \hat{u}_i^* \hat{u}_i + N^{-2} \hat{b}^* \hat{b} \tag{90}$$

The linear equation for $\hat{\mathbf{v}}$, derived from (86)- (88) is similar to (15), of the form

$$\frac{\partial \hat{\boldsymbol{v}}(\boldsymbol{k}, t)}{\partial t} + \mathbf{M}'(\boldsymbol{k}) \hat{\boldsymbol{v}}(\boldsymbol{k}, t) = 0$$

so that the RDT solution can be found in terms of the three eigenmodes of $\mathbf{M}'$, as

$$\hat{\boldsymbol{v}} = \xi^{(0)} \boldsymbol{N}^{(0)} + \xi^{(1)} \boldsymbol{N}^{(1)} + \xi^{(-1)} \boldsymbol{N}^{(-1)} \tag{91}$$

in which the eigenmodes $\boldsymbol{N}^{(s)}$, $s = 0, \pm 1$ are simple linear combinations of the vectors in (27). A Green's function similar to (74) is derived as

$$G_{ij}(\boldsymbol{k}, t, t_0) = \sum_{s=0,\pm 1} N_i^s(\boldsymbol{k}) N_j^{-s}(\boldsymbol{k}) \exp[\imath s \sigma_{\boldsymbol{k}}(t - t_0)], \tag{92}$$

but the essential difference with the case of pure rotation is the presence of the quasi-geostrophic mode $\boldsymbol{N}^{(0)}$ which involves a part of the toroidal velocity field and a part of the buoyancy field. In addition, the wavy modes $\boldsymbol{N}^{(\pm 1)}$ are also different from the helical modes $\boldsymbol{N}(\pm \boldsymbol{k})$, whereas

$$\sigma_k = \sqrt{N^2 \left(\frac{k_\perp}{k}\right)^2 + 4\Omega^2 \left(\frac{k_\parallel}{k}\right)^2} \tag{93}$$

holds for the absolute value of the frequency given by the dispersion law of inertia-gravity internal waves . Because of the form of the eigenvectors and of the dispersion law, the structure of $\mathbf{G}$ in (92) is consistent with axisymmetry around the axis of reference (chosen vertical here), without mirror symmetry, and $k_\parallel$ and $k_\perp$ hold for axial (along the axis) and transverse (normal to the axis) components of $\boldsymbol{k}$.

Anisotropy can be significantly broken through axisymmetrical response function for triple correlations only, or possibly for two time second order statistics (outside the scope of this chapter), but the linear limit exhibits no interesting creation of structural anisotropy in classic RDT for predicting second order single-point statistics. However in practice there is two-dimensionalisation in rotating turbulence and a horizontal layering tendency in the stably stratified case. In other words, RDT only alters phase dynamics, and conserves exactly the spectral density of typical modes (full kinetic energy for the rotating case, total energy and 'vortex', or potential vorticity, energy for the stably stratified case), so that two-dimensionalization or 'two-componentalization' (horizontal layering), which affect the distribution of this energy, are typically nonlinear effects.

Nevertheless, the eigenmodes of the linear regime, derived from RDT, form a useful basis for expanding the fluctuating velocity-temperature field, even when nonlinearity is present, and nonlinear interactions can be evaluated and discussed in terms of triadic interactions between these eigenmodes. Accordingly, the complete anisotropic description of two-point second order correlations, e.g. (24) for pure rotation, can be related to spectra and cospectra of these eigenmodes.

6.1 Pure Stratified Homogeneous Turbulence

The case of stably stratified turbulence is different from the one of pure rotation, even if the gravity waves present strong analogies with inertial waves. An additional element is the presence of the 'vortex', or potential vorticity (PV) mode, which is a particular case of the quasi-geostrophic mode $\mathbf{N}^0$, which is related to $s = 0$ steady motion in (92). Hence the motion in the linear limit is not completely dominated by wavy modes, in contrast with the case of pure rotation. Regarding dispersive gravity waves, the general relationship (93) reduces to

$$\sigma_k = N\frac{k_\perp}{k} \tag{94}$$

and cross-shaped structures can be found in a stratified tank (Mowbray and Rarity, 1967), from a local harmonic forcing, as discussed at length for pure rotation.

Focusing on nonlinear effects, interactions which do not involve the wavy mode, i.e. the part of the flow which is steady in the linear limit (PV mode + VSHF mode) have been found to be dominant in triggering the loss of isotropy, as a prerequisite to orient the evolution of the initially isotropic velocity field towards a two-component state. EDQNM2 (Godeferd and Cambon, 1994) and DNS results have shown that the spectral energy concentrates towards vertical wavenumbers $k_\perp \sim 0$. Because $\boldsymbol{k}$ and $\boldsymbol{u}$ are perpendicular, these wavenumbers correspond to predominantly horizontal, low-frequency motions, called VSHF after Smith and Waleffe (2002). As for the partial transition towards 2D structure shown in pure rotation, a new dynamical insight is given to the collapse of vertical motion expected in stably stratified turbulence, but the long-time behaviour essentially differs from a two-dimensionalisation.

A sketch of the different nonlinear effects of pure rotation and pure stratification is shown in Figure 12. Previous EDQNM studies focused on triple correlation characteristic times modified by wave frequencies, whereas wave-turbulence theories proposed scaling laws for wave-part spectra. None of them, however, was capable of connecting wave-vortex dynamics to the vertical collapse and layering. Only recently, by re-introducing a small but significant vortex part in their wave turbulence analysis, Caillol and Zeitlin (2000) found that: *'The vortex part obeys a limiting slow dynamics equation exhibiting vertical collapse and layering which may contaminate the wave-part spectra'.* This is in complete agreement with the main finding of Godeferd and Cambon (1994), where this result reflects a scrambling of any triadic interactions, including at least one wave mode, so that the pure PV (+VSHF) interaction becomes dominant. The corresponding 'wave-released' triadic energy transfer is strongly anisotropic. It does not

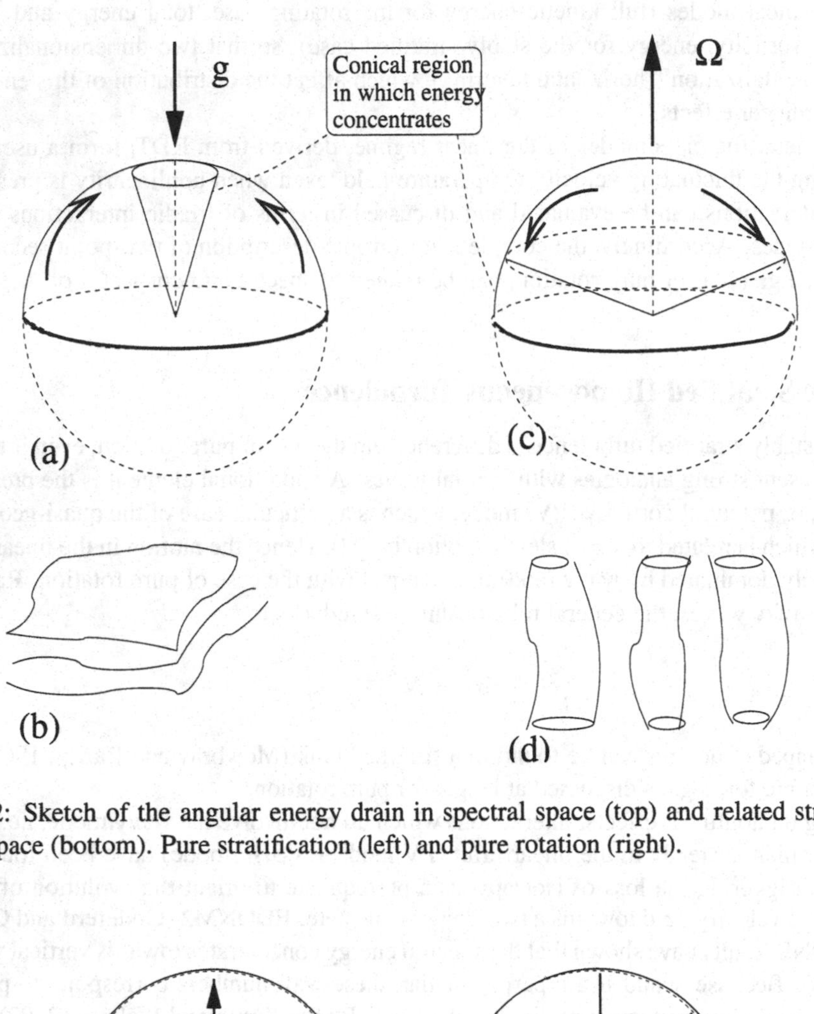

Figure 12: Sketch of the angular energy drain in spectral space (top) and related structure in physical space (bottom). Pure stratification (left) and pure rotation (right).

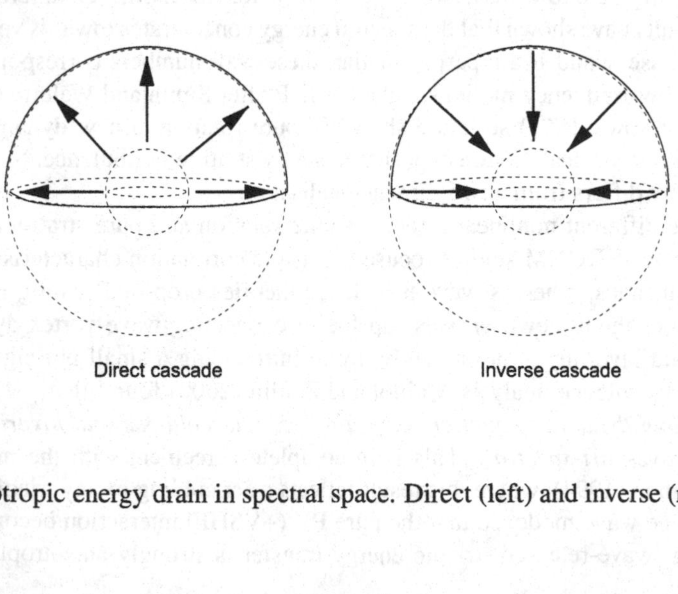

Figure 13: Isotropic energy drain in spectral space. Direct (left) and inverse (right) cascade.

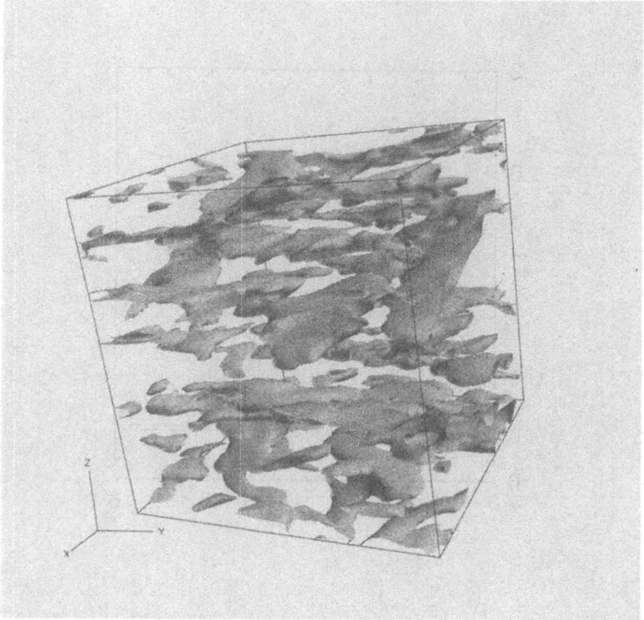

Figure 14: Isovalues of vertical gradient of horizontal velocity fluctuation. Pure stratification. 256^3 DNS with isotropic initial data. Courtesy of F. S. Godeferd and C. Staquet

yield a classic direct cascade, as sketched in Figure 13, (which would contribute to dissipate the energy) but instead yields the angular drain of energy which condenses the energy towards vertical wave-vectors, in agreement with vertical collapse and layering. The latter effect is reflected in physical space by a pancake structure, as illustrated in Figure 14, in which isovalues of velocity gradients are obtained from a snapshot of 256^3-DNS. This layering can be statisticaly quantified by the development of two different integral length scales, as shown in Figure 15 (from EDQNM2) and 16 (from DNS), with excellent agreement. The integral lengthscale related to horizontal velocity components and horizontal separation $L_{11}^{(1)}$ is shown to develop similarly to isotropic unstratified turbulence, whereas the one related to vertical separation $L_{11}^{(3)}$ is blocked. In the same conditions, with initial equipartition of potential and wave energy, linear calculation (RDT) exhibits no anisotropy, or $L_{11}^{(1)} = 2L_{11}^{(3)}$.

The mode related to vertical wave-vectors, which is linked to the VSHF mode in physical space, appears to be very important, since the concentration of spectral energy on it gives the most sensible identification of the development of vertical collapse and layering. It corresponds to the limit of the wavy mode, when the dispersion frequency tends to zero. Strictly speaking, this mode is a slow mode, which cannot be strictly referred to 'vortex' or 'wave'. It is absorbed in any decomposition based on the Craya-Herring frame (see equation (27) and Figure 4), provided that some care is taken to extend by continuity the definition of the unit vectors $(\mathbf{e}^{(1)}, \mathbf{e}^{(2)})$ towards $\boldsymbol{k}$ aligned with the polar (vertical here) axis of the frame of reference. In so doing, the

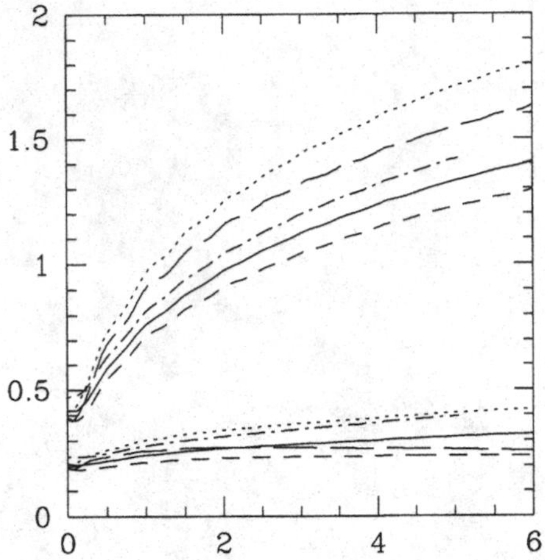

Figure 15: Development of typical integral lengthscales from EDQNM2. $L_{11}^{(1)}$ (top) and $L_{11}^{(3)}$ (bottom), with 1 and 3 horizontal and vertical directions, respectively. (Initial) isotropy implies $L_{11}^{(1)} = 2L_{11}^{(3)}$. Courtesy of F. S. Godeferd and C. Staquet.

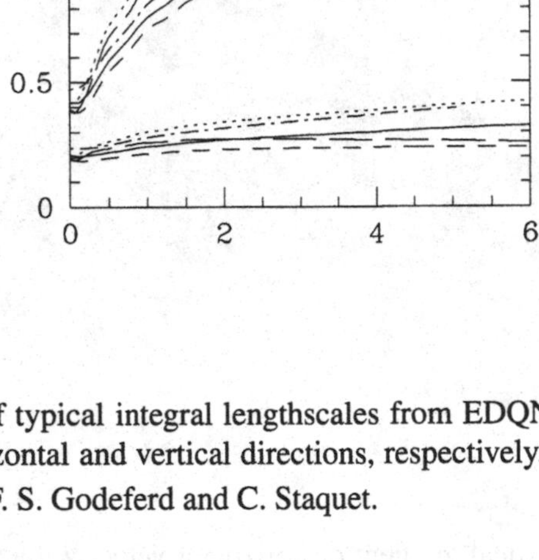

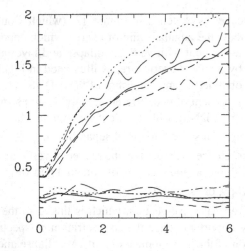

Figure 16: Same as Figure 15, from 256^3 DNS.

mode related to $\mathbf{e}^{(1)}$ coincides with a toroidal, or 'horizontal vortex', mode, but for vertical wave vectors, where it includes half the energy of the vertical slow (VSHF) mode. In the same way, the mode related to $\mathbf{e}^{(2)}$ coincides with a poloidal mode, affected by the wavy motion, but for vertical wave vectors, where it includes the other half of the energy of the VSHF mode.

6.2 Rotating Stratified Turbulence: Pancake versus Cigar Structuring

In agreement with the general RDT solution,

$$\boldsymbol{u}'(\boldsymbol{x},t) = \Sigma[\boldsymbol{A}^{(0)}(\boldsymbol{k}) + \boldsymbol{A}^{(1)}(\boldsymbol{k})\exp(-\sigma_k t) + \boldsymbol{A}^{(-1)}(\boldsymbol{k})\exp(+\sigma_k t)]\exp(\imath(\boldsymbol{k}\cdot\boldsymbol{x})), \quad (95)$$

with similar relationship for the buoyancy force, which is linked to temperature fluctuation. The $A^{(s)}, s = 0, \pm 1$ are projections of the initial field onto the base of eigenmodes, and are kept constant in the inviscid linear, or RDT, limit. This base of eigenmodes is used in different approaches, statistical theories (Cambon, 2001 and references therein) as well as full DNS (*e.g.* Smith and Waleffe, 2002). As seen before, dynamics from pure RDT is of poor interest in this case, if Eulerian single-time correlations are concerned, since strict conservation of the amplitudes $\boldsymbol{A}^{(s)}$ prevents interesting structuring to occur. These effects, two-dimensionalisation in pure rotation and two-componentalization in pure stratification, are thus ultimately controlled by nonlinear interactions, even weak. Typical 'cigar' and 'pancake' structure are shown in Figures 17 and 18, respectively. In order to account for nonlinearity, the $A^{(s)}$ have to be considered as

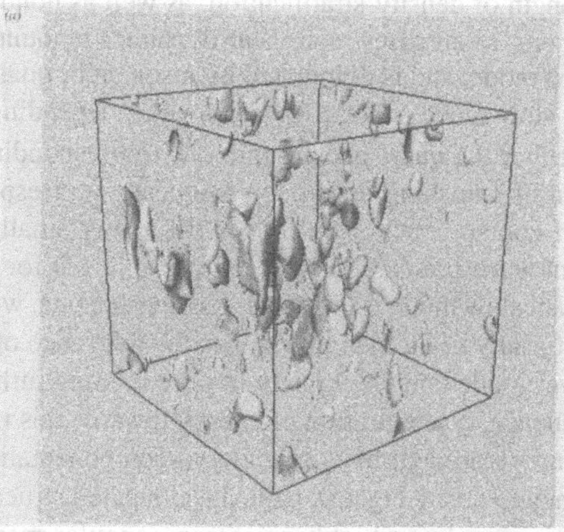

Figure 17: Isovalues of vorticity for dominant rotation. Courtesy of Kimura and Herring

slowly evolving amplitudes, for instance $\boldsymbol{A}^{(s)} = \boldsymbol{A}^{(s)}(\boldsymbol{k}, \epsilon t)$, and it is worthwhile to derive exact nonlinear equations for them from the background Navier Stokes equations for u_i and b, using (95). In other words, even if RDT is not relevant in itself, it suggest to substitute to primitive velocity-temperature variables a set of projections onto a convenient basis of eigenmodes. Analysis of long-time effects of nonlinear interactions is facilitated in terms of these eigenmodes. In addition to DNS, closure models can be constructed for predicting detailed energy distribution

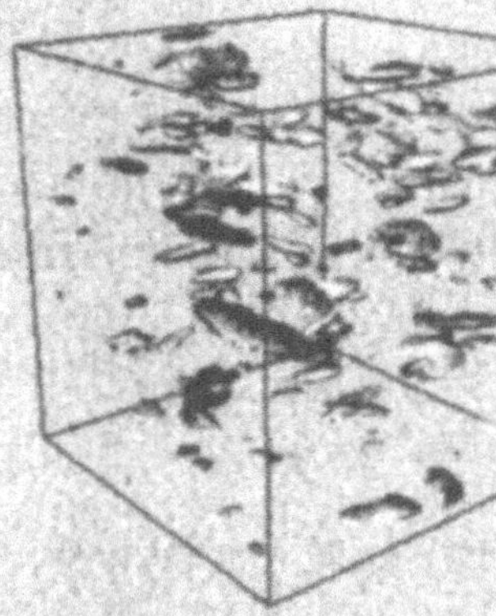

Figure 18: Isovalues of vorticity for dominant stratification. Courtesy of Kimura and Herring

in terms of different modes and different wavevectors, along the line of what was done in section 5. In order to give more physical meaning to the mathematical procedures introduced here, consistent results from (anisotropic multimodal) EDQNM and DNS are summarized as follows. Important parameters are the system vorticity of the rotating frame 2Ω, or the Coriolis parameter f in a geophysical context, the Brunt-Väisälä frequency N which characterizes the gravity wave frequency and the strength of density-stratification, as well as nondimensional Rossby (or Froude) and Reynolds numbers. In all cases, nonlinear dynamics amounts to a concentration of energy towards a typical slow mode, and is dominated by resonant or quasi-resonant triads in 3D wavespace. But the relevant slow mode is not always the same, depending on the ratio $N/(2\Omega)$. The relevant slow mode is either the quasi-geostrophic one (corresponding to $\boldsymbol{A}^{(0)}$ in (95) or a wavy one, found in the limit of vanishing dispersion frequency (corresponding to $\boldsymbol{A}^{(\pm 1)}$ if σ_k can reach a zero value). Of course, wavy slow modes fill a very small part of the configuration space, so as do resonant wave interactions ($\sigma_k \pm \sigma_p \pm \sigma_q = 0$ for a triad of wavevectors $\boldsymbol{k}+\boldsymbol{p}+\boldsymbol{q} = 0$). If the dominant slow mode is the quasi-geostrophic one, with all related nonlinear interactions being resonant for any triad, then this shows the relevance of the quasi-geostrophic model. Since the slow 'wave' mode is the 2D mode in pure rotating turbulence, and the VSHF mode in pure stratified turbulence, concentration of energy towards this mode is consistent with two-dimensionalisation in pure rotating turbulence and two-componentalisation in pure stratified turbulence, or related tendency to create cigar or pancake elongated structures in physical space, respectively. This concentration involves a highly anisotropic angular energy drain in wavespace. Another aspect is the presence of an inverse cascade for these typical slow modes, as evidenced by Smith and Waleffe (2002) using DNS forced isotropically at small scale.

Finally, it is important to point out that the building of large scale with an obvious, but very slow, concentration towards the VSHF mode in the latter DNS at dominant stratification was attributed by the authors to resonant gravity waves, and no significant creation of large scale PV mode was found. The mechanisms of concentration towards the VSHF mode are thus radically different in (Godeferd and Cambon, 1994, Caillol and Zeitlin, 2000), on the one hand, and in Smith and Waleffe (2002), on the other hand. Only in the case in which a significant large-scale PV part of the flow exists initially, the process of layering is relatively rapid and is mediated by those interactions which exclude wavy contributions. As for the relevance of the quasi-geostrophic model, The latter situation was shown by Smith and Waleffe (2002) to prevail when $1/2 < N/(2\Omega) < 2$, triadic 'wave' resonances being forbidden in this range of parameter.

As a conclusion of this subsection, a complete understanding of dominant nonlinear interactions in all these cases requires to have a combined description of the cascade (energy drain between different wave numbers) and of the angular energy drain (which is exclusively emphasized in Godeferd and Cambon (1994) and related Figure 12).

7 Towards Compressible Flows

7.1 Compressed Turbulence, RDT and Nonlinear Rescaling

An interesting class of solenoidal (divergence-free velocity fluctuations) homogeneous turbulent flows can be considered in the presence of a mean flow of type (7), which take into account a variation in mean volume. Provided that the Mach number is small enough, this set of assumptions is self-consistent, and it is possible to extend previous analysis, especially RDT, to 'compressed' turbulence: i.e divergence-free fluctuating velocity field in the presence of a mean dilatational flow, neglecting acoustics and thermal effects.

The mean flow is characterised by a volumetric ratio

$$J(t,t_0) = Det\mathbf{F}(t,t_0) = \exp\left[\int_{t_0}^{t} A_{ii}(t')dt'\right] \tag{96}$$

which differ from 1 when the constraint $A_{ii} = 0$ is relaxed in (8). For the sake of brevity, t_0 will be ommitted, so that abridged notations $\mathbf{F}(t)$, $J(t)$ will be used from now on in this subsection. Among different 'compressing' mean flows, the case of isotropic compression deserves particular attention. In this case, the matrices $\mathbf{A}$ and $\mathbf{F}$, and the trajectory equations write

$$A_{ij}(t) = S(t)\delta_{ij}, \quad F_{ij}(t) = J^{1/3}(t)\delta_{ij}, \quad x_i = J^{1/3}(t)X_j \tag{97}$$

in which $S = \frac{1}{3}\frac{1}{J}\frac{dJ}{dt}$. The fluctuating field is governed by

$$\frac{\partial u_i'}{\partial t} + Sx_j\frac{\partial u_i'}{\partial x_j} + Su_i' + \frac{1}{\rho}\frac{\partial p'}{\partial x_i} = -u_j'\frac{\partial u_i'}{\partial x_j} + \nu\nabla^2 u_i' \tag{98}$$

in which explicit nonlinear terms and viscous terms are gathered in the right-hand-side. Setting to zero this rigth-hand-side, the RDT solution is directly found in physical space [7] as

$$\boldsymbol{u}'(\boldsymbol{x},t) = J^{-1/3}(t)\boldsymbol{u}'(\boldsymbol{X},0)$$

More interesting is the possibility to derive a rescaling for full nonlinear equations (98), in terms of spatial coordinates, velocity and time, as follows

$$\boldsymbol{x}^* = J^{-1/3}\boldsymbol{x}, \quad \boldsymbol{u}^*(\boldsymbol{x}^*,t^*) = J^{1/3}\boldsymbol{u}'(\boldsymbol{x},t) \quad dt^* = J^{-2/3}(t)dt \tag{99}$$

[7] This is a very special case, in which the non-local potential term is zero in (30)

When substituting the previous relationship in the equation (98) that governs the primitive 'unstarred' variables, the 'starred' quantities are shown to satisfy the Navier-Stokes equations without the additional mean terms (with S) in the left-hand-side. For consistency, the pressure is rescaled as $p^* = J^{5/3}p$, and the only difference with freely (unstrained) decaying isotropic turbulence for the velocity field $\boldsymbol{u}^*(\boldsymbol{x}^*, t^*)$ is a possible impact of varying viscosity $\nu^*(t)$. The variation in Reynolds number follows directly since $u'L = u^*L^*$. If the Reynolds number is high enough, however, it is reasonable to stress that all classical spatio-temporal dynamics and statistics of isotropic freely decaying turbulence is valid for $(\boldsymbol{u}^*, \boldsymbol{x}^*, t^*)$, so that the corresponding laws for primitive variables $(\boldsymbol{u}', \boldsymbol{x}, t)$ can be readily derived using (99). The reader is referred to Cambon *et al.* (1992) for various applications.

This rescaling presents particular interest in this course, for two reasons. On the one hand, it illustrates a particular 'dynamical' version of the general scale invariance reported by Benzi (2002), or

$$\boldsymbol{x}^* = \lambda \boldsymbol{x}, \quad \boldsymbol{u}^* = \lambda^h \boldsymbol{u}, \quad t^* = t\lambda^{1-h}, \quad \nu^* = \lambda^{1+h}\nu \tag{100}$$

so that λ corresponds to the time-dependent mean density ratio $J^{-1/3}$, $h = -1$. In the latter invariance group, the viscosity would be unchanged if $h = -1$, but it can be noticed that the dynamical rescaling deals with a continuously time-varying parameter $J^{-1/3}(t)$ in contrast to λ.

On the other hand, it can suggest a new domain of 'homogeneous compressed turbulence' to extend the spatio-temporal scaling laws obtained by Oberlack (2002) in homogeneous turbulence.

7.2 Isentropic Turbulence in High Speed Flows

More generally, if the turbulent Mach number is not small, the effects of compressibility are much more complicated, since both acoustic and entropy modes are called into play, as well as the vortical mode inherited from the incompressible case (Kovasznay, 1953). Irrotational flows have been studied by Goldstein (1978) using an inhomogeneous RDT formulation (which can be based on (30)), while homogeneous RDT has been extended to quasi-isentropic compressible turbulence at significant Mach number, in the presence of either irrotational compression or mean shear flows (Simone *et al.* , 1997). For high speed compressible flows, it is no longer possible to consider the velocity field as divergence-free. Accordingly, the pressure disturbance can recover its role of thermodynamical variable, it is no longer a 'Lagrange multiplier', linked to divergence-free constraint, which can be eventually eliminated. The velocity field can be splitted into solenoidal (divergence-free) and dilatational parts, using a classic Helmholtz decomposition.

$$u'_i(\mathbf{x}, t) = u_i^{(s)} + u_i^{(d)} \tag{101}$$

Using a spectral description, this splitting is obtained when applying two projection operators, $\hat{u}_i^{(d)} = \frac{k_i k_j}{k^2}\hat{u}_j$, since the Fourier component related to the dilatational mode is aligned with $\boldsymbol{k}$, and $\hat{u}_i^{(s)} = \hat{u}_i - \frac{k_i k_j}{k^2}\hat{u}_i = P_{ij}\hat{u}_i$. In the presence of a distorting mean flow, it can be more interesting to display the two components which characterize the solenoidal part of the velocity field, using a poloidal-toroidal decomposition for instance. All can be gathered using the Craya-Herring frame of reference in (27), but now taking into account all three components

$$\widehat{u}_i(\mathbf{k}, t) = \underbrace{u^{(1)}.e_i^{(1)} + u^{(2)}.e_i^{(2)}}_{\widehat{u}^{(s)}} + \underbrace{u^{(3)}.e_i^{(3)}}_{\widehat{u}^{d}}$$

The classical description in terms of vorticity ω_i and divergence $d = \partial u'_i/\partial x_i$ is easily recovered as

$$\widehat{\omega}_i = \imath k \left(u^{(1)} e_i^{(2)} - u^{(2)} e_i^{(1)} \right)$$

and

$$\widehat{d} = \imath k u^{(3)}.$$

In addition, the pressure fluctuation p can be scaled as a fourth 'velocity' component, or

$$u^{(4)} = \imath \frac{\widehat{p}}{\overline{\rho} c},$$

with $\overline{\rho}$ and c being the mean density and the sonic speed, respectively. Homogeneous quasi-isentropic RDT for compressible flows can be developed as follows. For the mean flow with space-uniform gradients, a condition more restrictive than (8) is found as

$$A_{ij} + A_{ik} A_{kj} = 0 \tag{102}$$

As a simple explanation, the latter term comes from the contribution of mass-weighted acceleration in the equation which governs the fluctuating momentum, or $\overline{\rho}\gamma'_i + \rho'\overline{\gamma}$ with $\overline{\gamma}_i = (A_{ij} + A_{ik}A_{kj})x_j$. If the density fluctuation is not neglected, the nonzero value of the left-hand-side of (102) yields breaking the statistical homogeneity.

RDT solutions write

$$u^{(i)}\left(\boldsymbol{k}(t), t\right) = g_{ij}(\boldsymbol{k}, t, t_0) u^{(j)}\left(\boldsymbol{k}(t_0), t_0\right) \tag{103}$$

in which $i, j = 1, 4$, so that both velocity and pressure are involved. The time dependency of $\boldsymbol{k}$ is governed by the Cauchy matrix $\mathsf{F}(t, t')$ as in incompressible RDT. Typical results for irrotational compression and pure shear were obtained using (103). The matrix in the system of ODE that governs the set $(u^{(1)}, u^{(2)}, u^{(3)}, u^{(4)})$, with solution (103), depends on two different inverse time-scales, the mean velocity gradient $\mathbf{A}$, with magnitude S, and the sonic frequency kc. The ratio of these two parameters $S/(kc)$ is the key parameter of compressible RDT, and appears as a spectral counterpart of the gradient (or distortion) Mach number $M_g = SL/c$ used by Sarkar (1995), where L is a typical lengthscale of energetic structures. RDT results helped explain the systematic changes in energy growth rate with gradient Mach number found in numerical simulations (Sarkar, 1995 and Simone *et al.* , 1997). In the case of axial one-dimensional compression, the kinetic energy growth rate can increase with increasing compressibility, measured by M_g. This surprising 'destabilizing effect' of compressibility is due to the development of dilatational energy, in addition to solenoidal energy that develops as in the incompressible case. The same behaviour holds for the pure shear flow at small time ($St < 4$), but a classic 'stabilizing effect' is recovered at larger time. This 'cross-over' behaviour about $St = 4$, which is not present when the mean distortion is irrotational, is due to the linear coupling in (103) of dilatational and solenoidal modes, induced by the rotational part of the mean shear. These results also suggest that compressibility mainly alters the one-point properties of turbulence through the pressure-velocity correlation tensor, rather than via bulk viscous dissipation or other explicit compressible terms (e.g. the pressure-dilatation term) usually considered in compressible one-point modelling.

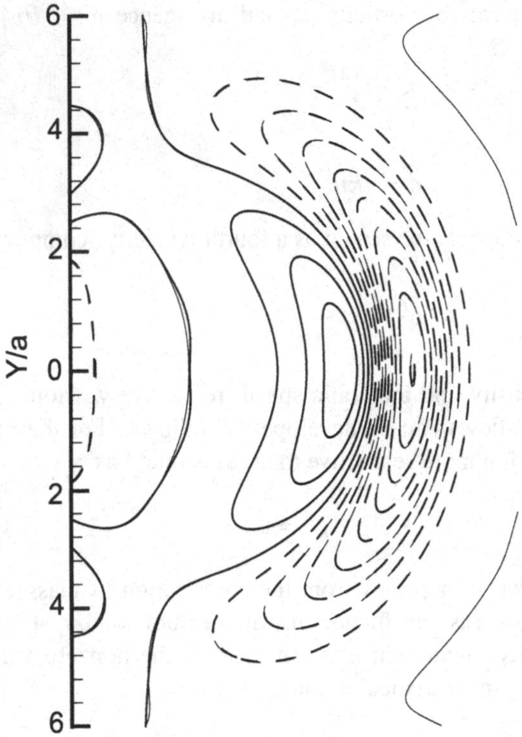

Figure 19: Isolines of pressure downstream of a shock wave, for the case where a Gaussian entropy spot has passed through the shock wave. The upstream Mach number is 4. The analytical linear solution (thin lines) is shown to coincide with its counterpart from DNS (thick lines). Courtesy Fabre, Jacquin and Sesterhenn (2001).

In the absence of mean velocity gradients, interactions between solenoidal, dilatational and pressure modes are purely nonlinear and can be analysed and modelled in pure isotropic homogeneous turbulence. In this context, the model by Fauchet *et al.* (1997) gave promising spectral information. Nonlinear transfer terms have a structure close to (59), with a Green's function, or response tensor, which gathers both a classic linear 'acoustic wave propagator' and a nonlinear damping factor which ensures a decorrelation time for triple correlations shorter than in classical EDQNM.

7.3 Overview of some Linear Theories

Continuing with pure linear theories, promising applications to compressible turbulence follow from RDT and short-wave 'geometric optics', as well as from Ribner's theory for shock-wave turbulence interaction. It seems interesting to explore the common background of these approaches, different though their applications are. In both RDT and stability analysis, distortion of the disturbance field is mediated by a large-scale base flow, whereas it is induced by linearised jump relationship across an infinitely thin shock wave in the Ribner's theory. The latter 'Linear Interaction Approximation' (LIA, from Sanjiva Lele), was recently revisited by Fabre *et al.* (2001) in order to complement analytical calculation of the linear transfer matrix, which connects any upstream disturbance mode to its downstream counterpart, when the disturbance field is passing through a shock wave. Any kind of upstream disturbance (input) can be constructed, and its output calculated, with an arbitrary combination of the three modes, solenoidal, acoustic and entropic, which are uncoupled in the absence of shock waves or mean distortion in the linear limit (e.g. Kovasznay, 1953). Previous LIA applications have dealt with homogeneous quasi-isotropic and incompressible upstream turbulence. Recent work by Fabre *et al.* (2001) shows the behaviour of an upstream spot of temperature, which turns into a pair of vortices, as shown in Figure 19 by results from LIA and DNS. This shows how powerful the method is, and illustrates the fact that Fourier synthesis for disturbances with a linear transfer function (as in homogeneous RDT!) is not an impediment to address strong inhomogeneity. Homogeneous RDT was generalized by Simone et al. (1997) towards isentropic compressible turbulence, using Fourier modes for disturbances to a mean flow with spatially-uniform velocity gradients, with emphasis on the stabilising-destabilising effects of compressibility on shear flows. This technique gave access to linear coupling of pressure, solenoidal and dilatational disturbance modes by strong mean velocity gradients, possibly including an acoustic mode. On the other hand, short-wave 'geometric optics' can also be applied to weakly compressible turbulence, for wavepackets that are convected along mean streamlines (Eckhoff and Storesletten, 1978), but the short-wave WKB development discards the acoustic mode (which consists of long waves). An entropic disturbance mode, however, can be called into play in a nontrivial way by the latter technique. Hence, homogeneous RDT and short-wave 'geometric optics' are no longer governed by similar disturbance equations in the weakly compressible case, as they are in the incompressible case (section 2).

Simple explanations for different types of compressible modes and their coupling, in the three theories, can be proposed as follows. In both LIA and homogeneous RDT, there is no physical cutoff due to the external distortion, because the mean velocity gradients are spatially-uniform in the former, and the shock-wave is infinitely thin in the latter. Hence, disturbances can be broadband or not, and long waves, such as acoustic waves, are not a priori forbidden. This is no longer valid for short-wave 'geometric optics', in which long waves are not allowed by the WKB development. In turn, the homogeneous assumption is not consistent with the presence of the entropic disturbance mode in RDT, since it actually consists of thermal inhomogeneities. It is not necessary to delve further into this area to be convinced that exciting issues remain for investigation. A more general theory has to be built in order to account for all the interactions between the three (solenoidal, acoustic, entropic) or four (solenoidal, dilatational, pressure and entropic) modes, for various external distortions.

8 The Challenge of Modelling Inhomogeneous Semi-Complex Flows

Only incompressible turbulence is considered in what follows. Generalisation of the Craya equation can be sought in the presence of an arbitrary mean flow, in deriving a complete equation for the two-point velocity correlation tensor, with centered position, or

$$R_{ij}(\boldsymbol{r},\boldsymbol{x},t) = \overline{u'_i(\boldsymbol{x}-\boldsymbol{r}/2)u'_j(\boldsymbol{x}+\boldsymbol{r}/2)} \tag{104}$$

from (3), as presented by Oberlack (2002). Fourier transformation can be performed with respect to the separation variable $\boldsymbol{r}$, so that the equation for the hybrid spectral-physical tensor $\hat{R}(\boldsymbol{k},\boldsymbol{x},t)$ can be displayed. Equations for both R and $\hat{\mathsf{R}}$ are very complicated. Correlations involving the pressure cannot be expressed in terms of velocity only, as in (47), especially if boundary conditions have to be accommodated. Hence, it is necessary to add some assumptions, or to introduce some multi-scale approach. The remaining necessary assumption is the separation of spectral and physical space dependencies of the correlations, for example by treating the statistical inhomogeneity as weak. Even for homogeneous turbulence, going beyond the isotropic case entails a high computational cost for two-point simulations using classical nonlinear closures, a cost which is not insignificant compared with that of direct numerical simulation. Thus, it is currently unattractive to solve the full set of equations resulting from closures such as DIA, TFM or EDQNM in the inhomogeneous case without simplifications.

An alternative approach can take inhomogeneity into account *via* the basis set of modes used to express the fluctuations, while, as far as possible, maintaining the structure of equations of the correlation matrix similar to that of the homogeneous case. The modes which substitute for Fourier components may, for instance, be chosen to satisfy the boundary and incompressibility conditions. Accordingly, strong inhomogeneity due to solid boundaries can be accommodated by the very definition of the fluctuation modes. This approach is illustrated by the recent work of Turner (1999), who considered the problem of channel flow using suitably chosen modes whose amplitude equations are analogous to those of Fourier modes in the homogeneous case and which were closed via a random phase approximation. The normal modes of the linear problem might well be good candidates in this type of approach.

8.1 Transport Models along Mean Trajectories

Simplified equations for $\hat{R}(\boldsymbol{k},\boldsymbol{x},t)$ are suggested by the short-wave analysis of subsection 2.2. In turbulent flows, the fluctuating field is not the single component (33), but instead consists of a random superposition of such components. As one might expect, given the behaviour of the underlying local Fourier components described above, it can be shown that, at leading order, weakly inhomogeneous turbulence evolves according to

$$\dot{\hat{R}}_{ij} + M_{ik}\hat{R}_{kj} + M_{jk}\hat{R}_{ik} = 0 \tag{105}$$

where the dot now represents the operator

$$\frac{\partial}{\partial t} + \overline{u}_i\frac{\partial}{\partial x_i} - \frac{\partial \overline{u}_j}{\partial x_i}k_j\frac{\partial}{\partial k_i} \tag{106}$$

and expresses both convection by the mean flow and evolution of the wavenumber of individual Fourier components according to (36). The spectral evolution equation (105) corresponds to the RDT limit of its homogeneous equivalent, Craya's equation, provided the dot operator is interpreted appropriately. Thus, following the mean flow, the leading-order, local spectral tensor $\hat{R}_{ij}(\mathbf{k},\mathbf{x},t)$ behaves as in homogeneous RDT, being given in terms of its initial values and the RDT Green's function. The obvious way to incorporate nonlinearity and viscosity into this description is by employing

$$\dot{\hat{R}}_{ij} + M_{ik}\hat{R}_{kj} + M_{jk}\hat{R}_{ik} = T_{ij} - D_{ij} - 2\nu k^2 \hat{R}_{ij} \quad (107)$$

rather than (105) to describe spectral evolution, where T_{ij} could be modelled using a homogeneous spectral closure. For the sake of completeness, the tensor D_{ij} would typically represent inhomogeneous diffusion across the mean streamline.

An interesting alternative, as proposed by Nazarenko *et al.* (1999), is to derive weakly inhomogeneous RDT using a Gabor transform and related WKB development. A small parameter as ϵ in (33) appears, it is the ratio of the wave-length of the Fourier mode to the length of its Gaussian envelope. The interest of the method is not to derive the equations for the wave vector and the amplitude of the fluctuating velocity field (the method in section 2.2 does the job in a simplest and more general way), but to calculate a space-dependent Reynolds stress tensor by integrating $\hat{R}_{ij}(\boldsymbol{x}, \boldsymbol{k}, t)$ as in (41). Consequently, the nonlinear term which expresses the feedback from the Reynolds stress tensor in (1) can be evaluated (it is zero in pure homogeneous RDT). An extension of this procedure is in progress with application to subgrid-scale modelling (Bérangère Dubrulle, private communication).

8.2 Transport Models along Characteristic Rays Related to Waves

The WKB method by Lifschitz and Hameiri (1991) is different from those developments, which lead to 'geometric optics' and 'physical optics'. Accordingly, the first one (Lifschitz and Hameiri, 1991) was referred to *short-wave* 'geometric optics' everywhere in the present chapter, whereas the second will be denoted *true* 'geometric optics' from now on. The starting point of 'true geometric optics' is similar to (33), but the spatio-temporal evolution is assumed to be slow, so that $\boldsymbol{x}$ and t in (33) ought to be replaced by $\epsilon\boldsymbol{x}$ and ϵt instead. In 'true geometric optics', the leading order approximation is $\epsilon^{(0)}$, and the inhomogeneous dispersion law is exhibited, for instance in injecting (33) with ($\boldsymbol{x} \to \epsilon\boldsymbol{x}$, $t \to \epsilon t$) in linearized equations (86)-(88), so that:

$$\dot{\Phi} = \pm\sigma(\boldsymbol{\nabla}\Phi) - \boldsymbol{\nabla}\Phi \cdot \bar{\boldsymbol{u}} \quad (108)$$

in which, for the sake of simplicity of notations, spatial and temporal operators concern 'slow' variables. Stressing, as before, that $\boldsymbol{k} = \boldsymbol{\nabla}\Phi$, an Hamiltonian function can be defined as

$$\dot{\Phi} = H(\boldsymbol{k}, \boldsymbol{x}) = \pm\sigma(\boldsymbol{k}) - \boldsymbol{k} \cdot \bar{\boldsymbol{u}} \quad (109)$$

Accordingly, an Hamiltonian dynamical system is derived, or

$$\dot{\boldsymbol{x}} = \frac{\partial H}{\partial \boldsymbol{k}} \quad (110)$$

$$\dot{\boldsymbol{k}} = -\frac{\partial H}{\partial \boldsymbol{x}} \quad (111)$$

Because H includes both the dispersion frequency and the doppler frequency due to convection by the mean flow, or $\boldsymbol{k} \cdot \bar{\boldsymbol{u}}$, the right-hand side of (110) is the sum of group and convection velocities, and the related characteristic curve is the ray along with energy propagates. Applications of the Hamiltonian dynamical system are used by Galmiche (1999), for instance, in the case of gravity waves propagating in an inhomogeneous medium. Note that the dispersion law cannot appear at the leading order in the Lifschitz-Hameiri WKB method, so that the previous system of Hamiltonian equations reduces to the trajectory equation (31) and to the Eikonal equation (36), respectively, with $H = -\boldsymbol{k}\cdot\bar{\boldsymbol{u}}$. The dispersion law can reappear in Lifschitz and Hameiri (1991) at the next order, via the solution for the amplitude equation (37), similarly to RDT solutions of sections 5 and 6. Using the development in terms of slow spatio-temporal variables, in 'true geometric optics', the next order (ϵ^1) gives the 'physical optics' step, which yields conservation of wave action.

In the same context of gravity waves, promising perspectives, with transport of statistical spectra with nonlinear effects and diffusion, are offered by Carnevale and Frederiksen (1983). In the latter work, the Hamiltonian function in (109) is affected by nonlinear dynamics in connection with a simplified version of DIA, and the role of resonant triad interactions is displayed.

8.3 Semi-Empirical Transport 'Shell' Models

This approach, discussed in Godeferd *et al.* (2001) is mainly illustrated by semi-empirical transport models, which treat the dependency with respect to the position variable by analogy with one-point modelling. These models cannot incorporate all the information from the general equation (107), but they retain some element of its structure. They are very far from the 'shell models' presented by, *e.g.* Benzi (2002), but they share with them the property that the spectral dependency is only retained through the modulus of the wave vector. Accordingly, it is assumed that primitive equations for $\hat{R}(\boldsymbol{k}, \boldsymbol{x}, t)$ are integrated over spherical shells of radius k. Because of spherical averaging, one has to forget the idea of recovering the asymptotic RDT limit, even in the homogeneous case, and needs to model the 'rapid' terms comprising distortion and pressure-strain correlations, modelling which is unnecessary in the fully anisotropic theory. Transport models for the joint physical/spectral space energy spectrum $E(k, \mathbf{x})$ have been developed, which describe inhomogeneity in a similar way to the diffusive terms in the $k-\varepsilon$ model, but allow a better treatment of dissipation, calculated from the energy spectrum. Examples include the inhomogeneous EDQNM model of Burden (1991), the SCIT (Simplified Closure for Inhomogeneous Turbulence) model developed at Lyon (Touil *et al.* , 2000) and the LWN (Local Wave Number) model developed at Los Alamos (Clark and Zemach, 1995). The SCIT model contains more of the underlying physics of turbulence, in particular the cascade, than do one-point models and first results for complex flows are promising.

9 General Conclusion

With respect to most present-day closure strategies, RDT and multi-point closures give a better access to two particular aspects of turbulence. The first is structural anisotropy, which is induced by mean velocity gradients, body forces, or the direct blocking effect of solid boundaries. The second is the energy cascade involving complex multi-scale interactions, which have to be accounted for in a more or less explicit way in the case of spectral imbalance. As an example of the former in the context of homogeneous turbulence subjected to space-uniform mean velocity gradients, RDT solutions exhibit strong anisotropy. In this case RDT has the same starting point as an initial value, linear stability analysis, and some mechanisms of instability (e.g. the elliptic flow instability) can be understood using RDT with full angular dependence in Fourier space. The latter type of instability cannot be accounted for by classical one-point closure methods, since the instability would affect the rapid pressure-strain term in a full Reynolds-stress models through the complex angular distribution of $\hat{\mathbf{R}}$ in wave-space. The challenge of reproducing RDT behaviour, by transporting extra tensors, was addressed by Cambon *et al.* (1992) and Kassinos *et al.* (2000), with some limited success. In the particular case of turbulence subjected to pure rotation, the complex structural anisotropy is created by the nonlinear cascade, with the angular dependence of energy in wave-space reflecting the loss of dimensionality. Such a behaviour occurs in other flow configurations in which the presence of dispersive waves is more important than the classic 'production' mechanisms.

9.1 Single or Multi-Point Closures

The terms present in the rate equations for Reynolds stress models in homogeneous turbulence can be exactly expressed as integrals over Fourier space of spectral contributions derived from the second order spectral tensor $\hat{R}_{ij}$, which is the Fourier transform of double correlations at two points, and from the third-order 'transfer' spectral tensor T_{ij}. All one-point quantities in the equation that governs $\overline{u'_i u'_j}$ can be expressed as integrals over wavenumber space, as for (41). The equation for the dissipation rate $\varepsilon = \nu\overline{\omega_i \omega_i}$ (in quasi-homogeneous and quasi-incompressible turbulence), can be derived from the exact equation that governs the fluctuating vorticity field ω_i. However, the practical procedure for deriving the ε-equation hardly uses the latter exact equation and consists of basing the equation for $\dot{\epsilon}/\epsilon$ on the equation for $\dot{k}/k$ with adjustable constants.

Advantages and drawbacks of single versus multi-point closure techniques can be briefly discussed as follows. Single-point closure models are much more economical and flexible, and can currently address anisotropic and inhomogeneous flows. Nevertheless, they can easily be questioned in the presence of complex anisotropising mechanisms, and in the presence of a modified cascade with spectral imbalance, even if one restricts the comparison to the pure homogeneous incompressible case. These weaknesses appear in predicting the dynamics of $\overline{u'_i u'_j}$, when looking at the 'rapid' pressure-strain correlation for complex anisotropisation processes, and when looking at the ε-equation and 'slow' pressure-strain tensor for the cascade, sophisticated though the single-point modelling may be. For the sake of brevity, only the anisotropy problem will be illustrated, by comparing RDT and single-point closures in the same 'rapid' limit of homogeneous turbulence in the presence of uniform mean velocity gradients. In this limit, $\overline{u'_i u'_j} - \varepsilon$ models seem to work satisfactorily in the presence of an *irrotational* mean flow, even for time-dependent pure straining processes, such as the successive plane strains addressed by Gence and Mathieu (1979), or, more recently, for cyclic irrotational compression (Le Penven, private communication and Hadzic *et al.* (2001)). In the same situation, $k - \varepsilon$ models give wrong results because of the instantaneous relationship between the deviatoric part of $\overline{u'_i u'_j}$ and the mean strain-rate tensor, a relationship that is usually known as the Boussinesq approximation.

The same contrast between $k - \varepsilon$ and full $\overline{u'_i u'_j} - \varepsilon$ (or Explicit Algebraic Stress Models, Johansson, 2002), with only the latter working satisfactorily, is found when looking at stabilising-destabilising effects of rotation in a plane channel (only the trends induced by terms present in homogeneous turbulence are analysed). A clue to understand why $R_{ij} - \epsilon$ can roughly mimic RDT in the cases mentioned above, is their ability to take into account the production term, in a way much more realistic than in $k - \epsilon$. Correct trends can also be captured by $\overline{u_i'u_j'} - \varepsilon$ models even if the rapid pressure-strain is only roughly modelled, *e.g.* as proportional to the deviatoric part of the production tensor. Regarding the relevance of rotational Bradshaw (or Richarson) numbers for stabilising-destabilising effects of rotation in a shear flow, Leblanc and Cambon (1997) have explained why an apparently two-dimensional and pressure-less analysis (Bradshaw, 1969) gave the same criterion as an 'exact' linear stability analysis (Pedley, 1969). The reason is the dominant role of pure spanwise modes, which are naturally unaffected by pressure fluctuations, yielding again a 'production dominated' mechanism. Things are completely different when rotation interplays with the straining process in a more subtle way, for instance by inducing inertial waves for which anisotropic dispersion relationships affect the 'rapid' pressure-strain term, not to mention slow terms and dissipation beyond the RDT limit. For instance, Townsend's equations for homogeneous RDT were shown (Cambon, 1982) to develop angular peaks of instability in Fourier space if the rotation rate (half the vorticity) of the mean flow is strictly larger than the strain rate, a fact which was recovered by Bayly (1986) in the different context of incisively revisiting the glorified 'elliptical flow instability'. The reader is referred to Salhi *et al.* (1997) for more details on the discussion which is touched upon here.

Even without additional mean strain, pure rotation induces complex 'rapid' and 'slow' effects, for which even the basic principles of single-point closures are questionable. Single-point closures look particularly poor since there is no production by the Coriolis force, whereas the dynamics is dominated by waves whose anisotropic dispersivity is induced by fluctuating pressure. This suggests discriminating 'turbulence dominated by production effects' from 'turbulence dominated by wavy effects'. In short, single-point closures are well adapted to simple turbulent flow patterns of the first class in rather complex geometry, whereas multi-point closures are more convenient for complex turbulent flows in simplified geometry, as illustrated by the second class.

9.2 Anisotropic Turbulence with Dispersive Waves

This case was illustrated by stratified and rotating turbulence, using multi-point closure (MPC). More generally, mathematical developments in the area of wave-turbulence theory (WT), have recently renewed interest in flows that consist of superimposed dispersive waves, in which nonlinear interactions drive the long time behaviour. Individual modes are of the kind

$$u_i'(\mathbf{x}, t) = a_i(t) \exp[\imath(\boldsymbol{k} \cdot \mathbf{x} - \sigma_K t)] \qquad (112)$$

with a known analytical dispersion law for $\sigma_k = \sigma(\boldsymbol{k})$. Similar averaged nonlinear amplitude equations can be found using either WT or MPC, the advantages and drawbacks of which are briefly discussed below.

In the case of wave-turbulence, statistical homogeneity and quasi-normal assumption have equivalent counterparts, obtained by assuming *a priori* Gaussian random phases for the wave fields. As a consequence, an isotropic version of the quasi-normal assumption (56) is derived as

$$\overline{a_i a_j a_k^* a_l^*} = \overline{a_i a_i^*}.\overline{a_j a_j^*}\,(\delta_{ik}\delta_{jl} + \delta_{il}\delta_{jk})\,,$$

as discussed in Staquet and Sommeria (2002). One should perhaps recall that a random phase approximation does not necessarily imply statistical homogeneity, if the basic modes differ from simple Fourier modes (Turner, 1999). In addition, *isotropic* dispersion laws such as $\sigma_k =$

$|\boldsymbol{k}|^\alpha$ in (112) are almost exclusively treated in WT for deriving Kolmogorov spectra, with the key hypothesis of constant and isotropic energy fluxes across different scales associated with a wavenumber $|\boldsymbol{k}|$ (Zakharov *et al.* , 1992). By contrast, in geophysical flows, dispersion laws are anisotropic, with for instance $\sigma = \pm\beta k_x/k^2$ in the case of Rossby waves, $\sigma = \pm 2\Omega k_\parallel/k$ for inertial waves and $\sigma = \pm N k_\perp/\kappa$ for gravity waves (k_x, $k_\parallel$ and $k_\perp$ are the components of the associated wavevector respectively in the zonal direction, and the directions parallel or perpendicular to the rotation/gravity vectors). In the latter two 3D cases, this anisotropy is reflected by the strange conical — 'St Andrew cross' — shape of iso-phase surfaces in typical experiments with a localized point forcing (see views of this type in Figures 6, and in Mc Ewan (1970), Mowbray and Rarity (1967), Godeferd and Lollini (1999)), and by angular-dependent energy drains when looking at nonlinear interactions, as illustrated in sections 5 and 6.

At least if Eulerian correlations are considered, The MPC and WT theories share in general an important background. Kinetic equations for mean spectral energy densities of waves are found in WT, similar to homogeneous MPC. Their slow evolution is governed by similar energy transfer terms, which are cubic in terms of wave amplitudes (triads). There is also a possibility that these transfers involve fourth-order interactions (quartets) in WT when triple resonances are forbidden by the dispersion laws and/or by geometric constraints (*e.g.* shallow waters). Resonant quartets seem to be particularly relevant when resonances are seen in a Lagrangian description. When triple resonances are allowed, for instance in cases of rotating, stably stratified and MHD turbulence (Caillol and Zeitlin, 2000, Galtier *et al.* , 2000), WT kinetic equations have exactly the same structure as their counterpart in elaborated MPC. Hence, WT and MPC have a common limit at very small interaction parameter (*e.g.* Rossby number, Froude number, magnetic number in MHD). The shape of the typical eddy damping parameter, which remains the heuristic correction to quasi-normal transfer in EDQNM, is unimportant in this limit. Its only role is to regularise the resonance operators. Beyond the weak nonlinearity assumption, the eddy damping, or more generally the nonlinear contribution to Kraichnan's response function, can regain some importance for moderate interaction parameters, in allowing extrapolation from WT through MPC towards a larger domain, until the case of strong interactions is reached (*e.g.* pure isotropic turbulence without external or wave effects, for which classic MPC models work satisfactorily). Proposals for renormalising such generalized eddy damping were offered by Carnevale and Rubinstein (private communications).

9.3 Anisotropy versus Structure

The two-point anisotropic description is more powerful, even if homogeneity is assumed, than is generally recognized. In rotating and stratified turbulence the anisotropic spectral description, with angular dependence of spectra and cospectra in Fourier space, allows quantification of columnar or pancake structuring in physical space. Among various indicators of the thickness and width of pancakes, which can be readily derived from anisotropic spectra, integral length scales $L_{ij}^{(n)}$ related to different components and orientations are the most useful (see Figures 14, 15 and 16). As another illustration (Lee *et al.* , 1990), the streak-like tendency in shear flows (see Figure 3) can be easily found in calculating both the $L_{11}^{(1)}$ component, which gives the streamwise length of the streaks, and $L_{11}^{(3)}$, which gives the spanwise separation length of the streaks (as usual, 1 and 3 refer to streamwise and spanwise coordinates, respectively). In pure homogeneous RDT at constant shear rate, both length scales can be calculated analytically and their ratio (elongation parameter) is found to increase as $(St)^2$, $S = \partial U_1/\partial x_2$ being the shear rate.

In order to take into account statistical inhomogeneity, the description in terms of Fourier modes for the turbulent field remains partially relevant, provided a convenient scale separation be assumed, as illustrated in section 8. If this is no longer possible, more appropriate sets of basis function other than sunusoidal ones should be used, such as wavelets, POD modes (Lumley, 1967), and inhomogeneous RDT modes.

References

Bardina, J., Ferziger, J. M., Rogallo, R. S. (1985) Effect of rotation on isotropic turbulence: computation and modelling. *J. Fluid Mech.*154:321–326.

Batchelor, G. K. (1953) *The theory of homogeneous turbulence.* Cambridge University Press.

Batchelor, G. K. and Proudman I. (1954) The effect of rapid distortion in a fluid in turbulent motion. *Q. J. Mech. Appl. Maths,* 7–83.

Bayly, B. J. (1986) Three-dimensional instability of elliptical flow. *Phys. Rev. Lett.*57:2160.

Bayly, B. J. Holm, D. D. and Lifschitz. (1996) Three-dimensional stability of elliptical vortex columns in external strain flows. *Phil. Trans. R. Soc. London* A354:895–926.

Benney, D. J., Saffman, P. G. (1966) Nonlinear interactions of random waves in a dispersive medium In *Proc. R. Soc. London, Ser. A* 289, 301–320.

Benzi, R. (2001) Chapter 1 of the present book

Beronov, K. and Kaneda, Y. (2001) RDT and long-time universality in uniformly sheared turbulence. Part 1:velocity correlations. *J. Fluid Mech.*, submitted.

Burden, A. D. (1991) Towards an EDQNM closure for inhomogeneous turbulence. in Johansson, A. V. and Alfredson, P. H., eds. *Advances in Turbulence III.* Springer Verlag. Berlin 387.

Busse, F. (2001) Present CISM Course

Caillol, P. and Zeitlin, W. (2000) Kinetic equations and stationary energy spectra of weakly nonlinear internal gravity waves. *Dyn. Atm. Oceans*32:81–112.

Cambon, C. (1982) Etude spectrale d'un champ turbulent incompressible soumis à des effets couplés de déformation et rotation imposés extérieurement. Thèse de Doctorat d'Etat, Université Lyon I, France.

Cambon, C. (2001) Turbulence and vortex structures in rotating and stratified flows. *Eur. J. Mech. B (fluids)*20:489–510.

Cambon C., Jacquin L. and Lubrano J-L. (1992) Towards a new Reynolds stress model for rotating turbulent flows. *Phys. Fluids A*4:812–824.

Cambon C., Mansour N. N. and Godeferd, F. S. (1997) Energy transfer in rotating turbulence. *J. Fluid Mech.*337:303–332.

Cambon C., Mao Y. and Jeandel D. (1992) On the application of time dependent scaling to the modelling of turbulence undergoing compression. *Eur. J. Mech, B/Fluids*6:683–703.

Cambon, C. and Scott, J. F. (1999) Linear and nonlinear models of anisotropic turbulence. *Ann. Rev. Fluid Mech.* 31:1–53.

Carnevale, G. F. and Frederiksen, J. S. (1983) A statistical dynamical theory of strongly nonlinear internal gravity waves. *Geophys. Astrophys. Fluid Dyn.*20,8:131–164.

Clark, T., T., and Zemach, C. (1995) A spectral model applied to homogeneous turbulence. *Phys. Fluids*7 (7):1674-1694.

Craya, A. (1958) Contribution à l'analyse de la turbulence associée à des vitesses moyennes. *P.S.T.* n^0 *345.* Ministère de l'air. France

Eckhoff, K. S. and Storesletten, L. (1978) A note on the stability of steady inviscid helical gas flows. *J. Fluid Mech.*89:401.

Fabre, D., Jacquin, L., and Sesterhenn, J. (2001) Linear interaction of a cylindrical entropy spot with a shock wave. *Phys. Fluid*13,8:1-20.

Fauchet, G., Shao, L., Wunenberger, R. and Bertoglio, J. P. (1997) An improved two-point closure for weakly compressible turbulence. *11-th Symp. Turb. Shear Flow*, Grenoble, Sept. 8-10.

Galmiche, M. (1999) Thèse de Doctorat, Université de Toulouse, France.

Galtier, S., Nazarenko, S., Newell, A. C. and Pouquet, A. (2000) A weak turbulence theory for incompressible MHD. *J. Plasma Phys.*63:447–488.

Gence J.-N. and Mathieu J. (1979) On the application of successive plane strains to grid-generated turbulence. *J. Fluid Mech.*93:501–513.

Godeferd F S. and Cambon C. (1994) Detailed investigation of energy transfers in homogeneous stratified turbulence. *Phys. Fluid*6:2084-2100.

Godeferd, F. S., Cambon, C., and Leblanc, S. (2001) Zonal approach to centrifugal, elliptic and hyperbolic instabilities in Suart vortices with external rotation. *J. Fluid Mech.*449:1–37.

Godeferd, F. S., Cambon, C. and Scott, J. F. (2001) Report on the workshop: Two-point closures and their application. *J. Fluid Mech.*436:393-407.

Godeferd, F. S., and Lollini, L. (1999) Direct numerical simulations of turbulence with confinment and rotation. *J. Fluid Mech.*393:257-308.

Goldstein M. E. (1978) Unsteady vortical and entropic distortions of potential flows round arbitrary obstacles. *J. Fluid Mech.*89:431.

Greenspan H. P. (1968) *The theory of rotating fluids.* Cambridge University Press.

Hadzic, I., Hanjalic, K. and Laurence, D. (2001) Modeling the response of turbulence subjected to cyclic irrotational strain. *Phys. Fluid*13 (6):1739–1747.

Hopf, E. (1952) Statistical hydrodynamics and functional calculus *J. Rat. Mech. Anal.*1:87.

Hunt J. C. R. and Carruthers D. J. (1990) Rapid distortion theory and the 'problems' of turbulence. *J. Fluid Mech.*212:497–532.

Jacquin L., Leuchter O., Cambon C., Mathieu J. (1990) Homogeneous turbulence in the presence of rotation. *J. Fluid Mech.*220:1–52.

Johansson, A. V. (2002) Chapter 5 of the present book

Kaneda, Y. (1981) Renormalized expansions in the theory of turbulence with the use of the Lagrangian position function. *J. Fluid Mech.*107:131–145.

Kassinos, S. C., Reynolds, W. C. and Rogers, M. M. (2000) One-point turbulence structure tensors. *J. Fluid Mech.*428:213–248.

Kovasznay, L. S. G. (1953) Turbulence in supersonic flow. *J. Aeronaut. Sci.*20:657–682.

Kraichnan R. H. (1959) The structure of turbulence at very high Reynolds numbers. *J. Fluid Mech.*5:497–543.

Leblanc, S. and Cambon, C. 1997 On the three-dimensional instabilities of plane flows subjected to Coriolis force. *Phys. Fluids*9 (5):1307–1316.

Lee, J. M., Kim, J. and Moin P. (1990) Structure of turbulence at high shear rate. *J. Fluid Mech.*216:561–583.

Lifschitz, A. and Hameiri, E. (1991) Local stability conditions in fluid dynamics. *Phys. Fluids* A 3:2644–2641.

Lighthill, M J. (1978) *Waves in Fluids*. Cambridge University Press.

Lumley J. L. (1967) The structure of inhomogeneous turbulence flows. In Yaglom, A M, Tatarsky, V I, eds., *Atmospheric Turbulence and Radio Wave Propagation.* Moscow: NAUCA. 166–167.

Lundgren, T. S. (1967) Distribution function in the statistical theory of turbulence. *Phys. Fluids*10 (5):969–975.

Mathieu, J. and Scott, J. F. (2000) *Turbulent Flows: An Introduction* Cambridge University Press.

Mc Ewan, A. D. (1970) Inertial oscillations in a rotating fluid cylinder. *J. Fluid Mech.*40:603–639.

Mc Comb, W. D. (1974) A local energy transfer theory of isotropic turbulence. *J. Phys. A*7 (5):632.

Mc Comb, W. D. (2002) Chapter 3 of the present book

Moffat, H. K. (1967) The interaction of turbulence with a strong shear. In Yaglom, A M, Tatarsky, V I, eds., *Atmospheric Turbulence and Radio Wave Propagation.* Moscow: NAUCA.

Monin, A. S. and Yaglom A. M. (1975) *Statistical Fluid Mechanics* I. Cambridge, MA. MIT Press.

Mowbray, D. E. and Rarity, B. S. H. (1967) A theoretical and experimental investigation of the phase configuration of internal waves of small amplitude in a density stratified liquid. *J. Fluid Mech.*28:1–16.

Nazarenko, S., Kevlahan, N. N., & Dubrulle, B. (1999) A WKB theory for rapid distortion of inhomogeneous turbulence. *J. Fluid Mech.*390:325.

Oberlack, M. (2002) Chapter 6 of the present book

Orszag, S. A. (1970) Analytical theories of turbulence. *J.Fluid Mech.*41:363–386.

Ribner, H. S. (1954) Convection of a pattern of vorticity through a shock wave. NACA REPORT, No 1164.

Riley, J. J., Metcalfe, R. W. and Weisman, M. A. (1981) DNS of homogeneous turbulence in density stratified fluids. In B. J. West, Ed., AIP, New York. *Proc. AIP conf. on nonlinear properties of internal waves* 79–112.

Rogers M. M. (1991) The structure of a passive scalar field with a uniform gradient in rapidly sheared homogeneous turbulent flow. *Phys. Fluids*A, 3:144-154.

Salhi A., Cambon C. and Speziale, C. G. (1997) Linear stability analysis of plane quadratic flows in a rotating frame. *Phys. Fluids*9 (8):2300–2309.

Sarkar S. (1995) The stabilizing effect of compressibility in turbulent shear flow. *J. Fluid Mech.*282:163–286.

Simone, A., Coleman G. N. and Cambon C. (1997) The effect of compressibility on turbulent shear flow: a RDT and DNS study. *J. Fluid Mech.*330:307–338.

Sipp, D., Lauga, E. and Jacquin, L. (1999) Vortices in rotating systems: Centrifugal, elliptic and hyperbolic type instabilities. *Phys. Fluids*11:3716–3728.

Smith, L. M. and Waleffe, F. (2002) Generation of slow, large scales in forced rotating, stratified turbulence. *J. Fluid Mech.* 451:145–168.

Staquet, C. and Sommeria, J. (2002) Internal gravity waves: from instabilities to turbulence. *Annu. Rev. Fluid Mech.*34:559–593.

Touil, H., Bertoglio, J-P., and Parpais, S. (2000) A spectral closure applied to anisotropic inhomogeneous turbulence. in Dopazo, C., ed. *Advances in Turbulence VIII*. CIMNE. Spain 689.

Townsend, A. A. (1956) *The structure of turbulent shear flow.* Revised version 1976. Cambridge University Press.

Turner, L. (1999) Macroscopic structures of inhomogeneous, Navier-Stokes turbulence. *Phys. Fluids*11:2367–2380.

Zakharov, V. E., L'vov, V. S. and Falkowich G. (1992) *Kolmogoroff spectra of turbulence I. Wave turbulence.* Springer series in nonlinear dynamics. Springer Verlag

Engineering Turbulence Models and their Development, with Emphasis on Explicit Algebraic Reynolds Stress Models

Arne Johansson *

Department of Mechanics, Royal Institute of Technology (KTH), Osquars Backe 18, SE-100 44 Stockholm, Sweden

Abstract

Single-point turbulence models will be discussed from a somewhat analytical point of view. The lowest level of modelling considered here is that of eddy-viscosity-based two-equation models, but particular attention is given to explicit algebraic Reynolds stress models (and explicit algebraic scalar flux models). Some new trends in models based directly on the Reynolds stress transport equations are also discussed.

1 Introduction

In January 2001, NASA arranged a workshop for the assessment of the direction of CFD research for the design of future generations of transport aircraft (Rubinstein et al., 2001). From the two days of discussion about the needs of aircraft manufacturers, the need for further developments of single-point turbulence models stood out in clear light. As an illustration we may cut a few key points from the executive summary of that report

- Advances in turbulence modeling are needed in order to calculate high Reynolds number flows near the onset of separation and beyond.

- NASA should support long-term research on Algebraic Stress Models and Reynolds Stress Models.

- ...a balanced effort in turbulence modeling development, validation, and implementation should include DNS, LES, and hybrid method approaches as well.

The lively discussions between model developers, aircraft designers, program managers etc revealed many of the difficulties involved in the development and validation of models for use

*The author wants to thank Dr Stefan Wallin for various valuable contributions, and Gustaf Mårtensson and professor Fritz Busse for many useful comments on the manuscript.

in the design process. Also evident is the aspiration to use CFD for more and more challenging problems covering many different flow features. The onset of separation is still a major challenge for CFD based on single-point turbulence models, although we see here a continuous improvement through the use of more complex models. One often refers to this type of CFD as RANS-methods (i.e. methods based on the Reynolds-averaged Navier-Stokes equations).

The complexity of engineering type, single-point turbulence models can be measured by the number of primary quantities carried in the closure, and thereby the number of transport equations for turbulence quantities carried in the closure. A dividing line in the set of single-point closures is represented by the use (or not) of a turbulent (eddy) viscosity. The eddy viscosity is a diffusivity coefficient of turbulent momentum transfer and can be seen essentially as a product of a characteristic velocity scale of the energetic eddies and the corresponding characteristic length scale. An eddy-viscosity-based model is usually said to be 'complete' if it includes transport equations for (at least) both characteristic scales. A major part of industrial CFD today is carried out with eddy-viscosity-based two-equation models, and a wide variety of such models have been developed. Generalizations of the eddy-viscosity concept have started to find their way into commercial CFD codes. In this type of approach, algebraic approximations of the Reynolds stress (or rather stress anisotropy) transport equations are typically used to replace the eddy-viscosity based relation to determine the energy distribution among the components.

In aircraft design simple models with zero or one transport equation have been used extensively, and with considerable success for attached fully turbulent flows. As the ambition has been increased to handle complex flow interactions, such as e.g. those involved in high-lift devices or compact air inlets for modern fighter aircraft, the need for models with a more general formulation has become accentuated. Also, as pointed out by A. Kumar (NASA) in the above-mentioned workshop report, in the historical perspective over the last few decades the pacing item in CFD has gone from numerics in the 1970's and 80's to grid-related issues in the 90's over to aspects of the actual physical modelling in the present decade.

We now also start to see a mixture of CFD with single-point models and large-eddy simulation (LES) (or similar) methods. For flows with inherent unsteadiness, that are usually associated with large scale separations, simulation methods are becoming a useful tool in the flow predictions, including noise predictions. LES as a complete replacement of CFD with single-point closures is still decades away, a conclusion also born out from the above-mentioned workshop.

Hybrid methods that combine RANS and LES methods in different regions of the same flow form an interesting possibility that still needs much development.

The general aspect that we shall focus on here is the generalization of single-point models to meet increasing demands for the prediction of complex flow phenomena such as onset of separation. Particular attention will be given to the category of explicit algebraic Reynolds stress models (EARSM) and comparisons with corresponding eddy-viscosity based two-equation models and with full Reynolds stress transport model formulations. In the latter category, we will take a closer look at realizability constraints and a formulation with non-linear models for the individual terms that gives reasonable near-wall behaviour without wall damping functions, wall-distance, wall-normal vectors, etc that are typical ingredients in most 'low Reynolds number' model formulations. Extensions of the EARSM-approach to handle also passive scalars will also be discussed briefly.

2 Basic Equations and Definitions for the Single-Point Turbulence Model Concept

Here we will restrict attention to incompressible flows and denote the instantaneous velocity and pressure fields by u_i and p, and assume that the fluid is Newtonian so that the stress tensor can be written

$$\sigma_{ij} = -p\delta_{ij} + 2\mu s_{ij} \tag{1}$$

where μ is the dynamic viscosity (the kinematic viscosity being $\nu = \mu/\rho$) and s_{ij} is the rate of strain

$$s_{ij} \equiv \frac{1}{2}\left(\frac{\partial u_i}{\partial x_j} + \frac{\partial u_j}{\partial x_i}\right). \tag{2}$$

Correspondingly, we define the rotation rate tensor (the antisymmetric part of the velocity gradient tensor) ω_{ij} as

$$\omega_{ij} \equiv \frac{1}{2}\left(\frac{\partial u_i}{\partial x_j} - \frac{\partial u_j}{\partial x_i}\right). \tag{3}$$

The rotation rate tensor is related to the vorticity, $\vec{\omega}(\vec{x},t) \equiv \text{curl}\,\vec{u}(\vec{x},t)$, as

$$\begin{aligned} \omega_k &= \epsilon_{kij}\omega_{ji} \\ \omega_{ij} &= \tfrac{1}{2}\epsilon_{kji}\omega_k \end{aligned} \tag{4}$$

where ϵ_{ijk} is the alternating permutation tensor.

We can then write the Navier-Stokes and continuity (incompressibility constraint) equations as

$$\frac{\partial u_i}{\partial t} + u_j\frac{\partial u_i}{\partial x_j} = -\frac{1}{\rho}\frac{\partial p}{\partial x_i} + 2\nu\frac{\partial s_{im}}{\partial x_m} + f_i \tag{5}$$

$$\frac{\partial u_i}{\partial x_i} = 0 \tag{6}$$

where the left-hand side represents the material derivative and f_i is a volume force.

2.1 The Reynolds Decomposition and Mean Flow Equation

In single-point turbulence modelling, we use a statistical approach with a closed set of transport equations for turbulence quantities in the form of statistical moments evaluated at a single point.

First we need to separate the flow field into mean and fluctuating parts by defining the mean value as the ensemble average over an infinite number of realizations

$$\overline{q} \equiv \lim_{N\to\infty}\frac{1}{N}\sum_{n=1}^{N} q_n \tag{7}$$

The Reynolds decomposition of the flow field is defined by

$$\begin{aligned} u_i(\vec{x},t) &\equiv \bar{u}_i(\vec{x},t) + u_i'(\vec{x},t) \\ p(\vec{x},t) &\equiv \bar{p}(\vec{x},t) + p'(\vec{x},t) \\ f_i(\vec{x},t) &\equiv \bar{f}_i(\vec{x},t) + f_i'(\vec{x},t) \end{aligned} \tag{8}$$

The mean flow equation, often referred to as the Reynolds equation, is then obtained by inserting the Reynolds decomposition into (5), and in the averaging the identity $\overline{u_i'} = 0$ is used. For the substantial derivative, where only advection by the mean flow is included, we will in the following use the notation

$$\frac{\bar{\mathrm{D}}}{\bar{\mathrm{D}}t} \equiv \frac{\partial}{\partial t} + \bar{u}_m \frac{\partial}{\partial x_m} \tag{9}$$

The Reynolds equation can then be written

$$\rho \frac{\bar{\mathrm{D}}\, \bar{u}_i}{\bar{\mathrm{D}}t} = -\frac{\partial \bar{p}}{\partial x_i} + \frac{\partial}{\partial x_m}\left(2\mu \bar{S}_{im} - \rho \overline{u_i' u_m'}\right) + \rho \bar{f}_i \tag{10}$$

$$\frac{\partial \bar{u}_i}{\partial x_i} = 0 \tag{11}$$

where $-\rho\overline{u_i' u_j'}$ is usually called the 'Reynolds stress tensor'. It acts as an additional stress and dominates the momentum transfer in most parts of fully turbulent flows.

$\bar{S}_{im}, s_{im}'$ represent the mean and fluctuating parts of the strain rate tensor, defined in analogy with (2). We will later also use the corresponding rotation rate tensor (mean: W_{ij}, fluctuating part: ω_{ij}').

Subtracting (10) from the Navier-Stokes equation for the instantaneous velocity (5) yields the equation for the fluctuating part of the velocity

$$\frac{\bar{\mathrm{D}}\, u_i'}{\bar{\mathrm{D}}t} = -u_m' \frac{\partial \bar{u}_i}{\partial x_m} + \frac{\partial}{\partial x_m}\left(\frac{p'}{\rho}\delta_{im} + 2\nu s_{im}' - u_i' u_m' + \overline{u_i' u_m'}\right) + f_i' \tag{12}$$

$$\frac{\partial u_i'}{\partial x_i} = 0. \tag{13}$$

Models of the Reynolds stress tensor are in single-point closures based on equations for statistical measures of various combinations of the fluctuating velocity and pressure fields. The starting point for constructing these equations is then (12). Also the transport equation for the dissipation rate may be derived by taking the gradient of equation (12) and multiplying that by $\nu \frac{\partial u_i'}{\partial x_m}$.

2.2 The Turbulent Kinetic Energy Equation

By multiplying (12) by u_i and averaging, we obtain the equation for the mean turbulent kinetic energy, K,

$$K \equiv \frac{1}{2}\overline{u_i' u_i'} \tag{14}$$

$$\frac{\bar{\mathrm{D}}\, K}{\bar{\mathrm{D}}t} = \mathcal{P} - \varepsilon - \frac{\partial}{\partial x_m}\left(J_m - \nu \frac{\partial K}{\partial x_m}\right) + \overline{u_i' f_i'} \tag{15}$$

where

$$\mathcal{P} \equiv -\overline{u_i' u_m'} S_{im} \tag{16}$$

$$\varepsilon \equiv \nu \overline{u_{i,m}' u_{i,m}'} \tag{17}$$

$$J_m \equiv \frac{1}{2}\overline{u_i' u_i' u_m'} + \frac{1}{\rho}\overline{u_m' p'} \tag{18}$$

On the right-hand side of (15), the first term represents turbulence energy production, i.e. exchange of energy between the mean flow and the turbulent fluctuations. This is practically always positive, but can under some conditions be negative temporarily and locally. The second term represents viscous dissipation into heat and the third is diffusion by viscous stresses and turbulent diffusion caused by fluctuating velocity and pressure. The last term represents the possible contribution from work by volume forces.

The complete viscous dissipation is actually $2\nu\overline{s'_{im}s'_{im}}$ of which ε as used above is the homogeneous part, but ε is for most practical purposes a good approximation of the total dissipation (for a more detailed discussion see Johansson & Burden, 1999).

2.3 The Dissipation Rate Equation

The transport equation for the (homogeneous part of the) dissipation rate, ε, of kinetic energy can be derived from (12) and may be written as

$$\frac{\bar{\mathrm{D}}\,\varepsilon}{\bar{\mathrm{D}}t} = \mathcal{P}^{\varepsilon} + T^{\varepsilon} - D^{\varepsilon} - \frac{\partial}{\partial x_m}\left(J^{\varepsilon}_m - \nu\frac{\partial\varepsilon}{\partial x_m}\right) + 2\nu\overline{u'_{i,m}f'_{i,m}} \tag{19}$$

where on the right-hand side we have grouped the terms in a way analogous to that for the K-equation

$$\mathcal{P}^{\varepsilon} = -2\nu S_{ij}\left(\overline{u'_{i,k}u'_{j,k}} + \overline{u'_{k,i}u'_{k,j}}\right) - 2\nu\bar{u}_{i,jk}\overline{u'_j u'_{i,k}} \tag{20}$$

$$T^{\varepsilon} = -2\nu\overline{u'_{i,j}u'_{i,k}u'_{j,k}} \tag{21}$$

$$D^{\varepsilon} = 2\nu^2\overline{u'_{i,jk}u'_{i,jk}} \tag{22}$$

$$J^{\varepsilon}_m = \nu\left\{\overline{u'_{i,j}u'_{i,j}u'_m} + \tfrac{2}{\rho}\overline{p'_{,i}u'_{m,i}}\right\} \tag{23}$$

The physical interpretation of the terms is mean flow related production ($\mathcal{P}^{\varepsilon}$), turbulence related production through vortex stretching (T^{ε}), viscous destruction (D^{ε}), turbulent transport flux (J^{ε}_m) and viscous diffusive flux ($-\nu\frac{\partial\varepsilon}{\partial x_m}$). All terms except the viscous diffusive flux term need to be modelled.

2.4 The Reynolds Stress Transport Equation

By multiplying (12) by u_j, averaging and adding the corresponding equation with switched indices i,j, we obtain an equation for the velocity correlation $\overline{u'_i u'_j}$ (and hence the Reynolds stress)

$$\frac{\bar{\mathrm{D}}\,\overline{u'_i u'_j}}{\bar{\mathrm{D}}t} = \mathcal{P}_{ij} - \varepsilon_{ij} + \Pi_{ij} - \frac{\partial}{\partial x_m}\left(J_{ijm} - \nu\frac{\partial\overline{u'_i u'_j}}{\partial x_m}\right) + G_{ij} \tag{24}$$

where

$$\mathcal{P}_{ij} \equiv -\overline{u_i'u_m'}\frac{\partial \bar{u}_j}{\partial x_m} - \overline{u_j'u_m'}\frac{\partial \bar{u}_i}{\partial x_m} \tag{25}$$

$$\varepsilon_{ij} \equiv 2\nu\overline{u_{i,m}'u_{j,m}'} \tag{26}$$

$$\Pi_{ij} \equiv \frac{2}{\rho}\overline{p's_{ij}'} \tag{27}$$

$$J_{ijm} \equiv \overline{u_i'u_j'u_m'} + \frac{1}{\rho}\left(\overline{u_j'p'}\delta_{im} + \overline{u_i'p'}\delta_{jm}\right) \tag{28}$$

$$G_{ij} \equiv \overline{u_i'f_j'} + \overline{u_j'f_i'} \tag{29}$$

Here we also obtain a molecular diffusion term that is explicit in the Reynolds stress tensor. All other viscous effects are lumped into one single term, denoted by ε_{ij}, normally referred to as the 'dissipation rate tensor'.

The production tensor represents the direct interaction between the turbulence and the mean flow, the trace being equal to $2\mathcal{P}$. In contrast to the energy production term, the individual $\mathcal{P}_{ij}$-terms are influenced both by the mean strain and mean rotation rate tensors. This is a reflection of the fact that three-dimensional turbulence does not satisfy material frame indifference, and is an important aspect to be considered in the formulation of models that should be expected to accurately capture effects of system rotation and mean flow curvature.

The pressure-strain rate correlation tensor, Π_{ij} has no counterpart in the kinetic energy equation, since it has zero trace, and is thereby associated with intercomponent energy redistribution. Since this correlation involves the pressure fluctuations, it represents a term associated with non-local interactions, and presents severe modelling difficulties. Its modelling is also a key element in capturing complex flow phenomena, such as effects of rotation, in single-point closures.

The spatial redistribution is represented by J_{ijm} and is driven both by turbulent velocity and pressure fluctuations. It involves third order correlations, revealing the well-known closure problem caused by the non-linearity of the Navier-Stokes equations.

We note that the volume force here contributes to the balance even in cases where it does not perform a net work. For instance, the Coriolis force in cases with system rotation is perpendicular to the velocity vector and thereby gives a zero contribution to the kinetic energy equation. In that case, the fluctuating part of the volume force is given by $f_i' = -2\epsilon_{ilm}\Omega_l^s u_m'$ and

$$G_{ij} = -2\Omega_l^s(\epsilon_{ilm}\overline{u_m'u_j'} + \epsilon_{jlm}\overline{u_m'u_i'}). \tag{30}$$

For the case of plane turbulent channel flow rotating about the spanwise axis ($\Omega_l^s = \Omega^s\delta_{3l}$) we see that

$$G_{11} = 4\Omega^s\overline{u_1'u_2'} \tag{31}$$

$$G_{22} = -4\Omega^s\overline{u_1'u_2'} \tag{32}$$

$$G_{12} = 2\Omega^s(\overline{u_2'u_2'} - \overline{u_1'u_1'}). \tag{33}$$

Hence, wall-normal mixing (through $\overline{u_2'u_2'}$) is enhanced on the side where $\overline{u_1'u_2'}$ is negative (the 'unstable' side) and damped on the other ('stable') side. This gives a more isotropic turbulence near the wall on the unstable side than is the case in the non-rotating case, a fact that has been well documented in the literature (see e.g. the DNS study of Kristoffersen & Andersson, 1993).

2.5 Passive Scalars in Turbulent Flow

A passive scalar is defined as a quantity that is carried by the turbulent flow, influenced by the turbulent fluctuations, but leaving the velocity field unchanged. It may typically be the concentration of a tracer (at low concentration) or temperature. The resulting density difference must be negligible.

The evolution of a passive scalar is influenced by the advection of the turbulent flow and by molecular diffusion. We here denote the molecular diffusivity by α.

We may divide the scalar value, θ, into a mean value, Θ and a fluctuation, θ'. The Reynolds averaged scalar equation can then be written

$$\frac{\bar{D}\Theta}{\bar{D}t} = \frac{\partial}{\partial x_j}\left(\alpha\frac{\partial\Theta}{\partial x_j} - \overline{u_j'\theta'}\right) \tag{34}$$

where the scalar flux vector $-\overline{u_j'\theta'}$ plays a role analogous to that of the Reynolds stress tensor in the mean flow equation.

A transport equation for the scalar flux vector can symbolically be written

$$\frac{\bar{D}\,\overline{u_j'\theta'}}{\bar{D}t} = \mathcal{P}_{\theta i} + \Pi_{\theta i} - \varepsilon_{\theta i} + \mathcal{D}_{\theta i} \tag{35}$$

where $\Pi_{\theta i}$ is the pressure-scalar gradient correlation, $\varepsilon_{\theta i}$ is the destruction rate vector, $\mathcal{D}_{\theta i}$ denotes the diffusion, and the production term is given by

$$\mathcal{P}_{\theta i} = -\overline{u_i'u_j'}\frac{\partial\Theta}{\partial x_j} - \overline{u_j'\theta'}\frac{\partial\bar{u}_i}{\partial x_j}. \tag{36}$$

In scalar flux transport modelling one sometimes also makes use of the transport equations for the scalar variance and its dissipation rate (see Wikström, Wallin & Johansson, 2000).

3 A Common Formulation of Single-Point Turbulence Models

A single-point closure is defined by its set of primary quantities, i.e. the statistical moments for which transport equations are solved. The Reynolds stress tensor, needed together with $\bar{u}_i, \bar{p}$, to close the Reynolds equation (10) is determined from these transport equations and one or more algebraic relations.

The Reynolds stress tensor depends on the dynamics of the turbulent flow, involving

- energy transfer between the mean flow and the fluctuations,
- intercomponent energy transfer,
- spatial redistribution of energy (and other quantities),
- dynamics of energy transfer between scales of different sizes and
- history effects.

In single-point closures, we cannot directly account for spectral dynamics features such as energy transfer between scales of different sizes. Instead we have to assume a self-similar type of development of the energy spectra, where the development is governed by a single lengthscale.

The primary scalar invariant of the Reynolds stress tensor is the trace, which equals twice the turbulence kinetic energy ($2K$). We can separate the information given by the Reynolds stress tensor into the scalar magnitude measure K and the anisotropy tensor defined as

$$a_{ij} \equiv \frac{\overline{u_i' u_j'}}{K} - \frac{2}{3}\delta_{ij}. \tag{37}$$

This symmetric tensor has zero trace ($a_{ii} = 0$), so that it has five nontrivial elements. Together with K it should of course carry the same information as $\overline{u_i' u_j'}$.

From the definition (37) we immediately see that the diagonal elements of a_{ij} attain values only in the interval from $-2/3$ to $4/3$. It can also readily be shown that the off-diagonal elements are restricted to the interval from -1 to 1.

In the literature we also commonly find the anisotropy tensor $b_{ij} = a_{ij}/2$.

3.1 The Anisotropy Invariant Map

Since the anisotropy tensor is symmetric we can always find a coordinate system in which it is in diagonal form. This, together with the fact that it has zero trace, means that it carries only two 'independent' pieces of information, i.e. has only two scalar invariants. We may take these to be

$$II_a \equiv a_{ij}a_{ji}, \qquad III_a \equiv a_{ij}a_{jk}a_{ki}. \tag{38}$$

Lumley (1978) introduced the anisotropy invariant map (AIM). We may depict all turbulence anisotropy states in a bounded region in the III_a, II_a plane (see Figure 1). One of the boundaries (the upper) represents the two-component limit where the energy content in one component is vanishingly small, and where $II_a = 8/9 + III_a$. The two other boundaries represent axisymmetric turbulence, where $II_a^{1/2} = 6^{1/6} \mid III_a \mid^{1/3}$. For the left boundary, the two components with equal energy dominate over the third component, whereas the opposite is true for the right boundary. Hence, the upper right corner represents the one-component limit.

For modelling purposes we also introduce

$$F = 1 - \frac{9}{8}\left(II_a - III_a\right) \tag{39}$$

as the 'degree of two-componentality' being zero at the two-component limit and unity at the origin $II_a = III_a = 0$.

3.2 The Reynolds Equation, Rewritten and a Common Formulation

Using the definition (37), we can rewrite the Reynolds equation as

$$\frac{\bar{\mathrm{D}}\bar{u}_i}{\bar{\mathrm{D}}t} = -\frac{\partial}{\partial x_i}\left(\frac{1}{\rho}P + \frac{2}{3}K\right) + \frac{\partial}{\partial x_m}\left(2\nu\bar{S}_{im} - Ka_{im}\right) + \overline{f}_i \tag{40}$$

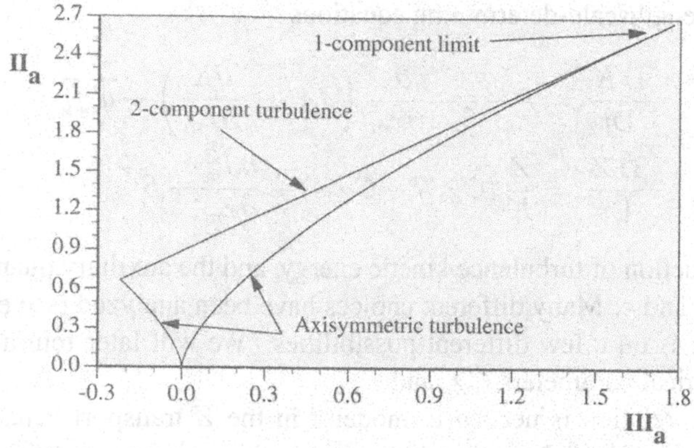

Figure 1: The anisotropy invariant map.

The closure problem is to form a closed set of equations where the unknown Reynolds stress in the mean flow equation is determined from transport and/or algebraic relations involving the set of primary quantities. We see from (40) that the Reynolds stress anisotropy plays a key role. In any single-point closure we need a relation for the determination of a_{ij}, or rather Ka_{ij}. In closures, where the turbulence kinetic energy is not determined explicitly, we cannot determine the mean pressure but only the modified pressure including the effects of the normal turbulent stresses.

Hence, a transport equation for the turbulence kinetic energy can be seen as a first natural ingredient in a single-point closure. The form of this equation is given by (15) and the exact expression for the individual terms are given by (16,18).

Any modelled form of the kinetic energy equation involves the dissipation directly or indirectly through some other lengthscale-determining quantity.

At high turbulence Reynolds numbers, one may under 'normal' conditions assume the viscous dissipation to be equal to the energy flux from the large energetic scales, through the spectrum down to the small dissipative scales. This argument gives us the relation

$$\varepsilon \simeq \frac{K^{3/2}}{\ell}, \tag{41}$$

between the dissipation and the lengthscale (ℓ) of the energetic scales. This relation is one of the underlying assumptions in most single-point models. One may also incorporate modifications due to a low Reynolds number, e.g. occurring near a wall (see Johansson & Burden, 1999).

One-equation models with a modelled K-equation complemented by an algebraic relation where the turbulence lengthscale is given directly by mean-flow quantities, have proven to be of limited generality and will not be discussed here.

The lowest level that we will analyze in the present work is the two-equation level with transport equations for the turbulence kinetic energy and a 'lengthscale-determining' quantity. To begin with, we will analyze 'standard' high-Reynolds number formulations without any near-wall corrections. Such formulations will be discussed separately.

Thus, in the high Reynolds number formulation of single-point models we close the set of equations by complementing the Reynolds equation (40) and continuity equation by a 'platform' of the velocity and lengthscale-determining equations

$$\frac{\bar{\mathrm{D}}K}{\bar{\mathrm{D}}t} = \mathcal{P} - \varepsilon - \frac{\partial}{\partial x_m}\left(J_m - \nu\frac{\partial K}{\partial x_m}\right) + \overline{u_i' f_i'} \tag{42}$$

$$\frac{\bar{\mathrm{D}}Z}{\bar{\mathrm{D}}t} = \frac{Z}{K}\left(C_{Z_1}\mathcal{P} - C_{Z_2}\varepsilon\right) - \frac{\partial J_m^Z}{\partial x_m} + S^Z \tag{43}$$

where $\mathcal{P}$ is the production of turbulence kinetic energy, and the auxiliary quantity Z is chosen as a combination of K and ε. Many different choices have been analyzed (see e.g. Piquet (1999)), but here we will focus on a few different possibilities. We will later return to the methods of calibrating the empirical parameters C_{Z_1} and C_{Z_2}.

When $Z \neq \varepsilon$, a relation is needed to model ε in the Z transport equation. This relation (normally) involves Z and K, but may also involve other primary quantities. If the auxiliary quantity is chosen to be ε the (exact) form of the source term (S) is given by $2\nu\overline{u'_{i,m} f'_{i,m}}$. This, as well as the source term in the K-equation, may need modelling, depending on the type of source term and level of closure.

The platform (42,43) is used directly in all forms of two-equation models, but may also be said to be used (directly or) indirectly in higher level closures, such as Reynolds stress transport (RST) models. The latter will mostly be referred to as differential Reynolds stress models (DRSM). In all these models or extensions thereof, such as in the ideas put forward by Reynolds and co-workers (see e.g. Poroseva et al., 2001), the model can be said to consist of

- the Reynolds equation and continuity,
- the platform (the K and Z equations),
- an a_{ij} relation

plus suitable boundary and initial conditions.

A DRSM is normally formulated in terms of a transport equation for the Reynolds stress tensor, in which case the a_{ij} relation is indirectly given by that equation. The DRSM can alternatively be formulated as a platform including the K-equation, and a transport equation for a_{ij}. The treatment of the boundary conditions will be somewhat different for the two cases (see Sjögren & Johansson, 2000).

In lower level models, the a_{ij} relation will be of algebraic form. We can divide two-equation models into two distinct categories, depending on the use or not of a linear Boussinesq hypothesis.

3.3 The Eddy-Viscosity Concept and the Boussinesq Hypothesis

In fully turbulent flow away from solid walls, the momentum transfer is dominated by the mixing caused by the large energetic turbulent eddies. The first ideas along these lines, i.e. to describe the turbulent momentum transfer by means of an equivalent eddy (or turbulent) viscosity were proposed by Boussinesq in the same period (1870's) as Reynolds made his pioneering studies of transition and turbulence in pipe flow.

The turbulent (kinematic) viscosity, ν_T, can essentially be seen as a product of the characteristic length and velocity scales of the large energetic eddies.

The simplest form of a momentum transfer hypothesis, linearly related to the mean strain in analogy with the Newtonian fluid hypothesis, can in coordinate invariant form be written as

$$\overline{u_i'u_j'} = \frac{2}{3}K\delta_{ij} - 2\nu_T\bar{S}_{ij}. \tag{44}$$

In contrast to the molecular viscosity, ν_T can normally not be assumed to be constant, but rather dependent on the dynamics of the flow. The very simplest assumption of a constant eddy-viscosity yields qualitatively correct predictions only in some basic free shear flows.

Since (44) is equivalent to five scalar relations for the relative energy distribution among the components of the Reynolds stress tensor, we may suitably rewrite it in terms of the anisotropy tensor a_{ij}

$$a_{ij} = -2\frac{\nu_T}{K}\bar{S}_{ij} \tag{45}$$

that more clearly illustrates the nature of the linear Boussinesq hypothesis. The relative distribution among the components of the Reynolds stress tensor is here assumed to be given directly by the mean strain field. This is of course a far-reaching assumption of a type of equilibrium between the turbulence state and the mean flow, and can only be expected to give reasonable results where the mean strain field generates a significant turbulence production.

In a physical situation where the mean strain field is abruptly changed or removed, the relation (45) cannot be expected to describe the Reynolds stress anisotropy state. For many cases of practical interest, though, a local relation between the anisotropy state and mean flow characteristics has proven to be a quite useful assumption.

The close analogy with the Newtonian fluid hypothesis also includes insensitivity to solid body rotation. This follows from the linearity of the relation which excludes any dependence on the vorticity tensor, which is antisymmetric in its indices (see (4)). This is here an artifact (see discussion in connection with (24)) in contrast to the situation for the Newtonian fluid hypothesis. In the latter case, the insensitivity to solid body rotations is explained by the large separation in scales between the molecular level where the mixing occurs and the macroscopic scales relevant for the rotation. In turbulent flows there is normally not a very large separation between the scales of rotation and the scales where the major turbulent mixing occurs. Hence, the insensitivity to rotation in (45) is a major weakness of this level of modelling, and is reflected in normally poor predictions of effects of rotation and/or effects of strong mean streamline curvature.

At this level of modelling the a_{ij} relation needed to complement the platform (42,43) is given by (45).

3.4 Linear Eddy-Viscosity Models

We will refer to models based on the linear Boussinesq hypothesis as 'linear eddy-viscosity models'. These are then given by the mean flow equation (40), the platform and a_{ij} relation (42,43,45), and an algebraic relation determining the eddy viscosity in terms of the primary quantities.

The mean flow equation

$$\frac{\bar{\mathrm{D}}\bar{u}_i}{\bar{\mathrm{D}}t} = -\frac{\partial}{\partial x_i}\left(\frac{1}{\rho}\bar{p} + \frac{2}{3}K\right) + \frac{\partial}{\partial x_m}\left(2\left(\nu + \nu_T\right)\bar{S}_{im}\right) + \bar{f}_i \tag{46}$$

here becomes very similar to the laminar form.

The standard form of the eddy-viscosity relation for $K-\varepsilon$ models is given by

$$\nu_T = C_\mu \frac{K^2}{\varepsilon} \tag{47}$$

where C_μ is an empirical parameter normally calibrated to give a correct turbulent shear stress in plane thin shear flows (where $\mathcal{P} \approx \varepsilon$).

For other choices of Z the corresponding relation is a straightforward transformation of (47) with ε given in terms of K and Z.

The modelling of the turbulence production ($\mathcal{P}$) is here determined by the Boussinesq hypothesis, and the (normal) form of the model for the turbulent diffusion terms is that of a simple gradient diffusion expression so that

$$J_m = -\frac{\nu_T}{\sigma_k}\frac{\partial K}{\partial x_m} \tag{48}$$

$$J_m^Z = -\frac{\nu_T}{\sigma_Z}\frac{\partial Z}{\partial x_m} \tag{49}$$

The spatial variation of the turbulent diffusivity coefficient is here assumed to be entirely described by the eddy-viscosity, and the magnitude is determined by the Schmidt number σ_k or σ_Z.

Hereby, the model in its high Reynolds number form is defined except for the calibration of the parameters involved for the respective choice of Z (plus boundary and initial conditions).

For clarity, we here also give the resulting K-equation (including molecular diffusion)

$$\frac{\bar{\mathrm{D}}K}{\bar{\mathrm{D}}t} = 2\nu_T \bar{S}_{ij}\bar{S}_{ij} - \varepsilon + \frac{\partial}{\partial x_i}\left[\left(\nu + \frac{\nu_T}{\sigma_k}\right)\frac{\partial K}{\partial x_i}\right] \tag{50}$$

This type of model is still dominating in commercial CFD codes. Here we will give some details about $K-\varepsilon$ and $K-\omega$ models, but many other possibilities have been tried and more details can be found in e.g. Speziale, Abid & Anderson (1992) or Wilcox (1994). The different choices tried for Z can be written as

$$Z = K^m \varepsilon^n \tag{51}$$

sometimes multiplied by a constant. The turbulence lengthscale (of the energetic scales) is then in principle extracted through the relation $l \sim K^{3/2}/\varepsilon$.

The ω quantity is proportional to ε/K and has been given several slightly different physical interpretations. Perhaps the most straightforward is that of an inverse timescale of the large energetic scales.

These two choices for the Z-quantity have quite different behaviours near solid walls and as a free-stream is approached. The limiting value of ε near a solid wall is finite (and non-zero), whereas ω becomes infinite at the wall. A third possibility that has a natural zero wall-boundary condition, and has attracted some recent attention, is the turbulence timescale itself, $\tau \equiv K/\varepsilon$. Near walls, the timescale usually has to be modified and is often set to be limited from below by the Kolmogorov timescale, or often six times that ($6\sqrt{\nu/\varepsilon}$) as proposed by Durbin (1996).

3.4.1 Analytical solutions for cases of homogeneous turbulence

To illustrate the character of the solutions of the model equations at this level of turbulence closure, we consider the situation of homogeneous turbulence. This can be homogeneous shear flow, plane strain, axisymmetric straining flow, pure rotation, etc.

The statistical moments (including the mean velocity gradient field) are then uniform in space, implying that the diffusion terms vanish. For brevity we introduce the notation

$$II_S \equiv \bar{S}_{ij}\bar{S}_{ij} \tag{52}$$

for the, in space, constant parameter that measures the strength of the mean strain. It is also the square of the inverse timescale of the mean strain and is thereby the natural parameter to which the turbulence timescale is compared.

It is easy to see that all models based on a Z-quantity defined by (51) have equivalent forms for the modelling of the production and destruction terms in the Z-equation. The differences only lie in the different approaches to the calibration of the model parameters.

The K and Z equations here take the form (in the absence of work-performing volume forces)

$$\frac{dK}{dt} = 2\nu_T II_S - \varepsilon \tag{53}$$

$$\frac{dZ}{dt} = \frac{Z}{K}\left(C_{Z_1} 2\nu_T II_S - C_{Z_2}\varepsilon\right) \tag{54}$$

With use of (51) and (47), they can be rewritten as

$$\frac{dK}{dt} = 2C_\mu II_S K^{2+\frac{m}{n}} Z^{-\frac{1}{n}} - K^{-\frac{m}{n}} Z^{\frac{1}{n}} \tag{55}$$

$$\frac{dZ}{dt} = 2C_\mu C_{Z_1} II_S K^{1+\frac{m}{n}} Z^{1-\frac{1}{n}} - C_{Z_2} K^{-\left(1+\frac{m}{n}\right)} Z^{1+\frac{1}{n}} \tag{56}$$

By identification with the corresponding K and ε equations (i.e. $m = 0, n = 1$), we find the relations

$$C_{Z_1} = m + nC_{\varepsilon 1} \tag{57}$$

$$C_{Z_2} = m + nC_{\varepsilon 2} \tag{58}$$

The equations (55,56) may even be solved analytically. We first notice that the only choices of Z for which the Z-equation becomes independent of K are those for which $m = -n$. This is satisfied for the turbulence timescale τ (or its inverse $\sim \omega$). The equation for τ becomes

$$\frac{d\tau}{dt} = -2C_\mu\left(C_{\varepsilon 1} - 1\right) II_S \tau^2 + C_{\varepsilon 2} - 1 \tag{59}$$

which has the solution

$$\frac{\tau}{\sqrt{II_S}} = a\,\frac{ce^{bt} - 1}{ce^{bt} + 1} \tag{60}$$

where

$$a = \sqrt{\frac{(C_{\varepsilon 2} - 1)}{2C_\mu\left(C_{\varepsilon 1} - 1\right)}}, \quad b = \frac{2}{\sqrt{II_S}}\sqrt{2C_\mu\left(C_{\varepsilon 2} - 1\right)\left(C_{\varepsilon 1} - 1\right)}, \quad c = \frac{(a + \tau_o^*)}{(a - \tau_o^*)} \tag{61}$$

and $\tau_o^* = \tau(t=0)/\sqrt{II_S}$.

We can readily see that the turbulence timescale, as solved from the model equation, will remain positive as long as $C_{\varepsilon 1}, C_{\varepsilon 2}$ are larger than unity and the initial value satisfies $\tau_o^* < a$, which for standard parameter values is about 3. In fact, the timescale is then monotonously growing for all times.

The timescale asymptotically tends to

$$\tau \rightarrow \frac{a}{\sqrt{II_S}} = \left(\frac{C_{\varepsilon 2} - 1}{2C_\mu II_S\,(C_{\varepsilon 1} - 1)}\right)^{1/2} \quad \text{as } t \rightarrow \infty \tag{62}$$

which also implies that the production to dissipation ratio

$$\frac{\mathcal{P}}{\varepsilon} \rightarrow \frac{C_{\varepsilon 2} - 1}{C_{\varepsilon 1} - 1} \quad \text{as } t \rightarrow \infty. \tag{63}$$

From the K-equation, we see that $K^{-1}\mathrm{D}K/\mathrm{D}t$ is expressed explicitly in τ. Hence, we readily see that the solution for the kinetic energy remains positive, which of course means that the solution for any choice of Z will remain positive for all times.

The above equations can also be of use as an intermediate step in the solution of the complete K, ε (or K, Z) equations for non-homogeneous flows. Such a solution strategy was devised by Mohammadi & Pironneau (1994) in order to guarantee positive solutions for K and ε for all times in complex flows. The reason for this is that in practical cases the occurrence of negative values may often occur temporarily for ordinary solution schemes, and can cause serious convergence problems.

The above asymptotic production to dissipation value was also used in the explicit algebraic Reynolds stress model of Gatski & Speziale (1993).

4 Two-Equation Models and the Calibration of Model Parameters

Here we will discuss some aspects of the high-Reynolds formulations of the standard $K-\varepsilon$ and $K-\omega$ models and some typical approaches for the parameter calibration (mostly) in the former case. The crucial differences between various two-equation models lie in the modelling of the diffusion terms (see 48,49) and near-wall treatments (including boundary conditions). For the former much attention has been paid to the possible need of introducing cross-diffusion terms in the model transport equations. The diffusion treatment is known to have a central importance near walls and in particular near boundary layer – free-stream interfaces.

In the $K-\varepsilon$-model, the kinetic energy equation (50) needs no further modelling, whereas for the $K-\omega$ and other two-equation models, the dissipation term needs to be modelled in terms of K and the auxiliary quantity.

4.1 The $K-\varepsilon$ Model

In the history of the $K-\varepsilon$ model much of the development was done in the seventies (e.g Launder & Spalding, 1972), although the origin may be traced back to Chou (1945), and there are still ongoing developments.

The kinetic energy equation is given by (50), and the ε-equation (we see from equation (43)) can be written as

$$\frac{\bar{\mathrm{D}}\varepsilon}{\bar{\mathrm{D}}t} = 2C_\mu C_{\varepsilon 1} K S_{ij} S_{ij} - C_{\varepsilon 2}\frac{\varepsilon^2}{K} + \frac{\partial}{\partial x_m}\left(\left(\frac{\nu_T}{\sigma_\varepsilon} + \nu\right)\frac{\partial \varepsilon}{\partial x_m}\right) \tag{64}$$

and the eddy-viscosity is given by (47).

A set of standard values for the model parameters may be said to be:

$$C_\mu = 0.09, \quad \sigma_k = 1.0, \quad \sigma_\varepsilon = 1.3, \quad C_{\varepsilon 1} = 1.44, \quad C_{\varepsilon 2} = 1.92. \tag{65}$$

The above set of parameters is arrived at by considering a few basic arguments and the model predictions for a set of 'corner-stone' flow cases, namely decaying isotropic turbulence, homogeneous shear flow and the log-layer.

We will revisit this approach below and arrive at a 'consistent' set of parameter values.

The traditional way of arriving at the above value of C_μ is to consider a thin shear flow with $\mathcal{P} \approx \varepsilon$. With y as the cross-stream coordinate we then get

$$C_\mu = \nu_T \frac{\varepsilon}{K^2} = \nu_T \frac{\nu_T \left(\frac{\partial \bar{u}}{\partial y}\right)^2}{K^2} = \left(\frac{-\overline{u'v'}}{K}\right)^2 \tag{66}$$

where use has been made of (47) and the Boussinesq hypothesis. The shear stress to kinetic energy ratio for thin shear flows is typically ≈ 0.3 (the so-called Bradshaw hypothesis) from which we get the above value of C_μ. This is probably a good compromise value for a range of shear flows. A closer look at the situation in the inner part of the log-layer where often the 'wall boundary condition' is applied yields that a suitable value of C_μ there should be lower, probably close to 0.07.

One should keep in mind that this calibration of C_μ, which is used in an equivalent manner in other two-equation models, is a rather serious restriction of the range of validity of this type of model. It obviously suggests that the primary target for this type of model should be turbulent shear flows without strong effects of rotation or other external body forces that may remove the turbulent state from this equilibrium value. We will return to more dynamic methods of determining an equivalent C_μ in connection with the discussion of algebraic Reynolds stress models.

The modelling of the other terms and the calibration of the corresponding parameter values are also of relevance when the ε equation is used as an auxiliary equation in Reynolds stress closures.

Since K is simply twice the trace of $\overline{u_i'u_j'}$ it is reasonable to take $\sigma_k = 1.0$.

4.1.1 Modelling the destruction term

The modelling of the dissipation rate destruction term can be motivated by considering the case of isotropic homogeneous turbulence decaying in the absence of a mean strain rate field (see e.g. Batchelor, 1953). It is reasonable to assume that under these circumstances turbulence should decay in a self-similar manner in the sense that the decay rates of K and ε differ at most by a constant factor

$$\frac{K/\frac{dK}{dt}}{\varepsilon/\frac{d\varepsilon}{dt}} = C_{\varepsilon 2}. \tag{67}$$

There are no production or diffusion terms in the K and ε equations (15,19) so the denominator in (67) becomes $-K/\varepsilon$, and it follows that in order to mimic the behaviour prescribed by (67), the ε equation should be modelled as

$$\frac{d\varepsilon}{dt} = -C_{\varepsilon 2}\frac{\varepsilon^2}{K} \tag{68}$$

in the case of isotropic turbulence. This modelling of the destruction term is also the standard one for general flow situations.

From the K and the above modelled ε equations, the evolution of K can then be obtained as

$$K(t) = K_0\left(1 + (C_{\varepsilon 2} - 1)\frac{\varepsilon_0}{K_0}t\right)^{-n}, \qquad n \equiv \frac{1}{C_{\varepsilon 2} - 1} \tag{69}$$

where the 0-subscripts denote initial values. This type of power-law decay is well documented in the literature on wind tunnel experiments of grid generated turbulence, *e.g.* in the study by Comte-Bellot & Corrsin (1993) where exponents of $n = 1.2$ - 1.3 were found.

The rate of decay may be expected to depend on the Reynolds number which is one of the parameters that are relevant in isotropic turbulence. Assuming a self-similar decay behaviour of high Reynolds number turbulence, described by the simple model spectrum of Figure 2, Comte-Bellot & Corrsin (1993) and Reynolds (1976) showed that $C_{\varepsilon 2}$ is related to the low wavenumber exponent a of the energy spectrum function. Evaluation of the kinetic energy from the model spectrum gives

$$K(t) = \int_0^\infty E(k,t)\,dk = \text{const} \times [\varepsilon(t)]^m, \qquad m = \frac{2(a+1)}{3a+5}$$

which is consistent with the first assumption (67) regarding the decay behaviour of K and ε if

$$C_{\varepsilon 2} = \frac{3a+5}{2(a+1)}. \tag{70}$$

With the ('standard') assumption of $a = 2$ (see Saffman, 1967 and Chasnov, 1994), we obtain $C_{\varepsilon 2} = 11/6 \approx 1.83$. A comparison with (69) gives a corresponding decay rate exponent of 1.2 for the kinetic energy, which lies in the range of results of Comte-Bellot & Corrsin (1993).

Reynolds number dependence may also be included in the above analysis. As shown by Aupoix et al. (1989) this effect is small for reasonably high Reynolds numbers, where the assumption of an extended inertial range can be assumed to be valid.

One can note in the above analysis that the quantity modelled is actually the energy drainage from the large scales rather than the viscous dissipation itself. Hence, we assume a turbulence in a state of equilibrium in the sense that the dissipation at small scales is balanced by the transfer through the spectrum. As we can see, although it may perhaps seem somewhat unintuitive, the large scale behaviour is of primary importance for the magnitude of the dissipation rate destruction term. The dependence of universality of the behaviour at large scales may be seen as somewhat of a weakness for single-point closures in general. These aspects were further discussed in Johansson & Burden (1999).

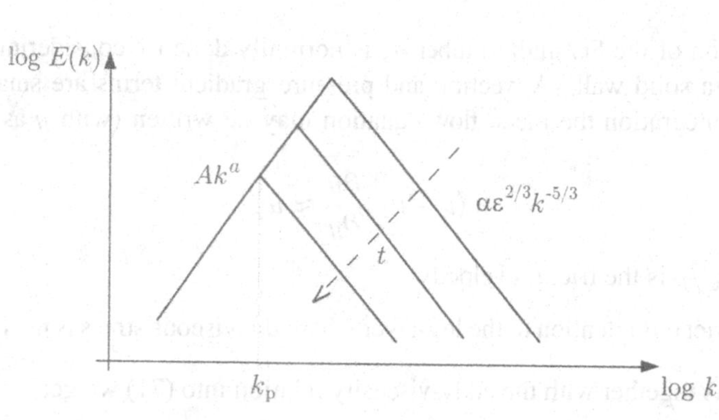

Figure 2: Self-similar decay of high Reynolds number model spectrum where A is assumed to be constant and $k_{\rm p}$ decreases with time.

4.1.2 Modelling the ε production term

The idea behind the standard modelling of the production term in the dissipation rate equation is based on the argument that the rate of energy drainage ($\varepsilon_f \sim K^{3/2}/l$) is intensified as K is increased by kinetic energy production. The simplest assumption that follows from this reasoning, which also follows from simple dimensional arguments, is to take $\mathcal{P}^\varepsilon$ to be proportional to $\varepsilon\, P/K$, reflecting a coupling between energetic and dissipative scales. This is exactly the type of production term modelling described in equation (43). This model for the dissipation production term can be seen as a serious weakness of this level of closure.

The third part of the mean flow related production term, associated with the mean flow second order derivative terms, (see equation 20) is negligibly small essentially everywhere in wall bounded shear flows (cf. *e.g.* Rodi & Mansour, 1993). Appreciable effects are found only in the buffer region (and viscous sublayer for cases with a strong pressure gradient). In some low-Reynolds number formulations of the $K - \varepsilon$ model, a term of this character is included to mimic the effects of this part of the production.

Homogeneous shear flow data may be used to assign $C_{\varepsilon 1}$ a numeric value. This type of flow reaches an equilibrium state with K and ε growing in a manner such that the turbulence time scale K/ε approaches an approximately constant value (see equation (62)) and the normalized Reynolds stresses, i.e. anisotropies, become nearly constant.

In the investigation of Tavoularis & Corrsin (1981), this type of behaviour was observed and the asymptotic value of the production to dissipation ratio was found to be about 1.8. A comparison with (63) yields $C_{\varepsilon 1} = 1.46$, if we assume $C_{\varepsilon 2} = 11/6 \approx 1.83$. These may be compared with the 'standard' values suggested by Launder et al. (1975), $C_{\varepsilon 1} = 1.44$ and $C_{\varepsilon 2} = 1.92$.

4.1.3 Calibration of the Schmidt number σ_ε

The complex diffusion, or spatial redistribution, term in the exact transport equation for the (homogeneous part of the) dissipation is described by a simple gradient diffusion expression. This approach has proven to be reasonably successful and is used in all the modelled transport equations.

The calibration of the Schmidt number σ_ε is normally done by considering the situation of a log-layer near a solid wall. Advection and pressure gradient terms are small up to the log-layer and after integration the mean flow equation may be written (with y as the wall-normal coordinate)

$$(\nu + \nu_T)\frac{\partial \bar{u}}{\partial y} \approx u_\tau^2 \tag{71}$$

where $u_\tau = \sqrt{\tau_w/\rho}$ is the friction velocity.

Next we restrict our attention to the log-layer where the viscous stress is negligible and $\dfrac{\partial U}{\partial y} = \dfrac{u_\tau}{\kappa y}$. Inserting this together with the eddy-viscosity relation into (71) we get

$$K^2 \approx \frac{u_\tau \kappa}{C_\mu} y \varepsilon \tag{72}$$

For the log-layer, we further assume that $\mathcal{P} \approx \varepsilon$, which gives

$$\varepsilon \approx \nu_T \frac{\partial \bar{u}}{\partial y} = C_\mu \frac{K^2}{\varepsilon}\left(\frac{u_\tau}{\kappa y}\right)^2 \tag{73}$$

Combining (72) and (73), we obtain the well known relations for K and ε in the log-layer

$$\begin{aligned} \varepsilon &\approx \frac{u_\tau^3}{\kappa y} \\ K &\approx \frac{u_\tau^2}{\sqrt{C_\mu}} \end{aligned} \tag{74}$$

Since K is constant and production is balanced by dissipation, the K equation is trivially satisfied. Using $\mathcal{P} \approx \varepsilon$ and the assumption of negligible advection, we can write the ε equation as

$$0 = (C_{\varepsilon 1} - C_{\varepsilon 2})\frac{\varepsilon^2}{K} + \frac{\partial}{\partial y}\left(\frac{\nu_T}{\sigma_\varepsilon}\frac{\partial \varepsilon}{\partial y}\right) \tag{75}$$

Combining (75) and (74) we finally obtain

$$\sigma_\varepsilon = \frac{\kappa^2}{\sqrt{C_\mu}} / (C_{\varepsilon 2} - C_{\varepsilon 1}) \tag{76}$$

Hence, with the above 'standard' values $\kappa = 0.41, C_\mu = 0.09, C_{\varepsilon 2} = 1.92, C_{\varepsilon 1} = 1.44$, we would get $\sigma_\varepsilon \approx 1.2$ to be compared with the 'standard' value of 1.3. Alternatively, with $C_{\varepsilon 2} = 1.83, C_{\varepsilon 1} = 1.46$ we would get $\sigma_\varepsilon \approx 1.5$.

Hence, the set of parameters consistent with the present calibration becomes:

$$C_\mu = 0.09, \quad \sigma_k = 1.0, \quad \sigma_\varepsilon = 1.5, \quad C_{\varepsilon 1} = 1.46, \quad C_{\varepsilon 2} = 1.83. \tag{77}$$

It may be interesting to make a closer comparison with the asymptotic homogeneous shear flow data of Tavoularis & Corrsin (1981). The asymptotic time-scale and $\mathcal{P}/\varepsilon$ predicted by the model are given by (62,63). The Boussinesq hypothesis yields the values of the anisotropies ($a_{12} = -2C_\mu(K/\varepsilon)\bar{S}_{12}$) and we may summarize the comparison with the asymptotic values as

	a_{11}	a_{22}	a_{12}	$2S_{12}K/\varepsilon$	$\mathcal{P}/\varepsilon$
TC81-exp	0.40	-0.29	-0.30	6	1.8
$K-\varepsilon$-pred.	0	0	-0.40	4.5	1.8

Table 1: Asymptotic values for homogeneous shear flow, comparison with the experimental data of Tavoularis & Corrsin (1981)

4.1.4 Near-wall treatment and low Reynolds number formulations

Still the most common way to treat wall boundary conditions in commercial CFD applications is to use so-called log-layer boundary conditions. This means that the ε equation is retained in its high Reynolds number form (64) and the equations are integrated only down to the inner portion of the log-layer. In this way one strongly reduces numerical problems and the need for dense computational meshes in the areas where viscous diffusion is strong. One also avoids modification of the model to account for the low Reynolds numbers near the wall. If the first node, y_1 is chosen to lie in the (inner portion) of the log-layer (usually chosen in the range $50 < y_1^+ < 100$) we have

$$\bar{u}(y_1) = u_\tau \left(\frac{1}{\kappa} \ln y_1^+ + B \right) \tag{78}$$

The values of K and ε are then given by (74).

Normally the friction velocity is not known, which means that it has to be determined from some iterative process, or simply as

$$u_\tau = \frac{\sqrt{K(y = y_1)}}{C_\mu^{1/4}}$$

in which case we use

$$\frac{\partial K}{\partial y} = 0 \text{ at } y = y_1$$

as boundary condition.

In separated regions or for flow near separation, the above approach is highly questionable. This is also true in a number of other situations, *e.g.* in cases with strong effects of rotation, or for unsteady flows.

For flows where the existence of universal wall functions is not established, a low Reynolds number formulation of the turbulence model equations must be formulated. There are a number of low *Re* versions of the K-ε model suggested in the literature (see e.g. Rodi & Mansour, 1993) which for two-dimensional shear layers can be written on the form

$$\frac{\bar{\mathrm{D}} K}{\bar{\mathrm{D}} t} = \frac{\partial}{\partial y}\left[\left(\nu + \frac{\nu_t}{\sigma_K}\right)\frac{\partial K}{\partial y}\right] + \nu_t \left(\frac{\partial \bar{u}}{\partial y}\right)^2 - \varepsilon \tag{79}$$

$$\frac{\bar{\mathrm{D}} \tilde{\varepsilon}}{\bar{\mathrm{D}} t} = \frac{\partial}{\partial y}\left[\left(\nu + \frac{\nu_t}{\sigma_\varepsilon}\right)\frac{\partial \tilde{\varepsilon}}{\partial y}\right] + C_{\varepsilon 1} f_1 \frac{\tilde{\varepsilon}}{K} \nu_t \left(\frac{\partial \bar{u}}{\partial y}\right)^2 - C_{\varepsilon 2} f_2 \frac{\tilde{\varepsilon}^2}{K} + E \tag{80}$$

$$\nu_t = C_\mu f_\mu \frac{K^2}{\tilde{\varepsilon}}, \qquad \tilde{\varepsilon} \equiv \varepsilon - D$$

The low Reynolds number corrections to the high Reynolds number model form are found in the damping functions f_μ, f_1 and f_2 and in the extra terms D and E. The various proposed models differ through the choice of these quantities. In the fully turbulent region away from the wall all three damping functions approach unity and the two extra terms approach zero. The damping function f_μ is introduced to make the model predict a shear stress $-\overline{u'v'} \sim y^3$ in agreement with the expected limiting behaviour. An often used form of D is such that

$$\tilde{\varepsilon} = \varepsilon - 2\nu \left(\frac{\partial K^{1/2}}{\partial y} \right)^2 \sim y^2 \quad \text{as} \quad y \to 0$$

This redefined dissipation rate has the advantage of approaching zero at the wall, as opposed to ε, whence $\tilde{\varepsilon}_w = 0$ may be used as a boundary condition. Away from the wall $\varepsilon \to \tilde{\varepsilon}$. The role of the extra term E (or alternatively f_1) is to increase the production of $\tilde{\varepsilon}$ near the wall. Yet, most models predict the maximum of ε away from the wall in disagreement with DNS data (see So et al., 1991). The damping functions are usually expressed as functions of either y^+ or $Re_{\tilde{\varepsilon}} \equiv K^2/\nu\tilde{\varepsilon}$, or both. For instance, in the Launder-Sharma model the low Reynolds number corrections are given by

$$f_\mu = \exp\left(\frac{-3.4}{(1 + Re_{\tilde{\varepsilon}}/50)^2} \right), \quad f_1 = 1.0, \quad f_2 = 1 - 0.3 \exp(-Re_{\tilde{\varepsilon}}^2)$$

$$D = 2\nu \left(\frac{\partial K^{1/2}}{\partial y} \right)^2, \quad E = 2\nu\nu_t \left(\frac{\partial^2 \bar{u}}{\partial y^2} \right)^2$$

As boundary conditions at the wall, most low Reynolds number K-ε models use

$$K = \tilde{\varepsilon} = 0 \quad \text{at} \quad y = 0$$

Exceptions are models that use (*e.g.* the Lam-Bremhorst model, Lam & Bremhorst, 1981)

$$\varepsilon = \nu \frac{\partial^2 K}{\partial y^2} \quad \text{or} \quad \frac{\partial \varepsilon}{\partial y} = 0 \quad \text{at} \quad y = 0$$

There is no a priori reason why the latter condition should hold. Yet another boundary condition for ε was suggested by Lindberg (1994) who used

$$\varepsilon^+ = \langle a_1^{+2} \rangle + \langle c_1^{+2} \rangle \approx 0.17$$

which is justifiable insofar as one has reason to believe that this value is reasonably constant with respect to flow conditions.

4.2 The $K - \omega$ Model

The $K - \omega$ model is mostly known from the works of Wilcox, see e.g. Wilcox (1994), but goes back to the study of Kolmogorov (1942) which was the first real work on two-equation models. It was later revived and extended by Saffman (1970), and in many contributions by Wilcox.

The quantity ω has been given several different interpretations, but in essence it may be taken as an inverse time scale of the large eddies. The dissipation is consequently modelled through the following relation with ω

$$\varepsilon = C_\mu \omega K \tag{81}$$

The 'standard' model, is defined by the modelled mean flow equation (46) (and continuity), the kinetic energy equation (50) with ε as above, and, for high *Re*, the following ω-equation

$$\frac{\bar{\mathrm{D}}\omega}{\bar{\mathrm{D}}t} = 2\alpha \bar{S}_{ij}\bar{S}_{ij} - \beta\omega^2 + \frac{\partial}{\partial x_i}\left[\left(\nu + \frac{\nu_T}{\sigma_\omega}\right)\frac{\partial \omega}{\partial x_i}\right] \tag{82}$$

together with the eddy-viscosity relation

$$\nu_T = \frac{K}{\omega} \tag{83}$$

The model parameter values given by Wilcox (1994) are:

$$C_\mu = 0.09, \quad \sigma_k = 2.0, \quad \sigma_\omega = 2.0, \quad \alpha = 5/9 \approx 0.56, \quad \beta = 3/40 = 0.075. \tag{84}$$

The choice of σ_ω is consistent with the log-layer condition discussed in conjunction with the calibration of σ_ε, but one may note, perhaps somewhat surprisingly, that here the Schmidt number σ_k is given a value twice that in the standard $K-\varepsilon$-model.

In a comparison with the general relations (57,58), taking into account that here $Z = \varepsilon/(C_\mu K)$, we find that the above choice of model parameters would correspond to a $K-\varepsilon$ model (in the homogeneous case) with $C_{\varepsilon 1} = 1 + \alpha = 1.56$ and $C_{\varepsilon 2} = 1 + \beta/C_\mu = 1.83$. It is noteworthy that these values are quite close to those arrived at with the above model parameter calibration for the $K-\varepsilon$-model.

Inserting $\varepsilon = C_\mu \omega K$ into the ε transport equation (64), we find that the turbulent diffusion term in the transformed ω equation reads

$$\frac{\partial}{\partial x_i}\left[\frac{\nu_T}{\sigma_\varepsilon}\frac{\partial \omega}{\partial x_i}\right] + \frac{\omega}{K}\frac{\partial}{\partial x_i}\left[\nu_T\left(\frac{1}{\sigma_\varepsilon} - \frac{1}{\sigma_k}\right)\frac{\partial K}{\partial x_i}\right] + \frac{2\nu_T}{K\sigma_\varepsilon}\frac{\partial K}{\partial x_i}\frac{\partial \omega}{\partial x_i}.$$

and can be seen to involve terms that are not of gradient diffusion type. The original $K-\omega$ model is known to have problems near free-stream boundaries (see Wilcox, 1994). To remedy this behaviour Menter (1992) has proposed the inclusion of a cross-diffusion term that is equivalent to the last term of the above expression of the diffusion term in the $K-\omega$ model. This means that the similarity between the two models becomes even closer. A further modification that has been proposed is a formulation where the influence of the cross-diffusion term is weighted such that it will only influence the flow far away from walls. The generality of such formulations for complex flows can perhaps be doubted.

The $K-\omega$ model has been used extensively for boundary layer problems with different types of pressure gradients. For instance, in boundary layers with a local separation bubble, the behaviour after reattachment has been found to be improved over typical $K-\varepsilon$ models. As seen from the transformation of the ε equation into an ω equation, the qualitative difference between the two models are few, at least in their high Reynolds number formulations.

The near-wall behaviour of the two quantities is quite different though, and can be an important factor in explaining the noted prediction differences.

4.3 Realizability Considerations for Two-Equation Models

A general requirement in constructing models of various terms in the exact dynamic equations of different statistical moments, such as the Reynolds stresses, is that the modelled equations should not give rise to physically unrealizable solutions, such as negative energies. When dealing with two-equation closures based on the linear Boussinesq hypothesis, we should first of all demand that K and ε (or other Z-quantity) remain non-negative. Secondly, the anisotropies predicted by the Boussinesq hypothesis should attain physically realizable values, i.e. the diagonal components should satisfy, $-2/3 \leq a_{\alpha\alpha} \leq 4/3$ (no summation over Greek indices) and the off-diagonal components should be restricted by $\mid a_{\alpha\beta} \mid \leq 1.0 \ \ (\alpha \neq \beta)$.

It is easy to see, however, that the linear, local modelling of the Reynolds stress in terms of the mean strain rate in the form of the Boussinesq hypothesis does not satisfy the realizability constraint for large values of the (normalized) mean strain rate.

This would typically occur when the variation in strain is large or a large strain rate is suddenly imposed on the turbulence. For such cases, the turbulence timescale $\frac{K}{\varepsilon}$ does not have time to adjust to the change in mean strain.

For instance, with the $K - \varepsilon$ model description of the eddy viscosity, the anisotropy tensor can be written as

$$a_{ij} = -2C_\mu \frac{K}{\varepsilon} \bar{S}_{ij} \equiv -2C_\mu \bar{S}^*_{ij}$$

where $\bar{S}^*_{ij}$ denotes the mean strain rate normalized by the turbulence time scale.

A quantity of central interest in model predictions is the production to dissipation ratio, $\mathcal{P}/\varepsilon \equiv -a_{ij}\bar{S}^*_{ij}$. Since the anisotropy components in the real physical situation are always bounded, the true production to dissipation ratio for large strain rate values will be of the order of the non-dimensional mean strain rate. In a $K - \varepsilon$ model, on the other hand, this quantity will be of the order of the square of the strain rate (see 50).

The prediction of the anisotropy components can be illustrated by the following examples. For a parallel shear flow with a non-dimensional mean strain, $\sigma = \dfrac{1}{2}\dfrac{K}{\varepsilon}\dfrac{d\bar{u}}{dy}$, the eddy viscosity model will predict a shear stress anisotropy component

$$a_{12} = -2C_\mu \sigma$$

Hence, the Boussinesq hypothesis will here generate unphysical results for $\sigma > 1/2C_\mu \approx 5.6$. This is obviously a primary reason for the strong need for near-wall damping functions in this type of model. A simple remedy for this problem would be use of a limiter (equal to unity).

We can find a similar approach in the SST model of Menter (1993), which adopts the Bradshaw assumption, $a_{12} = -0.3$ for $\mathcal{P}/\varepsilon$ ratios greater than unity. This model has also been motivated by the observation that in flows with an adverse pressure gradient, the production to dissipation ratio is greater than one and eddy viscosity models with constant C_μ overestimate the turbulent viscosity or the a_{12} anisotropy.

For flows with large irrotational strain, e.g. flows in nozzles etc, the inability of the Boussinesq hypothesis to satisfy realizability may cause severe problems. We may, for instance, consider axisymmetric strain with $\bar{S}^*_{11} = \sigma$ $(\bar{S}^*_{22} = \bar{S}^*_{33} = -\sigma/2)$. Here the Boussinesq hypothesis yields (with $\nu_t = C_\mu \frac{K^2}{\varepsilon}$)

$$a_{11} = -2C_\mu \sigma$$

which shows that unphysical results are produced for $\sigma > 1/3C_\mu \approx 3.7$.

5 Differential Reynolds Stress Models – Simplest High-Re Form

In the previous section, we noted some of the shortcomings in the prediction capacity at the two-equation modelling level when a linear Boussinesq hypothesis is used. This implies that the turbulence anisotropy state is assumed to be in balance with the mean strain field. The linearity also excludes any dependence on system rotation in the anisotropy relation. A natural remedy for these shortcomings is to remove the eddy-viscosity hypothesis altogether, and formulate modelled transport equations for the Reynolds stress tensor (24). As pointed out in section 3, the six scalar equations for the stress components must still be complemented by (at least) one equation for a length-scale-determining quantity, such as ε. This approach is then equivalent to that described in section 3, in terms of a platform with equations for K and ε (or e.g. ω), complemented by a a_{ij} relation in the form of a modelled transport equation.

We will later return to this approach, but start here with a short description of the simplest form of differential Reynolds stress models (DRSM) formulated directly with a transport equation for $\overline{u_i'u_j'}$, i.e.

$$\frac{\bar{\mathrm{D}}\,\overline{u_i'u_j'}}{\bar{\mathrm{D}}t} = \mathcal{P}_{ij} - \varepsilon_{ij} + \Pi_{ij} - \frac{\partial}{\partial x_m}\left(J_{ijm} - \nu\frac{\partial \overline{u_i'u_j'}}{\partial x_m}\right) + G_{ij}. \tag{85}$$

We note that major advantages at this level of modelling are the improved description of history effects, especially in rapidly varying flow situations, and the fact that the production tensor $\mathcal{P}_{ij}$, and thereby also the kinetic energy production, is described without need for any modelling, i.e.

$$\mathcal{P}_{ij} \equiv -\overline{u_i'u_m'}\frac{\partial \bar{u}_j}{\partial x_m} - \overline{u_j'u_m'}\frac{\partial \bar{u}_i}{\partial x_m} = \tag{86}$$

$$= -K\left(\frac{4}{3}\bar{S}_{ij} + a_{im}\bar{S}_{mj} + \bar{S}_{im}a_{mj} - \left(a_{im}\bar{W}_{mj} - \bar{W}_{im}a_{jm}\right)\right). \tag{87}$$

The latter rewritten form of the production tensor illustrates clearly how the rotational part of the mean flow naturally enters in the production term, in contrast to eddy-viscosity based models. Also, an imposed system rotation enters explicitly through the body force term via the Coriolis term (see discussion in conjunction with equation 24). One should mention though, that even at this level of modelling the description of rotational effects forms a major challenge and difficulty.

One should note that in the kinetic energy equation, which is simply half the trace of (85), only the turbulent diffusion term needs explicit modelling. The mean flow (or Reynolds) equation needs no modelling at this level.

The first model based on the Reynolds stress transport equation was devised 1951 by Rotta (1951), although much of the development of Reynolds stress models occurred during the 1970's with e.g. the landmark paper of Launder et al. (1975) (see also Naot et al., 1973). Continued development of more advanced forms of this type of model is still ongoing. See also e.g. the review of Speziale (1991). Some details on boundary conditions and near-wall treatment can be found in Johansson & Burden (1999).

The natural inclusion of the rotational effects through the Coriolis term in the Reynolds stress (or a_{ij}) transport equation means, for instance, that the well known asymmetric mean velocity profile in plane turbulent channel flow rotating around the spanwise axis, can be predicted at this level with relatively simple models for the various terms involved in equation (85). This was shown first by Launder et al. (1987). A standard two-equation model yields results that are insensitive to the rate of rotation in this case. For similar reasons, effects of mean streamline curvature are also better described here than at the two-equation modelling level.

The unknown terms on the right-hand-side of (85) have to be described in terms of the primary quantities, i.e. K, ε (or other Z), a_{ij} (unless we introduce further primary quantities for which transport equations have to be solved) and the mean flow field quantities. Geometrical constraints and the kinematic viscosity, ν, may enter the problem through a parametric dependence.

In this section, we will give a brief description of DRSM's, in high Reynolds number formulation, and with models for the right-hand-side terms that are linear in the anisotropy tensor a_{ij}. Hence, we here have seven transport equations for turbulence quantities.

5.1 The Dissipation Rate Tensor

The trace of the dissipation rate tensor equals twice the dissipation rate $\varepsilon_{ii} = 2\varepsilon$. We will return to the modelling of the ε transport equation in the DRSM context.

We may here introduce an anisotropy measure for the ε_{ij} tensor

$$e_{ij} \equiv \frac{\varepsilon_{ij}}{\varepsilon} - \frac{2}{3}\delta_{ij}. \tag{88}$$

The simplest form of modelling of this term can then be written as

$$\varepsilon_{ij} = \tfrac{2}{3}\varepsilon\delta_{ij} \quad \text{or} \quad e_{ij} = 0. \tag{89}$$

This simply means that the dissipation rate tensor is assumed to be isotropic, i.e. equal amounts of energy are drained from the three (diagonal) components. The general motivation for this model would be that the dissipation is dominated by contributions from scales that are much smaller than the large energetic ones, and thereby more isotropic (cf. Kolmogorov's theory of small scale universal equilibrium). In more extreme situations where the Reynolds stress anisotropy is very large, and the turbulence is close to a two-component state (e.g. near solid walls), this is a rather poor approximation, but one may then interpret the dissipation rate anisotropy contributions in (85) to be lumped together with the pressure-strain rate term. Such situations are often also associated with relatively low turbulence Reynolds numbers.

A model assumption that would still lie in the linear modelling context would be to assume the dissipation rate anisotropy to be proportional to a_{ij} with a proportionality coefficient dependent on the turbulence Reynolds number.

5.2 The Pressure-Strain Rate Term

The pressure strain-rate tensor has zero trace and, hence, gives no net contribution to the kinetic energy equation. Instead, it has a purely redistributive effect among the components of the Reynolds stress tensor.

The equation for the pressure itself may be derived by taking the divergence of the equation for the fluctuating velocity (13). This results in a Poisson equation (for a divergence-free body force term)

$$-\frac{1}{\rho}p'_{,kk} = 2\bar{u}_{i,j}u'_{j,i} + u'_{i,j}u'_{j,i} - \overline{u'_{i,j}u'_{j,i}} \tag{90}$$

with the wall boundary condition

$$\frac{1}{\rho}p'_{,2} = \nu u'_{2,22} \tag{91}$$

where x_2 is the wall-normal coordinate.

The source term in the Poisson equation suggests that there are three different types of contribution to the pressure, when written as a formal integral solution to (90). The contribution associated with the satisfaction of the boundary condition can normally be assumed to be negligibly small compared with the other two parts. The part associated with the interaction with the mean velocity gradient field is usually referred to as the rapid part, in that it responds directly to changes in the mean field.

The part associated with the last source term is consequently referred to as the slow part of the pressure.

The solution of the rapid part, can for instance, be written as

$$p'^{(r)}(\vec{x}) = \frac{1}{4\pi}\int_V 2\bar{u}^{(1)}_{k,l}u'^{(1)}_{l,k}\left(\frac{1}{|\vec{x}^{(1)} - \vec{x}|} + \frac{1}{|\vec{x}^{(1)} - \vec{x}^{(*)}|}\right) dV^{(1)}$$

and correspondingly for the slow term. The effect of the wall condition is here included in the form of Green's function, where $\vec{x}^{(*)}$ is the mirror image point of $\vec{x}$.

In forming the expressions for the rapid and slow pressure-strain rate correlations, it is common to impose three local homogeneity assumptions (cf. Chou, 1945), viz.

- no influence of boundary conditions (i.e. applied at infinity),
- assume the mean velocity gradient field to vary slow enough so that it can extracted from the Poisson integral,
- two-point correlations are assumed to vary slowly enough with position $\vec{x}$ to be approximated by two-point velocity-correlation derivatives with respect to separation $\vec{r}$ only.

A more detailed discussion of the derivation of the model expressions for the pressure strain rate terms can e.g. be found in Johansson & Burden (1999).

5.2.1 The rapid part

The homogeneous approximation of the rapid pressure strain rate term can, with the above approximations, be written as

$$\Pi_{ij}^{(r)} = 4K\bar{u}_{p,q}\{M_{iqpj} + M_{jqpi}\} \tag{92}$$

where we have introduced the fourth rank tensor $\mathbf{M}$

$$M_{ijpq} \equiv -\frac{1}{8\pi K}\int \frac{\partial^2 R_{ij}}{\partial r_p \partial r_q}\frac{dV}{|\vec{r}|} = \frac{1}{2K}\int \frac{k_p k_q}{k^2}\widehat{R}_{ij} d^3k \tag{93}$$

where $R_{ij}(\vec{r})$ is the two-point double velocity correlation and $\widehat{R}_{ij}$ its spectral counterpart.

The problem is thus to model the dimensionless tensor $\mathbf{M}$. The $\mathbf{M}$-tensor satisfies two symmetry conditions, a continuity condition and the so-called Green's condition

$$\begin{aligned} M_{ijpq} &= M_{jipq} \\ M_{ijpq} &= M_{ijqp} \\ M_{ijjq} &= 0 \\ M_{ijpp} &= \frac{\overline{u_i' u_j'}}{2K} = \tfrac{1}{2}a_{ij} + \tfrac{1}{3}\delta_{ij} \end{aligned} \tag{94}$$

The second symmetry condition is obvious and the first one follows from the fact that the anti-symmetric part of the spectrum tensor integrates to zero.

The standard approach is to model $\mathbf{M}$ in terms of a. A complete linear expansion satisfying the symmetry conditions of (94) is given by

$$\begin{aligned} M_{ijpq} = {} & A_1\,\delta_{ij}\delta_{pq} + A_2\,(\delta_{ip}\delta_{jq} + \delta_{iq}\delta_{jp}) + A_3\,\delta_{ij}a_{pq} \\ & + A_4\,a_{ij}\delta_{pq} + A_5\,(\delta_{ip}a_{jq} + \delta_{iq}a_{jp} + \delta_{jp}a_{iq} + \delta_{jq}a_{ip}) \end{aligned} \tag{95}$$

Imposing the last two conditions of (94) yields all of the coefficients expressed in terms of only one;

$$A_1 = \frac{4}{15}, \quad A_2 = -\frac{1}{15}, \quad A_3 = \frac{C_2}{2}, \quad A_4 = \frac{5+2C_2}{11}, \quad A_5 = -\frac{2+3C_2}{11}$$

Insertion of the linear ansatz (95) and the above coefficients into (92) yields the linear model of Launder et al. (1975)

$$\begin{aligned} \Pi_{ij}^{(r)} = {} & \frac{4}{5}K\bar{S}_{ij} + \frac{9C_2+6}{11}K(a_{ik}\bar{S}_{jk} + a_{jk}\bar{S}_{ik} - \frac{2}{3}a_{kl}\bar{S}_{kl}\delta_{ij}) \\ & + \frac{-7C_2+10}{11}K(a_{ik}\bar{W}_{jk} + a_{jk}\bar{W}_{ik}) \end{aligned} \tag{96}$$

where $C_2 = 0.4$ was found to give the best fit in calibration against the normal stresses in the nearly homogeneous shear flow of Champagne et al. (1970). Calibration against RDT would give $C_2 = 10/7$, as a consequence of which, as pointed out by Shih et al. (1990), the mean rotation related term of (96) would vanish. The best compromise value, for a wide range of flow situations, seems to be somewhat higher than that proposed by Launder et al. A value of (or close to) 5/9 has been proposed by Lumley (1978) and used in algebraic Reynolds stress modelling by Taulbee (1992) and Wallin & Johansson (2000).

A further truncated version of (96) known as the 'isotropization of production' model is given by

$$\begin{aligned}\Pi_{ij}^{(r)} &= \frac{4}{3}\gamma K\bar{S}_{ij} + \gamma K(a_{ik}\bar{S}_{jk} + a_{jk}\bar{S}_{ik} - \frac{2}{3}a_{kl}\bar{S}_{kl}\delta_{ij}) + \gamma K(a_{ik}\bar{W}_{jk} + a_{jk}\bar{W}_{ik}) \\ &= -\gamma(\mathcal{P}_{ij} - \frac{2}{3}\mathcal{P}\delta_{ij}) \end{aligned} \quad (97)$$

where $\gamma = 0.6$ was chosen by Launder et al. (1975) in order to give the correct initial response to mean strain of isotropic turbulence ($\Pi_{ij}^{(r)} = 4/5K\bar{S}_{ij}$). The simplified version (97) does not satisfy the symmetry conditions of the **M**-tensor.

5.2.2 The slow part

The coupling between the pressure-strain rate term and the two-point correlation spectrum, which integrated over wavenumber space gives the Reynolds stresses, is less clear for the slow term than for the rapid term. The justification for modelling the slow pressure-strain-rate correlation in terms of the Reynolds stresses is thus weaker than for the rapid part.

Even though not rigorously shown, it is regarded as conventional wisdom, based on experience, that the slow pressure-strain rate term plays the role of redistributing energy among velocity components towards isotropy. The simplest possible way to model this tendency of isotropization mathematically is to apply dimensional analysis and assume that the pressure-strain rate is proportional to the deviation from isotropy as was suggested by Rotta (1951)

$$\overline{\frac{p'^{(s)}}{\rho}\left(\frac{\partial u_i'}{\partial x_j} + \frac{\partial u_j'}{\partial x_i}\right)} = -C_{\mathrm{R}}\frac{K^{3/2}}{l}\left(\frac{\overline{u_i'u_j'}}{K} - \frac{2}{3}\delta_{ij}\right) \quad (98)$$

In isotropic turbulence, the pressure has been shown to scale with the kinetic energy, $\sqrt{\overline{(p'/\rho)^2}} \sim K$ (Batchelor, 1953). On the other hand, this is far from true at a wall where $K = 0$, but $p'^{(s)}_{\mathrm{rms}} \neq 0$, reflecting the non-local character of the pressure field. Usually, the high Reynolds number estimate $l \sim K^{3/2}/\varepsilon$ is used to eliminate the turbulence length scale from (98) and to reformulate the model as

$$\Pi_{ij}^{(s)} = -C_1\varepsilon a_{ij} \quad (99)$$

where C_1 is known as the "Rotta constant", usually assigned a value in the range 1.5 - 1.8. This type of modelling has been adopted by most modellers although C_1 is in general not a constant, but rather a function of the turbulence state.

It is notable that a linear model cannot satisfy the strong realizability condition unless the parameter C_1 is allowed to depend on scalar measures of the turbulence state, such as the two-componentality parameter F or the Reynolds number Re_T. Indeed, a rather strong Re_T-dependence of C_1 has been observed (see Sjögren & Johansson, 1998).

There are different methods to incorporate effects of the presence of a solid wall for positions in the close vicinity of the boundary. Pressure reflection effects incorporated into linear models were discussed in e.g. Launder et al. (1975), Gibson & Launder (1978) and Shih & Lumley

(1986). An alternative approach can be taken by use of nonlinear models satisfying strong realizability, as will be discussed later.

5.3 Formulation in Rotating Systems

Deriving a Poisson equation from equation (12) for the fluctuating velocity component in a rotating frame, we see that the 'homogeneous' expression for the rapid part of the pressure strain (92) should be altered to

$$\Pi_{ij}^{(\mathrm{r})} = 4K\left(\bar{u}_{p,q} + \epsilon_{plq}\Omega_l^{\mathrm{s}}\right)\{M_{iqpj} + M_{jqpi}\} \tag{100}$$

i.e. the rotation tensor should be replaced by the absolute rotation tensor (where the contribution from the system rotation is added).

The direct contributions from the Coriolis term in the Reynolds stress transport equation have been discussed earlier. Further details concerning formulation of the DRSM model equations can be found in e.g. Sjögren & Johansson (2000).

5.4 The Turbulent Stress Diffusion Term

The turbulent flux term J_{ijm} is normally modelled by a gradient diffusion expression in analogy with those in two-equation models. Here though, it is possible to describe the diffusion coefficient with a tensor, giving it properties that account for the spreading effects of different turbulence intensities in the different components (Daly & Harlow (1970))

$$J_{ijm} = -c_s \frac{K}{\varepsilon} \overline{u_l' u_m'} \frac{\partial \overline{u_i' u_j'}}{\partial x_l} \tag{101}$$

with $c_s = 0.25$ as the 'standard' value.

5.5 The ε Equation in DRSM Closures

The modelled DRSM equations are usually complemented by a model equation for the dissipation rate. To the lowest order of refinement, the latter equation is typically given by

$$\frac{\bar{\mathrm{D}}\varepsilon}{\bar{\mathrm{D}}t} = C_{\varepsilon 1}\frac{\mathcal{P}}{K}\varepsilon - C_{\varepsilon 2}\frac{\varepsilon^2}{K} - \frac{\partial}{\partial x_m}\left(J_m^{\varepsilon} - \nu \frac{\partial \varepsilon}{\partial x_m}\right) \tag{102}$$

where

$$\frac{\mathcal{P}}{K} = -a_{lm}\bar{S}_{lm}$$

$$C_{\varepsilon 1} \approx 1.5\,, \quad C_{\varepsilon 2} \approx 1.9$$

and the transport term is again described by a gradient diffusion hypothesis. The Daly & Harlow (1970) model reads

$$J_l^{\varepsilon} = -c_{\varepsilon}\frac{K}{\varepsilon}\overline{u_l' u_m'}\frac{\partial \varepsilon}{\partial x_m} \tag{103}$$

with $c_{\varepsilon} = 0.15$ (Launder et al., 1975).

6 Nonlinear DRSM's Satisfying Realizability

It is a major advantage if the models for the terms in the Reynolds stress transport equation can be ensured to satisfy strong realizability, by which we mean that the models should allow the two-component limit to be accessed but not exceeded. This can be illustrated with the Reynolds stress anisotropy map, where the anisotropy state should be allowed to approach the boundaries but not cross them.

A basic tool in the derivation of nonlinear realizable models is the Caley-Hamilton theorem which states that, e.g., a second rank symmetric tensor (matrix) satisfies its own characteristic equation. For such a tensor, say B_{ij}, we may express the Caley-Hamilton theorem as

$$B_{ik}B_{kl}B_{lj} = I_B B_{ik}B_{kj} + \frac{1}{2}\left(II_B - I_B^2\right)B_{ij} + \frac{1}{6}\left(2III_B - 3II_B I_B + I_B^3\right)\delta_{ij}\,. \quad (104)$$

The normal modelling assumption for the dissipation rate anisotropy tensor and the slow pressure strain rate tensor, is that they do not explicitly depend on the mean strain and rotation tensor. Instead we normally assume that their tensorial form is given solely by the Reynolds stress anisotropy tensor. We may then use the Caley-Hamilton theorem to find the most general form of the models for these terms as

$$e_{ij} = f_1 a_{ij} + f_2\left(a_{ik}a_{kj} - \frac{1}{3}II_a\delta_{ij}\right), \quad (105)$$

$$\Pi_{ij}^{(s)}/\varepsilon = g_1 a_{ij} + g_2\left(a_{ik}a_{kj} - \frac{1}{3}II_a\delta_{ij}\right), \quad (106)$$

where f_1, f_2, g_1, g_2 are scalar functions of the invariants of a_{ij} and possibly other scalar parameters, such as the turbulence Reynolds number ($Re_T = K^2/\nu\varepsilon$).

Sjögren & Johansson (1998) studied the above forms in detail and carried out experiments in axisymmetric turbulence at a relatively high Reynolds number. As an example of a model for the dissipation rate anisotropy tensor we may take a closer look at the one suggested by Sjögren & Johansson (1998)

$$e_{ij} = \left(1 - \frac{1}{2}F\right)a_{ij} \quad (107)$$

where F is the degree of two-componentality (given by 39).

Since $F = 1$ for vanishing anisotropy the model predicts $e_{ij} = a_{ij}/2$ as the initial response when isotropic turbulence is subjected to a sudden distortion (strain or shear). This is in agreement with rapid distortion theory (RDT). Also, at a solid wall we approach the two-component limit ($F = 0$) and the model predicts $e_{ij} = a_{ij}$, which is in accordance with the asymptotic behaviour in the vicinity of a solid wall (see Sjögren & Johansson, 2000).

Komminaho & Skote (2001) used DNS-data to compute dissipation rate anisotropies in the immediate vicinity of the wall in plane turbulent Couette flow and turbulent boundary layer flow. It was shown that the near-wall region in plane turbulent Couette flow corresponds well to a high Reynolds number limit of zero-pressure-gradient turbulent boundary layer flow, since it satisfies the condition of constant total stress.

The limiting values of $(e_{11}, e_{22}, e_{33}, e_{12})$ (and corresponding a_{ij}-values) in Couette flow was found to be $(0.72, -2/3, -0.05)$. This is predicted by the model (107), whereas the third order model by Hallbäck, Groth & Johansson (1990), although satisfying realizability, predicts, e.g., a too low e_{11} (0.48).

With the homogeneity assumptions discussed in section 5.2 we can write the rapid pressure-strain rate term as $\Pi_{ij}^{(r)} = 4K\bar{u}_{p,q}\{M_{iqpj} + M_{jqpi}\}$. The standard approach to modelling the M-tensor is then to assume that it is a function of the Reynolds stress anisotropy tensor (and scalar parameters). We discussed earlier the complete linear expansion (see equation 95) of the **M**-tensor. Using the Cayley-Hamilton theorem we can find the most general tensorial form as expressed in combinations of the a_{ij}-tensor. This contains 15 terms with non-linearity up to fourth order (see Johansson & Hallbäck, 1994), each multiplied by a function of scalar parameters such as the invariants of a_{ij}. Using index symmetry, continuity and Green's conditions (see section 5.2), reduces the number of independent functions to nine.

The most general form of the **M**-tensor yields that any model (based on the above assumptions) for the rapid pressure-strain rate can be expressed with the aid of the following eight terms,

$$G_{ij}^{(1)} = a_{ij} \tag{108}$$

$$G_{ij}^{(2)} = \bar{S}_{ij} \tag{109}$$

$$G_{ij}^{(3)} = a_{ik}\bar{S}_{kj} + \bar{S}_{ik}a_{kj} - \frac{2}{3}I_{aS}\delta_{ij} \tag{110}$$

$$G_{ij}^{(4)} = a_{ik}\bar{W}_{kj} - \bar{W}_{ik}a_{kj} \tag{111}$$

$$G_{ij}^{(5)} = a_{ik}a_{kj} - \frac{1}{3}II_a\delta_{ij} \tag{112}$$

$$G_{ij}^{(6)} = a_{ik}\bar{S}_{kl}a_{lj} - \frac{1}{3}I_{aaS}\delta_{ij} \tag{113}$$

$$G_{ij}^{(7)} = a_{ik}a_{kl}\bar{W}_{lj} - \bar{W}_{ik}a_{kl}a_{lj} \tag{114}$$

$$G_{ij}^{(8)} = a_{ik}a_{kl}\bar{W}_{lm}a_{mj} - a_{ik}\bar{W}_{kl}a_{lm}a_{mj} \tag{115}$$

where $I_{aS} = a_{ik}\bar{S}_{ki}$ and $I_{aaS} = a_{ik}a_{kl}\bar{S}_{li}$.

The nine independent functions appearing in the **M**-tensor expression are further reduced to six after insertion into the expression for the rapid pressure-strain (see Sjögren & Johansson, 2000), and we obtain the following form for the most general $\Pi_{ij}^{(r)}$-model that conserves linearity in the mean strain and rotation tensors,

$$\begin{aligned}\frac{\Pi_{ij}^{(r)}}{K} &= (q_1 I_{aS} + q_9 I_{aaS})\, G_{ij}^{(1)} + q_2 G_{ij}^{(2)} + q_3 G_{ij}^{(3)} + q_4 G_{ij}^{(4)} \\ &+ (q_5 I_{aS} + q_{10} I_{aaS})\, G_{ij}^{(5)} + q_6 G_{ij}^{(6)} + q_7 G_{ij}^{(7)} + q_8 G_{ij}^{(8)}\end{aligned} \tag{116}$$

where only six of the scalar functions are independent,

$$q_2 = \frac{4}{5} - \frac{1}{10}(4q_1 - 3q_7)II_a - \frac{2}{5}q_9 III_a \tag{117}$$

$$q_3 = \frac{12}{7} + \frac{9}{7}q_4 - \frac{1}{7}(3q_8 + 2q_9)II_a - \frac{2}{7}q_{10}III_a \tag{118}$$

$$q_5 = q_9 \tag{119}$$

$$q_6 = 6q_1 - 9q_7 - q_{10}II_a\,. \tag{120}$$

6.1 A Nonlinear DRSM without Wall-Damping Functions

The recently developed model of Sjögren & Johansson (2000) is constructed to be integrated all the way to the wall, but completely avoids damping functions and near-wall corrections. Instead it is based on non-linear models for the two parts of the pressure strain and the dissipation rate anisotropy, that all satisfy the condition of strong realizability. For instance, the form of the rapid pressure strain-rate model is given by (116) with the scalar functions taken as expansions in terms of the invariants II_a, III_a. The final model tested contained a $\Pi_{ij}^{(r)}$-model that is of fifth order in the anisotropy. The model constants for this and the other terms involved were mainly calibrated against a set of different homogeneous turbulent flows.

Figure 3 shows a comparison between the model predictions and the DNS data of Kim et al. (1987) for channel flow at a friction Reynolds number of 395. Considering the fact that no damping functions are used, and no explicit use is made of the wall normal vector or wall-normal distance, the agreement is more than satisfactory.

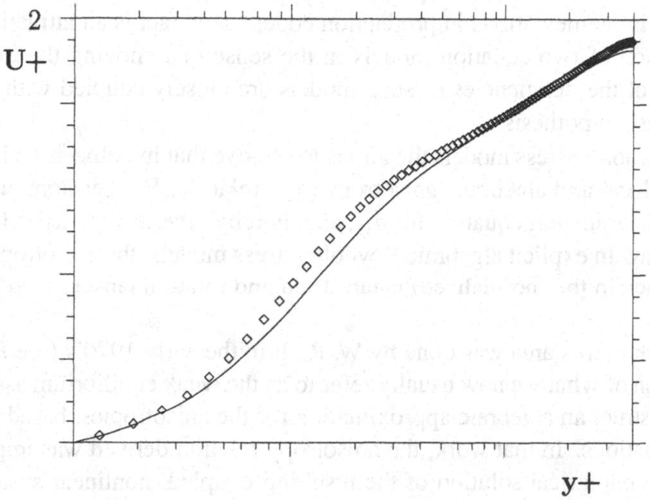

Figure 3: Comparison between the Sjögren & Johansson (2000) DRSM prediction and DNS data (Kim et al. 1987) for the mean velocity in turbulent channel flow.

An extension of the standard non-linear rapid pressure strain model is also given in that study which is aimed at capturing effects of strong rotation (or curvature). With the proposed modification it was shown that even phenomena such as damped oscillations of the anisotropy measures in pure rotation of homogeneous turbulence could be captured.

7 Explicit Algebraic Reynolds Stress Models

It was noted in section 1 that there is considerable interest in the further development of so-called explicit algebraic Reynolds stress models (EARSM) in order to meet increasing demands on generality of turbulence models in commercial CFD codes.

So far, eddy-viscosity-based two-equation models have been dominating in the context of industrial flow computations, but with higher demands on prediction accuracy in increasingly challenging flow situations, the need for more complex models has become more pressing. These situations include onset of separation, combinations with flow control in highly curved flows, rapidly rotating flows etc. It is well known that standard eddy-viscosity models tend to underpredict separation tendency. The modelling of the production that results from the Boussinesq hypothesis is indeed rather crude. The effect of 'decorrelation' caused by rotation typically causes the production to dissipation ratio to decrease with increasing rate of rotation. Such effects can, of course, not be predicted by eddy-viscosity models where the production model is for instance insensitive to system rotation.

The level of DRSM undoubtedly includes much more of the flow physics in a natural way, but the implementation in codes for design work of complex industrial flows have proven to still be a considerable challenge. We are undoubtedly going to see more work done on these aspects in the near future. For standard two-equation models, on the other hand, there are numerous well-tested, robust implementations in production codes. This fact is an attractive starting point for the generalization of two-equation models in the sense of removing the linear Boussinesq hypothesis. Many of the deficiencies of such models are closely coupled with the character of the linear Boussinesq hypothesis.

In algebraic Reynolds stress models the aim is to remove that hypothesis and replace it with a more general, still local and algebraic, anisotropy (a_{ij}) relation. Furthermore, it is derived as an approximation of the transport equation for a_{ij}, and thereby inherits properties from the DRSM-level in a natural way. In explicit algebraic Reynolds stress models, the anisotropy is described as an explicit expression in the (normalized) mean strain and rotation tensors (and possibly various scalar parameters).

Pioneering work in this area was done by W. Rodi in the early 1970's (see Rodi, 1976) who formulated a version of what we now usually refer to as the weak equilibrium assumption, which is necessary to construct an algebraic approximation for the anisotropies, based on the Reynolds stress transport equations. In that work, the anisotropy relation derived was implicit, and it was found later that the numerical solution of the resulting coupled, nonlinear system of equations could cause serious difficulties in complex flows. The lack of viscous damping in the equations was shown to be a source of numerical stiffness that was not easy to handle.

This type of closure was shown to retain a number of features from the Reynolds stress transport models, but the numerical difficulties form a major drawback. Based on the Rodi approach, Pope (1975) studied an explicit form of the anisotropy relation for two-dimensional mean flows. This pioneering work was practically forgotten until the 1990's when also more general forms for three-dimensional flows were derived. It can still be regarded as an open field for research, although much progress has already been obtained.

With an explicit anisotropy relation, the numerical robustness becomes equivalent to that for eddy-viscosity-based two-equation models, and the additional work to evaluate the more complicated anisotropy is quite limited. A property that it shares with the DRSM level is that the Reynolds equation (40) is left without any explicit modelling. Since the algebraic Reynolds stress model is derived as an algebraic approximation of a DRSM, no additional model constants are needed and some of the basic behaviour and experiences of the particular DRSM will be inherited. The generality of the EARSM closures is of course still somewhat limited by the locality of the anisotropy relation.

7.1 The EARSM Platform

The EARSM concept is in itself independent of the choice of an auxiliary quantity Z. Typically, one would choose a low-Reynolds number version of the equation for that quantity. We will here focus on the principles and choose the basic high Reynolds form of the K and ε equations

$$\frac{\bar{\mathrm{D}}K}{\bar{\mathrm{D}}t} = \mathcal{P} - \varepsilon - \frac{\partial J_m}{\partial x_m} \tag{121}$$

$$\frac{\bar{\mathrm{D}}\varepsilon}{\bar{\mathrm{D}}t} = \frac{\varepsilon}{K}(C_{\varepsilon 1}\mathcal{P} - C_{\varepsilon 2}\varepsilon) - \frac{\partial J^{\varepsilon}_m}{\partial x_m} \tag{122}$$

An important difference between this level of modelling and the standard eddy-viscosity approach is that the production of kinetic energy does not need to be modelled in the platform equations for the EARSM.

In the description of the EARSM we will need rather extensive algebra and for brevity we shall introduce the mean strain and rotation rate tensors normalized by the turbulence timescale $\tau \equiv K/\varepsilon$

$$\bar{S}^*_{ij} = \frac{\tau}{2}(\bar{u}_{i,j} + \bar{u}_{j,i})$$

$$W^*_{ij} = \frac{\tau}{2}(\bar{u}_{i,j} - \bar{u}_{j,i}). \tag{123}$$

We shall later also refer to these tensors in a matrix type of notation

$$\mathbf{S}, \mathbf{W}$$

The production to dissipation ratio is a key parameter in the following description and can now be expressed as

$$\frac{\mathcal{P}}{\varepsilon} = -a_{ij}\bar{S}^*_{ij}. \tag{124}$$

In the following we will drop the superscript $*$.

An advantage over eddy-viscosity models is also that the transport terms can be modelled by a gradient diffusion with a tensor diffusivity coefficient, such as in the model of Daly & Harlow (1970). The flux terms can then be expressed as

$$J_m = -c_s\frac{K^2}{\varepsilon}\left(a_{ml} + \frac{2}{3}\delta_{ml}\right)\frac{\partial K}{\partial x_l} \qquad J^{\varepsilon}_m = -c_{\varepsilon}\frac{K^2}{\varepsilon}\left(a_{ml} + \frac{2}{3}\delta_{ml}\right)\frac{\partial \varepsilon}{\partial x_l}. \tag{125}$$

Launder et al. (1975) recommend $c_s = 0.25$ and $c_{\varepsilon} = 0.15$.

7.2 The Weak Equilibrium Assumption

The basic idea behind the EARSM concept can be expressed as the goal to obtain the anisotropy relation as an algebraic approximation of the transport equation for a_{ij}. Originally, Rodi (see

Rodi, 1976) expressed this assumption as

$$\frac{\bar{\mathrm{D}}\,\overline{u_i'u_j'}}{\bar{\mathrm{D}}t} + \frac{\partial J_{ijm}}{\partial x_m} = \frac{\overline{u_i'u_j'}}{K}\left(\frac{\bar{\mathrm{D}}K}{\bar{\mathrm{D}}t} + \frac{\partial J_m}{\partial x_m}\right). \tag{126}$$

Using the symbolic expression for the right-hand-side of the Reynolds stress transport equation (85) and the right-hand-side of (121), we then obtain

$$\frac{\overline{u_i'u_j'}}{K}(\mathcal{P} - \varepsilon) = \mathcal{P}_{ij} - \varepsilon_{ij} + \Pi_{ij}. \tag{127}$$

Alternatively, we may transform (85) into a transport equation for a_{ij}, and equivalently write the 'weak equilibrium assumption' as

$$\frac{\bar{\mathrm{D}}\,a_{ij}}{\bar{\mathrm{D}}t} - \mathcal{D}_{ij}^{(a)} = 0, \tag{128}$$

where $\mathcal{D}_{ij}^{(a)}$ is the diffusion term resulting from the transformation of the Reynolds stress transport equation. Hence, the weak equilibrium assumption amounts to neglecting advection and diffusion of a_{ij}. The advection term is indeed exactly zero for all stationary parallel mean flows, such as fully developed channel and pipe flows. This gives a local description of the anisotropy, but one should keep in mind that history effects enter through the K and ε equations.

The dissipation rate tensor and the pressure strain rate tensor need to be modelled. Inserting the choice of models for these terms results in an implicit, non-linear, coupled set of equations for the anisotropy components. The resulting relation involves both the mean strain and rotation rate tensors, and replaces the Boussinesq hypothesis. The linearity of the Boussinesq hypothesis excludes any dependence of a_{ij} on the rotational (antisymmetric) part of the mean velocity gradient tensor. Here an ARSM approach represents a systematic method to construct a non-linear stress relationship that includes effects of the rotational part of the mean velocity gradient tensor.

The goal of the EARSM approach is to derive an explicit anisotropy relation

$$a_{ij} = a_{ij}\left(\bar{\mathbf{S}}, \bar{\mathbf{W}}\right) \tag{129}$$

from the implicit relation (128).

7.3 The Choice of Model for $\Pi_{ij} - \varepsilon_{ij}$

The transport equation for a_{ij} may be written

$$\frac{\bar{\mathrm{D}}\,a_{ij}}{\bar{\mathrm{D}}t} - \mathcal{D}_{ij}^{(a)} = \mathcal{P}_{ij}^{(a)} + \frac{\Pi_{ij}}{K} - \frac{\varepsilon}{K}\left(\frac{\varepsilon_{ij}}{\varepsilon} - \frac{2}{3}\delta_{ij} - a_{ij}\right) + C_{ij}^{(a)} \tag{130}$$

where $\mathcal{P}_{ij}^{(a)}$ is the resulting production tensor and $C_{ij}^{(a)}$ here denotes the Coriolis term.

In the regular EARSM approach, the left-hand-side is neglected (the weak equilibrium assumption). We will later return to this issue and discuss how the model may be extended to incorporate effects of strong streamline curvature (and thereby also rotation).

The presence of the Coriolis term implies that the ARSM will inherit the natural description of effects of system rotation from the DRSM (see e.g. Wallin & Johansson, 2000). We shall first, however, consider the case without system rotation ($C_{ij}^{(a)} = 0$).

The ARSM approximation of the a_{ij} transport equation with this approach is equivalent to neglecting the advective term (and diffusion) in the chosen coordinate system. Hence, the adequacy of the ARSM approach is coupled to the choice of a coordinate system where the neglection of advection terms in the a_{ij} equation can be justified.

We need to specify the model choice for the dissipation and pressure strain rate tensor. We first note that the production tensor in (130) is given explicitly and can be written as

$$\mathcal{P}_{ij}^{(a)} \equiv -\frac{4}{3}\bar{S}_{ij} - \left(a_{ik}\bar{S}_{kj} + \bar{S}_{ik}a_{kj} - \left(a_{ij} + \frac{2}{3}\delta_{ij} \right) a_{kl}\bar{S}_{lk} \right) + a_{ik}\bar{W}_{kj} - \bar{W}_{ik}a_{kj} \quad (131)$$

Alternatively, we may use the more compact matrix notation, where we denote a matrix product as e.g. $\mathbf{a}\bar{\mathbf{W}}$ (corresponding to $a_{ik}\bar{W}_{kj}$) and $\mathbf{I}$ is the identity matrix. The trace of a matrix is denoted by tr. With this notation the production tensor $\mathcal{P}_{ij}^{(a)}$ may be written

$$-\frac{4}{3}\bar{\mathbf{S}} - \left(\mathbf{a}\bar{\mathbf{S}} + \bar{\mathbf{S}}\mathbf{a} - \left(\mathbf{a} + \frac{2}{3}\mathbf{I} \right) \mathbf{a}\bar{\mathbf{S}} \right) + \mathbf{a}\bar{\mathbf{W}} - \bar{\mathbf{W}}\mathbf{a}$$

In order to allow an exact solution of the resulting implicit ARSM relation the pressure strain model has to be linear or quasi-linear, meaning that it must be tensorially linear in $\mathbf{a}$ but may depend nonlinearly on the production to dissipation ratio ($-\mathbf{a}\bar{\mathbf{S}}$).

The general quasi-linear model for $\Pi_{ij} - \varepsilon e_{ij}$, where effects of the dissipation rate anisotropy e_{ij} and slow pressure strain have been lumped together, can be written

$$\frac{\Pi_{ij}}{\varepsilon} - e_{ij} = -\frac{1}{2}\left(C_1^0 + C_1^1 \frac{\mathcal{P}}{\varepsilon} \right) a_{ij} + C_2\bar{S}_{ij} + \frac{C_3}{2}\left(a_{ik}\bar{S}_{kj} + \bar{S}_{ik}a_{kj} - \frac{2}{3}a_{kl}\bar{S}_{lk}\delta_{ij} \right)$$
$$-\frac{C_4}{2}\left(a_{ik}\bar{W}_{kj} - \bar{W}_{ik}a_{kj} \right). \quad (132)$$

For further discussions of anisotropic dissipation modelling in this context see e.g. Jongen et al. (1998).

7.4 The Implicit a_{ij} Relation Resulting from ARSM

Insertion of (131) and (132) into the anisotropy transport equation (130) yields (Wallin & Johansson, 2001)

$$\frac{1}{A_0}\left(\frac{\bar{\mathrm{D}}\, a_{ij}}{\bar{\mathrm{D}}t} - \mathcal{D}_{ij}^{(a)} \right) = \left(A_3 + A_4 \frac{\mathcal{P}}{\varepsilon} \right) a_{ij} + A_1\bar{S}_{ij} +$$
$$A_2\left(a_{ik}\bar{S}_{kj} + \bar{S}_{ik}a_{kj} - \frac{2}{3}a_{kl}\bar{S}_{lk}\delta_{ij} \right) - \left(a_{ik}\bar{W}_{kj} - \bar{W}_{ik}a_{kj} \right). \quad (133)$$

The A coefficients can be expressed in terms of the C coefficients as

$$A_0 = \frac{C_4}{2} - 1, \quad A_1 = \frac{3C_2 - 4}{3A_0}, \quad A_2 = \frac{C_2 - 2}{2A_0}$$
$$A_3 = \frac{2 - C_1^0}{2A_0}, \quad A_4 = \frac{-2 - C_1^1}{2A_0} \quad (134)$$

	A_0	A_1	A_2	A_3	A_4
Lin-SSG	-0.80	1.22	0.47	0.88	2.37
WJ2000	-0.44	1.20	0	1.80	2.25
WJ2001	-0.72	1.20	0	1.80	2.25

Table 2: A-coefficients corresponding to the linearized SSG-model of Gatski & Speziale (1993), the recalibrated LRR-model of Wallin & Johansson (2000) and the curvature-corrected dito of Wallin & Johansson (2001)

The models described by (132) include e.g. the general linear (LRR) model for the rapid pressure strain rate of Launder et al. (1975) and the linearized (SSG) model of Speziale et al. (1991). The model for the slow part of the pressure strain rate essentially corresponds to a classical Rotta-model (Rotta, 1951).

Taulbee (1992) and Wallin & Johansson (2000) based their EARSM's on a somewhat recalibrated LRR model, whereas Gatski & Speziale (1993) based their EARSM on a linearized SSG-model (see table 2). For future reference, we also give the coefficients of the Wallin & Johansson (2001) curvature corrected EARSM in table 2.

For the basic situation where no extra corrections for streamline curvature are included, i.e. when the left-hand-side of (133) is neglected, we see that the resulting implicit algebraic anisotropy relation (ARSM) is tensorially linear in $\mathbf{a}$ and we may express it with the aid of the matrix notation in the following form valid for all quasi-linear pressure strain models

$$N\mathbf{a} = -A_1\bar{\mathbf{S}} + (\mathbf{a}\bar{\mathbf{W}} - \bar{\mathbf{W}}\mathbf{a}) - A_2\left(\mathbf{a}\bar{\mathbf{S}} + \bar{\mathbf{S}}\mathbf{a} - \frac{2}{3}\mathrm{tr}\{\mathbf{a}\bar{\mathbf{S}}\}\right) \tag{135}$$

where

$$N = A_3 + A_4\frac{\mathcal{P}}{\varepsilon} . \tag{136}$$

Note that relation (135) contains a scalar nonlinearity since N contains the production to dissipation ratio, $\mathcal{P}/\varepsilon = -\mathrm{tr}\{\mathbf{a}\bar{\mathbf{S}}\}$.

Gatski & Speziale (1993) imposed an additional approximation in order to avoid the nonlinearity in the ARSM equation system. The approximation entailed using the asymptotic value $(C_{\varepsilon 2} - 1)/(C_{\varepsilon 1} - 1)$ for the production to dissipation ratio as a universal constant. Various improvements on this model have subsequently been proposed by Gatski, Speziale and their co-workers.

The key element of the recalibration of the LRR model that was applied by both Taulbee (1992) and Wallin & Johansson (2000) was to set the original c_2-coefficient (note that c_2 and C_2 denote different parameters) in the rapid pressure strain model to $5/9$. It was originally set to 0.4 but as noted in section 5 it has later been found that a higher value, close to $5/9$ should be more appropriate. The specific choice of $c_2 = 5/9 \Leftrightarrow C_3 = 2$, which gives $A_2 = 0$, results in a major simplification of the equations, as we shall see in the following. With the choice of A-coefficients of Wallin & Johansson (2000) the ARSM relation can be written

$$N\mathbf{a} = -\frac{6}{5}\bar{\mathbf{S}} + (\mathbf{a}\bar{\mathbf{W}} - \bar{\mathbf{W}}\mathbf{a}) \tag{137}$$

with

$$N = c_1' + \frac{9}{4}\frac{\mathcal{P}}{\varepsilon}\,, \qquad c_1' = \frac{9}{4}\left(c_1 - 1\right). \tag{138}$$

The coefficient c_1 is the standard Rotta constant in the model for the slow part of the pressure strain rate. It was chosen as 1.8 in the study of Wallin & Johansson (2000).

7.5 Deriving an Explicit a_{ij} Relation from ARSM

Direct numerical solution of the implicit **a**-relation (135) is not a feasible alternative in complex flows. The difficulties are associated with the lack of diffusion or damping in the equation system. One instead wishes to construct an explicit **a**-relation $\mathbf{a} = \mathbf{a}(\bar{\mathbf{S}}, \bar{\mathbf{W}})$. Models based on such an explicit anisotropy relation, referred to as EARSM, i.e. explicit algebraic Reynolds stress models, essentially avoid these numerical difficulties. The computational cost of evaluating the explicit a_{ij} relation is also low.

For the construction of an explicit anisotropy relation, we return to the basic linear algebra theory, as found e.g. in Spencer & Rivlin (1959). Here we wish to express the second rank symmetric tensor a_{ij} in terms of two second rank tensors, one that is symmetric ($\bar{S}_{ij}$) and one antisymmetric ($\bar{W}_{ij}$). From the general results of Spencer & Rivlin (1959), we can see that the most general form for such an expression consists of ten tensorially independent groups to which all higher order tensor combinations can be reduced with the aid of the Caley-Hamilton theorem.

$$\begin{aligned}
\mathbf{a} = {} & \beta_1 \bar{\mathbf{S}} + \beta_3 \left(\bar{\mathbf{W}}^2 - \frac{1}{3} I\!I_W\, \mathbf{I}\right) + \beta_4 \left(\bar{\mathbf{S}}\bar{\mathbf{W}} - \bar{\mathbf{W}}\bar{\mathbf{S}}\right) \\
& + \beta_6 \left(\bar{\mathbf{S}}\bar{\mathbf{W}}^2 + \bar{\mathbf{W}}^2\bar{\mathbf{S}} - \frac{2}{3} I\!V\, \mathbf{I}\right) + \beta_9 \left(\bar{\mathbf{W}}\bar{\mathbf{S}}\bar{\mathbf{W}}^2 - \bar{\mathbf{W}}^2\bar{\mathbf{S}}\bar{\mathbf{W}}\right) \\
& + \beta_2 \left(\bar{\mathbf{S}}^2 - \frac{1}{3} I\!I_S\, \mathbf{I}\right) + \beta_5 \left(\bar{\mathbf{S}}^2\bar{\mathbf{W}} - \bar{\mathbf{W}}\bar{\mathbf{S}}^2\right) \\
& + \beta_7 \left(\bar{\mathbf{S}}^2\bar{\mathbf{W}}^2 + \bar{\mathbf{W}}^2\bar{\mathbf{S}}^2 - \frac{2}{3} V\, \mathbf{I}\right) + \beta_8 \left(\bar{\mathbf{S}}\bar{\mathbf{W}}\bar{\mathbf{S}}^2 - \bar{\mathbf{S}}^2\bar{\mathbf{W}}\bar{\mathbf{S}}\right) + \\
& + \beta_{10} \left(\bar{\mathbf{W}}\bar{\mathbf{S}}^2\bar{\mathbf{W}}^2 - \bar{\mathbf{W}}^2\bar{\mathbf{S}}^2\bar{\mathbf{W}}\right)
\end{aligned} \tag{139}$$

The β coefficients may be functions of the five independent invariants of the combinations of $\bar{\mathbf{S}}$ and $\bar{\mathbf{W}}$, which can be written as

$$I\!I_S = \mathrm{tr}\{\bar{\mathbf{S}}^2\}, I\!I_W = \mathrm{tr}\{\bar{\mathbf{W}}^2\}, I\!I\!I_S = \mathrm{tr}\{\bar{\mathbf{S}}^3\}, I\!V = \mathrm{tr}\{\bar{\mathbf{S}}\bar{\mathbf{W}}^2\}, V = \mathrm{tr}\{\bar{\mathbf{S}}^2\bar{\mathbf{W}}^2\}. \tag{140}$$

Also other scalar parameters may be involved. For a further discussion of the completeness of the expression (139) and the independence of the groups involved the reader is referred to Wallin & Johansson (2000) and Taulbee et al. (1994).

In two-dimensional mean flows, one can show that there are only three independent tensor groups, e.g., the $\beta_{1,2,4}$ groups (i.e. the first line of 139), and two independent invariants, $I\!I_S$ and $I\!I_W$. As a comparison, we should note that the Boussinesq hypothesis only corresponds to the first term with $\beta_1 = 2C_\mu$. In the EARSM context, the tensorial dependence is much more complex and we do not restrict β_1 to be constant. Instead, already that coefficient, which is a counterpart to the C_μ-coefficient, will partly reflect the model's response to magnitude of rotation and mean strain.

The determination of the ten β coefficients in general three-dimensional mean flows is very complex for the general form of a quasi-linear pressure strain rate model. The strategy is to insert the general anisotropy relation (139) into the ARSM-relation (135). This generates a complex system of equations for the ten β-coefficients (see Gatski & Speziale, 1993).

An alternative to deriving the coefficients from the ARSM relation, i.e. constructing a direct approximation of the Reynolds stress transport equations, could be to choose to calibrate the β-coefficients against some chosen set of 'basic flows'. This approach has been investigated by e.g. Shih et al. (1992) and could be referred to as a nonlinear eddy-viscosity approach.

In the following, we will present results for the recalibrated LRR model with $c_2 = 5/9$, as used by Taulbee (1992) and Wallin & Johansson (2000) (and Johansson & Wallin, 1996). We will describe the specific choice of parameters of Wallin & Johansson (2000), where the ARSM relation is given by (137), (138). This model will be referred to as the WJ-model in the following.

An important aspect of this choice of model is the absence of the A_2 term. This has the important effect that a set of five of the groups in (139) will map back onto itself. The complete solution for the WJ model for general three-dimensional mean flows hence only contains five non-zero coefficients, namely, $\beta_1, \beta_3, \beta_4, \beta_6, \beta_9$. Thus, the model is described by the first two rows in (139).

A complicating factor is the (scalar) nonlinearity of equation (137). Wallin & Johansson (2000) solved the nonlinear system of equations in the form of a linear system of (five) equations complemented by a nonlinear scalar equation for $\mathcal{P}/\varepsilon$. For two-dimensional mean flows, Johansson & Wallin (1996) and Girimaji (1995), (1996) independently showed that this equation has a closed and fully explicit solution that can be expressed in a compact form.

The removal of the need for ad-hoc relations for $\mathcal{P}/\varepsilon$ represents a substantial improvement over eddy-viscosity-based two-equation models.

A constant $\mathcal{P}/\varepsilon$ gives an erroneous asymptotic behaviour for large strain rates, as also noticed by Speziale & Xu (1996), while the fully consistent solution of the non-linear equation system automatically fulfills the correct asymptotic behaviour. Also, in the very near-wall region, the correct asymptotic behaviour significantly improves the predictions.

After inserting the general expression for $\mathbf{a}$ in (137), we obtain a system of five linear equations for $\beta_1, \beta_3, \beta_4, \beta_6, \beta_9$, although it still contains the unknown N (related to $\mathcal{P}/\varepsilon$).

A nonlinear scalar equation for N can then be obtained in the following way. We first multiply the solution for $\mathbf{a}$ (containing the solutions for the β-coefficients) by $\bar{\mathbf{S}}$ and then take the trace. This gives us an expression for $\mathcal{P}/\varepsilon$, and hence N.

7.6 The WJ Model for two-dimensional Mean Flows

For two-dimensional mean flows, the WJ-solution is reduced to only two non-zero coefficients β_1, β_4, and the anisotropy tensor can be expressed as

$$\mathbf{a} = -\frac{6}{5}\frac{1}{N^2 - 2II_W}\left(N\bar{\mathbf{S}} + \bar{\mathbf{S}}\bar{\mathbf{W}} - \bar{\mathbf{W}}\bar{\mathbf{S}}\right) \tag{141}$$

It is clearly seen that the solution cannot become singular since II_W in the denominator is always negative.

The non-linear equation for N in two-dimensional mean flow can be derived by multiplying (141) by $\bar{\mathbf{S}}$ and thereafter taking the trace. This readily yields a cubic equation for N

$$N^3 - c_1' N^2 - \left(\frac{27}{10} II_S + 2 II_W\right) N + 2 c_1' II_W = 0. \tag{142}$$

which can be solved in a closed form with the solution for the positive root being

$$N = \begin{cases} \frac{c_1'}{3} + \left(P_1 + \sqrt{P_2}\right)^{1/3} + \operatorname{sign}\left(P_1 - \sqrt{P_2}\right) \mid P_1 - \sqrt{P_2} \mid^{1/3}, & P_2 \geq 0 \\ \frac{c_1'}{3} + 2\left(P_1^2 - P_2\right)^{1/6} \cos\left(\frac{1}{3} \arccos\left(\frac{P_1}{\sqrt{P_1^2 - P_2}}\right)\right) & , P_2 < 0 \end{cases} \tag{143}$$

with

$$P_1 = \left(\frac{c_1'^2}{27} + \frac{9}{20} II_S - \frac{2}{3} II_W\right) c_1', \; P_2 = P_1^2 - \left(\frac{c_1'^2}{9} + \frac{9}{10} II_S + \frac{2}{3} II_W\right)^3. \tag{144}$$

N remains real and positive for all possible values of II_S and II_W.

The resulting solution for $\mathcal{P}/\varepsilon$ is illustrated in Figure 4, where σ and ω are defined as $\sigma \equiv \sqrt{II_S/2}$ and $\omega \equiv \sqrt{-II_W/2}$. These are direct measures of the strength of the mean strain and rotation rate, respectively. The model has the attractive feature of reproducing a monotonously decreasing $\mathcal{P}/\varepsilon$ ratio with increasing influence of rotation. For all parallel shear flows, $\sigma = \omega$.

The WJ-model replicates the results for the asymptotic value of 1.8 for $\mathcal{P}/\varepsilon$ (see Tavoularis & Corrsin, 1981) for homogeneous shear flow, where the asymptotic normalized timescale corresponds to $\sigma = \omega = 3$.

Furthermore, in the log-layer of a boundary layer we know that the production balances the dissipation rate, $\mathcal{P} = \varepsilon$, which is obtained with the WJ model when the strain rate $\sigma = 1.69$. This is within the range of σ values found in the log-layer of the DNS data for channel flow (Kim, 1989). This is also consistent with an effective $C_\mu = 0.09$ which gives $\sigma = 1.67$, also marked as a circle in the figure.

The anisotropies predicted with the WJ-model for the two cases are compared in Tables 3 and 4. In plane shear flows the a_{12} anisotropy component can be said to be the most important one to predict accurately. The a_{22} component also plays an important role in determining the magnitude of the diffusivity coefficient in this case. The tables show that a_{12} and a_{22} are well predicted for the two different cases. The a_{11} and a_{33} components are not as well predicted, however, due to the simplification of setting $c_2 = 5/9$ since this implies $a_{33} = 0$.

For moderate to large strain rates (in parallel flows), the WJ model predicts an a_{12} anisotropy that is nearly constant. This range includes values relevant for the log-layer and the asymptotic homogeneous shear flow. This also means that it also avoids unrealizable results for large strain rates, unlike standard eddy viscosity models. Bradshaw's assumption, that is adopted by Menter (1993) in the SST model, forces the a_{12} anisotropy to be constant for $\mathcal{P}/\varepsilon$ ratios greater than unity, which gives that $\beta_1 \sim 1/\sigma$. Hence, this is also fulfilled for the WJ model in the limit of large strain rates.

An example where the benefits of a direct, explicit solution of the production to dissipation ratio is further illustrated is rotating plane channel flow. For this case, the effective II_W in the

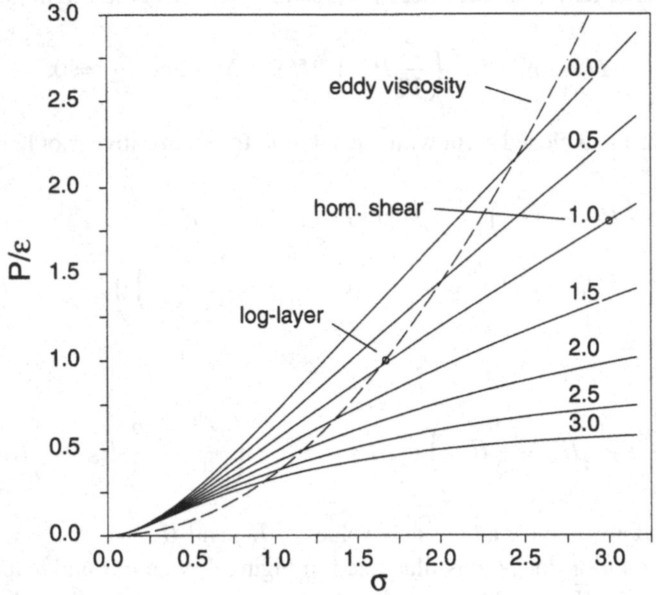

Figure 4: Production to dissipation ratio versus strain rate σ for different rotation ratios ω/σ. The current model (—-) compared to an eddy viscosity model (---).

	a_{12}	a_{11}	a_{22}	a_{33}	σ
DNS	−0.29	0.34	−0.26	−0.08	1.65
WJ-EARSM model	−0.30	0.25	−0.25	0.00	1.69

Table 3: The computed anisotropy in the log-layer using the current model assuming balance between turbulence production and dissipation compared to channel DNS data (Kim 1989).

rotating system is decreased on one side and increased on the other, which results naturally in a decreased production on the stabilized side and an increased production on the destabilized side.

Furthermore, for the case of irrotational straining mean flow, the cubic equation degenerates to a quadratic equation with the following solution for the production to dissipation ratio

$$\frac{\mathcal{P}}{\varepsilon} = \left(\frac{27}{10} II_S + \left(\frac{c_1'}{2} \right)^2 \right)^{1/2} - \frac{2}{9} c_1' \tag{145}$$

which for large strain rates thus tends to $\sigma\sqrt{27/5}$, with the correct linear increase with increasing strain rate.

	a_{12}	a_{11}	a_{22}	a_{33}	$\mathcal{P}/\varepsilon$
expr	−0.30	0.40	−0.28	−0.12	1.8
WJ-EARSM model	−0.30	0.31	−0.31	0.00	1.8

Table 4: The anisotropy in asymptotic homogeneous shear flow using the current model with $\sigma = 3.0$ compared to measurements by Tavoularis & Corrsin (1981).

7.7 The WJ Model for three-dimensional Mean Flow

For three-dimensional mean flows the WJ-model gives the following non-zero β-coefficients

$$\begin{aligned} \beta_1 &= -\frac{N\left(2N^2 - 7II_W\right)}{Q} \qquad \beta_3 = -\frac{12N^{-1}IV}{Q} \\ \beta_4 &= -\frac{2\left(N^2 - 2II_W\right)}{Q} \qquad \beta_6 = -\frac{6N}{Q} \qquad \beta_9 = \frac{6}{Q} \end{aligned} \tag{146}$$

where the denominator

$$Q = \frac{5}{6}\left(N^2 - 2II_W\right)\left(2N^2 - II_W\right) \tag{147}$$

is also clearly seen to remain always positive since $II_W \leq 0$.

The non-linear equation for N or the corresponding equation for $\mathcal{P}/\varepsilon$ is of sixth order. Wallin & Johansson (2000) proposed an approximative solution of this equation in terms of a linear expansion around the solution for the corresponding cubic equation valid for two-dimensional flows.

Fully developed turbulent flow in a circular pipe rotating around its length axis represents a three-dimensional flow that can be described with only one spatial coordinate in a cylindrical coordinate system. If the flow is laminar, the tangential velocity, U_θ, varies linearly with the radius, r, like a rigid body rotation. In turbulent flow, on the other hand, the tangential velocity is nearly parabolic (Imao et al., 1996), which cannot be described with an eddy-viscosity-based turbulence model. Wallin & Johansson (2000), among others, have demonstrated that cubic terms are needed in the EARSM to capture this feature. One may note that such terms are not present in two-dimensional formulations of EARSM and not in most of the EARSM:s proposed in the literature.

7.8 Near Wall Treatments

When EARSM is used with integration all the way down to solid walls near-wall damping functions are needed in a similar fashion as with standard two-equation models. Wallin & Johansson (2000) showed, however, that a correct asymptotic behaviour for large strain rates which is possible to achieve with EARSM improves the near-wall situation substantially over that for eddy-viscosity models. They showed by comparison with DNS channel flow data that a simple van Driest damping function was quite sufficient to obtain excellent agreement for the shear stress anisotropy. For further details, the reader is referred to that paper.

7.9 A Model for Curvature Corrections in the WJ EARSM

The adequacy of the ARSM approximation is coupled to the choice of the coordinate system, and more precisely to what extent we can justify the neglection of the advection terms in the a_{ij} transport equation (133) as described in that coordinate system. For flow situations with strong streamline curvature, we may attempt to improve the approximation by introducing an algebraic approximation of the advection term in a streamline-based coordinate system. The first steps in this manner of relaxing the weak equilibrium assumption were taken by Girimaji (1997) and Sjögren (1997). They showed that in order to relax the weak equilibrium assumption to neglect only the advection in the local streamline-based coordinate system, we need to approximate the anisotropy advection term by

$$-\left(\mathbf{a}\bar{\mathbf{W}}^{(r)} - \bar{\mathbf{W}}^{(r)}\mathbf{a}\right).$$

where $\bar{\mathbf{W}}^{(r)}$ is essentially the rotation rate of the coordinate system following an ensemble averaged fluid particle (see Wallin & Johansson (2001)). $\bar{\mathbf{W}}^{(r)} = -\epsilon_{ijk}\omega_k/\tau$ where ω_k is the coordinate system rotation vector (the normalization by $\tau = K/\varepsilon$ gives a non-dimensional $\bar{\mathbf{W}}$). The exact determination of this quantity for general three-dimensional flows is quite complicated. Wallin & Johansson (2001) proposed the use of the following approximation

$$\omega_i^{\text{approx.}} = \epsilon_{ijk} V_j \frac{\bar{\mathrm{D}} V_k}{\bar{\mathrm{D}} t} / V^2 \tag{148}$$

where V_i is the acceleration $\frac{\bar{\mathrm{D}} \bar{u}_i}{\bar{\mathrm{D}} t}$, and V is its magnitude. This approximation is relatively easy to compute since it uses only first and second order time derivatives of the mean velocity. It thereby also ensures Galilean invariance.

This curvature correction is easy to implement. One needs only to replace $\bar{\mathbf{W}}$ by

$$\bar{\mathbf{W}}^* = \bar{\mathbf{W}} - \frac{1}{A_0}\bar{\mathbf{W}}^{(r)} \tag{149}$$

in the original ARSM-relation (or EARSM solution). It also has the important property that it leaves the original (WJ) EARSM unaltered in flows without curvature (or system rotation) effects. In order to avoid problems where the acceleration becomes small, a limiter of the numerator in (148) was used in Wallin & Johansson (2001).

They showed that considerable improvements could be achieved with this correction in strongly curved or rotating flows. Figure 5 illustrates the improvement obtained in a plane turbulent channel flow rotating around the spanwise axis. The rotation number (Ro) based on the channel height and angular rotation rate of the system is here as high as 0.77, indicating strong effects of rotation.

In (the journal version of) Wallin & Johansson (2001) the limitations of the above approximation (148) are discussed and a more generally valid (although somewhat more complicated) approximation is introduced.

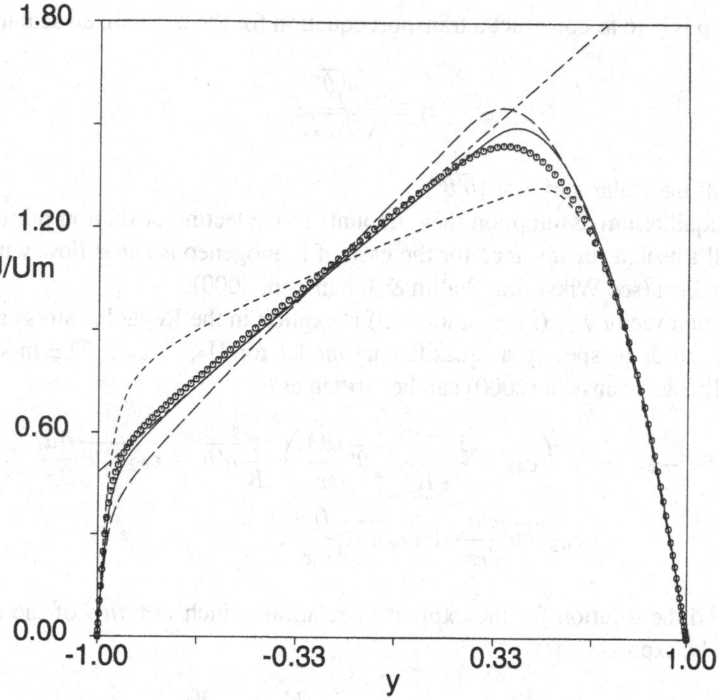

Figure 5: Mean velocity profile in rotating plane channel flow. Symbols represent DNS data (Re based on bulk velocity and half channel height = 3446) of Alvelius & Johansson (1999), solid curve: curvature corrected WJ EARSM, Chain-dashed curve: slope of $2\omega^s$, fine dashed curve: original WJ EARSM.

8 Explicit Algebraic Scalar Flux Models

In analogy with the eddy-viscosity assumption, the most widely used model for passive scalar transport modelling is based on an eddy-diffusivity concept, so that

$$-\overline{u_j'\theta'} = \frac{\nu_T}{\mathrm{Pr}_T}\frac{\partial\Theta}{\partial x_j} \tag{150}$$

where Pr_T represents a turbulent Prandtl number (i.e. the ratio between the eddy viscosity and the scalar eddy diffusivity).

The assumption of alignment between the scalar flux vector and the mean scalar gradient is often not well satisfied in real flows. The idea of constructing a more generally valid expression for the scalar flux vector in a manner similar to that for EARSM has been studied by several investigators, see e.g. Adumitroaie et al. (1997) and Girimaji & Balachandar (1997). Wikström, Wallin & Johansson (2000) and Högström, Wallin, & Johansson (2001) studied the possibility to construct a simple explicit algebraic scalar flux model (EASFM) to complement the WJ EARSM.

The first step is here to construct a transport equation for the normalized scalar flux vector

$$\xi_j \equiv \frac{\overline{u'_j\theta'}}{\sqrt{K_\theta K}} \tag{151}$$

where K_θ is half the scalar variance $\left(\overline{\theta'\theta'}\right)$.

The weak equilibrium assumption then amounts to neglecting advection minus diffusion of ξ_j. This is well satisfied for instance for the case of homogeneous shear flow with an imposed mean scalar gradient (see Wikström, Wallin & Johansson, 2000).

The production vector $\mathcal{P}_{\theta i}$ (in equation (35) is explicit in the Reynolds stress and scalar flux vector, but one needs to specify a (quasilinear) model for $\Pi_{\theta i} - \varepsilon_{\theta i}$. The model chosen by Wikström, Wallin & Johansson (2000) can be written as:

$$\begin{aligned}\Pi_{\theta i} - \varepsilon_{\theta i} = & -\left(c_{\theta 1} + \frac{1}{2}\frac{K}{\varepsilon K_\theta}\overline{u'_k\theta'}\frac{\partial\Theta}{\partial x_k}\right)\frac{\varepsilon}{K}\overline{u'_i\theta'} + c_{\theta 2}\overline{u'_j\theta'}\frac{\partial\bar{u}_i}{\partial x_j} + \\ & c_{\theta 3}\overline{u'_j\theta'}\frac{\partial\bar{u}_j}{\partial x_i} + c_{\theta 4}\overline{u'_iu'_j}\frac{\partial\Theta}{\partial x_j}.\end{aligned} \tag{152}$$

They derived the solution for the explicit ξ_j-relation, which in terms of the original scalar flux vector can be expressed as

$$\overline{u'_i\theta'} = -(1 - c_{\theta 4})B_{ij}\frac{K}{\varepsilon}\overline{u'_ju'_k}\frac{\partial\Theta}{\partial x_k} \tag{153}$$

with (in matrix notation)

$$\mathbf{B} = \frac{\left(G^2 - \frac{1}{2}Q_1\right)\mathbf{I} - G(c_s\bar{\mathbf{S}} + c_W\bar{\mathbf{W}}) + (c_s\bar{\mathbf{S}} + c_W\bar{\mathbf{W}})^2}{G^3 - \frac{1}{2}GQ_1 + \frac{1}{2}Q_2} \tag{154}$$

where $c_s = 1 - c_{\theta 2} - c_{\theta 3}$, $c_W = 1 - c_{\theta 2} + c_{\theta 3}$ and

$$G = \frac{1}{2}\left(2c_{\theta 1} - 1 - \frac{1}{r} + \frac{\mathcal{P}}{\varepsilon}\right) \tag{155}$$

where r is the timescale ratio

$$r \equiv \frac{K_\theta/\varepsilon_\theta}{K/\varepsilon}. \tag{156}$$

Also involved are the invariant measures

$$Q_1 = c_s^2 II_S + c_W^2 II_W, \qquad Q_2 = \frac{2}{3}c_s^3 III_S + 2c_s c_W^2 IV \tag{157}$$

Good predictions were obtained for homogeneous shear flow with an imposed mean scalar gradient in different directions, turbulent channel flow with an imposed temperature difference between the plates, and the wake downstream a heated cylinder, with the following set of parameters

$$c_{\theta 1} = 1.6\frac{r+1}{r}, \quad c_{\theta 2} = c_{\theta 3} = c_{\theta 4} = 0.$$

The presence of the timescale ratio that is allowed to vary in the flow also requires the solution of transport equations for K_θ and ε_θ.

Högström, Wallin, & Johansson (2001) later pursued the investigation of this model and proposed a simplified version with a constant timescale ratio, removing the need for solution of the K_θ and ε_θ equations. A good compromise was found to be $r = 0.55$. The same expression for $c_{\theta 1}$ was used and the other parameters were modified to give the best compromise for the chosen set of test cases,

$$c_{\theta 1} = 4.51, \quad c_{\theta 2} = -0.47 \quad c_{\theta 3} = 0.02 \quad c_{\theta 4} = 0.08.$$

One should, of course, use such a model with assumed constant timescale ratio, with caution in flows with (molecular) Prandtl numbers (ν/α) much smaller or much larger than unity.

Remaining problems were found in regions of low shear. A simple remedy for this was found in Högström, Wallin, & Johansson (2001) to be an inclusion of a diffusion correction that simply amounts to a redefinition of G,

$$G = \frac{1}{2}\left(2c_{\theta 1} - 1 - \frac{1}{r} + \frac{\mathcal{P}}{\varepsilon}\right) + 8.0 \max\left(1 - \frac{\mathcal{P}}{\varepsilon}, 0\right). \tag{158}$$

9 Concluding Remarks

Although it has been more than 50 years since the first steps were taken in the realm of single-point turbulence eddy-viscosity based models with two transport equations, we have seen a continuous development and in more recent years a penetration of ideas that removes the eddy viscosity type of modelling in order to enable the capturing of complex flow phenomena. The EARSM concept is finding its way into commercial CFD codes, but is also still under development. The curvature correction discussed above is one important aspect in the work to establish this level of modelling for complex flows, including onset of separation on curved surfaces etc.

The numerical implementation of robust nonlinear DRSM:s that satisfy strong realizability still needs considerable work, but may extend the generality of engineering turbulence models even further, not least in rapidly rotating flows. New efforts in highly automated code generation tools are presently starting to produce model testers and flow prediction codes for this level of complex models with much reduced coding effort in comparison with traditional approaches. This will facilitate the introduction and validation of complex models in future CFD codes.

References

Adumitroaie, V., Taulbee, D.B. & Givi, P. (1997). Explicit algebraic scalar flux models for turbulent reacting flows. *A.i.ch.e.J.* 43:1935-2147.

Alvelius, K. & Johansson, A.V. (1999). Direct numerical simulation of rotating channel flow at various Reynolds numbers and rotation numbers. In PhD thesis of K. Alvelius, Dept. of Mechanics, KTH, Stockholm, Sweden

Aupoix, B., Cousteix, J. & Liandrat, J. (1989). MIS: A way to derive the dissipation equation Turbulent Shear Flow vol. 6 (ed. J.-C. André), Springer

Batchelor, G.K. (1953). The theory of homogeneous turbulence. Cambridge University Press.

Champagne, F.H., Harris, V.G. & Corrsin, S. (1970). Experiments on nearly homogeneous turbulent shear flow. *J. Fluid Mech.* 41:81-139.

Chasnov, J.R. (1994). Similarity states of passive scalar transport in isotropic turbulence. *Phys. Fluids* 6:1036.

Chou, P.Y. (1945). On velocity correlations and the solutions of the equations of turbulent motion. *Quart. of Applied Math.* 3:38-54.

Comte-Bellot, G. & Corrsin, S. (1966). The use of a contraction to improve the isotropy of grid-generated turbulence. *J. Fluid Mech.* 25:657-682.

Daly, B.J. & Harlow, F.H. (1970). Transport equations in turbulence. *Phys. Fluids* 13:2634-2649.

Durbin, P.A. (1996). On the $k - \epsilon$ stagnation point anomaly. *Int. J. Heat and Fluid Flow* 17:89-90.

Gatski, T.B. & Speziale, C.G. (1993). On explicit algebraic Reynolds stress models for complex turbulent flows. *J. Fluid Mech.* 254:59-78.

Gibson, M.M. & Launder, B.E. (1978). Ground effects on pressure fluctuations in the atmospheric boundary layer. *J. Fluid Mech.* 86:491-511.

Girimaji, S.S. (1995). Fully-explicit and self-consistent algebraic Reynolds stress model. *ICASE Report No. 95-82.*

Girimaji, S.S. (1996). Improved algebraic Reynolds stress model for engineering flows. In Engineering Turbulence Modelling and Experiments 3 pp 121-129. Eds W. Rodi and G. Bergeles. Elsevier. Science B.V.

Girimaji, S.S. (1997). A Galilean invariant explicit Reynolds stress model for turbulent curved flows. *Phys. Fluids* 9:1067-1077.

Girimaji, S.S. & Balachandar, S. (1997). Analysis and modeling of buoyancy-generated turbulence using numerical data. *Int. J. Heat and Mass Transfer* 41:915-929.

Hallbäck, M., Groth, J. & Johansson, A.V. (1990). An algebraic model for nonisotropic turbulent dissipation rate in Reynolds stress closures. *Phys Fluids A* 2:1859-1866.

Högström, C.M., Wallin, S. & Johansson, A.V. (2001). Passive scalar flux modelling for CFD. In *Proceedings of Turbulence and Shear Flow Phenomena II, Stockholm, June 27-29, 2001*, II:383-388.

Imao, S., Itoh, M. & Harada, T. (1996). Turbulent characteristics of the flow in an axially rotating pipe. *Int. J. Heat and Fluid Flow* 17:444-451.

Johansson, A.V. & Burden, A.D. (1999). An introduction to turbulence modelling. Chapter 4 in Transition, Turbulence and Combustion Modelling, ERCOFTAC Series vol. 6, Kluwer. pp 159-242.

Johansson, A.V. & Hallbäck, M. (1994). Modelling of rapid pressure-strain rate in Reynolds tress closures. *J. Fluid Mech.* 269:143-168 (see also *J. Fluid Mech.* 290:405 (1995)).

Johansson, A.V. & Wallin, S. (1996). A new explicit algebraic Reynolds stress model. In *Proc. Sixth European Turbulence Conference*, Lausanne, July 1996, Ed. P. Monkewitz, 31–34.

Johansson, A.V. & Wikström, P.M. (2000). DNS and modelling of passive scalar transport in turbulent channel flow with a focus on scalar dissipation rate modelling. *Flow, Turbulence and Combustion* 63:223–245.

Jongen, T., Mompean, G. & Gatski, T.G. (1998). Accounting for Reynolds stress and dissipation rate anisotropies in inertial and noninertial frames. *Phys. Fluids* 10:674-684.

Kim, J. (1989). Turbulence statistics in fully developed channel flow at low Reynolds number. *J. Fluid Mech.* 177:133-166.

Kim, J., Moin, P. & Moser, R. (1987). On the structure of pressure fluctuations in simulated turbulent channel flow. *J. Fluid Mech.* 205:421-451.

Kolmogorov, A.N. (1942). Equations of turbulent motion of an incompressible fluid. *Izvestia Academy of Sciences, USSR; Physics* 6:56-58.

Komminaho, J. & Skote, M. (2001). Reynolds stress budgets in Couette and boundary layer flows. Submitted to *Flow, Turbulence and Combustion.*

Kristoffersen, R. & Andersson, H. (1993). Direct simulation of low-Reynolds-number turbulent flow in a rotating channel. *J. Fluid Mech.* 256:163-197.

Lam, C.K.G. & Bremhorst, K.A. (1981). Modified form of K-ϵ model for predicting wall turbulence. *ASME, J. Fluids Engineering* 103:456-460.

Launder, B.E., Reece, G.J. & Rodi, W. (1975). Progress in the development of a Reynolds-stress turbulence closure. *J. Fluid Mech.* 41:537-566.

Launder, B.E. & Spalding, D.B. (1972). Mathematical models of turbulence. Academic Press.

Launder, B.E., Tselepidakis, D.P. & Younis, B.A. (1987). A second-moment closure study of rotating channel flow. *J. Fluid Mech.* 183:63-75.

Lindberg, P.Å. (1994). Near-wall turbulence models for 3D boundary layers. *Appl. Sci. Res.* 53:139-162.

Lumley, J.L. (1978). Computational modeling of turbulent flows. *Adv. Appl. Mech.* 18:123-177.

Menter, F.R. (1992). Improved Two-Equation $k - \omega$ Turbulence Model for Aerodynamic Flows. *NASA TM-103975*

Menter, F.R. (1993). Zonal Two Equation $k - \omega$ Turbulence Models for Aerodynamic Flows. *AIAA 93-2906*

Mohammadi, B. & Pironneau, O. (1994). Analysis of the $K - \epsilon$ model Wiley/Masson.

Naot, D., Shavit, A. & Wolfstein, M. (1973). Two-point correlation model and the redistribution of Reynolds stresses *Phys. Fluids* 16:738-743.

Piquet, J. (1999). Turbulent flows, models and physics Springer.

Pope, S.B. (1975). A more general effective-viscosity hypothesis. *J. Fluid Mech.* 72:331-340.

Poroseva, S.V., Kassinos,S.C., Langer, C.A. & Reynolds, W.C. (2001). Simulation of turbulent flow in a rotating pipe using the structure-based model. In *Proceedings of Turbulence and Shear Flow Phenomena II, Stockholm, June 27-29, 2001*, II:223-228.

Reynolds, W.C. (1976). Computation of turbulent flows. *Ann. Rev. Fluid Mech.* 8:183-208.

Rodi, W. (1976). A new algebraic relation for calculating the Reynolds stresses. *Z. angew. Math. Mech.* 56:T219–221.

Rodi, W & Mansour, N.N. (1993). Low Reynolds number K-ϵ modelling with the aid of direct simulation data. *J. Fluid Mech.* 250:509-529.

Rotta, J (1951). Statistische theorie nichthomogener turbulenz I. *Zeitschrift für Physik* 129:547-572.

R. Rubinstein, C.L. Rumsey, M.D. Salas, and J.L. Thomas, editors (2001). Turbulence modelling workshop ICASE Interim Report No. 37 NASA/CR-2001-210841.

Saffman, P.G. (1970). A model for inhomogeneous turbulent flow. *Proc. Roy. Soc. London A* 317:417-433.

Saffman, P.G. (1967). The large-scale structure of homogeneous turbulence. *J. Fluid Mech.* 27:581-593.

Shih, T.-H. & Lumley, J.L. (1986). Second-order modeling of near-wall turbulence. *Phys. Fluids* 29:971-975.

Shih, T.-H., Reynolds, W.C. & Mansour, N.N. (1990). A spectrum model for weakly anisotropic turbulence. *Phys. Fluids A* 2:1500-1502.

Shih, T.H., Zhu, J. & Lumley, J.L. (1992). A Realizable Reynolds Stress Algebraic Equation Model NASA TM 105993, ICOMP-92-27, CMOTT-92-14.

Sjögren, T.I.Å (1997). Development and Validation of Turbulence Models Through Experiment and Computation. Doctoral thesis, Dept. of Mechanics, KTH, Stockholm, Sweden.

Sjögren, T.I.Å & Johansson, A.V. (1998). Measurement and modelling of homogeneous axisymmetric turbulence. *J. Fluid Mech.* 374:59-90.

Sjögren, T.I.Å & Johansson, A.V. (2000). Development and calibration of algebraic non-linear models for terms in the Reynolds stress transport equations. *Phys. Fluids* 12:1554-1572.

So, R.M.C., Zhang, H.S. & Speziale, C.G. (1991). Near-wall modeling of the dissipation rate equation. *AIAA J.* 29:2069-2076.

Spencer, A.J.M. & Rivlin, R.S. (1959). The theory of matrix polynomials and its application to the mechanics of isotropic continua. *Arch. Rat. Mech. Anal.* 2:309-336.

Speziale, C.G. (1991). Analytical methods for the development of Reynolds-stress closures in turbulence. *Ann. Rev. Fluid Mech.* 23:107-157.

Speziale, C.G. Abid, R. & Anderson, E.C. (1992). Critical evaluation of two-equation models for near-wall turbulence. *AIAA J.* 30:324-331.

Speziale, C.G., Sarkar, S. & Gatski, T.B. (1991). Modelling the pressure-strain correlation of turbulence: an invariant dynamical systems approach. *J. Fluid Mech.* 227:245-272.

Speziale, C.G. & Xu, X.-H. (1996). Towards the development of second-order closure models for nonequilibrium turbulent flows. *Int. J. Heat and Fluid Flow* 17:238-244.

Taulbee, D.B. (1992). An improved algebraic Reynolds stress model and corresponding nonlinear stress model. *Phys. Fluids* 4:2555-2561.

Taulbee, D.B., Sonnenmeier, J.R. & Wall, K.M. (1994). Stress relation for three-dimensional turbulent flows. *Phys. Fluids* 6:1399-1401.

Tavoularis, S. & Corrsin, S. (1981). Experiments in nearly homogeneous turbulent shear flow with a uniform mean temperature gradient. Part I. *J. Fluid Mech.* 104:311-347.

Wallin, S. & Johansson, A.V. (2000). An explicit algebraic Reynolds stress model for incompressible and compressible turbulent flows. *J. Fluid Mech.* 403:89–132.

Wallin, S. & Johansson, A.V. (2001). Modelling of streamline curvature effects on turbulence in explicit algebraic Reynolds stress turbulence models. In *Proceedings of Turbulence and Shear Flow Phenomena II, Stockholm, June 27-29, 2001*, II:223-228. Also to appear in revised form in *Int. J. Heat and Fluid Flow*.

Wikström, P.M., Wallin, S. & Johansson, A.V. (2000). Derivation and investigation of a new explicit algebraic model for the passive scalar flux. *Phys. Fluids* 12:688–702.

Wilcox, D.C. (1994). Turbulence modeling for CFD DCW Industries Inc. ISBN 0-9636051-0-0, USA 1994.

Symmetries and Invariant Solutions of Turbulent Flows and their Implications for Turbulence Modelling

Martin Oberlack
Hydromechanics and Hydraulics Group, Darmstadt University of Technology
Petersenstrasse 13, 64287 Darmstadt, Germany

Abstract

First a short introduction to the notion of symmetries of differential equations is given including infinitesimal transformations, invariant functions and invariant solutions. Then it is shown that the symmetry properties i.e. invariant transformations of the Navier-Stokes equations are pivotal to understand the physics of fluid flow. We demonstrate that all common symmetries "transfer" to the statistical equations such as the Reynolds stress transport equations or the multi-point correlation equations. From the knowledge of the symmetries we derive from the latter equations a broad variety of invariant solutions (scaling laws) using only first principles. These solutions comprise classical results such as the logarithmic-law-of-the-wall and other wall bounded shear flows. Also homogeneous and inhomogeneous time-dependent flows are analyzed and solutions are discussed. Since the symmetries of fluid motion are admitted by all statistical quantities of turbulent flows we give necessary conditions on turbulence models such that they "capture" the proper physics i.e. the symmetries and their corresponding invariant solutions. Particularly we will investigate two-equation models such as the k-ϵ model as well as Reynolds stress transport models with respect to their symmetry properties. Finally we give conditions for the sub-grid scale model in large-eddy simulation of turbulence to obey the proper symmetries. For all of the latter turbulence models it is demonstrated that symmetry violation gives rather disadvantageous prediction capabilities of the model under investigation.

1 Introduction to Symmetry Group Methods

1.1 Introduction

Symmetry group analyses of differential equations (DE) on the basis of continuous transformation groups (Lie groups) unify a wide variety of *ad hoc* methods to analyze and solve DE's. These methods cover ideas such as:

a) rigorous finding of symmetries of DE's,

b) algorithmic reduction and solution of ordinary differential equations (ODE),

c) algorithmic derivation of all invariant (similarity) solutions of partial differential equations (PDE),

d) the decision whether a non-linear PDE may be transformed to a linear PDE,

e) the decision whether a linear non-constant coefficient PDE may be transformed to a linear PDE with constant coefficients and,

f) derivation of conservation laws.

Symmetries are pivotal to a profound understanding of the underlying physics associated with the problem under investigation. The basic idea of symmetries, and of continuous symmetries in particular, will be explained in detail and elementary examples will be given. In particular it will be shown how invariant solutions can be obtained from symmetries. Detailed descriptions of the methods may be found in e.g. Bluman and Kumei (1989), Olver (1986) or Ibragimov (1995a, 1995b,1996).

1.2 Symmetries of Differential Equations and their Global and Infinitesimal forms

The concept of symmetries may be approached from the question on how to extend a given solution $\boldsymbol{y} = \boldsymbol{\Omega}(\boldsymbol{x})$ of the differential equation

$$\boldsymbol{F}(\boldsymbol{x}, \boldsymbol{y}, \underset{1}{\boldsymbol{y}}, \underset{2}{\boldsymbol{y}}, \ldots) = 0 \tag{1}$$

to a new solution $\boldsymbol{y}^* = \boldsymbol{\Omega}^*(\boldsymbol{x}^*)$ by means of transformations. Here $\boldsymbol{x}, \boldsymbol{y}$ and $\underset{i}{\boldsymbol{y}}$ respectively denote the vector of independent variables, the vector of dependent variables and the vector of all i^{th} order derivatives with respect to $\boldsymbol{y}$. By definition $\boldsymbol{y}^* = \boldsymbol{\Omega}^*(\boldsymbol{x}^*)$ is a solution of the equation

$$\boldsymbol{F}(\boldsymbol{x}^*, \boldsymbol{y}^*, \underset{1}{\boldsymbol{y}}^*, \underset{2}{\boldsymbol{y}}^*, \ldots) = 0 \tag{2}$$

written in the "*" variables. In order to accomplish the above task of finding new solutions from known ones we have to find a transformation

$$\boldsymbol{x}^* = \boldsymbol{\phi}(\boldsymbol{x}, \boldsymbol{y}), \;\; \boldsymbol{y}^* = \boldsymbol{\psi}(\boldsymbol{x}, \boldsymbol{y}) \tag{3}$$

such that the following equivalence holds for (3)

$$\boldsymbol{F}(\boldsymbol{x}, \boldsymbol{y}, \underset{1}{\boldsymbol{y}}, \underset{2}{\boldsymbol{y}}, \ldots) = 0 \quad \Leftrightarrow \quad \boldsymbol{F}(\boldsymbol{x}^*, \boldsymbol{y}^*, \underset{1}{\boldsymbol{y}}^*, \underset{2}{\boldsymbol{y}}^*, \ldots) = 0 \;, \tag{4}$$

i.e. equation (1) does not change its functional form written in the new variables $\boldsymbol{x}^*$ and $\boldsymbol{y}^*$. If such a transformation is known it is called a symmetry transformation or simply symmetry . A symmetry of a DE always constitutes a transformation group i.e. it admits group properties.

For the present purpose we are primarily interested in symmetries (3) which constitute Lie groups i.e. continuous transformation groups which allow for the construction of analytic solutions. Such transformations depend on a continuous parameter ϵ of the form

$$S_\epsilon : \;\; \boldsymbol{x}^* = \boldsymbol{\phi}(\boldsymbol{x}, \boldsymbol{y}; \epsilon) \quad \text{and} \quad \boldsymbol{y}^* = \boldsymbol{\psi}(\boldsymbol{x}, \boldsymbol{y}; \epsilon) \tag{5}$$

obeying the requirement to have group properties: closure, containing an unitary element, containing an inverse element and associativity.

An example of the latter Lie group is the scaling group

$$S_\epsilon^{(1)} : \ t^* = \mathrm{e}^{2\epsilon}t, \ x^* = \mathrm{e}^{\epsilon}x, \ u^* = u \tag{6}$$

admitted by the heat equation

$$\frac{\partial u}{\partial t} = \frac{\partial^2 u}{\partial t^2}. \tag{7}$$

It may readily be shown that (7) can be transformed to

$$\frac{\partial u^*}{\partial t^*} = \frac{\partial^2 u^*}{\partial t^{*2}} \tag{8}$$

by means of (6). It is obvious that equations (7) and (8) are fully equivalent, i.e. they obey the condition (4). The above mentioned group properties of (6), which are in the fact the theoretical basis, need not be considered any further.

Another group admitted by equation (7) is that of the scaling group only involving the dependent variable

$$S_\epsilon^{(2)} : \ t^* = t, \ x^* = x, \ u^* = \mathrm{e}^{\epsilon_1}u. \tag{9}$$

It is an easy matter to show that the latter transformation group also admits an equivalence between (7) and (8).

Most important for the results to be presented in the sections 2-6 is the fact that once two or more distinct symmetries $S_\epsilon^{(i)}$ of a DE are known their combination is a new symmetry. This may be written as

$$S_{\epsilon_1,\dots,\epsilon_n} = S_{\epsilon_1}^{(1)} \circ S_{\epsilon_2}^{(2)} \circ \dots \circ S_{\epsilon_n}^{(n)} \ . \tag{10}$$

The above notation means that transformations are successively introduced one after the other.

As an example we may consider the combination of the two scaling groups (6) and (9):

$$S_{\epsilon,\epsilon_1} = S_\epsilon^{(1)} \circ S_{\epsilon_1}^{(2)} : \ t^* = \mathrm{e}^{2\epsilon}t, \ x^* = \mathrm{e}^{\epsilon}x, \ u^* = \mathrm{e}^{\epsilon_1}u \ . \tag{11}$$

Clearly (11) is also a symmetry of (7).

At this point it is interesting to note that as well as continuous groups many DEs also admit finite groups . The heat equation (7) e.g. admits the reflection group

$$R_1 : \ t^* = t, \ x^* = -x, \ u^* = u. \tag{12}$$

In the following we restrict our analysis to continuous transformation groups . In the context of this restriction it is of considerable importance to introduce infinitesimal transformations . I.e. we do a Taylor series expansion in ϵ of the transformation groups (5) to obtain

$$S_\epsilon : \quad \begin{aligned} \boldsymbol{x}^* &= \boldsymbol{x} + \epsilon \left.\frac{\partial \boldsymbol{\phi}}{\partial \epsilon}\right|_{\epsilon=0} + O(\epsilon^2) \\ &= \boldsymbol{x} + \epsilon \boldsymbol{\xi}(\boldsymbol{x}, \boldsymbol{y}) + O(\epsilon^2) \end{aligned} \quad \text{and} \quad \begin{aligned} \boldsymbol{y}^* &= \boldsymbol{y} + \epsilon \left.\frac{\partial \boldsymbol{\psi}}{\partial \epsilon}\right|_{\epsilon=0} + O(\epsilon^2) \\ &= \boldsymbol{y} + \epsilon \boldsymbol{\eta}(\boldsymbol{x}, \boldsymbol{y}) + O(\epsilon^2) \end{aligned} \tag{13}$$

where $\boldsymbol{\xi}$ and $\boldsymbol{\eta}$ are called infinitesimals . In (13) we have implicitly used the identity transformation of (5), corresponding to $\epsilon = 0$, i.e.

$$S_{\epsilon=0}: \ \boldsymbol{x}^* = \boldsymbol{\phi}(\boldsymbol{x}, \boldsymbol{y}; \epsilon = 0) = \boldsymbol{x} \quad \text{and} \quad \boldsymbol{y}^* = \boldsymbol{\psi}(\boldsymbol{x}, \boldsymbol{y}; \epsilon = 0) = \boldsymbol{y}, \tag{14}$$

which is one of the defining properties of symmetry groups (see e.g. Bluman and Kumei, 1989).

As an example of obtaining $\boldsymbol{\xi}$ and $\boldsymbol{\eta}$ for the scaling group from the definition of the infinitesimals we readily obtain by introducing (6) into (13)

$$\xi_t = 2t, \ \ \xi_x = x, \ \ \eta_u = 0. \tag{15}$$

Accordingly we find the infinitesimals from (9)

$$\xi_t = 0, \ \ \xi_x = 0, \ \ \eta_u = u. \tag{16}$$

It can be proved rigorously that if the transformations (5) are restricted to continuous groups only terms up to the order ϵ need to be kept, as can be taken from Lie's first theorem to be given below.

The key property of the Lie group method is that the continuous transformation group (5) and its infinitesimal form (13) are fully equivalent.

This is in fact what is stated in Lie's first theorem (see e.g. Bluman and Kumei, 1989). Once the infinitesimals of a transformation are known the global form of the transformation (5) can be uniquely determined by integrating the first order system

$$\frac{\mathrm{d}\boldsymbol{x}^*}{\mathrm{d}\epsilon} = \boldsymbol{\xi}(\boldsymbol{x}^*, \boldsymbol{y}^*) \quad \text{and} \quad \frac{\mathrm{d}\boldsymbol{y}^*}{\mathrm{d}\epsilon} = \boldsymbol{\eta}(\boldsymbol{x}^*, \boldsymbol{y}^*) \tag{17}$$

furnished by the initial conditions

$$\epsilon = 0 \ : \ \boldsymbol{x}^* = \boldsymbol{x} \quad \text{and} \quad \boldsymbol{y}^* = \boldsymbol{y}. \tag{18}$$

Employing again the latter example of the scaling group (6) we introduce the infinitesimals (15) into (17) by invoking (18) to obtain

$$\frac{\mathrm{d}t^*}{\mathrm{d}\epsilon} = 2t^*, \ \ \frac{\mathrm{d}x^*}{\mathrm{d}\epsilon} = x^*, \ \ \frac{\mathrm{d}u^*}{\mathrm{d}\epsilon} = 0 \quad \text{with} \quad \epsilon = 0 \ : \ t^* = t, \ \ x^* = x, \ \ u^* = u. \tag{19}$$

The latter system of ODE's may easily be integrated to yield (6). The equations (6) and (15) are respectively the *global* and the *infinitesimal* form of the same transformation group .

The property (10) which states that two or more groups admitted by a DE can be combined may also be transferred to the infinitesimal form of a symmetry . Without any proof we state that, once the infinitesimals $\boldsymbol{\xi}^{(i)}$ and $\boldsymbol{\eta}^{(i)}$ of the groups are given, any linear combination of these groups is a new group, i.e.

$$\boldsymbol{\xi} = k_1\boldsymbol{\xi}^{(1)} + k_2\boldsymbol{\xi}^{(2)} + \cdots + k_n\boldsymbol{\xi}^{(n)}, \ \ \boldsymbol{\eta} = k_1\boldsymbol{\eta}^{(1)} + k_2\boldsymbol{\eta}^{(2)} + \cdots + k_n\boldsymbol{\eta}^{(n)}. \tag{20}$$

The latter superposition principle rests on the fact, that the infinitesimals are solutions of a set of linear homogeneous PDEs to be shown below.

For the above given scaling groups in infinitesimal form (15) and (16) we find

$$\xi_t = k_{s_1} 2t, \quad \xi_x = k_{s_1} x, \quad \eta_u = k_{s_2} u. \tag{21}$$

Utilizing Lie's first theorem we find, without proof, the equivalence between (20) and (10). Applying this to the latter infinitesimals we obtain the global group after implementing (20) into (17) and actuating (18)

$$t^* = \mathrm{e}^{2k_{s_1}\epsilon} t, \quad x^* = \mathrm{e}^{k_{s_1}\epsilon} x, \quad u^* = \mathrm{e}^{k_{s_2}\epsilon} u. \tag{22}$$

This group is equivalent to (11) since the contained parameters are arbitrary.

Though in case of the heat equation some of the symmetries such as (6) may be easily obtained by inspection, others may be very difficult to guess. Hence in the following we very briefly sketch how to rigorously obtain all point symmetries of a DE in infinitesimal form. They may in turn be transformed into a global transformation .

In order to write the symmetry condition (4) into infinitesimal form we implement (13) into the right hand side of (4) and expand with respect to ϵ to obtain

$$\begin{aligned} &\boldsymbol{F}(\boldsymbol{x}, \boldsymbol{y}, \underset{1}{\boldsymbol{y}}, \underset{2}{\boldsymbol{y}}, \ldots, \underset{p}{\boldsymbol{y}}) + \epsilon \mathrm{X}^{(p)} \boldsymbol{F}(\boldsymbol{x}, \boldsymbol{y}, \underset{1}{\boldsymbol{y}}, \underset{2}{\boldsymbol{y}}, \ldots, \underset{p}{\boldsymbol{y}}) \\ &+ \frac{\epsilon^2}{2} \left[\mathrm{X}^{(p)}\right]^2 \boldsymbol{F}(\boldsymbol{x}, \boldsymbol{y}, \underset{1}{\boldsymbol{y}}, \underset{2}{\boldsymbol{y}}, \ldots, \underset{p}{\boldsymbol{y}}) + O(\epsilon^3) = 0 \end{aligned} \tag{23}$$

where the generator $\mathrm{X}^{(p)}$ is given by

$$\mathrm{X}^{(p)} = \xi_i \frac{\partial}{\partial x_i} + \eta_j \frac{\partial}{\partial y_j} + \zeta_{j;i_1} \frac{\partial}{\partial y_{j,i_1}} + \cdots + \zeta_{j;i_1 i_2 \ldots i_p} \frac{\partial}{\partial y_{j,i_1 i_2 \ldots i_p}}. \tag{24}$$

and $\zeta_{j;i_1 i_2 \ldots i_p}$ is defined according to

$$\begin{aligned} \zeta_{k;i} &= \frac{\mathcal{D}\eta_k}{\mathcal{D}x_i} - y_{k,m} \frac{\mathcal{D}\xi_m}{\mathcal{D}x_i} \quad \text{and} \\ \zeta_{k;i_1 \ldots i_s} &= \frac{\mathcal{D}\zeta_{k;i_1 \ldots i_{s-1}}}{\mathcal{D}x_{i_s}} - y_{k,m i_1 \ldots i_{s-1}} \frac{\mathcal{D}\xi_m}{\mathcal{D}x_{i_s}} \quad \text{for} \quad s > 1 \end{aligned} \tag{25}$$

with

$$\frac{\mathcal{D}}{\mathcal{D}x_i} = \frac{\partial}{\partial x_i} + y_{k,i} \frac{\partial}{\partial y_k} + y_{k,ij} \frac{\partial}{\partial y_{k,j}} + \ldots. \tag{26}$$

Those terms in $\mathrm{X}^{(p)}$ containing ζ are called prolongations of the generator

$$\mathrm{X} = \xi_i \frac{\partial}{\partial x_i} + \eta_j \frac{\partial}{\partial y_j} \tag{27}$$

of order p.

The power of the operator $\left[\mathrm{X}^{(p)}\right]^n$ denotes the successive application of $\mathrm{X}^{(p)}$ in the form $\underbrace{\mathrm{X}^{(p)} \ldots \mathrm{X}^{(p)}}_{n}$. The term ζ is the infinitesimal form of the derivatives. Details of the derivation of (25) and (26) may be taken from Bluman and Kumei (1989).

Recalling the example of the scaling group (6) and (9) with its infinitesimal form (15) and (16) we may respectively write the generators of these groups as

$$\mathrm{X}_{s_1} = 2t\frac{\partial}{\partial t} + x\frac{\partial}{\partial x} \tag{28}$$

and

$$\mathrm{X}_{s_2} = u\frac{\partial}{\partial u}. \tag{29}$$

Prolonging these groups to the first order, i.e. considering only first order derivatives, we obtain from (24) by invoking (25) and (26)

$$\mathrm{X}_{s_1}^{(1)} = 2t\frac{\partial}{\partial t} + x\frac{\partial}{\partial x} - 2u_t\frac{\partial}{\partial u_t} - u_x\frac{\partial}{\partial u_x} \tag{30}$$

and

$$\mathrm{X}_{s_2}^{(1)} = u\frac{\partial}{\partial u} + u_t\frac{\partial}{\partial u_t} + u_x\frac{\partial}{\partial u_x}. \tag{31}$$

Since the infinitesimals admit the superposition principle given in (20) we may readily show that also the corresponding set of n generators X_i possess this theorem of linearity, i.e.

$$\mathrm{X} = k_1\mathrm{X}_1 + k_2\mathrm{X}_2 + \cdots + \mathrm{X}_n. \tag{32}$$

In case of the two scaling groups (28) and (29) we have

$$\mathrm{X}_s = k_{s_1}\mathrm{X}_{s_1} + k_{s_2}\mathrm{X}_{s_2} = k_{s_1}\left[2t\frac{\partial}{\partial t} + x\frac{\partial}{\partial x}\right] + k_{s_2}u\frac{\partial}{\partial u}. \tag{33}$$

Using the first equation of (4) we may eliminate the first term in the first line of (23). For the remaining terms it is a necessary and sufficient condition if the following holds

$$\mathrm{X}^{(p)}\boldsymbol{F} = 0. \tag{34}$$

It is apparent that any successive application of $\mathrm{X}^{(p)}$ onto the latter equation may not lead to additional constraints.

Since the original equation (1) may always be taken into account for further computations of symmetries we write the final symmetry condition in infinitesimal form as

$$\left[\mathrm{X}^{(p)}\boldsymbol{F}\right]\Big|_{\boldsymbol{F}=0} = 0. \tag{35}$$

The usual way of finding Lie symmetry groups is to introduce arbitrary $\boldsymbol{\xi}$ and $\boldsymbol{\eta}$ into condition (35) by employing (25) and (24). The resulting condition constitutes an overdetermined set of linear homogeneous differential equations, the determining equations , which have to be solved for the infinitesimals $\boldsymbol{\xi}$ and $\boldsymbol{\eta}$. They can in turn be used to derive the global transformations of the form (5) from Lie's differential equations (17) and (18).

Since for most equations of mathematical physics, and particularly for the equations of fluid mechanics, the symmetry transformations are already known we will in the subsequent sections only state their forms and not show their derivation.

1.3 Invariant Functions and Invariant Solutions

In full analogy to the invariance of differential equations under certain transformation groups we also define the invariance of a function. A function is called invariant under a transformation (3) if it admits the property

$$f(\boldsymbol{x},\boldsymbol{y}) = f(\boldsymbol{x}^*,\boldsymbol{y}^*). \tag{36}$$

Again we limit our definition of an invariant function by only considering Lie groups. Hence, we may adopt the expansion according to (23) to obtain from (36)

$$\begin{aligned} f(\boldsymbol{x},\boldsymbol{y}) &= f[\boldsymbol{x}+\boldsymbol{\xi}(\boldsymbol{x},\boldsymbol{y})\epsilon + O(\epsilon^2), \boldsymbol{y}+\boldsymbol{\eta}(\boldsymbol{x},\boldsymbol{y})\epsilon + O(\epsilon^2)] \\ \Leftrightarrow f(\boldsymbol{x},\boldsymbol{y}) &= f(\boldsymbol{x},\boldsymbol{y}) + \epsilon \mathrm{X} f + \frac{\epsilon^2}{2}\mathrm{X}^2 f + O(\epsilon^3), \end{aligned} \tag{37}$$

where the generator X is given by (27).

The first term on the left and on the right hand side cancel and we obtain the condition for an invariant function

$$\mathrm{X} f = \xi_k \frac{\partial f}{\partial x_k} + \eta_l \frac{\partial f}{\partial y_l} = 0 \tag{38}$$

in infinitesimal form. Again the higher order terms such as $\mathrm{X}^2 f$ also cancel due to $\mathrm{X} f = 0$ because $\mathrm{X}^2 f$ is equivalent to $\mathrm{X}(\mathrm{X} f)$. The latter hyperbolic DE may be solved with the method of characteristics, composed by the m independent and n dependent variables

$$\frac{\mathrm{d}x_1}{\xi_1} = \frac{\mathrm{d}x_2}{\xi_2} = \cdots = \frac{\mathrm{d}x_m}{\xi_m} = \frac{\mathrm{d}y_1}{\eta_1} = \frac{\mathrm{d}y_2}{\eta_2} = \cdots = \frac{\mathrm{d}y_n}{\eta_n}. \tag{39}$$

The solution to this system possesses $m+n-1$ constants of integration C_i, which, employed as new independent variables of f, form a complete solution of equation (38). Consequentially we have defined a function $f(C_1,\ldots,C_{m+n-1})$ which is invariant under a given group of transformation, defined by its infinitesimals .

As an illustration we may again employ the example of the scaling group (6) in infinitesimal form (15). Implementing the latter infinitesimals into (38) provided f depends on x, t and u we obtain

$$\mathrm{X}_{s_1} f = 2t\frac{\partial f}{\partial t} + x\frac{\partial f}{\partial x} = 0. \tag{40}$$

The latter hyperbolic equation possesses, respectively, the characteristics and the general solution

$$\frac{\mathrm{d}t}{2t} = \frac{\mathrm{d}x}{x} = \frac{\mathrm{d}u}{0} \quad \text{and} \quad f = f\left(u, \frac{x}{\sqrt{t}}\right). \tag{41}$$

It is a simple matter to show that the latter result (41) obeys the condition of an invariant function (36) under the symmetry scaling group (6). In its global form this may readily be verified by implementing (6) into (41) to obtain

$$f\left(u, \frac{x}{\sqrt{t}}\right) = f\left(u^*, \frac{\mathrm{e}^{-\epsilon}x^*}{\sqrt{\mathrm{e}^{-2\epsilon}t^*}}\right) = f\left(u^*, \frac{x^*}{\sqrt{t^*}}\right). \tag{42}$$

At this point it is instructive to define the invariance under multiple symmetry groups. Classically invariance of a function under n symmetry groups means the consecutive application of these groups to a given function, i.e.

$$\mathrm{X}_1 f = 0 \ \wedge \ \mathrm{X}_2 f = 0 \ \wedge \cdots \wedge \wedge \mathrm{X}_n f = 0. \tag{43}$$

Suppose we consider the invariance of $f(t,x,u)$ under both scaling groups (6) and (9). In addition to the already known invariant function (41) we also add

$$\mathrm{X}_{s_2} f = u \frac{\partial f}{\partial u}. \tag{44}$$

As a result we immediately obtain

$$f = f\left(\frac{x}{\sqrt{t}}\right) \tag{45}$$

since the u dependence has been lost due to (44).

A very different meaning of the invariance of a function under multiple symmetry groups is intended if all or several groups are combined to a single group as is defined in (32). In case of the combined two scaling groups of the heat equation (33) we get

$$\mathrm{X}_s f = k_{s_1} \left[2t \frac{\partial f}{\partial t} + x \frac{\partial f}{\partial x} \right] + k_{s_2} u \frac{\partial f}{\partial u} = 0. \tag{46}$$

The latter invariance condition, respectively, admits the characteristic equation, also sometimes referred to as invariant surface condition ,

$$\frac{\mathrm{d}t}{k_{s_1} 2t} = \frac{\mathrm{d}x}{k_{s_1} x} = \frac{\mathrm{d}u}{k_{s_2} u} \tag{47}$$

with the two fundamental invariants and the general solution

$$\gamma = \frac{x}{\sqrt{t}}, \quad \Delta = \frac{u}{t^{\frac{k_{s_2}}{2k_{s_1}}}} \quad \text{and} \quad f = f(\gamma, \Delta). \tag{48}$$

It is apparent that the two invariant solutions (45) and (48) have a very different structure.

A key property of an invariant function under a certain group admitted by a differential equation and most important for what follows is the fact that the fundamental invariants may be used as new variables leading to a similarity reduction (invariant solution). In the latter case of the heat equation we employ the fundamental invariants Δ and γ under the combined scaling groups admitted by the heat equation as new dependent and independent variables. Implementing this into the equation (7) we find

$$-\frac{k_{s_2}}{2k_{s_1}} \Delta + \frac{1}{2} \gamma \frac{\mathrm{d}\Delta}{\mathrm{d}\gamma} + \frac{\mathrm{d}^2\Delta}{\mathrm{d}\gamma^2} = 0, \tag{49}$$

which is the reduced heat equation.

An important result to be taken from the latter example is that once more than one group is used to derive invariants, one or more free parameters appear in the reduced equation which allows for construction of a family of solutions.

The idea of an invariant solution may be written in general terms. In accordance to the latter example we define an invariant solution as follows. If a PDE such as (2) admits the symmetry given in form of a generator X in (27) we call $\boldsymbol{y} = \boldsymbol{\Theta}(\boldsymbol{x})$ an invariant solution of the PDE if the following two conditions hold:

(i) $\boldsymbol{y} - \boldsymbol{\Theta}(\boldsymbol{x})$ is invariant under X and

(ii) $\boldsymbol{y} = \boldsymbol{\Theta}(\boldsymbol{x})$ is a solution of the PDE (2).

According to the definition of an invariant function (38) the condition (i) yields the equation

$$\mathrm{X}\left[\boldsymbol{y} - \boldsymbol{\Theta}(\boldsymbol{x})\right] = 0 \quad \text{with} \quad \boldsymbol{y} = \boldsymbol{\Theta}(\boldsymbol{x}). \tag{50}$$

Expanding the derivatives in (50) by employing X according to (27) we obtain the hyperbolic system

$$\xi_k(\boldsymbol{x}, \boldsymbol{\Theta}) \frac{\partial \Theta_l}{\partial x_k} = \eta_l(\boldsymbol{x}, \boldsymbol{\Theta}). \tag{51}$$

The corresponding characteristic equation reads

$$\frac{\mathrm{d}x_1}{\xi_1(\boldsymbol{x}, \boldsymbol{y})} = \frac{\mathrm{d}x_2}{\xi_2(\boldsymbol{x}, \boldsymbol{y})} = \cdots = \frac{\mathrm{d}x_m}{\xi_m(\boldsymbol{x}, \boldsymbol{y})} = \frac{\mathrm{d}y_1}{\eta_1(\boldsymbol{x}, \boldsymbol{y})} = \frac{\mathrm{d}y_2}{\eta_2(\boldsymbol{x}, \boldsymbol{y})} = \cdots = \frac{\mathrm{d}y_n}{\eta_n(\boldsymbol{x}, \boldsymbol{y})}, \tag{52}$$

where $\boldsymbol{\Theta}$ has been replaced by $\boldsymbol{y}$.

As above the system admits $m + n - 1$ solutions, which may be taken as new variables. Generally it is advised to take the $m-1$ solutions of the m equations on the left hand side as new independent variables. The n terms on the right hand side may be set equal to an arbitrary term of the m terms on the left hand side. The solution of the resulting system may be employed as new dependent variables. As a result of this process we find that the set of independent variables in the original PDE has been reduced by at least one. The latter process of deriving invariant solutions under a given symmetry is called similarity reduction or simply reduction. For most cases the derivative order of the PDE is unaltered after the reduction.

2 Symmetries of the Euler and Navier-Stokes Equations

The starting point of all analysis and results to be presented in the subsequent sections are the Navier-Stokes equations for an incompressible fluid under the restriction of constant density and viscosity. In a cartesian coordinate system the continuity and momentum equation are respectively

$$\frac{\partial u_k}{\partial x_k} = 0 \tag{53}$$

and

$$\frac{\mathrm{D}u_i}{\mathrm{D}t} = -\frac{\partial p}{\partial x_i} + \nu \frac{\partial^2 u_i}{\partial x_k \partial x_k} - 2\,\Omega_k\, e_{ikl}\, u_l \tag{54}$$

where

$$\frac{\mathrm{D}}{\mathrm{D}t} = \frac{\partial}{\partial t} + u_k \frac{\partial}{\partial x_k} \tag{55}$$

defines the substantial derivative. t, $\boldsymbol{x}$, $\boldsymbol{u}$, p and ν define the usual quantities, $\boldsymbol{\Omega}$ constitutes a constant rotation vector and e_{ijk} is the alternating tensor

$$e_{ijk} = \begin{cases} 1 & \text{for} \quad ijk = 123, 312, 231 \\ -1 & \text{for} \quad ijk = 321, 132, 213\,. \\ 0 & \text{otherwise} \end{cases} \tag{56}$$

Since for the turbulence dynamics of all cases to be discussed in the present work centrifugal forces and the value of the fluid density do not matter in equation (54), these terms have been absorbed into the pressure term according to the form

$$p^* = \frac{p}{\rho} + \frac{1}{2}\left(\Omega_k\,\Omega_l\,x_k\,x_l - \Omega_k\,\Omega_k\,x_l\,x_l\right), \tag{57}$$

where "$*$" of "p^*" in (54) has been omitted.

The foundation of all subsequent results on turbulence rely on the symmetries of the Euler equation and the Navier-Stokes equation. Rotation is only considered for some results in sections 3 and 4. Hence, if not stated otherwise we set

$$\boldsymbol{\Omega} = 0. \tag{58}$$

The Euler equations in a non-rotating frame, i.e. equation (53) and (54) with

$$\nu = 0 \tag{59}$$

admit the ten-parameter symmetry group

$$\begin{aligned}
X_t &= \frac{\partial}{\partial t}, \\
X_{r_3} &= -x_2\frac{\partial}{\partial x_1} + x_1\frac{\partial}{\partial x_2} - u_2\frac{\partial}{\partial u_1} + u_1\frac{\partial}{\partial u_2} \\
X_{r_1} &= -x_3\frac{\partial}{\partial x_2} + x_2\frac{\partial}{\partial x_3} - u_3\frac{\partial}{\partial u_2} + u_2\frac{\partial}{\partial u_3} \\
X_{r_2} &= -x_3\frac{\partial}{\partial x_1} + x_1\frac{\partial}{\partial x_3} - u_3\frac{\partial}{\partial u_1} + u_1\frac{\partial}{\partial u_3} \\
X_{u_1} &= f_1(t)\frac{\partial}{\partial x_1} + \frac{\mathrm{d}f_1(t)}{\mathrm{d}t}\frac{\partial}{\partial u_1} - x_1\frac{\mathrm{d}^2 f_1(t)}{\mathrm{d}t^2}\frac{\partial}{\partial p}, \\
X_{u_2} &= f_2(t)\frac{\partial}{\partial x_2} + \frac{\mathrm{d}f_2(t)}{\mathrm{d}t}\frac{\partial}{\partial u_2} - x_2\frac{\mathrm{d}^2 f_2(t)}{\mathrm{d}t^2}\frac{\partial}{\partial p}, \\
X_{u_3} &= f_3(t)\frac{\partial}{\partial x_3} + \frac{\mathrm{d}f_3(t)}{\mathrm{d}t}\frac{\partial}{\partial u_3} - x_3\frac{\mathrm{d}^2 f_3(t)}{\mathrm{d}t^2}\frac{\partial}{\partial p}, \\
X_p &= f_4(t)\frac{\partial}{\partial p}, \\
X_{s_1} &= x_i\frac{\partial}{\partial x_i} + u_j\frac{\partial}{\partial u_j} + 2p\frac{\partial}{\partial p}, \\
X_{s_2} &= t\frac{\partial}{\partial t} - u_i\frac{\partial}{\partial u_i} - 2p\frac{\partial}{\partial p},
\end{aligned} \tag{60}$$

where $f_1(t)$-$f_3(t)$ are at least twice differentiable function of time and $f_4(t)$ is an arbitrary function of time.

The symmetries X_{u_1}-X_{u_3} are called generalized Galilean invariance and comprise the usual translation invariance if f_1-f_3 are constants and the classical Galilean invariance if f_1-f_3 are linear functions of time. The generalized Galilean invariance and the pressure invariance X_p are an immediate consequence of incompressibility and have no correspondence in the Euler equation for compressible flows. Details are given below in equation (70).

Employing Lie's differential equations (17)/(18) which constitute the equivalence between the global and the infinitesimal form of a transformation group we obtain from (60)

$$T_t:\ t^* = t + a_1\ ,\ \boldsymbol{x}^* = \boldsymbol{x}\ ,\ \boldsymbol{u}^* = \boldsymbol{u}\ ,\ p^* = p, \tag{61}$$

$$T_{r_1} - T_{r_3}:\ t^* = t\ ,\ \boldsymbol{x}^* = \mathbf{a}\cdot\boldsymbol{x}\ ,\ \boldsymbol{u}^* = \mathbf{a}\cdot\boldsymbol{u}\ ,\ p^* = p, \tag{62}$$

$$T_{u_1} - T_{u_3}:\ t^* = t\ ,\ \boldsymbol{x}^* = \boldsymbol{x} + \boldsymbol{f}(t)\ ,\ \boldsymbol{u}^* = \boldsymbol{u} + \frac{\mathrm{d}\boldsymbol{f}}{\mathrm{d}t}\ ,\ p^* = p - \boldsymbol{x}\cdot\frac{\mathrm{d}^2\boldsymbol{f}}{\mathrm{d}t^2}, \tag{63}$$

$$T_p:\ t^* = t\ ,\ \boldsymbol{x}^* = \boldsymbol{x}\ ,\ \boldsymbol{u}^* = \boldsymbol{u}\ ,\ p^* = p + a_4 f_4(t), \tag{64}$$

$$T_{s_1}:\ t^* = t\ ,\ \boldsymbol{x}^* = \mathrm{e}^{a_2}\boldsymbol{x}\ ,\ \boldsymbol{u}^* = \mathrm{e}^{a_2}\boldsymbol{u}\ ,\ p^* = \mathrm{e}^{2a_2}p, \tag{65}$$

$$T_{s_2}:\ t^* = \mathrm{e}^{a_3}t\ ,\ \boldsymbol{x}^* = \boldsymbol{x}\ ,\ \boldsymbol{u}^* = \mathrm{e}^{-a_3}\boldsymbol{u}\ ,\ p^* = \mathrm{e}^{-2a_3}p, \tag{66}$$

where the constants a_1-a_4 are the corresponding group parameters, a is a constant rotation matrix with the properties $\mathbf{a}\cdot\mathbf{a}^{\mathsf{T}} = \mathbf{a}^{\mathsf{T}}\cdot\mathbf{a} = \mathbf{I}$ and $|\mathbf{a}| = 1$ and $\boldsymbol{f}(t) = (a_4 f_1(t), a_5 f_2(t), a_6 f_2(t))^{\mathsf{T}}$ respectively, $f_4(t)$ meets the restrictions given below (60).

Introducing the restriction of a two-dimensional flow , i.e.

$$\boldsymbol{u} = \boldsymbol{u}(x_1, x_2, t) \text{ and } p = p(x_1, x_2, t), \tag{67}$$

we find that the Euler equations admit an additional symmetry, which in the turbulence literature is sometimes referred to as "2D material frame indifference" (2DMFI) (see e.g. Speziale, 1981). In infinitesimal form we have

$$\mathrm{X}_{2D} = tx_2\frac{\partial}{\partial x_1} - tx_1\frac{\partial}{\partial x_2} + (x_2 + u_2 t)\frac{\partial}{\partial u_1} - (x_1 + u_1 t)\frac{\partial}{\partial u_2} + 2\psi\frac{\partial}{\partial p}$$
$$\text{with}\quad u_1 = \frac{\partial\psi}{\partial x_2}\ \text{ and }\ u_2 = -\frac{\partial\psi}{\partial x_1}, \tag{68}$$

where ψ is the two-dimensional stream function . At this point it may be interesting to note that the two-dimensionality in the x_1-x_2-plane does not exclude a u_3 velocity component.

In primitive variables the symmetry (68) denotes a non-local transformation since the stream-function is a line integral of the velocities.

$$\psi = \int_Q (u_1 dx_2 - u_2 dx_1)\,. \tag{69}$$

With the aid of Lie's theorem the symmetry (68) may be given in global form

$$\begin{aligned} T_{2D}: t^* &= t,\ x_1^* = x_1\cos(a_5 t) - x_2\sin(a_5 t),\ x_2^* = x_1\sin(a_5 t) + x_2\cos(a_5 t), \\ u_1^* &= u_1\cos(a_5 t) - u_2\sin(a_5 t) - a_5 x_1\sin(a_5 t) - a_5 x_2\cos(a_5 t), \\ u_2^* &= u_1\sin(a_5 t) + u_2\cos(a_5 t) + a_5 x_1\cos(a_5 t) - a_5 x_2\sin(a_5 t), \\ p^* &= p + 2\,a_5\int_Q (u_2 dx_1 - u_1 dx_2) + \frac{1}{2}a_5^2(x_1^2 + x_2^2)\ . \end{aligned} \tag{70}$$

In their global form (61)-(66) and (70) admit immediate interpretation from a physical point of view. Being autonomous in t, i.e. independent of time, is the basis for the symmetry T_t. The transformation $T_{r_1} - T_{r_3}$ is often called covariance principle (see e.g. Cantwell, 1997) and reflects the homogeneity of space with respect to rotation. As mentioned earlier T_{u_1}-T_{u_3} comprise the two symmetry translations in space for $\boldsymbol{f} = \boldsymbol{a}$ and classical Galilean invariance for $\boldsymbol{f} = \boldsymbol{b}t$ where $\boldsymbol{a}$ and $\boldsymbol{b}$ are constant vectors. The invariance T_p states that a uniform but time-dependent background pressure does not change the dynamics of a given flow and is an artifact of incompressibility. The symmetries T_{s_1} and T_{s_2} have their roots in the fact that in classical inviscid mechanics space and time can be scaled independently since no external length or time scale is symmetry breaking. We will later see that this is not true for Navier-Stokes equations .

The global form of the symmetry T_{2D} illuminates the fact that rotation of a two-dimensional flow at a constant rotation rate does not alter the flow if the axis of rotation is aligned with the axis of independence. This symmetry has a close relation to the Taylor-Proudman theorem (see e.g. Greenspan, 1990). The theorem states that a three-dimensional flow exposed to rotation in the limit of infinite rotation rate becomes two-dimensional. Though neither a two-dimensional flow exists in practice nor an infinite rotation rate can be achieved the symmetry T_{10} still has some importance for turbulence. In flows with sufficient large rotation rates such as in recirculation regions or due to forced rotation we find extended elongated coherent structures with their axes aligned with the rotation.

The complete set of all classical point symmetries (60) has originally been derived by Pukhnachev (1972). The first derivation of (68) in its global form that the author could find is in Batchelor (1967). Employing symmetry methods (68) has been derived by Cantwell (1978) and Andreev and Rodionov (1988) independent of its first derivation.

There is considerable uncertainty with respect to the latter invariance and completely contradictory or wrong conclusions have been drawn from it. The phrase "2DMFI" in the turbulence literature is confusing. "Material frame indifference" is a concept taken from classical mechanics, essentially from the theory of constitutive equations, and refers to the invariance of a material property under arbitrary and time-dependent translation and rotation. This is also known as Euclidean transformation

$$x_i^* = p_{ij}(t)x_j + q_i(t), \tag{71}$$

where $\mathbf{p}(t)$ is a real orthonormal and time dependent rotation matrix. Constitutive equations describing material properties such as the Stokes law of viscosity are invariant under the latter transformation.

In contrast the symmetry (70) is not a material property but indeed a dynamical feature. It is restricted to purely two-dimensional flows at a constant rate of rotation. Also there is no three-dimensional correspondence of the latter invariance since three-dimensional arbitrary flows are very sensitive to rotation as is known from DNS and experiments.

This cardinal misunderstanding has its roots in the analogy between molecular and Reynolds stresses. However, both stresses have a very different foundation and may under no circumstances be compared. In fact, turbulence models which wrongly impose "material frame indifference" onto the Reynolds stresses have very adverse properties such as insensitivity with respect to rotation, strong stream-line curvature etc. as will be pointed out in detail in section 6.

Additionally very special symmetries of the Euler-equations may be derived for the case of axisymmetric flow and plane stationary flow (see e.g. Andreev et al., 1998).

If we drop the restriction (59), i.e. we move on from the Euler to the Navier-Stokes equations, slightly different symmetry properties are observed. Except for the two scaling groups X_{s_1} and X_{s_2} the Navier-Stokes equations admit exactly all of the former symmetries recognized by the Euler equations. However, the two scaling groups X_{s_1} and X_{s_2} combine to

$$X_s = 2t\frac{\partial}{\partial t} + x_i\frac{\partial}{\partial x_i} - u_j\frac{\partial}{\partial u_j} - 2p\frac{\partial}{\partial p} \tag{72}$$

or in global form

$$T_s : t^* = \mathrm{e}^{2a}t \ , \ \boldsymbol{x}^* = \mathrm{e}^{a}\boldsymbol{x} \ , \ \boldsymbol{u}^* = \mathrm{e}^{-a}\boldsymbol{u} \ , \ p^* = \mathrm{e}^{-2a}p. \tag{73}$$

It should be noted that the symmetry (68) is unaltered by a non-zero viscosity i.e. (68) is also admitted by the 2D Navier-Stokes equations.

It is particularly the difference in scaling symmetries between the Euler and the Navier-Stokes equations which is important for the understanding of turbulent solutions to be presented below.

Beside the continuous symmetries the Euler and Navier-Stokes equations also admit important discrete symmetries, which do not depend on a continuous parameter. Both Euler and Navier-Stokes equations admit the spatial reflection symmetries

$$T_{R1} : t^* = t, \ x_a^* = -x_a, \ u_a^* = -u_a, \ x_\beta^* = x_\beta, \ u_\beta^* = u_\beta \text{ with } \beta \neq a, \ p^* = p, \tag{74}$$

where the index a refers to the values 1, 2 or 3 and β denotes the remaining two indices a time. On top of this the Euler equations admit time reflection

$$T_{R2} : t^* = -t, \ \boldsymbol{x}^* = \boldsymbol{x}, \ \boldsymbol{u}^* = -\boldsymbol{u}, \ p^* = p. \tag{75}$$

From the knowledge of all the above symmetries of the Euler and Navier-Stokes equations we will subsequently derive a broad variety of results for turbulent flows. This is in fact due to the important property that all of the above symmetries of the instantaneous equations transfer to the statistical equations of turbulence theory to be shown in the subsequent sections.

3 Multi-point Correlation Equations in Turbulence and their Symmetries

In this chapter we deal with the derivation and the properties of the multi-point correlation equations. The pivotal advantage of the multi-point approach compared to the one-point equations is that on each level of correlation order only *one unclosed term* is present. Since the structure of each multi-point correlation equation on each level of correlation (tensor) order is equivalent it is sufficient to deal only with the two-point correlation equations .

3.1 Reynolds Decomposition and Correlation Equations

For the derivation of the correlation equations we employ the usual Reynolds decomposition

$$z = \bar{z} + z' \tag{76}$$

where z, $\bar{z}$ and z' are generic parameters denoting the instantaneous, the mean and the fluctuating quantities for velocity and pressure.

Implementing this decomposition into the Navier-Stokes equations (53) and (54) and applying the usual Reynolds averaging we obtain

$$\frac{\partial \bar{u}_k}{\partial x_k} = 0, \tag{77}$$

and

$$\frac{\bar{\mathrm{D}}\bar{u}_i}{\bar{\mathrm{D}}t} = -\frac{\partial \bar{p}}{\partial x_i} + \nu \frac{\partial^2 \bar{u}_i}{\partial x_k \partial x_k} - \frac{\partial \overline{u_i' u_k'}}{\partial x_k} - 2\,\Omega_k\, e_{ikl}\, \bar{u}_l \tag{78}$$

where $\overline{u_i' u_k'}$ is the Reynolds stress tensor and

$$\frac{\bar{\mathrm{D}}}{\bar{\mathrm{D}}t} = \frac{\partial}{\partial t} + \bar{u}_k \frac{\partial}{\partial x_k} \tag{79}$$

denotes the mean substantial derivative.

Subtracting (77) and (78) from (53) and (54) respectively we obtain the continuity and momentum equations for the fluctuating quantities

$$\frac{\partial u_k'}{\partial x_k} = 0. \tag{80}$$

and

$$\mathcal{N}_i(\boldsymbol{x}) = \frac{\bar{\mathrm{D}}u_i'}{\bar{\mathrm{D}}t} + u_k' \frac{\partial \bar{u}_i}{\partial x_k} - \frac{\partial \overline{u_i' u_k'}}{\partial x_k} + \frac{\partial u_i' u_k'}{\partial x_k} + \frac{\partial p}{\partial x_i} - \nu \frac{\partial^2 u_i'}{\partial x_k \partial x_k} + 2\,\Omega_k\, e_{ikl}\, u_l' = 0 \tag{81}$$

Equations (77) and (78) constitute an unclosed set of equations since an evolution equation for the Reynolds stress tensor is required. One way of treating the classical closure problem of turbulence for practical applications is to derive a transport equation for $\overline{u_i' u_j'}$ from (80) and (81) and to introduce empirical closure assumptions for the five unknown terms appearing in the resulting so-called Reynolds stress transport equation .

Presently we shall introduce a multi-point correlation approach relying on the infinite set of correlation functions. The concept of two- and multi-point correlation functions was born out of the necessity to obtain length-scale information on turbulent flows. At the same time the resulting correlation equations have considerably less unknown terms at the expense of additional dimensions in the equations. In each of the correlation equations of tensor order n an additional tensor of the order $n+1$ appears as unknown term. The first of the infinite sequence of correlation functions is the two-point correlation tensor defined as

$$R_{ij}(\boldsymbol{x}, \boldsymbol{r}; t) = \overline{u_i'(\boldsymbol{x}; t) u_j'(\boldsymbol{x} + \boldsymbol{r}; t)} \tag{82}$$

which is closely related to the Reynolds stress tensor by the identity

$$\overline{u_i' u_j'}(\boldsymbol{x};t) = R_{ij}(\boldsymbol{x}, \boldsymbol{r}=0;t). \tag{83}$$

The infinite set of correlation equations is derived in Oberlack (2000b) in compact form. From there we find that the form of each correlation equation of different order is essentially equivalent in structure.

As a conclusion, and in order to keep the mathematical formalism to a minimum, we only consider the two-point correlation equation in the following analysis

$$\begin{aligned}
&\frac{\bar{\mathrm{D}} R_{ij}}{\bar{\mathrm{D}} t} + R_{kj}\frac{\partial \bar{u}_i(\boldsymbol{x},t)}{\partial x_k} + R_{ik}\left.\frac{\partial \bar{u}_j(\boldsymbol{x},t)}{\partial x_k}\right|_{\boldsymbol{x}+\boldsymbol{r}} + \left[\bar{u}_k(\boldsymbol{x}+\boldsymbol{r},t) - \bar{u}_k(\boldsymbol{x},t)\right]\frac{\partial R_{ij}}{\partial r_k} \\
&+\frac{\partial \overline{p'u'}_j}{\partial x_i} - \frac{\partial \overline{p'u'}_j}{\partial r_i} + \frac{\partial \overline{u_i'p'}}{\partial r_j} - \nu\left[\frac{\partial^2 R_{ij}}{\partial x_k \partial x_k} - 2\frac{\partial^2 R_{ij}}{\partial x_k \partial r_k} + 2\frac{\partial^2 R_{ij}}{\partial r_k \partial r_k}\right] \\
&+\frac{\partial R_{(ik)j}}{\partial x_k} - \frac{\partial}{\partial r_k}\left[R_{(ik)j} - R_{i(jk)}\right] + 2\,\Omega_k\left[e_{kli}R_{lj} + e_{klj}R_{il}\right] = 0.
\end{aligned} \tag{84}$$

The terms $\overline{p'u'}_j$, $\overline{u_i'p'}$ and $R_{(ik)j}$, $R_{i(jk)}$ are defined according to

$$\overline{p'u'}_j(\boldsymbol{x},\boldsymbol{r},t) = \overline{p'(\boldsymbol{x},t)\,u_j'(\boldsymbol{x}+\boldsymbol{r},t)} \quad , \quad \overline{u_i'p'}(\boldsymbol{x},\boldsymbol{r},t) = \overline{u_i'(\boldsymbol{x},t)\,p'(\boldsymbol{x}+\boldsymbol{r},t)}. \tag{85}$$

and

$$R_{(ik)j} = \overline{u_i'(\boldsymbol{x},t)\,u_k'(\boldsymbol{x},t)\,u_j'(\boldsymbol{x}+\boldsymbol{r},t)} \quad , \quad R_{i(jk)} = \overline{u_i'(\boldsymbol{x},t)\,u_j'(\boldsymbol{x}+\boldsymbol{r},t)\,u_k'(\boldsymbol{x}+\boldsymbol{r},t)}. \tag{86}$$

All of the vectors and tensors in (84) obey additional constraints derived from continuity given by

$$\frac{\partial R_{ij}}{\partial x_i} - \frac{\partial R_{ij}}{\partial r_i} = 0\,, \quad \frac{\partial R_{ij}}{\partial r_j} = 0, \tag{87}$$

$$\frac{\partial \overline{p'u'}_i}{\partial r_i} = 0\,, \quad \frac{\partial \overline{u_j'p'}}{\partial x_j} - \frac{\partial \overline{u_j'p'}}{\partial r_j} = 0. \tag{88}$$

and

$$\frac{\partial R_{i(jk)}}{\partial r_i} = 0\,, \quad \frac{\partial R_{(ik)j}}{\partial x_j} - \frac{\partial R_{(ik)j}}{\partial r_j} = 0. \tag{89}$$

At this point it is very important to note that any result to be derived in the following chapters is consistent with all correlation equations up to any arbitrary correlation tensor of order n.

The non-local property of the two- and multi-point correlation equation may be most impressively demonstrated considering the commutation of the two-point velocity tensor. We readily recognize the identity $\overline{u_i'(\boldsymbol{x})u_j'(\boldsymbol{x}+\boldsymbol{r})} = \overline{u_j'(\boldsymbol{x}+\boldsymbol{r})u_i'(\boldsymbol{x})}$. From this we derive, together with (82), the functional relation

$$R_{ij}(\boldsymbol{x},\boldsymbol{r};t) = R_{ji}(\boldsymbol{x}+\boldsymbol{r},-\boldsymbol{r};t) \tag{90}$$

and

$$\overline{p'u'}_j(\boldsymbol{x},\boldsymbol{r};t) = \overline{u_j'p'}(\boldsymbol{x}+\boldsymbol{r},-\boldsymbol{r};t)\ . \tag{91}$$

Analogous identities may be derived for all other two- and multi-point correlation equations.

We can show that the correlation equations admit all symmetries of the Navier-Stokes equations in a modified format. In global form we find from (61)-(66) and (73)

$$\begin{aligned} \bar{T}_t: \; & t^* = t + a_1 \,, \quad \boldsymbol{x}^* = \boldsymbol{x} \,, \quad \bar{\boldsymbol{u}}^* = \bar{\boldsymbol{u}} \,, \quad \bar{p}^* = \bar{p} \,, \\ & \boldsymbol{r}^* = \boldsymbol{r} \,, \quad R^* = R \,, \quad \ldots \end{aligned} \tag{92}$$

$$\begin{aligned} \bar{T}_{r_1} - \bar{T}_{r_3}: \; & t^* = t \,, \quad \boldsymbol{x}^* = \mathbf{a} \cdot \boldsymbol{x} \,, \quad \bar{\boldsymbol{u}}^* = \mathbf{a} \cdot \bar{\boldsymbol{u}} \,, \quad \bar{p}^* = \bar{p} \,, \\ & \boldsymbol{r}^* = \mathbf{a} \cdot \boldsymbol{r} \,, \quad R^* = \mathbf{a} \cdot R \cdot \mathbf{a} \,, \quad \ldots \end{aligned} \tag{93}$$

$$\begin{aligned} \bar{T}_{\bar{u}_1} - \bar{T}_{\bar{u}_3}: \; & t^* = t \,, \quad \boldsymbol{x}^* = \boldsymbol{x} + \boldsymbol{f}(t) \,, \quad \bar{\boldsymbol{u}}^* = \bar{\boldsymbol{u}} + \frac{\mathrm{d}\boldsymbol{f}}{\mathrm{d}t} \,, \quad \bar{p}^* = \bar{p} - \boldsymbol{x} \cdot \frac{\mathrm{d}^2 \boldsymbol{f}}{\mathrm{d}t^2} \,, \\ & \boldsymbol{r}^* = \boldsymbol{r} \,, \quad R^* = R \,, \quad \ldots \end{aligned} \tag{94}$$

$$\begin{aligned} \bar{T}_{\bar{p}}: \; & t^* = t \,, \quad \boldsymbol{x}^* = \boldsymbol{x} \,, \quad \bar{\boldsymbol{u}}^* = \bar{\boldsymbol{u}} \,, \quad \bar{p}^* = \bar{p} + f_4(t) \,, \\ & \boldsymbol{r}^* = \boldsymbol{r} \,, \quad R^* = R \,, \quad \ldots \end{aligned} \tag{95}$$

$$\begin{aligned} \bar{T}_s: \; & t^* = \mathrm{e}^{2a_2} t \,, \quad \boldsymbol{x}^* = \mathrm{e}^{a_2} \boldsymbol{x} \,, \quad \bar{\boldsymbol{u}}^* = \mathrm{e}^{-a_2} \bar{\boldsymbol{u}} \,, \quad \bar{p}^* = \mathrm{e}^{-2a_2} \bar{p} \,, \\ & \boldsymbol{r}^* = \mathrm{e}^{a_2} \boldsymbol{r} \,, \quad R^* = \mathrm{e}^{-2a_2} R \,, \quad \ldots \end{aligned} \tag{96}$$

while in infinitesimal form we have time invariance

$$\bar{\mathrm{X}}_t = \frac{\partial}{\partial t}, \tag{97}$$

the three rotation groups are

$$\begin{aligned} \bar{\mathrm{X}}_{r_3} = & x_2 \frac{\partial}{\partial x_1} - x_1 \frac{\partial}{\partial x_2} + \bar{u}_2 \frac{\partial}{\partial \bar{u}_1} - \bar{u}_1 \frac{\partial}{\partial \bar{u}_2} + r_2 \frac{\partial}{\partial r_1} - r_1 \frac{\partial}{\partial r_2} \\ & + (R_{21} + R_{12}) \frac{\partial}{\partial R_{11}} + (R_{22} - R_{11}) \frac{\partial}{\partial R_{12}} + R_{23} \frac{\partial}{\partial R_{13}} + (R_{22} - R_{11}) \frac{\partial}{\partial R_{21}} \\ & - (R_{21} + R_{12}) \frac{\partial}{\partial R_{22}} - R_{13} \frac{\partial}{\partial R_{23}} + R_{32} \frac{\partial}{\partial R_{31}} - R_{31} \frac{\partial}{\partial R_{32}} + \ldots \,, \end{aligned} \tag{98}$$

$$\begin{aligned} \bar{\mathrm{X}}_{r_1} = & x_3 \frac{\partial}{\partial x_2} - x_2 \frac{\partial}{\partial x_3} + \bar{u}_3 \frac{\partial}{\partial \bar{u}_2} - \bar{u}_2 \frac{\partial}{\partial \bar{u}_3} + r_3 \frac{\partial}{\partial r_2} - r_2 \frac{\partial}{\partial r_3} \\ & + R_{13} \frac{\partial}{\partial R_{12}} - R_{12} \frac{\partial}{\partial R_{13}} + R_{31} \frac{\partial}{\partial R_{21}} + (R_{32} + R_{23}) \frac{\partial}{\partial R_{22}} + (R_{33} - R_{22}) \frac{\partial}{\partial R_{23}} \\ & - R_{21} \frac{\partial}{\partial R_{31}} + (R_{33} - R_{22}) \frac{\partial}{\partial R_{32}} - (R_{32} + R_{23}) \frac{\partial}{\partial R_{33}} + \ldots \,, \end{aligned} \tag{99}$$

$$\begin{aligned} \bar{\mathrm{X}}_{r_2} = & x_1 \frac{\partial}{\partial x_3} - x_3 \frac{\partial}{\partial x_1} + \bar{u}_1 \frac{\partial}{\partial \bar{u}_3} - \bar{u}_3 \frac{\partial}{\partial \bar{u}_1} + r_1 \frac{\partial}{\partial r_3} - r_3 \frac{\partial}{\partial r_1} \\ & - (R_{31} + R_{13}) \frac{\partial}{\partial R_{11}} - R_{32} \frac{\partial}{\partial R_{12}} - (R_{33} - R_{11}) \frac{\partial}{\partial R_{13}} - R_{23} \frac{\partial}{\partial R_{21}} \\ & + R_{21} \frac{\partial}{\partial R_{23}} - (R_{33} - R_{11}) \frac{\partial}{\partial R_{31}} + R_{12} \frac{\partial}{\partial R_{32}} + (R_{31} + R_{13}) \frac{\partial}{\partial R_{33}} + \ldots \,, \end{aligned} \tag{100}$$

the three generalized Galilean groups are given by

$$\begin{aligned}\bar{X}_{\bar{u}_1} &= f_1(t)\frac{\partial}{\partial x_1} + \frac{\mathrm{d}f_1(t)}{\mathrm{d}t}\frac{\partial}{\partial \bar{u}_1} - x_1\frac{\mathrm{d}^2 f_1(t)}{\mathrm{d}t^2}\frac{\partial}{\partial \bar{p}},\\ \bar{X}_{\bar{u}_2} &= f_2(t)\frac{\partial}{\partial x_2} + \frac{\mathrm{d}f_2(t)}{\mathrm{d}t}\frac{\partial}{\partial \bar{u}_2} - x_2\frac{\mathrm{d}^2 f_2(t)}{\mathrm{d}t^2}\frac{\partial}{\partial \bar{p}},\\ \bar{X}_{\bar{u}_3} &= f_3(t)\frac{\partial}{\partial x_3} + \frac{\mathrm{d}f_3(t)}{\mathrm{d}t}\frac{\partial}{\partial \bar{u}_3} - x_3\frac{\mathrm{d}^2 f_3(t)}{\mathrm{d}t^2}\frac{\partial}{\partial \bar{p}},\end{aligned} \tag{101}$$

while the scaling group for $\nu \neq 0$ yields

$$\bar{X}_s = 2t\frac{\partial}{\partial t} + x_i\frac{\partial}{\partial x_i} - \bar{u}_i\frac{\partial}{\partial \bar{u}_i} - 2\bar{p}\frac{\partial}{\partial \bar{p}} + r_i\frac{\partial}{\partial r_i} - 2R_{ij}\frac{\partial}{\partial R_{ij}} + \dots \,. \tag{102}$$

The pressure invariance $\bar{T}_8$ has been omitted since it is not of any relevance for the results below.

The crucial point for the understanding of large Reynolds number turbulence is that viscosity is only significant for small scale turbulence at the order of the Kolmogorov length-scale, which will be defined later. These eddies have a negligible amount of energy but provide the necessary dissipation for the energy balance equation. In contrast the energy containing large scale eddies determine the mean velocity, the Reynolds stress tensor and similar variables.

It is this distinction and the corresponding difference in symmetries which is the basis for the understanding of the invariant solutions of turbulent flows to be derived in sections 4 and 5. The disentangling of these regions leads to a singular asymptotic expansion of the two-point correlation equations in correlation space $\boldsymbol{r}$ to be derived in the subsequent subsection. The latter expansion is very similar to Prandtl's boundary layer expansion for laminar flow in physical space where the influence of viscosity is limited to a thin region close to the wall scaled by the Reynolds number.

3.2 Small-scale Asymptotic Expansion of the Correlation Equation in Correlation Space

In order to clarify the difference between small- and large-scale properties and hence understand the asymptotic expansion of equation (84) below we may first analyze the two-point correlation function in the light of Kolmogorov's theory of small scale isotropic turbulence (Kolmogorov, 1941b, 1941c). For this purpose we consider the longitudinal normalized two-point correlation function f in r-space (correlation space) taken from the isotropic divergence-free two-point correlation tensor (see von Kármán and Howarth, 1938)

$$R_{ij} = \overline{u'^2}\left[\delta_{ij}\left(f + \frac{r}{2}\frac{\partial f}{\partial r}\right) - \frac{r_i r_j}{r^2}\frac{r}{2}\frac{\partial f}{\partial r}\right] \,. \tag{103}$$

A sketch of $f(r)$ is given in Figure 1. For very small correlation distances $r = |\boldsymbol{r}|$ the isotropic longitudinal two-point function scales as

$$\mathcal{F}(r) = \overline{u'^2} f(r) = \overline{u'^2} - \frac{1}{30}\frac{r^2}{\nu} + O(r^4) \tag{104}$$

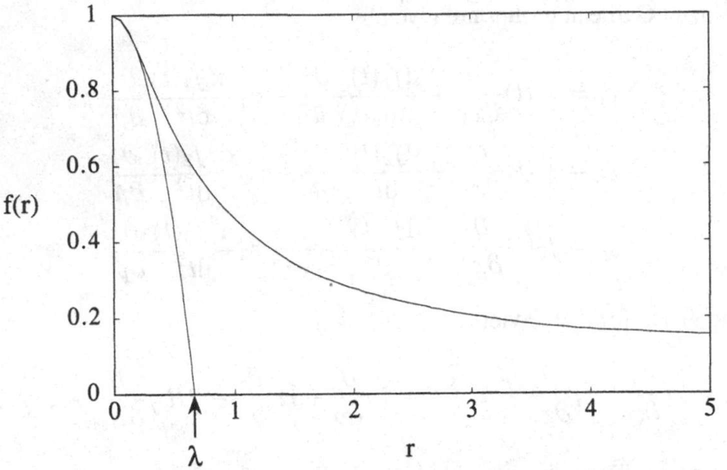

Figure 1: Schematic sketch of the two-point correlation function including the Taylor length-scale λ.

where $\overline{u'^2}$ is the rms-value of the turbulent fluctuations and is the dissipation , given by:

$$= \nu \overline{\frac{\partial u'_i}{\partial x_j} \frac{\partial u'_i}{\partial x_j}} = -\nu \frac{\partial^2 R_{ii}}{\partial r_j \partial r_j}. \tag{105}$$

In the limit $r \to 0$ locally isotropic turbulence is essentially determined by viscous forces. The region where (104) is valid is limited to the order of the Kolmogorov length-scale

$$\eta_K = \left(\frac{\nu^3}{\varepsilon} \right)^{\frac{1}{4}} . \tag{106}$$

The quadratic behavior (104) at $r = 0$ determines the Taylor micro-scale

$$\lambda^{-2} = - \left. \frac{\partial^2 f}{\partial r^2} \right|_{r=0} \tag{107}$$

as is sketched in Figure 1.

According to Kolmogorov's sub-range theory there is a region in correlation space obeying the limits $\eta_K \ll r \ll \ell_t$ where viscosity is negligible and large-scale influences are also asymptotically small. ℓ_t is the integral length-scale

$$\ell_t(t) = \int_0^\infty f(r,t) \mathrm{d}r \ . \tag{108}$$

In this domain which is independent of viscosity Kolmogorov (1941c) proposed the inertial sub-range

$$\mathcal{F}(r) = \overline{u'^2} f(r) = \overline{u'^2} - \frac{C}{2} (\varepsilon r)^{\frac{2}{3}} , \tag{109}$$

where C is a universal constant.

In Oberlack and Peters (1993) an asymptotic equation for the correlation function $\mathcal{F}$ was derived such that the viscous regime (104) as well as the inertial sub-range domain (109) are included in the resulting asymptotic equation. The following boundary layer type of expansion for small r based on the turbulent Reynolds number

$$Re_t = \frac{\overline{u'^2}\ell_t}{\nu} \tag{110}$$

was introduced for the isotropic correlation function

$$\mathcal{F}(r) = \overline{u'^2}[1 - Re_t^{-\frac{1}{2}} f^{(1)}(\hat{r}) - O(Re_t^{-\frac{3}{4}})] \tag{111}$$

where

$$\hat{r} = Re_t^{\frac{3}{4}} r \ . \tag{112}$$

is a scaled correlation distance covering the region of the Kolmogorov length-scale up to the inertial sub-range.

It turned out from the resulting asymptotic equation in the limit $\hat{r} \to \infty$ that the correlation function varies as (109) for $\hat{r} \to \infty$ while for small scales, i.e. $r = O(\eta_K)$, it has functional behavior like (104).

The expansion (111)/(112) is in fact the inner region of a boundary layer type asymptotic in r-space while the outer part is obtained in the limit of $Re_t \to \infty$, i.e. viscosity is essentially set to zero for the correlation equations for isotropic turbulence. Due to the matching process of the inner region in the limit $\hat{r} \to \infty$ and the outer region $r \to 0$ the dissipation ε was obtained in Oberlack (2000a). From a physical point of view the matching process may be interpreted as such that the dissipation ε is a small scale quantity but is driven by the large scales.

The matched asymptotic expansion for locally isotropic turbulence in correlation space in Oberlack and Peters (1993) may readily be extended to inhomogeneous flows such as wall-bounded shear flows. The outer part of the asymptotic expansion in r-space, i.e. the domain $\eta_K \ll r$, is obtained by taking the limit $\nu \to 0$ respectively $Re_t \to \infty$ in the equations (84) yielding

$$\begin{aligned}
&\frac{\bar{\mathrm{D}}R_{ij}}{\bar{\mathrm{D}}t} + R_{kj}\frac{\partial \bar{u}_i(\boldsymbol{x},t)}{\partial x_k} + R_{ik}\left.\frac{\partial \bar{u}_j(\boldsymbol{x},t)}{\partial x_k}\right|_{\boldsymbol{x}+\boldsymbol{r}} + \left[\bar{u}_k\left(\boldsymbol{x}+\boldsymbol{r},t\right) - \bar{u}_k\left(\boldsymbol{x},t\right)\right]\frac{\partial R_{ij}}{\partial r_k} \\
&+\frac{\partial \overline{p'u'}_j}{\partial x_i} - \frac{\partial \overline{p'u'}_j}{\partial r_i} + \frac{\partial \overline{u'_i p'}}{\partial r_j} + \frac{\partial R_{(ik)j}}{\partial x_k} - \frac{\partial}{\partial r_k}\left[R_{(ik)j} - R_{i(jk)}\right] \\
&+2\Omega_k\left[e_{kli}R_{lj} + e_{klj}R_{il}\right] = 0.
\end{aligned} \tag{113}$$

It is apparent that the latter equation is not valid in the limit $r \to 0$ since no dissipation is contained which becomes important for $r = O(\eta_K)$.

From a symmetry point of view the key difference between the original two-point correlation equation (84) and the large-scale equation (113) is, that the latter equation admits two scaling symmetry groups given by

$$\begin{aligned}\bar{T}_{s_1}: \ t^* &= t \quad , \ \boldsymbol{x}^* = \mathrm{e}^{a_2}\boldsymbol{x} \quad , \ \bar{\boldsymbol{u}}^* = \mathrm{e}^{a_2}\bar{\boldsymbol{u}} \ , \ \bar{p}^* = \mathrm{e}^{2a_2}\bar{p}, \\ \boldsymbol{r}^* &= \mathrm{e}^{a_2}\boldsymbol{r}, \ R^* = \mathrm{e}^{2a_2}R \ , \ldots \ ,\end{aligned} \tag{114}$$

$$\begin{aligned}\bar{T}_{s_2}: \ t^* &= \mathrm{e}^{a_3}t \ , \ \boldsymbol{x}^* = \boldsymbol{x} \quad , \ \bar{\boldsymbol{u}}^* = \mathrm{e}^{-a_3}\bar{\boldsymbol{u}}, \ \bar{p}^* = \mathrm{e}^{-2a_3}\bar{p}, \\ \boldsymbol{r}^* &= \boldsymbol{r} \quad , \ R^* = \mathrm{e}^{-2a_3}R, \ \ldots \ .\end{aligned} \tag{115}$$

or in form of generators

$$\bar{\mathrm{X}}_{s_1} = x_i\frac{\partial}{\partial x_i} + \bar{u}_i\frac{\partial}{\partial \bar{u}_i} + 2\bar{p}\frac{\partial}{\partial \bar{p}} + r_i\frac{\partial}{\partial r_i} + 2R_{ij}\frac{\partial}{\partial R_{ij}} + \ldots \ , \tag{116}$$

$$\bar{\mathrm{X}}_{s_2} = t\frac{\partial}{\partial t} - \bar{u}_i\frac{\partial}{\partial \bar{u}_i} - 2\bar{p}\frac{\partial}{\partial \bar{p}} - 2R_{ij}\frac{\partial}{\partial R_{ij}} + \ldots . \tag{117}$$

Apart from this distinction equation (113) recognizes all symmetry groups of equation (84), i.e. (92)-(95). It should be noted that system rotation ($\boldsymbol{\Omega} \neq 0$) breaks scaling of time and hence $a_3 = 0$. This reduced number of symmetries is important for some invariant solutions to be shown below.

The inner part of the asymptotic expansion of the correlation function corresponding to (111) may be obtained by introducing

$$\begin{aligned}R_{ij}(\boldsymbol{x},\boldsymbol{r}) &= \overline{u_i'u_j'}(\boldsymbol{x}) - Re_t^{-\frac{1}{2}}R_{ij}^{(1)}(\boldsymbol{x},\hat{\boldsymbol{r}}) - O(Re_t^{-\frac{3}{4}}), \\ R_{(ik)j}(\boldsymbol{x},\boldsymbol{r}) &= \overline{u_i'u_j'u_k'}(\boldsymbol{x}) + Re_t^{-\frac{3}{4}}R_{(ik)j}^{(1)}(\boldsymbol{x},\hat{\boldsymbol{r}}) - O(Re_t^{-1}), \\ R_{i(jk)}(\boldsymbol{x},\boldsymbol{r}) &= \overline{u_i'u_j'u_k'}(\boldsymbol{x}) + Re_t^{-\frac{3}{4}}R_{i(jk)}^{(1)}(\boldsymbol{x},\hat{\boldsymbol{r}}) - O(Re_t^{-1}) \quad \text{with} \quad \hat{\boldsymbol{r}} = Re_t^{\frac{3}{4}}\boldsymbol{r}\end{aligned} \tag{118}$$

into (84) leading to

$$\begin{aligned}&\frac{\bar{\mathrm{D}}\overline{u_i'u_j'}}{\bar{\mathrm{D}}t} + \overline{u_j'u_k'}\frac{\partial \bar{u}_i(\boldsymbol{x},t)}{\partial x_k} + \overline{u_i'u_k'}\frac{\partial \bar{u}_j(\boldsymbol{x},t)}{\partial x_k} - \frac{\partial \bar{u}_k(\boldsymbol{x},t)}{\partial x_l}\hat{r}_l\frac{\partial R_{ij}^{(1)}}{\partial \hat{r}_k} \\ &+ \left[\frac{\partial \overline{p'u'}_j}{\partial x_i} - \frac{\partial \overline{p'u'}_j}{\partial r_i} + \frac{\partial \overline{u_i'p'}}{\partial r_j}\right]\Bigg|_{r=0} + 2\,\frac{\partial^2 R_{ij}^{(1)}}{\partial \hat{r}_k\partial \hat{r}_k} \\ &+ \frac{\partial \overline{u_i'u_j'u_k'}}{\partial x_k} - \frac{\partial}{\partial \hat{r}_k}\left[R_{(ik)j}^{(1)} - R_{i(jk)}^{(1)}\right] + 2\,\Omega_k\left[e_{kli}\overline{u_j'u_l'} + e_{klj}\overline{u_i'u_l'}\right] = 0.\end{aligned} \tag{119}$$

The pressure-velocity correlations $\overline{p'u'}_j$ and $\overline{u_i'p'}$ are determined by the Poisson equation and hence are not independent of the velocity correlations.

Comparing equations (113) and (119) with Prandtl's boundary layer theory for laminar flows we find that (113) corresponds to the inviscid outer flow while (119) is the analog of the boundary layer equation. In complete analogy to boundary layer theory where the pressure-gradient in stream-wise direction is determined by the outer inviscid flow we may compute $R_{ij}^{(1)}(\boldsymbol{x},\hat{\boldsymbol{r}})$ in (119). Therein the quantities $\overline{u_i'u_j'}$, $\overline{u_i'u_j'u_k'}$, $\overline{p'u'}_j$ and $\overline{u_i'p'}$ are determined by the outer equations

(113) by invoking the appropriate limit $r = 0$. The only term that has no counterpart in equation (113) is the last term in the second line of equation (119) which denotes dissipation.

Though there is a strong similarity between laminar and turbulent boundary layers from an asymptotic point of view we have to note that there are two fundamental differences for the solution of the equations. First, the equations for the laminar flow are a closed set with a finite number of variables. In contrast both equation (113) and (119) are in principle an infinite set of equations. This immediately limits our ability to construct solutions for these equations both with analytic and numerical tools.

Second and most important, it is the set of equations for the inner solution, namely the boundary layer equations, which determine the near-wall behavior of laminar flows. The outer solution is fixed by the Euler equations for inviscid flows. This is in clear contrast to the turbulent flows where the mean velocity of wall-bounded flows is largely determined by inviscid equations. This is demonstrated in the next subsection.

Finally we have to note that within the present analysis the very near-wall region , the viscous sub-layer , is not included. This region possesses a linear mean-velocity profile (see e.g. Rotta, 1972), which cannot be bridged to the solutions to be derived in the next subsection.

4 Plane and Round Turbulent Shear Flows

The first application to which the ideas of symmetry groups will be applied is that of wall-bounded shear flows. This is in fact also the classical case where turbulent scaling laws originated.

4.1 Plane Parallel Shear Flows

Within this subsection we consider flows where all statistical one-point quantities depend only on the wall-normal coordinate and in particular the mean velocity obeys

$$\bar{u}_1 = \bar{u}_1(x_2). \tag{120}$$

Under this assumption the two-point correlation equation (113) for the large-scale quantities admits the following four symmetries consisting of two scaling symmetries

$$\bar{\mathrm{X}}_{s_1} = x_2\frac{\partial}{\partial x_2} + \bar{u}_1\frac{\partial}{\partial \bar{u}_1} + r_i\frac{\partial}{\partial r_i} + 2R_{ij}\frac{\partial}{\partial R_{ij}} + \dots \tag{121}$$

and

$$\bar{\mathrm{X}}_{s_2} = -\bar{u}_1\frac{\partial}{\partial \bar{u}_1} - 2R_{ij}\frac{\partial}{\partial R_{ij}} + \dots, \tag{122}$$

Galilean invariance in x_1-direction

$$\bar{\mathrm{X}}_{\bar{u}_1} = \frac{\partial}{\partial \bar{u}_1} \tag{123}$$

as well as translation invariance in x_2-direction

$$\bar{\mathrm{X}}_{x_2} = \frac{\partial}{\partial x_2}\ . \tag{124}$$

At this point Galilean invariance (123) appears in a somewhat unusual form since the time and spatial coordinate do not appear explicitly. The classical form of Galilean invariance will appear again in later sections where unsteady flows will be discussed.

There exist four symmetries for the present flow which may be used to derive invariant solutions. Each of the symmetries (121)-(124) are independent invariant transformations of the two-point correlation equations for plane shear flows .

It has been pointed out in sub-section 1.2 in equation (32) that generators admit the superposition principle. Hence we also invoke this theorem for the above four symmetries to obtain

$$\bar{\mathrm{X}} = k_{s_1}\bar{\mathrm{X}}_{s_1} + k_{s_2}\bar{\mathrm{X}}_{s_2} + k_{\bar{u}_1}\bar{\mathrm{X}}_{\bar{u}_1} + k_{x_2}\bar{\mathrm{X}}_{x_2} \tag{125}$$

where k_i are constants.

It should be noted that rotational symmetries may not be implemented owing to the restriction to a plane shear flow .

At the very beginning it was mentioned that due to system rotation , i.e.

$$\boldsymbol{\Omega} \neq 0 \ , \tag{126}$$

we have symmetry breaking of the scaling symmetry for time (122). Since we have restricted the present flow to be steady the time coordinate does not appear explicitly. Still, if we compare (122) with the full form of the scaling symmetry (117), we recognize (122) as being the scaling symmetry of time. An analysis of the rotating shear flow appears at the very end of this section.

In full analogy to the example of the heat equation with the invariant surface condition (47) in section 1 we may derive the invariant solution from condition (52). In the present case of plane shear flows we have:

$$\frac{\mathrm{d}x_2}{k_{s_1}x_2 + k_{x_2}} = \frac{\mathrm{d}r_{[k]}}{k_{s_1}r_{[k]}} = \frac{\mathrm{d}\bar{u}_1}{(k_{s_1} - k_{s_2})\bar{u}_1 + k_{\bar{u}_1}} = \frac{\mathrm{d}R_{[ij]}}{2(k_{s_1} - k_{s_2})R_{[ij]}} = \dots \ , \tag{127}$$

where the square brackets refer to a suppression of the summation convention.

For the different combination of parameter k_{s_1} and k_{s_2} we will subsequently derive a variety of different flows. At this point it is important to note that $\bar{u}_1$ solely depends on x_2 and has no dependence on $\boldsymbol{r}$.

The importance of the two scaling symmetries (121) and (122) may best be interpreted from its combination in global form

$$x_2^* = \mathrm{e}^{k_{s_1}} x_2 \quad \text{and} \quad \bar{u}_1^* = \mathrm{e}^{k_{s_1} - k_{s_2}} \bar{u}_1. \tag{128}$$

As mentioned above k_{s_2} is the group parameter representing scaling of time.

4.1.1 Logarithmic-law-of-the-wall

The first case to be investigated is that of the classical logarithmic-law-of-the-wall. von Kármán's key assumption was, that close to the wall, just beyond the viscous sub-layer, the friction velocity u_τ is the only flow determining parameter. The friction velocity is defined by the integrated form of the leading order of the momentum equation (78) in stream-wise direction according to

$$\nu \frac{\partial \bar{u}_1}{\partial x_2} - \overline{u_1' u_2'} = \frac{\tau_w}{\rho} = u_\tau^2 \ . \tag{129}$$

At this point it should be pointed out, that the density appearing in the Navier-Stokes equations has been absorbed into the pressure. The wall-friction τ_w given above however is a "real" friction according to its dimension.

u_τ may be considered a given external parameter, such as a boundary condition. Hence we have a symmetry breaking due to the friction velocity such that the arbitrary scaling factor of $\bar{u}_1$ in (128) vanishes. As a result we obtain

$$k_{s_1} = k_{s_2} . \tag{130}$$

With this key restriction in mind we obtain for the mean velocity from (127)

$$\bar{u}_1 = \frac{k_{\bar{u}_1}}{k_{s_1}} \ln\left(x_2 + \frac{k_{x_2}}{k_{s_1}}\right) + C_{log} \tag{131}$$

and the corresponding reduced variables read

$$\tilde{r}_k = \frac{r_k}{x_2 + \frac{k_{x2}}{k_{s_1}}} \quad , \quad R_{ij} = \tilde{R}_{ij} \; , \quad \ldots \tag{132}$$

In this context we should mention that C_{log} in (131) and those variables denoted by "~" in (132) are integration constants. These constants are, as has been explained in Chapter 1.3, the new set of reduced variables (similarity variables) which only depend on $\tilde{r}$. The only exception is C_{log}, since $\bar{u}_1$ according to (120) only depends on x_2. As a result, C_{log} may not depend on $\tilde{r}$ defined in (132). The only possibility is that C_{log} is a constant.

(131) is a somewhat modified form of the classical logarithmic-law-of-the-wall since due to the term $\frac{k_{x_2}}{k_{s_1}}$ a shifting of the origin is possible. In modification of the classical dimensionless form of the logarithmic-law-of-the-wall we find

$$u^+ = \frac{1}{\kappa} \ln(x_2^+ + A^+) + C \; , \tag{133}$$

where the quantities denoted by "+" are non-dimensionalized by u_τ and ν/u_τ and A^+ has its origin in $\frac{k_{x_2}}{k_{s_1}}$. The reduced variables (132) are given by

$$\tilde{r}_k = \frac{r_k^+}{x_2^+ + A^+} \quad , \quad R_{ij}^+ = \tilde{R}_{ij}^+ \; , \quad \ldots \tag{134}$$

Implementing (133) and (134) into the two-point correlation equation (84) by invoking the limit of plane shear flow we obtain

$$\begin{aligned} 0 = &- \frac{1}{\kappa}\left[\delta_{i1} R_{2j}^+ + \delta_{1j} R_{i2}^+ \frac{1}{1+\tilde{r}_2} + \ln(1+\tilde{r}_2)\frac{\partial R_{ij}^+}{\partial \tilde{r}_1}\right] \\ &+ \left[\delta_{i2}\tilde{r}_k \frac{\partial \overline{p' u'_j}^+}{\partial \tilde{r}_k} + \frac{\partial \overline{p' u'_j}^+}{\partial \tilde{r}_i} - \frac{\partial \overline{u'_i p'}^+}{\partial \tilde{r}_j}\right] \\ &+ \left[\delta_{i2}\tilde{r}_k \frac{\partial R^+_{(i2)j}}{\partial \tilde{r}_k} + \frac{\partial R^+_{(ik)j}}{\partial \tilde{r}_k} - \frac{\partial R^+_{i(jk)}}{\partial \tilde{r}_k}\right] , \end{aligned} \tag{135}$$

where also the correlation tensors have been non-dimensionalized by u_τ.

An experimental validation of the logarithmic-law-of-the-wall may not be given here, since it has been confirmed by a very large number of publications in the literature. In the context of the turbulent pipe flow in subsection 4.2 we will re-consider the logarithmic-law-of-the-wall again and it will be verified there in Figure 12 on page 337.

For all of the subsequent laws for the mean velocity we can do an analogous reduction of the two-point correlation equations. Overall we will find five different cases including the latter logarithmic-law-of-the-wall depending on different combinations of the two scaling groups k_{s_1} and k_{s_2}. In a fully analogous way we will obtain from (127) the new set of variables and may use those to obtain the reduction of the two-point correlation equation. Due to the limitation of space we will only give the different reduced variables stemming from (127); however we will not give the reduced correlation equations for all cases.

4.1.2 Algebraic scaling law

The maximum number of combined symmetries is admitted if we integrate (127) under the assumption that all group parameters in (125) are non-zero and particularly $k_{s_1} \neq k_{s_2}$. As a result we obtain the algebraic law

$$\bar{u}_1 = C_{alg}\left(x_2 + \frac{k_{x2}}{k_{s_1}}\right)^{1-\frac{k_{s2}}{k_{s1}}} + \frac{k_{\bar{u}_1}}{k_{s_2} - k_{s_1}} \quad . \tag{136}$$

The corresponding set of reduced variables is:

$$\tilde{r}_k = \frac{r_k}{x_2 + \frac{k_{x2}}{k_{s_1}}} \quad , \quad R_{ij} = \left(x_2 + \frac{k_{x2}}{k_{s_1}}\right)^{2\left(1-\frac{k_{s2}}{k_{s1}}\right)} \tilde{R}_{ij} \ , \ldots \ . \tag{137}$$

Though the form is also algebraic, there are major differences from the laws derived in Barenblatt (1993), George et al. (1996) and (136). In both references it is stated that the algebraic law is applicable to the near-wall region.

Since the present case is identified by a maximum of symmetries, it may be localized in regions where any symmetry breaking influence, such as a solid wall, is negligible. Hence we propose the algebraic law to be located in the center of a pressure driven turbulent channel flow .

Since the flow configuration between the two parallel walls admits a reflection symmetry with respect to the center-line (see equation 74) we find that the term $\frac{k_{\bar{u}_1}}{k_{s_2} - k_{s_1}}$ in (136) may only represent the maximum velocity $\bar{u}_{max}$ on the center line. In normalized and dimensionless form we obtain the classical law-of-the-wake in the modified form

$$\frac{\bar{u}_{max} - \bar{u}_1}{u_\tau} = a_{alg}\left(\frac{y}{h}\right)^{\beta_{alg}} \quad . \tag{138}$$

The variables h, a_{alg} and β_{alg} respectively denote the channel width and two parameters which in the limit $Re \rightarrow \infty$ become constants. Figures 2 and 3 show experimental and DNS data for turbulent channel flow. In order to see the algebraic scaling law (138) the data are plotted on a double logarithmic scale, such that a linear region corresponds to (138). In all data sets it is

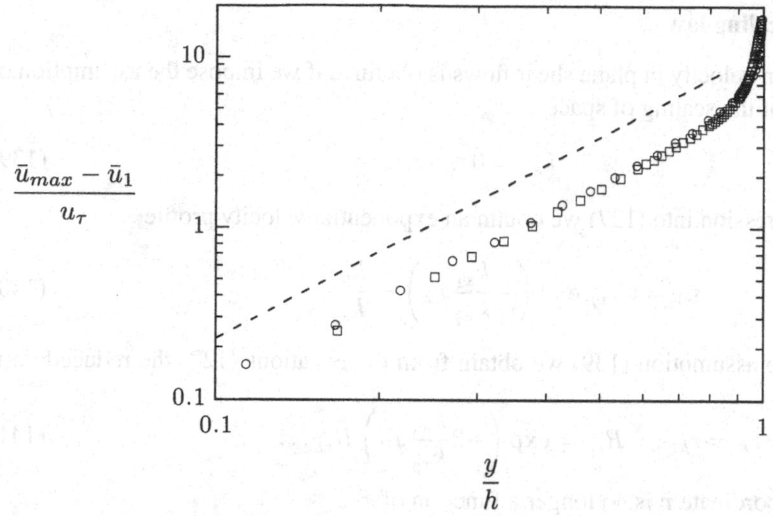

Figure 2: Mean velocity of the experimental data of a turbulent channel flow: ○, $Re_b = 18000$, Niederschulte (1996); □, $Re_b = 12200$, Fischer et al. (1999); $- - -$, parallel line to the algebraic function (138).

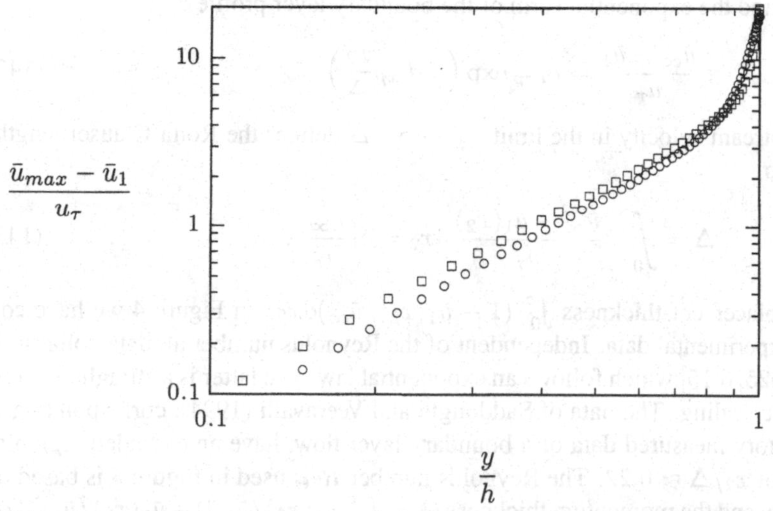

Figure 3: Mean velocity of the DNS data of a turbulent channel flow from Kim et al. (1987): □, $Re_b = 7900$; ○, $Re_b = 3300$.

clearly visible that in the center of the channel in the region $y/h \in [0.1, 0.8]$ we find an algebraic law. Fitting the parameter in (138) to the data in Figure 2 we find $a_{alg} \approx 5.8$ and $\beta_{alg} \approx 1.69$. The Reynolds number in Figures 2 and 3 is defined using the bulk velocity and the channel width.

4.1.3 Exponential scaling law

A new law for the mean velocity in plane shear flows is obtained if we impose the assumption of a symmetry breaking of the scaling of space

$$k_{s_1} = 0 \ . \tag{139}$$

Implementing this expression into (127) we obtain an exponential velocity profile

$$\bar{u}_1 = C_{exp} \exp\left(-\frac{k_{s_2}}{k_{x_2}} x_2\right) + \frac{k_{\bar{u}_1}}{k_{s_2}} \ . \tag{140}$$

Accordingly, under the assumption (139) we obtain from the equations (127) the reduced variables

$$\tilde{r}_k = r_k \ , \quad R_{ij} = \exp\left(-2\frac{k_{s_2}}{k_{x_2}} x_2\right) \tilde{R}_{ij} \quad \dots \ . \tag{141}$$

Apparently, the new coordinate $\tilde{\boldsymbol{r}}$ is no longer a function of x_2.

For positive values of $\frac{k_{s_2}}{k_{x_2}}$ the velocity law (140) converges for $x_2 \to \infty$ to a constant velocity. It is obvious that this may only be applicable to an infinite or semi-infinite domain such as a boundary-layer type of flow. The implicit assumption (139) imposes a symmetry breaking on (128) such that the length-scale may not be scaled. In a boundary layer flow the symmetry breaking length-scale may only be the boundary layer thickness itself. In normalized and non-dimensional form we find the exponential form of the boundary layer profile

$$\frac{\bar{u}_\infty - \bar{u}_1}{u_\tau} = a_{exp} \exp\left(-\beta_{exp} \frac{x_2}{\Delta}\right) \ , \tag{142}$$

where $\bar{u}_\infty$ is the free stream velocity in the limit $x_2 \to \infty$. Δ defines the Rotta-Clauser length-scale, corresponding to

$$\Delta = \int_0^\infty \frac{\bar{u}_\infty - \bar{u}_1(x_2)}{u_\tau} \mathrm{d}x_2 = \delta_1 \frac{\bar{u}_\infty}{u_\tau} \ , \tag{143}$$

in which δ_1 is the displacement-thickness $\int_0^\infty (1 - \bar{u}_1(x_2)/\bar{u}_\infty)\mathrm{d}x_2$. In Figure 4 we have collected a diversity of experimental data. Independent of the Reynolds number all data collapse in the range $x_2/\Delta \in [0.025, 0.15]$ which follows an exponential law. The latter is a straight line due to the semi-logarithmic scaling. The data of Saddoughi and Veeravalli (1994), corresponding to the largest in a laboratory measured data of a boundary layer flow, have an extended region of validity up to a value of $x_2/\Delta \approx 0.22$. The Reynolds number Re_Θ used in Figure 4 is based on the free-stream velocity and the momentum thickness $\Theta = \int_0^\infty \bar{u}_1(x_2)/\bar{u}_\infty[1 - \bar{u}_1(x_2)/\bar{u}_\infty]\mathrm{d}x_2$.

The extension of the region of validity with increasing Reynolds number , as is known for the logarithmic-low-of-the-wall, is also visible for the exponential scaling law in Figure 4. A simple explanation for this effect is obtained if the reduced set of variables (141) and the exponential law (142) is implemented into the full two-point correlation equations (84) also containing viscosity. Cancelling all factors in the inviscid terms the following x_2-dependent factor $\frac{\nu}{\Delta u_\tau} \exp\left(\beta_{exp} \frac{x_2}{\Delta}\right)$ remains in front of the viscous term.

At this point it becomes apparent that the previous expression may only be considered a small parameter if the following holds true: $Re_{\Delta u_\tau} = \frac{\Delta}{\nu} \gg \exp\left(\beta_{exp} \frac{x_2}{\Delta}\right)$ where $\beta_{exp} > 0$.

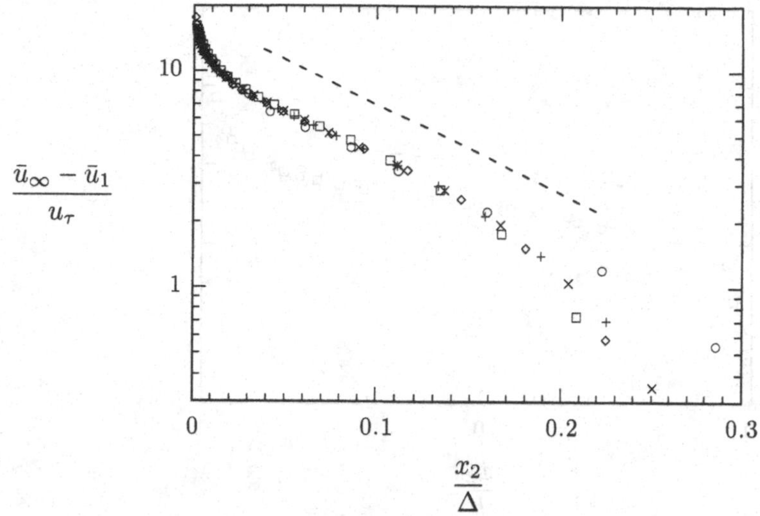

Figure 4: Mean velocity of the experimental data of a turbulent boundary layer flow: ○, $Re_\theta = 370000$, Saddoughi and Veeravalli (1994); □ and ◇, $Re_\theta = 60000$, Fernholz et al. (1995); +, $Re_\theta = 15000$ and ×, $Re_\theta = 20000$, DeGraaff et al. (1999); − − −, line parallel to the exponential function (142).

With increasing Reynolds number the right hand side may also increase while the inequality is still valid. The right hand side may also increase due to a larger value of x_2. Hence we may conclude that the outer part of the turbulent boundary layer is considerably influenced by the Reynolds number.

It should be noted that the logarithmic-law-of-the-wall in Figure 4 is valid up to a value of $x_2/\Delta = 0.025$. Hence the range of the exponential law is about a factor of six to eight times larger than the logarithmic region.

A simple relation between a_{exp} and β_{exp} may be derived from the simplified assumption that the exponential velocity profile is valid in the entire semi-infinite domain $x_2 \in [0, \infty]$, i.e. also the region down to the solid wall. Employing (142) in the definition of the Rotta-Clauser length-scale in (143) we find the relation

$$a_{exp} = \beta_{exp} \,. \tag{144}$$

A fit of the data in Figure 4 to the profile (142) in the region $x_2/\Delta \in [0.025, 0.15]$ leads to the values $a_{exp} \approx 10.34$ and $\beta_{exp} \approx 9.46$, which is a good approximation to the previous relation.

4.1.4 Linear scaling law

A classical plane shear flow is given by the symmetry breaking assumptions

$$k_{s_1} = 0 \quad \text{und} \quad k_{s_2} = 0 \,. \tag{145}$$

For this case we obtain from (127) the mean velocity

$$\bar{u}_1 = \frac{k_{\bar{u}_1}}{k_{x_2}} x_2 + C_{lin} \tag{146}$$

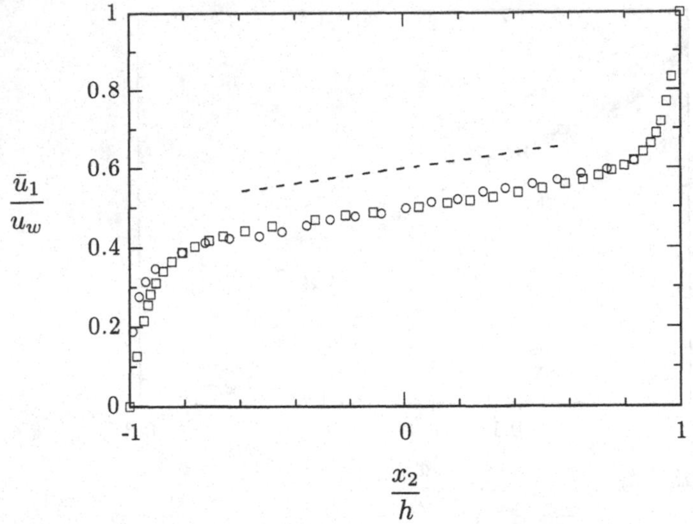

Figure 5: Mean velocity of a turbulent Couette flow: ∘, experimental data from El Telbany and Reynolds (1980); □, DNS data from Lee and Kim (1991); − − −, parallel line to the linear function (148).

and the corresponding reduced set of variables

$$\tilde{r}_k = r_k \ , \quad R_{ij} = \tilde{R}_{ij} \ , \dots \ , \tag{147}$$

which, as for the exponential law, does not depend on x_2.

A plane turbulent shear flow, which is influenced both by an external velocity and length-scale is the turbulent Couette flow . According to the assumption that a turbulent flow has the tendency to maximize the number of symmetries we find this case in the core of the flow.

In normalized form the linear mean velocity profile of the turbulent Couette flow reads

$$\bar{u}_1 = u_w \left[a_{lin} \frac{x_2}{h} + \beta_{lin} \right] \ , \tag{148}$$

where h and u_w are the channel width and the velocity of the pulled wall respectively.

In Figure 5 the linear velocity profile (148) is verified by experimental and DNS data.

4.1.5 Plane shear flow in a rotating frame

In contrast to all previously discussed plane shear flows the last case to be discussed is in a rotating and hence a non-inertial frame of reference. The flow may rotate about one of the three coordinate axes ($\mathbf{\Omega} \neq 0$). The expression $1/|\mathbf{\Omega}|$ denotes a time scale which acts as a symmetry breaking quantity on the flow. Though the time coordinate t does not explicitly appear in the two-point correlation for plane shear flows, we find from the two scaling groups in (128) that k_{s_2} refers to the scaling of time.

It has been pointed out above that the turbulent channel flow possesses a maximum of symmetries in the core region of the pressure driven channel flow i.e. no scaling symmetries are broken. For the present case the rotation about one coordinate axis, namely the time scale $1/|\boldsymbol{\Omega}|$, is symmetry breaking in the core region which leads to a new velocity profile.

In the following we consider two different rotation axes. The classical case is a pressure driven rotating channel flow rotating about the span-wise direction. In the literature experiments of this case are referenced by Johnston et al. (1972) while a corresponding DNS has been conducted by Kristoffersen and Andersson (1993). The axis of rotation is parallel to the mean vorticity vector.

The DNS of the second test case of a rotating channel flow rotating about its stream-wise direction was first published in Oberlack et al. (1998). In fact, it was the Lie group approach which gave rise to the idea to conduct the DNS.

$\boldsymbol{\Omega}$ is a new parameter in the two-point correlation equation. In extension of the symmetry analysis used for the last four cases we now introduce $\boldsymbol{\Omega}$ as a new independent parameter into the symmetry analysis. The purpose of this step is to obtain a new set of reduced variables such that $\boldsymbol{\Omega}$ disappears from the two-point correlation equation. In the large Reynolds number limit we find the following set of variables

$$x_2 = \acute{x}_2\gamma(\Omega),\ r_i = \acute{r}_i\gamma(\Omega)\ ,\quad \bar{u}_i = \acute{u}_i\gamma(\Omega)\Omega\ , R_{ij} = \acute{R}_{ij}\gamma(\Omega)^2\Omega^2\ ,\ldots\ . \tag{149}$$

Due to the introduction of the "´" variables $\boldsymbol{\Omega}$ does not longer appear in two-point correlation equations as if $\boldsymbol{\Omega}$ were set to 1. Ω denotes any one of the rotation rates Ω_1 or Ω_3 depending on the flow case. $\gamma(\Omega)$ refers to a free and so far unknown function only depending on Ω.

An analysis of the resulting two-point correlation equations written in the "´" variables leads to the symmetry breaking

$$k_{s_2} = 0\ , \tag{150}$$

induced by the external time scale $1/|\boldsymbol{\Omega}|$. The differential equations (127) for the reduced variables admit an identical form for the "´" variables. The resulting mean velocity is derived from (127) under consideration of the symmetry breaking (150)

$$\acute{u}_1 = C_{rot}\left[\acute{x}_2 + \frac{k_{x_2}}{k_{s_1}}\right] - \frac{k_{\bar{u}_1}}{k_{s_1}}\ , \tag{151}$$

and the reduced variables are

$$\tilde{r}_k = \frac{\acute{r}_k}{\acute{x}_2 + \frac{k_{x_2}}{k_{s_1}}}\ ,\quad \acute{R}_{ij} = \tilde{R}_{ij}\left(\acute{x}_2 + \frac{k_{x_2}}{k_{s_1}}\right)^2\ ,\ldots\ . \tag{152}$$

Converting (151) and (152) by employing (149) into the original variables we obtain

$$\bar{u}_1 = C_{rot}x_2\Omega + \gamma(\Omega)\left[\frac{k_{x_2}}{k_{s_1}} - \frac{k_{\bar{u}_1}}{k_{s_1}}\right]\ . \tag{153}$$

The functional form of $\gamma(\Omega)$ may not be derived by the symmetry analysis. However, due to the physical reason of dimensional homogeneity, γ may in general not depend on Ω. To finally bring (153) into a form which only contains parameters of the rotating channel flow we have the friction velocity u_τ, the bulk velocity $\bar{u}_m$ and the velocity at the center line $\bar{u}_{cl}$ at our disposal.

From an analysis of the two-point correlation equation for the case of a rotating channel flow rotating about the span-wise direction x_3 we find that there is no reflection symmetry about the

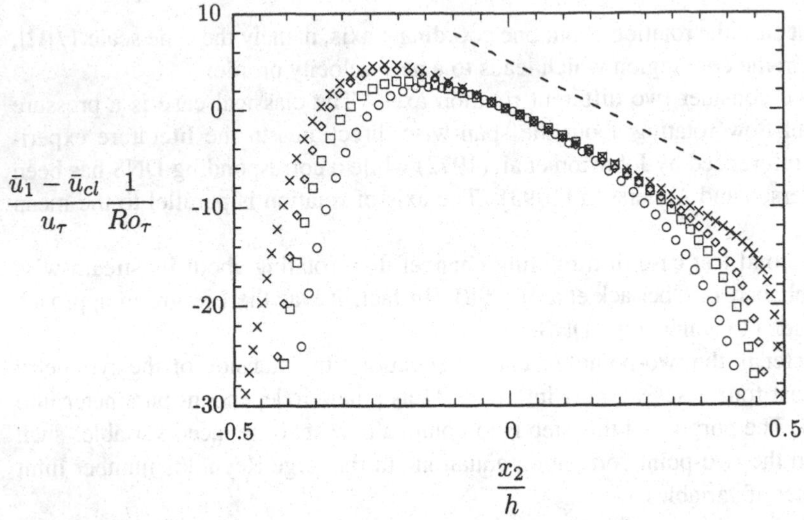

Figure 6: Mean velocity of a turbulent channel flow rotating about the x_3 direction. DNS data from Kristoffersen and Andersson (1993): ○, $Ro_m = 0.1$; □, $Ro_m = 0.15$; ◇, $Ro_m = 0.2$; ×, $Ro_m = 0.5$; — — —, parallel line to the linear function (154).

center line. As a result an asymmetry of both the mean velocity as well as of the statistical quantities is observed. The maximum velocity of the channel flow in an inertial system loses its meaning and will for the present case be replaced by $\bar{u}_{cl}$.

With this consideration in mind we derive from (153) the new scaling law

$$\bar{u}_1 = a_{rot}\, \Omega\, x_2 + \bar{u}_{cl} \tag{154}$$

where the origin of the coordinate system for x_2 lies in the center of the channel. From the DNS data of Kristoffersen and Andersson (1993) we may approximate the numerical value of a_{rot} to 2. In order to validate (154) we use the data of Kristoffersen and Andersson (1993) in Figure 6 in normalized form. For this reason we define the two rotation numbers:

$$Ro_m = \frac{\Omega h}{\bar{u}_m} \quad \text{and} \quad Ro_\tau = \frac{\Omega h}{u_\tau} \ . \tag{155}$$

In the core of the channel flow, namely in the region where we expect the maximum number of symmetries, all DNS data nicely collapse to a single line. With increasing rotation number the linear region increases.

A modification of the classical rotating channel flow is that with rotation about the streamwise direction, i.e. about the x_1 axis. This configuration has first been suggested and investigated in Oberlack et al. (1998) by DNS and different turbulence models.

Three major differences between the two cases of rotating channel flow are conspicuous and will subsequently be analyzed:

- different reflection symmetries with respect to the center line.
- in the new case a cross-stream velocity is induced.

- in the new case all six Reynolds stresses are non-zero.

Analyzing each single equation of the two-point correlation equations we find that both the "13" as well as the "23" equations contain Coriolis terms. These terms may not be set to zero since they depend on the non-zero terms R_{12}, R_{22} and R_{33}. From the structure of the equation we may readily see that as a result also the corresponding elements R_{13} and R_{23} are non-zero. Since both elements R_{22} and R_{33} are non-zero we have also the Coriolis term in "23" which to leading order may only be balanced by a production term . However, this term may only be non-zero due to a non-zero $\bar{u}_3$ velocity. As a result the rotation about the x_1 axis induces the mean velocity $\bar{u}_3$.

In contrast to the classical rotating channel flow about the x_3 axis the new rotating channel flow admits a reflection symmetry about the centerline. In terms of symmetry transformations this means that the flow is invariant under the transformation

$$\begin{gathered} x_2^* = -x_2 \ , \quad r_1^* = r_1 \ , \quad r_2^* = -r_2 \ , \quad r_3^* = -r_3 \ , \\ \bar{u}_1^* = \bar{u}_1 \ , \quad \bar{u}_3^* = -\bar{u}_3 \ , \quad {u'_1}^* = u'_1 \ , \quad {u'_2}^* = -u'_2 \ , \quad {u'_3}^* = -u_3 \ , \quad p'^* = p' \ . \end{gathered} \tag{156}$$

It is important to note at this point that the boundary conditions of the channel are also invariant under (156).

From the transformation properties of the fluctuating quantities we may derive all symmetry transformations for the statistical variables. E.g. for the two-point correlation tensor we find

$$\begin{pmatrix} R_{11}^* & R_{12}^* & R_{13}^* \\ R_{21}^* & R_{22}^* & R_{23}^* \\ R_{31}^* & R_{32}^* & R_{33}^* \end{pmatrix} = \begin{pmatrix} R_{11} & -R_{12} & -R_{13} \\ -R_{21} & R_{22} & R_{23} \\ -R_{31} & R_{32} & R_{33} \end{pmatrix} \ . \tag{157}$$

From the transformation (156) it may be taken that $\bar{u}_1$ is symmetric about the centerline. In contrast the induced cross flow velocity $\bar{u}_3$ is skew-symmetric about the center line. The symmetries of the Reynolds stress tensor originate from (157). The $\overline{u'_1u'_3}$ and $\overline{u'_2u'_3}$ stresses which are zero for the rotating channel flow about the span-wise direction are respectively skew-symmetric and symmetric.

Due to the induction of the cross flow the two scale symmetries (121) and (122) are extended and a term with $\bar{u}_3$ appears which is in full analogy to $\bar{u}_1$. In addition, and also in full analogy to $\bar{u}_1$ in (123) the symmetry

$$X_{\bar{u}_3} = \frac{\partial}{\partial \bar{u}_3} \tag{158}$$

appears, which corresponds to the Galilean invariance in x_3 direction.

The derivation of the mean velocity $\bar{u}_3$ is similar to the derivation of $\bar{u}_1$ from (127), (150) and (153). We finally obtain the mean velocity profile

$$\bar{u}_3 = C_{rot_3} x_2 \Omega_1 + \gamma_3(\Omega_1) \left[\frac{k_{x_2}}{k_{s_1}} - \frac{k_{\bar{u}_3}}{k_{s_1}} \right] . \tag{159}$$

Due to homogeneity in x_1- and x_3-direction the net mass flux in x_3 direction is zero. Choosing the origin of the coordinate system on the centerline there will be no additive constant in (159).

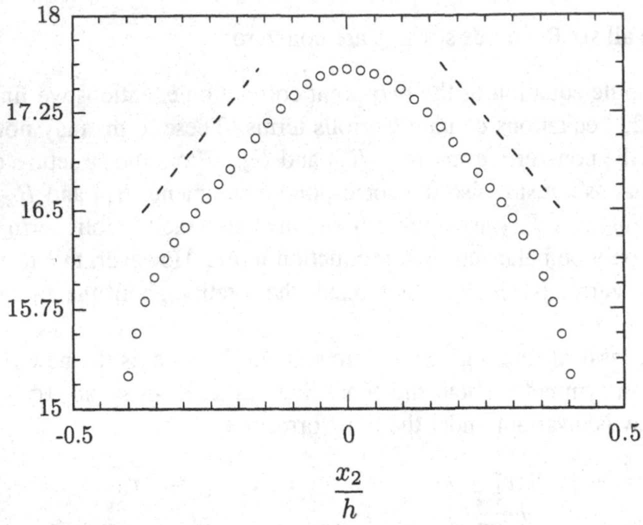

Figure 7: Mean velocity in x_1 direction of a turbulent channel flow rotating about the x_1 direction taken from the DNS data of Oberlack et al. (1998): $Ro_\tau = 20$; $- - -$, parallel line to the linear function (154).

From the knowledge of the above discussed symmetry properties a cone-like symmetric mean velocity profile for $\bar{u}_1$ is to be anticipated which has linear flanks on both sides. For the cross flow $\bar{u}_r$ a skew-symmetric profile with linear flanks as well is to be expected. The Figures 7 and 8 verify these expectations from the symmetry analysis. In order to better see the linear regions in 7, only the upper part of the mean velocity is shown.

Both profiles have been taken from a low Reynolds number DNS. Details may be taken from Oberlack et al. (1998).

4.2 Axially Symmetric Parallel Shear Flows

In the present chapter we will investigate axially symmetric parallel shear flows, such as pipe flows, with respect to their symmetry properties including rotation about the axis of symmetry. Hence, in the subsequent analysis we assume a radial dependence of the axial and azimuthal velocities

$$\bar{u}_z = \bar{u}_z(r) \quad \text{and} \quad \bar{u}_\phi = \bar{u}_\phi(r) \tag{160}$$

respectively. All of the following calculations will be presented in a cylindrical coordinate system where z, r and ϕ are respectively the axial, radial and azimuthal coordinates. The velocities in the respective directions have corresponding indices. Since only parallel flows are considered we have

$$\bar{u}_r = 0 \tag{161}$$

In order to avoid any confusion of the radial coordinate in physical space with the correlation distance $\boldsymbol{r}$, we name the correlation distances in the cylindrical coordinate system r_r, ϕ_r and z_r. Rotation about the z axis is denoted by Ω_z.

Using this presumption a Lie group analysis of the two-point correlations has been conducted leading to the subsequent maximum number of symmetry groups (see e.g. Oberlack, 1999).

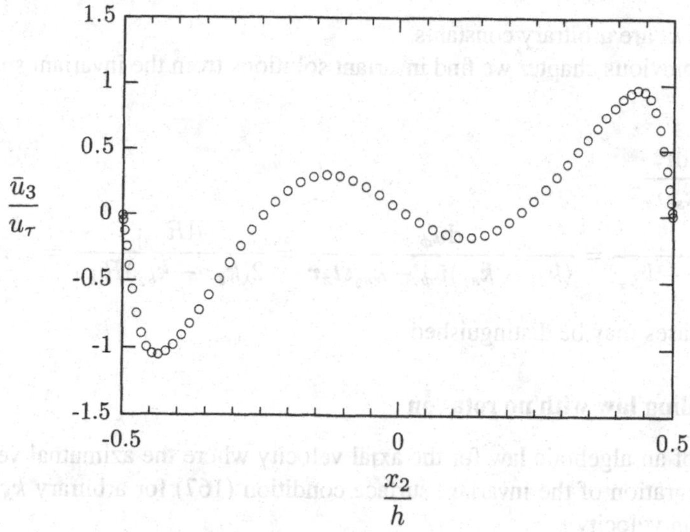

Figure 8: Mean velocity in x_3 direction of a turbulent channel flow rotating about the x_1 direction taken from the DNS data of Oberlack et al. (1998): $Ro_\tau = 20$.

Beside the two scaling groups

$$\bar{X}_{s_1} = r\frac{\partial}{\partial r} + \bar{u}_z\frac{\partial}{\partial \bar{u}_z} + \bar{u}_\phi\frac{\partial}{\partial \bar{u}_\phi} + z_r\frac{\partial}{\partial z_r} + r_r\frac{\partial}{\partial r_r} + 2R_{ij}\frac{\partial}{\partial R_{ij}} + \ldots \tag{162}$$

and

$$\bar{X}_{s_2} = -\bar{u}_z\frac{\partial}{\partial \bar{u}_z} - (\bar{u}_\phi + \Omega_z r)\frac{\partial}{\partial \bar{u}_\phi} - 2R_{ij}\frac{\partial}{\partial R_{ij}} - \ldots \ , \tag{163}$$

we also find the Galilean group in the z-direction

$$\bar{X}_{\bar{u}_z} = \frac{\partial}{\partial \bar{u}_z} \ . \tag{164}$$

The present axially symmetric case does not admit a translational invariance in the r direction nor Galilean invariance in the ϕ direction. For this reason rotationally symmetric flows are distinguished from plane flows as has been discussed in the previous sub-section. In particular the rotation Ω_z does not lead to a symmetry as we can see from $\bar{X}_{s_2}$. Important consequences are to be derived from this.

In cylindrical coordinates the two-point correlation tensor reads as

$$R = \begin{pmatrix} R_{zz} & R_{zr} & R_{z\phi} \\ R_{rz} & R_{rr} & R_{r\phi} \\ R_{\phi z} & R_{\phi r} & R_{\phi\phi} \end{pmatrix} \ . \tag{165}$$

Since all symmetries (162)-(164) are linearly independent, we find new symmetries from linear combination given by

$$\bar{X} = k_{s_1}\bar{X}_{s_1} + k_{s_2}\bar{X}_{s_2} + k_{\bar{u}_z}\bar{X}_{\bar{u}_z} \tag{166}$$

where the coefficients k_i are arbitrary constants.

According to the previous chapter we find invariant solutions from the invariant surface condition

$$\frac{\mathrm{d}r}{k_{s_1} r} = \frac{\mathrm{d}z_r}{k_{s_1} z_r} = \frac{\mathrm{d}r_r}{k_{s_1} r_r}$$
$$= \frac{\mathrm{d}\bar{u}_z}{(k_{s_1} - k_{s_2})\bar{u}_z + k_{\bar{u}_z}} = \frac{\mathrm{d}\bar{u}_\phi}{(k_{s_1} - k_{s_2})\bar{u}_\phi - k_{s_2}\Omega_z r} = \frac{\mathrm{d}R_{[ij]}}{2(k_{s_1} - k_{s_2})R_{[ij]}} = \dots \quad (167)$$

Three different flow cases may be distinguished.

4.2.1 Algebraic scaling law with no rotation

The first case is that of an algebraic law for the axial velocity where the azimuthal velocity has been set to zero. Integration of the invariant surface condition (167) for arbitrary k_{s_1}, k_{s_2} and $k_{\bar{u}_z}$ leads to axial mean velocity

$$\bar{u}_z = C_{algPipe}\, r^{1-\frac{k_{s2}}{k_{s1}}} + \frac{k_{\bar{u}_z}}{k_{s_2} - k_{s_1}} \quad , \quad (168)$$

and the corresponding forms of the reduced correlation functions are

$$\tilde{z}_r = \frac{z_r}{r} \; , \quad \tilde{r}_r = \frac{r_r}{r} \; , \quad R_{ij} = r^{2\left(1-\frac{k_{s2}}{k_{s1}}\right)} \tilde{R}_{ij} \quad \dots \; , \quad (169)$$

In view of the results of the turbulent channel flow where the maximum number of combined symmetries has been observed in the core of the flow we also expect the region of validity of (168) in the center of the turbulent pipe flow.

Due to the rotation symmetry of the pipe flow the term $\frac{k_{\bar{u}_z}}{k_{s_2} - k_{s_1}}$ in (168) may only represent the maximum velocity $\bar{u}_{max}$ on the pipe axis. Hence we rewrite (168) in physical coordinates to obtain

$$\frac{\bar{u}_{max} - \bar{u}_z}{u_\tau} = a_{algPipe} \left(\frac{r}{R}\right)^{\beta_{algPipe}} \quad , \quad (170)$$

where R is the pipe radius. There is a fundamental difference between (170) and the classical empirical scaling laws for turbulent pipe flow such as the "1/7"-law (see e.g. Schlichting, 1982). The latter is based on the distance from the wall. However, the group analysis clearly shows that the distance from the axis of symmetry is the only "natural" coordinate in an axially symmetric geometry.

An experimental verification of the latter may be found in Figure 9. The "1/7"-law is only an approximation of a very small range of Reynolds numbers ($Re \approx 10^5$) and is only applicable to the near-wall region. In contrast (170) admits a large range of applicability covering several orders of magnitude of Reynolds numbers. For the sake of clarity the different data sets in Figure 9 are shifted relative to each other in vertical direction.

An important indication for the validity of the algebraic law (168) in the core region of a turbulent pipe flow may be taken from the DNS data of Wagner and Friedrich (1998). Besides the classical boundary condition of an impermeable wall they also conducted a DNS with a porous

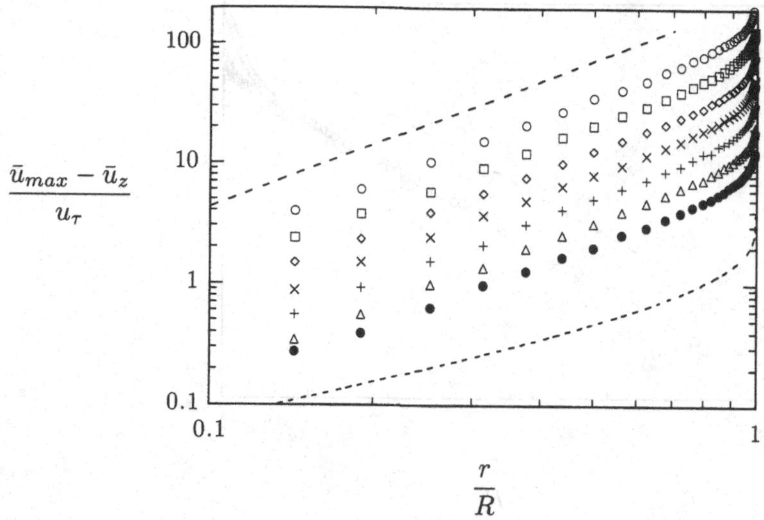

Figure 9: Mean velocity of a turbulent pipe flow experimentally derived by Zagarola (1996): •, $Re_m = 3.5 \cdot 10^7$; △, $Re_m = 1.4 \cdot 10^7$; +, $Re_m = 4.4 \cdot 10^6$; ×, $Re_m = 1.3 \cdot 10^6$; ◇, $Re_m = 4.1 \cdot 10^5$; □, $Re_m = 1.5 \cdot 10^5$; ○, $Re_m = 4.2 \cdot 10^4$; − − −, line parallel to the algebraic function (170), (beginning with $Re_m = 1.4 \cdot 10^7$ all data profiles are shifted against each other by the factor $10^{0.2}$.); - - - - -, line parallel to the "1/7"-law.

wall. The boundary condition is implemented as such that there is no net mass flux through the wall. They found two remarkable results.

First, at a constant pressure-gradient the mass flux through the pipe is drastically reduced. Second and most importantly the functional dependence in the core region of the flow is not affected. It is only the near-wall region which drastically changes its behavior. In Figure 10 the mean velocity is plotted with the classical and the modified boundary condition at equal Reynolds number. It is apparent that the modified boundary condition has essentially no influence on the core region of the flow. This region closely follows the law (170) which is dictated by the maximum number of combined symmetries.

4.2.2 Near-wall region of a turbulent pipe flow

As for the turbulent channel flow, the algebraic law (9) is only applicable to the core region of a turbulent pipe flow. Nevertheless, Zagarola et al. (1997) as well as many other researchers have found a logarithmic-law-of-the-wall in the near-wall region of a pipe flow. In contrast to the algebraic results above their results are based on a coordinate originating at the wall rather than in the center of the pipe.

The basis of this is the fact, that the near-wall region of the pipe flow may be considered a plane problem. This may be analyzed by introducing a coordinate system attached to the contour of the pipe wall into the Navier-Stokes equations

$$z = z \; , \quad y = R - r \; , \quad s = r\phi \; , \quad u_z = u_z \; , \quad u_y = -u_r \; , \quad u_s = u_\phi \; , \tag{171}$$

where y and s are respectively the wall-based coordinate pointing towards the center of the pipe

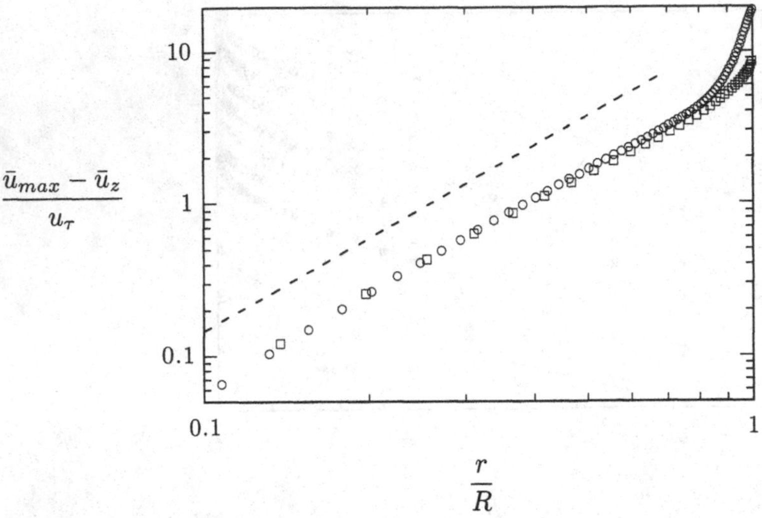

Figure 10: Mean velocity of a turbulent pipe flow taken from the DNS data in Wagner and Friedrich (1998) at $Re_R = 180$: □, porous pipe; ○, impermeable wall; $- - -$, line parallel to the algebraic function (170).

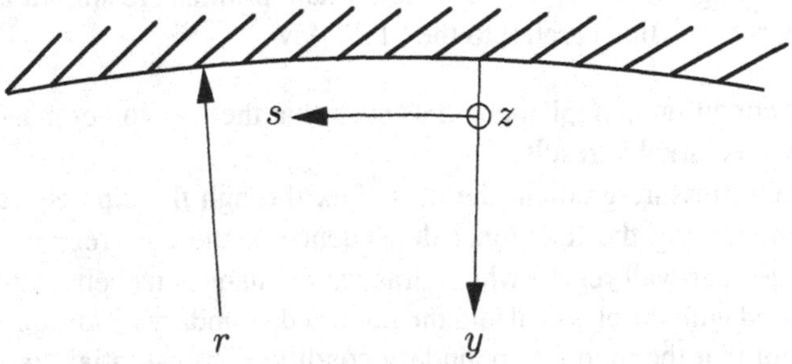

Figure 11: Sketch of the wall-based coordinate system as adopted for the near-wall scaling laws.

and the azimuthal coordinate along the circumference of the pipe (see Figure 11).

In order to identify the leading order terms in the resulting equations, we non-dimensionalize all variables with the classical wall parameters namely the friction velocity and viscous length-scale

$$u_\tau = \left(\nu \left. \frac{\partial \bar{u}_z}{\partial y} \right|_{y=0} \right)^{\frac{1}{2}} \quad \text{and} \quad l^+ = \nu / u_\tau \tag{172}$$

respectively.

Reformulating the Navier-Stokes equations (53) and (54) in cylinder coordinates and introducing (171) as well as the normalization (172) into the resulting system we obtain

$$\frac{\partial u_z}{\partial z} + \frac{\partial u_y}{\partial y} + \frac{\partial u_s}{\partial s} + \frac{1}{y - Re_R} \frac{\partial}{\partial s}(s u_y) = 0 \tag{173}$$

and

$$\frac{\partial u_z}{\partial t} + u_z \frac{\partial u_z}{\partial z} + u_y \frac{\partial u_z}{\partial y} + u_s \frac{\partial u_z}{\partial s} + \frac{s}{y - Re_R} u_y \frac{\partial u_z}{\partial s} = -\frac{\partial p}{\partial z}$$
$$+ \frac{\partial^2 u_z}{\partial z^2} + \frac{\partial^2 u_z}{\partial y^2} + \frac{\partial^2 u_z}{\partial s^2} + \frac{2s^{\frac{1}{2}}}{y - Re_R} \frac{\partial}{\partial s} \left(s^{\frac{1}{2}} \frac{\partial u_z}{\partial y} \right) + \frac{s}{(y - Re_R)^2} \frac{\partial}{\partial s} \left(s \frac{\partial u_z}{\partial s} \right) , \tag{174}$$

$$\frac{\partial u_y}{\partial t} + u_z \frac{\partial u_y}{\partial z} + u_y \frac{\partial u_y}{\partial y} + u_s \frac{\partial u_y}{\partial s} + \frac{1}{y - Re_R} \left(\frac{s}{2} \frac{\partial u_y^2}{\partial s} - u_s^2 \right) = -\frac{\partial p}{\partial y} - \frac{s}{y - Re_R} \frac{\partial p}{\partial s}$$
$$+ \frac{\partial^2 u_y}{\partial z^2} + \frac{\partial^2 u_y}{\partial y^2} + \frac{\partial^2 u_y}{\partial s^2} + \frac{1}{y - Re_R} \left[2s^{\frac{1}{2}} \frac{\partial}{\partial s} \left(s^{\frac{1}{2}} \frac{\partial u_y}{\partial y} \right) - 2 \frac{\partial u_s}{\partial s} \right]$$
$$+ \frac{1}{(y - Re_R)^2} \left(s^2 \frac{\partial^2 u_y}{\partial s^2} + 2s \frac{\partial u_y}{\partial s} - u_y \right) , \tag{175}$$

$$\frac{\partial u_s}{\partial t} + u_z \frac{\partial u_s}{\partial z} + u_y \frac{\partial u_s}{\partial y} + u_s \frac{\partial u_s}{\partial s} + \frac{1}{y - Re_R} \left(s \frac{\partial u_s}{\partial s} + u_s u_y \right) = -\frac{\partial p}{\partial s}$$
$$+ \frac{\partial^2 u_s}{\partial z^2} + \frac{\partial^2 u_s}{\partial y^2} + \frac{\partial^2 u_s}{\partial s^2} + \frac{1}{y - Re_R} \left[2s^{\frac{1}{2}} \frac{\partial}{\partial s} \left(s^{\frac{1}{2}} \frac{\partial u_s}{\partial y} \right) + 2 \frac{\partial u_y}{\partial s} \right]$$
$$+ \frac{1}{(y - Re_R)^2} \left(s^2 \frac{\partial^2 u_s}{\partial s^2} + 2s \frac{\partial u_s}{\partial s} - u_s \right) . \tag{176}$$

The Reynolds number is defined as

$$Re_R = \frac{u_\tau R}{\nu} . \tag{177}$$

For asymptotically large differences between Re_R and y we may consider $1/(y - Re_R)$ a small perturbation parameter. In the limit $(y - Re_R) \to \infty$ the equations (173)-(176) converge to the Navier-Stokes equations in cartesian coordinates. On the basis of cartesian equations we derived in subsection 4.1 the classical logarithmic-law-of-the-wall for plane shear flows.

There is also an additional indication for the validity of the above argument. For a given sufficient large difference of y and Re_R we find that with increasing Reynolds number also the value for y may increase. It may be concluded from this that the region of validity increases with increasing Reynolds number. A good experimental verification of this may be found in Figure 12 where the region of the logarithmic-law-of-the-wall increases with the Reynolds number. Hence in the limit of large Reynolds number and small wall distance turbulence in a pipe may be considered a plane problem.

4.2.3 Algebraic scaling law including rotation

In extension of the classical pipe flow we will consider the rotation of a turbulent pipe flow about its axis of symmetry z in the present subsection. For this case we find from (167) an invariant solution for the azimuthal velocity

$$\bar{u}_\phi = C_{rotPipe} \, r^{1 - \frac{k_{s2}}{k_{s1}}} - \Omega_z r , \tag{178}$$

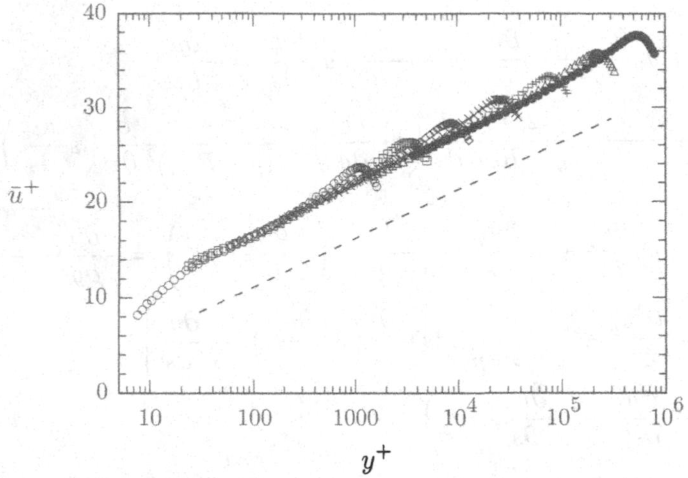

Figure 12: Experimental validation of the logarithmic-law-of-the-wall in a turbulent pipe flow taken from Zagarola (1996): •, $Re_m = 3.5 \cdot 10^7$; △, $Re_m = 1.4 \cdot 10^7$; +, $Re_m = 4.4 \cdot 10^6$; ×, $Re_m = 1.3 \cdot 10^6$; ◇, $Re_m = 4.1 \cdot 10^5$; □, $Re_m = 1.5 \cdot 10^5$; ○, $Re_m = 4.2 \cdot 10^4$; $- - -$, line parallel to the logarithmic function (133).

and the corresponding correlation functions may be taken from (169). Apparently in (178) a solid body rotation is induced by the rotation of the system. An algebraic mean velocity remains. Physically this may be interpreted that it is fully equivalent to rotate the solid wall in an inertial system or to consider the entire pipe in a rotating frame. Hence without loss of generality we may set Ω_z equal to zero and therefore in the following we consider rotation effects due to the motion of the solid wall.

Employing physical flow parameters we find from (178)

$$\frac{\bar{u}_\phi}{u_w} = a_{rotPipe} \left(\frac{r}{R}\right)^{\beta_{algRohr}} . \tag{179}$$

u_w denotes the wall velocity in circumferential direction. The key result is that the mean azimuthal velocity in a rotating turbulent pipe flow does not follow solid body rotation but rather follows an algebraic law as the axial velocity does. An experimental validation of this result may be taken from Figure 13.

Independent of the rotation number

$$N = \frac{u_w}{\bar{u}_m} \tag{180}$$

all data collapse for a major part of the pipe radius on an algebraic function which, in double-log scaling in Figure 13, is a straight line. u_w and $\bar{u}_m$ denote the azimuthal velocity at the wall and bulk velocity in axial direction respectively. Only the inner part of the pipe is dominated by viscosity which leads to solid body rotation. This region decreases with increasing Reynolds number.

It is very important to note that the mean velocity in axial and azimuthal in direction have a close link with each other given by an identical exponent $\beta_{algRohr}$ in (170) and (179) stemming

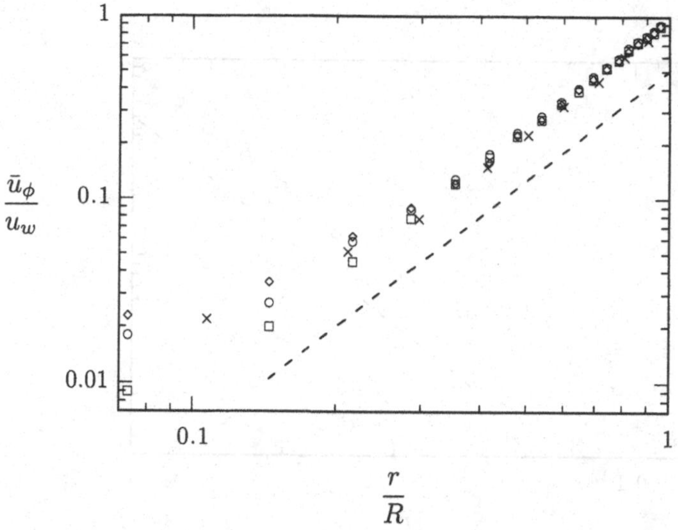

Figure 13: Azimuthal mean velocity of the experimental data of a turbulent pipe flow: ∘, $Re_m = 20000$, $N = 0.5$; □, $Re_m = 20000$, $N = 1.0$; ⋄, $Re_m = 50000$, $N = 0.5$ Reich (1988): ×, $Re_m = 50000$, $N = 1.0$ Kikuyama et al. (1983); – – –, parallel line to the algebraic function (179).

from the group parameter k_{s_1} and k_{s_2}. This result is experimentally confirmed in the work by Kikuyama et al. (1983) in Figure 14. Parallel lines in double-logarithmic presentation indicate identical scaling properties.

For a considerable larger rotation number N we find from the DNS data of Orlandi and Fatica (1997) in Figure 15 a comparable confirmation of the existence for the latter two scaling laws. However, two important differences are striking.

First, in the core region of the azimuthal velocity we find a rather large region of solid body rotation due to the small Reynolds number of the DNS. A second noticeable difference may be taken from the axial velocity in the near-wall region beginning at $r/R \approx 0.6$. From this value on we find a considerable deviation from the algebraic law. Since influence of the Reynolds number on the mean flow is usually limited to a region up to $y^+ \approx 100$ the difference to the experimental results in Figure 14 at $N = 1$ is believed to be caused by the considerably larger rotation number of $N = 2$. The wall velocity u_w in azimuthal direction implies the existence of a symmetry breaking velocity scale such as the friction velocity u_τ for logarithmic-law-of-the-wall. For small N the rotation number may play a less dominant role. However, with increasing N the rotation velocity u_w becomes the dominant parameter, leading to a symmetry breaking of the velocity scaling.

As for the logarithmic-law-of-the-wall the symmetry breaking velocity leads to $k_{s_1} = k_{s_2}$ in the condition of an invariant solution (167). With this premise we obtain the axial velocity

$$\bar{u}_z = \frac{k_{\bar{u}_z}}{k_{s_1}} \ln(r) + C_{logPipe} \; . \tag{181}$$

Evidently we find for this case of large rotation rates that the wall velocity u_w is the dominating parameter which has been introduced to non-dimensionalize (181). Hence we re-write (181) in

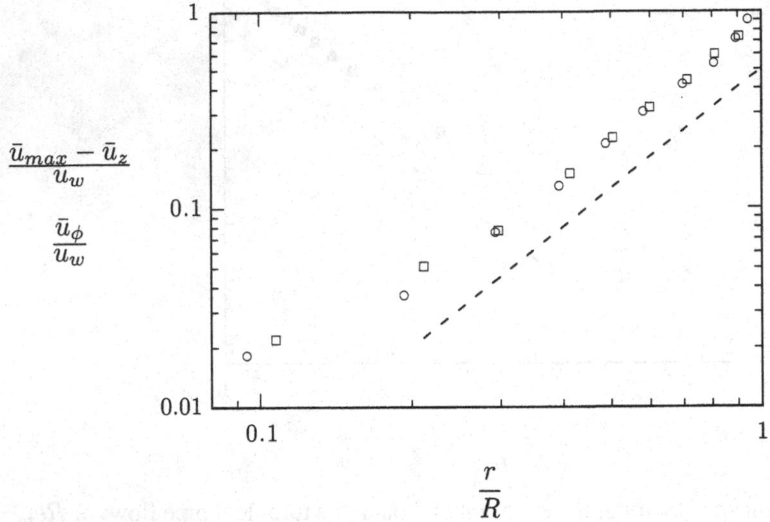

Figure 14: Axial and azimuthal mean velocity of the experimental data of a turbulent pipe flow with rotating walls from Kikuyama et al. (1983) where $Re_m = 50000$ and $N = 1.0$: ○, $(\bar{u}_{max} - \bar{u}_z)/u_w$; □, $\bar{u}_\phi/u_w$; — — —, parallel line to the algebraic function according to (170) and (179).

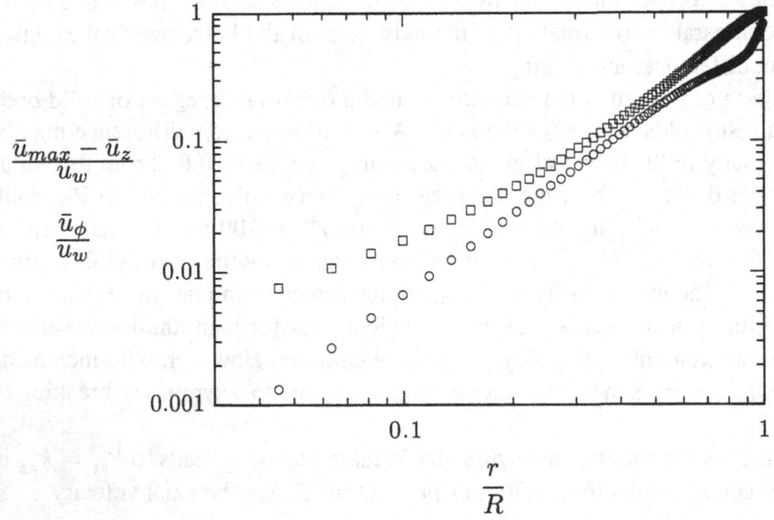

Figure 15: Axial and azimuthal mean velocity of the DNS data of a turbulent pipe flow with rotating walls from Orlandi and Fatica (1997) where $Re_m = 4900$ and $N = 2.0$: ○, $(\bar{u}_{max} - \bar{u}_z)/u_w$; □, $\bar{u}_\phi/u_w$.

physical coordinates as

$$\bar{u}_z = u_w \left[\alpha_{logPipe} \ln\left(\frac{r}{R}\right) + \beta_{logPipe} \right] . \tag{182}$$

A validation of the new log scaling law may be determined by a different plotting of the axial velocity. In the upper graph of Figure 16 we have a linear plot of the DNS data of Orlandi and Fatica (1997) at $N = 2$. Two turning points are visible. In the lower graph of Figure 16 identical data are plotted on a semi-logarithmic scale. In the range $r/R \in [0.55, 0.8]$ a logarithmic range according to (182) is apparent. It is important to note that the new log law should not be confused with the classical logarithmic-law-of-the-wall.

Three key differences are evident. First, the new log law (182) is based on the pipe radius coordinate and not on the distance from the wall. Second, the range of applicability is much further towards the center of the pipe and, third, has with $r/R \in [0.55, 0.8]$ a much larger extent of about 25% of the pipe radius. The classical logarithmic-law-of-the-wall is limited to about 10% of the pipe radius.

5 Homogeneous and Inhomogeneous Time-dependent Turbulent Shear Flows

5.1 Homogeneous Turbulent Flows

From experiments and DNS data it is known that the turbulent kinetic energy or the Reynolds normal stresses decay according to t^{-n} with $n > 0$ in the limit of homogeneous isotropic turbulence as long as the Reynolds number is large enough. This property has been implemented in essentially all Reynolds averaged turbulence models. If, however, the homogeneous flow is sheared or strained by a constant gradient of the mean velocity we obtain very different behavior. In both cases we find exponential behavior for $t \to \infty$.

5.1.1 Decay of homogeneous or isotropic turbulence

Classical work on homogeneous isotropic turbulence dates back to the work of von Kármán and Howarth (1938) who were the first to show that the decay of turbulence is according to the above power law. The latter work was extended by Batchelor (1946) and Chandrasekhar (1950) to axisymmetric turbulence.

Both theories are essentially based on the rotation symmetries (93) or rather on their infinitesimal form (98)-(100). For the case of isotropic turbulence the von Kármán-Howarth equation is invariant under all rotation groups and hence all statistical quantities depend only on a single spherical correlation coordinate.

Theories of axisymmetric turbulence have been derived under the assumption of invariance under one rotation group. From this it follows that all statistical quantities depend on only two independent variables.

Without loss of generality we set $\bar{\boldsymbol{u}} = 0$. As a result we find for the large-scale two-point correlation equation the following set of combined symmetries

$$\mathrm{X}_{HT} = k_t\,\mathrm{X}_t + k_{r_1}\,\mathrm{X}_{r_1} + k_{r_2}\,\mathrm{X}_{r_2} + k_{r_3}\,\mathrm{X}_{r_3} + k_{s_1}\,\mathrm{X}_{s_1} + k_{s_2}\,\mathrm{X}_{s_2} \tag{183}$$

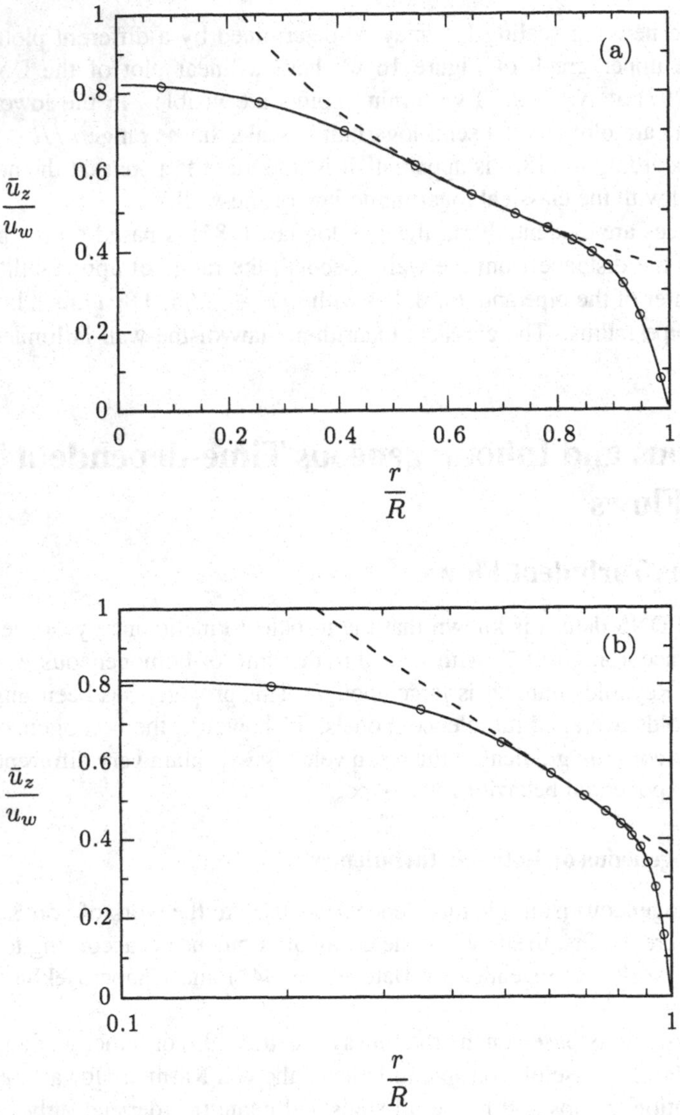

Figure 16: Axial mean velocity of the DNS data of a turbulent pipe flow with rotating walls at $Re_m = 4900$ and $N = 2.0$: ———, DNS data from Orlandi and Fatica (1997); — — —, logarithmic law according to (182).

for an arbitrary set of constants $k_\varkappa$, where $\varkappa$ represents any index. The above symmetries are taken from (97)-(100), (114) and (115). From these symmetries we may derive the following reductions, i.e. the derivation of similarity solutions of the two-point correlation equations:

(i) Independent use of the rotation groups, i.e. not in combined form with the remaining groups in (183), leads to isotropic turbulence (see von Kármán and Howarth, 1938). As a result the von-Kármán-Howarth equation was derived which is a scalar equation for the correlation function. A single rotation group leads to axisymmetric turbulence (see Batchelor, 1946, Chandrasekhar, 1950). In this case the two-point correlation equations reduce to two equations.

(ii) From the two scaling groups we may derive the decay law of homogeneous turbulence. For isotropic turbulence this was already done by von Kármán and Howarth (1938) in an *ad hoc* manner. In the limit of infinite Reynolds number we find identical results.

In the following we will only deal with the temporal evolution of turbulence rather than repeating the classical results of spatial symmetries. For this reason we will only employ the remaining groups. These are the group of time invariance and the two scaling groups corresponding to k_t, k_{s_1} and k_{s_2} respectively. Employing this combined group to give the condition for an invariant solution (52) we find

$$\frac{\mathrm{d}t}{k_{s_2}t+k_t}=\frac{\mathrm{d}r_{[k]}}{k_{s_1}r_{[k]}}=\frac{\mathrm{d}R_{[ij]}}{2(k_{s_1}-k_{s_2})R_{[ij]}}=\ldots\,. \tag{184}$$

It should be noted here that the rotation groups do not have any influence on the temporal evolution of turbulence. The integration of (184) leads to

$$\tilde{r}_k=\frac{r_k}{\left(t+\frac{k_t}{k_{s_2}}\right)^{\frac{k_{s_1}}{k_{s_2}}}}\,,\quad R_{ij}=\left(t+\frac{k_t}{k_{s_2}}\right)^{2\left(\frac{k_{s_1}}{k_{s_2}}-1\right)}\tilde{R}_{ij}\,,\quad\ldots \tag{185}$$

where the constants of integration $\tilde{r}_k$, $\tilde{R}_{ij}$ are to be taken as the new variables.

Since the reduced correlation functions $\tilde{R}_{ij}$ are solely a function of $\tilde{r}_k$ we find in the limit of vanishing correlation distance $\boldsymbol{r}\to 0$ the long time evolution of the Reynolds stress tensor

$$\overline{u_i'u_j'}\propto\left(t+\frac{k_t}{k_{s_2}}\right)^{2\left(\frac{k_{s_1}}{k_{s_2}}-1\right)}\,, \tag{186}$$

where the parameter in the exponent has to obey the condition $\frac{k_{s_1}}{k_{s_2}}<1$. Analogous decay laws are obtained for other one-point correlations.

For the special case of isotropic turbulence a large variety of experimental and DNS data are available. In essentially all experiments and DNS data the law (186) has been observed with great accuracy. As an example DNS data of Iida and Kasagi (1993) have been plotted in Figure 17 in double logarithmic scaling. After an initial transient period of time almost perfect agreement with the algebraic law is visible. From the scaled variable $\tilde{r}_k$ we may also derive the temporal development of the integral length-scale. In extension of the usual definition of the integral length-scale in (108) we define

$$\ell_{ijk}=\frac{1}{\overline{u_m'u_m'}}\left(\int_{-\infty}^{\infty}R_{ij}\mathrm{d}r_k\right)_{r_l=0}\qquad\text{for all}\quad k\neq l\,, \tag{187}$$

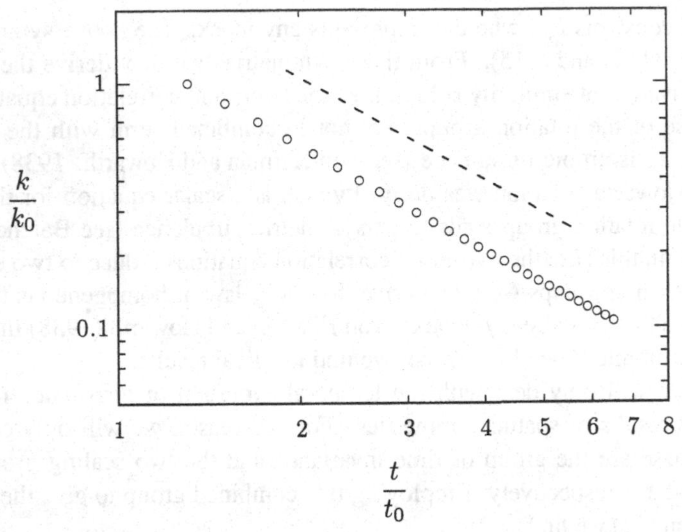

Figure 17: Temporal evolution of the turbulent kinetic energy in decaying isotropic turbulence: o, DNS data from Iida and Kasagi (1993); — — —, parallel line according to the algebraic function (186).

(see Rotta, 1972). Substituting the similarity variables (185) into (187) we find

$$\ell_{ijk} \propto \left(t + \frac{k_t}{k_{s_2}}\right)^{\frac{k_{s_1}}{k_{s_2}}} . \tag{188}$$

In the limit of isotropic turbulence von Kármán and Howarth (1938) established a relation between the two-point tensor in 3D space with the isotropic correlation function $f(r,t)$ according to (103). As a result we obtain from (185)

$$\tilde{r} = \frac{r}{t^{\frac{2}{\sigma+3}}} \quad , \quad \overline{u'^2} \sim t^{-2\frac{\sigma+1}{\sigma+3}} \quad \text{where} \quad \sigma = \frac{2 - 3\frac{k_{s_1}}{k_{s_2}}}{\frac{k_{s_1}}{k_{s_2}}} . \tag{189}$$

The parameter σ has been introduced since it gives a link between the temporal evolution and the spatial decay of the correlation function for large distances r (large eddies) according to

$$\lim_{r \to \infty} f(r) \sim r^{-\sigma} \quad \text{or} \quad \lim_{k \to 0} E(k) \sim k^{\sigma} , \tag{190}$$

where $E(k)$ is the isotropic energy spectrum and the constants of proportionality are independent of time.

During the last decades a variety of invariants have been proposed which will be discussed from a symmetry point of view. For this purpose we recall the integral length in (108). Introducing $\tilde{r}$ from (189) into (108) we find

$$\ell(t) = t^{\frac{2}{\sigma+3}} \int_{\tilde{r}=0}^{\infty} f(\tilde{r}) \mathrm{d}\tilde{r} \sim t^{\frac{2}{\sigma+3}} \quad \text{and} \quad Re_t = \frac{\sqrt{\overline{u'^2}(t)}\,\ell(t)}{\nu} \sim t^{-\frac{\sigma-1}{\sigma+3}} . \tag{191}$$

Loitsyansky (1939) proposed the invariant Λ based on an integrated form of the von Kármán-Howarth equation

$$\Lambda = \overline{u'^2} \int_0^\infty r^4 f(r) \mathrm{d}r \quad \Rightarrow \quad \sigma = 4, \quad \overline{u'^2} \sim t^{-10/7}, \quad \ell \sim t^{2/7}, \quad Re \sim t^{-3/7}, \tag{192}$$

while Kolmogorov (1941a) recognized the latter time dependence which also corresponds to the low wave number energy spectrum $E(k) = \frac{\Lambda}{3\pi} k^4$.

Later Birkhoff (1954) suggested

$$B = \overline{u'^2} \int_0^\infty r^2 \left(3f(r) + r \frac{\partial f}{\partial r} \right) \mathrm{d}r = \overline{u'^2} \lim_{r \to \infty} [r^2 f(r)] \tag{193}$$

as another invariant. Saffman (1967) derived the corresponding time dependence

$$\sigma = 2, \quad \overline{u'^2} \sim t^{-6/5}, \ell \sim t^{2/5}, \quad Re \sim t^{-1/5}, \tag{194}$$

with the low wave number energy spectrum $E(k) = \frac{B}{\pi} k^2$. In Oberlack (2001) it has been shown that the latter case corresponds to the conservation of energy for the large scales.

Accordingly we may also suggest other more commonly known physical quantities as invariants, i.e. a quantity independent of time, such as the integral length-scale

$$l = \int_0^\infty f(r) \mathrm{d}r = const. \quad \Rightarrow \quad \overline{u'^2} \sim t^{-2}, \quad Re \sim t^{-1} . \tag{195}$$

Physically the latter case corresponds to high Reynolds number turbulence in a box of length L where the integral length-scale has reached the box size $\ell \sim L$. The imposed length-scale of the finite box is symmetry breaking for the scaling of space.

For the following final case it is unnecessary to invoke the large Reynolds number limit since it is the only case which allows for full similarity of all terms in the two-point correlation equation or rather the von Kármán-Howarth equation. Since the Reynolds number is assumed to be an invariant, i.e. independent of time, we have

$$Re = const. \quad \Rightarrow \quad \sigma = 1, \quad \overline{u'^2} \sim t^{-1}, \quad \ell \sim t^{\frac{1}{2}} . \tag{196}$$

From a similarity point of view this is the only case being fully consistent with the viscous terms. Hence it allows similarity of all scales down to the Kolmogorov length-scale. However we find that the decay law (196) is essentially never observed in experimental or DNS data (see Mohamed and LaRue, 1990). This supports the asymptotic expansion in correlation space implying that viscosity has very little influence on the large scale quantities.

5.1.2 Evolution of homogeneous turbulence with a constant value of the mean velocity

Within this subsection we consider homogeneous flows subjected to a constant mean velocity gradient

$$\frac{\partial \bar{u}_k}{\partial x_l} = \bar{A}_{kl} = const. \quad \text{and} \quad \frac{\partial \bar{u}_k}{\partial x_k} = \bar{A}_{kk} = 0 \tag{197}$$

where the latter is due to continuity.

In the general case $\bar{A}$ does not admit any symmetry condition. It may be separated in a unique manner into a symmetric and anti-symmetric part $\bar{S}$ and $\overline{W}$, respectively

$$\bar{A} = \bar{S} + \overline{W} \ . \tag{198}$$

Since in homogeneous turbulent flows we do not need to consider boundary conditions we may rotate the coordinate system such that $\bar{S}$ is always in normal form i.e. has only diagonal elements. Hence we find $\bar{S}$ and $\overline{W}$ have the form

$$\bar{S} = \begin{pmatrix} \bar{S}_1 & 0 & 0 \\ 0 & \bar{S}_2 & 0 \\ 0 & 0 & -(\bar{S}_1+\bar{S}_2) \end{pmatrix} \quad \text{and} \quad \overline{W} = \begin{pmatrix} 0 & \overline{W}_{12} & -\overline{W}_{31} \\ -\overline{W}_{12} & 0 & \overline{W}_{23} \\ \overline{W}_{31} & -\overline{W}_{23} & 0 \end{pmatrix} \ . \tag{199}$$

If not specified otherwise we assume in the following that $\bar{A}$ possesses the tensor symmetry properties implied by (199). As a result we have that $\bar{A}$ is defined by five parameters.

Physically speaking $\bar{S}$ and $\overline{W}$ may be assigned typical flows, i.e. $\bar{S}$ corresponds to stretching while $\overline{W}$ characterizes rotation (see Oberlack, 1994).

The key information for the classification of the different flows is obtained from the determining equations (see page 306). Those parts of the determining equations (not shown here in total) which are independent of the elements of $\bar{A}$ have been solved until only eight equations are left. As a result we obtain all the symmetries (183) which have to be obeyed by the remaining parts of the determining equations as follows

$$k_{s_2}\,\bar{S}_1 = 0 \ , \tag{200}$$
$$k_{s_2}\,\bar{S}_2 = 0 \ , \tag{201}$$
$$k_{r_3}\,(\bar{S}_1 - \bar{S}_2) = 0 \ , \tag{202}$$
$$k_{r_1}\,(\bar{S}_1 + 2\bar{S}_2) = 0 \ , \tag{203}$$
$$k_{r_2}\,(2\bar{S}_1 + \bar{S}_2) = 0 \ , \tag{204}$$

$$k_{s_2}\,\overline{W}_{12} + k_{r_1}\,\overline{W}_{31} - k_{r_2}\,\overline{W}_{23} = 0 \ , \tag{205}$$
$$k_{s_2}\,\overline{W}_{23} - k_{r_3}\,\overline{W}_{31} + k_{r_2}\,\overline{W}_{12} = 0 \ , \tag{206}$$
$$k_{s_2}\,\overline{W}_{31} + k_{r_3}\,\overline{W}_{23} - k_{r_1}\,\overline{W}_{12} = 0 \ . \tag{207}$$

Since homogeneous turbulence with zero mean velocity has been studied in the previous subsection we assume that at least one element of $\bar{B}$ is non-zero. Since the systems (200)-(204) and (205)-(207) formally decouple they may be treated separately.

(205)-(207) admits the unique solution

$$\begin{pmatrix} k_{r_3} \\ k_{r_1} \\ k_{r_2} \\ k_{s_2} \end{pmatrix} = C_1 \begin{pmatrix} \overline{W}_{12} \\ \overline{W}_{23} \\ \overline{W}_{31} \\ 0 \end{pmatrix} \ , \tag{208}$$

where C_1 is an arbitrary constant. Apparently any combination of $\overline{W}_{12}$, $\overline{W}_{23}$ and $\overline{W}_{31}$ leads to symmetry breaking of k_{s_2}. We also obtain the same from the first two equations of (200)-(204). Since k_{s_2} corresponds to the scaling of time we find that there is a symmetry breaking of time due to the external time scale given by

$$t_{\bar{A}} = \frac{1}{\|\bar{A}\|} \ . \tag{209}$$

Still there may be some rotational symmetries in the equations (202)-(207).

Several cases have to be distinguished. If $\overline{W} = 0$ and $\bar{S} \neq 0$ there may be at most one of the coefficients k_{r_1}, k_{r_2} or k_{r_3} in (202)-(204) non-zero if due to a special combination of $\bar{S}_1$ and $\bar{S}_2$ one of the terms in brackets vanishes. In the remaining two equations the terms in brackets are then non-zero. Hence the corresponding group parameter in front of these terms must be zero. If e.g. $\bar{S}_1 = \bar{S}_2$ we may have a k_{r_3} non-zero. As a result we may have at most a rotation symmetry in the r_1-r_2-plane. Furthermore, if we have $\bar{S}_1 = \bar{S}_2$ in the equations (203) and (204) we find that k_{r_1} and k_{r_2} must vanish. Analogous to the previous case we may also investigate the remaining two cases. All three cases correspond to a homogeneous strain or compression flow along the corresponding coordinates (see Oberlack, 1994). For the case $\bar{S}_1 = \bar{S}_2$ we get the compression part of $\bar{A}$ to

$$\bar{S} = \operatorname{diag}(\bar{S}_1, \bar{S}_1, -2\bar{S}_1) \ . \tag{210}$$

This corresponds to a homogeneous strain and compression along the x_3-axis respectively in the x_1-x_2-plane. The remaining two cases correspond to strain and compression along the x_1- and x_2-axis.

The second class of flows is characterized by $\bar{S} = 0$ and $\overline{W} \neq 0$. Since equations (205)-(207) always have a solution for arbitrary $\overline{W}$ as a general solution there is always one rotation symmetry admitted. However, we should note that even in the case when all rotation group parameters are non-zero they combine into a single rotation group. The reason for this is that for a given $\overline{W}$ the group parameters have fixed numbers. As a result the rotation group is determined by the combination of the groups (98)-(100)

$$\mathrm{X}_{\overline{W}} = C_1 \left(\overline{W}_{12}\, \mathrm{X}_{r_3} + \overline{W}_{23}\, \mathrm{X}_{r_1} + \overline{W}_{31}\, \mathrm{X}_{r_2} \right) \ . \tag{211}$$

The direction of the rotation axis of the previous group coincides with the direction of the vector

$$\check{\overline{W}}_i = \frac{1}{2} e_{ijk}\, \overline{W}_{jk} \tag{212}$$

which is equivalent to the rotation tensor $\overline{W}$.

Finally we consider the combination of strain and rotation given by $\bar{S} \neq 0$ and $\overline{W} \neq 0$. As has been discussed above, a possible rotation symmetry exists for only three combinations of $\bar{S}_1$ and $\bar{S}_2$. This symmetry is only preserved if the superimposed rotation $\overline{W}$ is parallel to the axis of strain.

In summary we obtain the following complete classification for constant $\bar{A}$:

a.) No scaling symmetry with respect to time exists for any combination of $\bar{A}$ and hence $k_{s_2} = 0$.

b.) For $\bar{S} \neq 0$ and $\overline{W} = 0$ we only find an axisymmetric strain or compression which allows for a rotation symmetry about the axis of strain or compression.

c.) For $\bar{S} = 0$ and $\overline{W} \neq 0$ we always find the axis of symmetry parallel to the vector .

d.) For $\bar{S} \neq 0$ and $\overline{W} \neq 0$ we find at most one rotation symmetry if the axis of rotation is parallel to the axis of strain or compression.

e.) For none of the parameter combinations $\bar{A}$ there is more than one rotation symmetry.

The latter result is obvious from a geometrical point since for a given non-zero $\bar{A}$ at most one symmetry is achievable.

In contrast, a reduction due to the remaining scaling group may always be achieved. The condition for an invariant solution for homogeneous turbulence with zero mean velocity is given by (184), where in the present case we have the symmetry breaking of scaling of time which imposes the constraint $k_{s_2} = 0$

$$\frac{\mathrm{d}t}{k_t} = \frac{\mathrm{d}r_{[k]}}{k_{s_1} r_{[k]}} = \frac{\mathrm{d}R_{[ij]}}{2k_{s_1} R_{[ij]}} = \ldots \quad . \tag{213}$$

The resulting similarity coordinates are

$$\tilde{r}_k = \frac{r_k}{\exp\left(\frac{k_{s_1}}{k_t} t\right)} \quad , \qquad R_{ij} = \exp\left(2\frac{k_{s_1}}{k_t} t\right) \tilde{R}_{ij} \ , \tag{214}$$

where the integration constants $\tilde{r}_k$, $\tilde{R}_{ij}$ may be taken as new variables. Introducing them into the large-scale two-point correlation equation we immediately obtain a reduction to the latter variables. Invoking the limit $\boldsymbol{r} \longrightarrow 0$ we obtain the classical result for the Reynolds stress tensor

$$\overline{u_i' u_j'} \propto \exp\left(2\frac{k_{s_1}}{k_t} t\right) \quad . \tag{215}$$

In Figure 18 we have the DNS results of a homogeneous shear flow taken from Rogers and Moin (1987) where the temporal development of the turbulent kinetic energy is shown. After an initial transient of time we find that the exponential behavior according to (215) has been established.

In analogy to the evolution of the integral length-scale for homogeneous turbulence with zero mean velocity we find from (187) for the present case

$$\ell_{ijk} \propto \exp\left(\frac{k_{s_1}}{k_t} t\right) \ . \tag{216}$$

5.2 Time-dependent Turbulent Shear Flows

In contrast to the previous sections where we only investigated steady inhomogeneous shear flows or time-dependent homogeneous shear we analyze time-dependent inhomogeneous shear

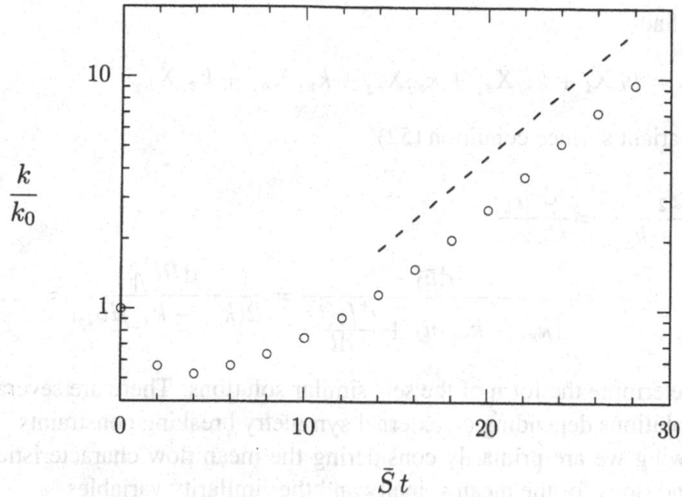

Figure 18: Temporal development of the turbulent kinetic energy for homogeneous shear from Rogers and Moin (1987); $- - -$, parallel line according to the exponential function (215).

flows in the present subsection. Hence we consider a one-dimensional time-dependent Reynolds averaged mean velocity $\bar{\boldsymbol{u}}$ in the x_1-direction

$$\bar{\boldsymbol{u}} = \begin{pmatrix} \bar{u}_1(x_2,t) \\ 0 \\ 0 \end{pmatrix} . \tag{217}$$

Taking the latter assumption into account the subsequent symmetries have to be considered. Scaling of space and time are respectively

$$\bar{\mathrm{X}}_{s_1} = x_2\frac{\partial}{\partial x_2} + \bar{u}_1\frac{\partial}{\partial \bar{u}_1} + r_i\frac{\partial}{\partial r_i} + 2R_{ij}\frac{\partial}{\partial R_{ij}} + \dots \tag{218}$$

and

$$\bar{\mathrm{X}}_{s_2} = t\frac{\partial}{\partial t} - \bar{u}_1\frac{\partial}{\partial \bar{u}_1} - 2R_{ij}\frac{\partial}{\partial R_{ij}} + \dots , \tag{219}$$

The generalized Galilean invariance in the x_1-direction gives

$$\bar{\mathrm{X}}_{\bar{u}_1} = \frac{\mathrm{d}f_1(t)}{\mathrm{d}t}\frac{\partial}{\partial \bar{u}_1} - x_1\frac{\mathrm{d}^2 f_1(t)}{\mathrm{d}t^2}\frac{\partial}{\partial \bar{p}_1} , \tag{220}$$

the translation invariance in x_2-direction is defined by

$$\bar{\mathrm{X}}_{x_2} = \frac{\partial}{\partial x_2} . \tag{221}$$

and invariance with respect to time yields

$$\bar{\mathrm{X}}_{t} = \frac{\partial}{\partial t} . \tag{222}$$

In combined form we find

$$\bar{\mathrm{X}} = k_t \bar{\mathrm{X}}_t + k_{s_1}\bar{\mathrm{X}}_{s_1} + k_{s_2}\bar{\mathrm{X}}_{s_2} + k_{\bar{u}_1}\bar{\mathrm{X}}_{\bar{u}_1} + k_{x_2}\bar{\mathrm{X}}_{x_2} \ . \tag{223}$$

Employing the invariant surface condition (52)

$$\begin{aligned} \frac{\mathrm{d}t}{k_{s_2}t + k_t} = \frac{\mathrm{d}x_2}{k_{s_1}x_2 + k_{x_2}} &= \frac{\mathrm{d}r_{[k]}}{k_{s_1}r_{[k]}} \\ &= \frac{\mathrm{d}\bar{u}_1}{(k_{s_1} - k_{s_2})\bar{u}_1 + \dfrac{\mathrm{d}f_1(t)}{\mathrm{d}t}} = \frac{\mathrm{d}R_{[ij]}}{2(k_{s_1} - k_{s_2})R_{[ij]}} = \dots \ , \end{aligned} \tag{224}$$

the latter generators determine the form of the self-similar solutions. There are several different types of self-similar solutions depending on external symmetry breaking constraints.

Since in the following we are primarily considering the mean flow characteristics we only investigate the two conditions for the mean velocity and the similarity variables

$$\frac{\mathrm{d}x_2}{k_{s_1}x_2 + k_{x_2}} = \frac{\mathrm{d}t}{k_{s_2}t + k_t} \quad \text{and} \quad \frac{\mathrm{d}\bar{u}_1}{(k_{s_1} - k_{s_2})\bar{u}_1 + \dfrac{\mathrm{d}f_1(t)}{\mathrm{d}t}} = \frac{\mathrm{d}t}{k_{s_2}t + k_t} \tag{225}$$

which lead to the following self-similar solutions.

For arbitrary k_{s_1} and k_{s_2}, we find from (225) the similarity coordinate

$$\tilde{x}_{2_{alg}} = \frac{k_{s_1}x_2 + k_{x_2}}{(k_{s_2}t + k_t)^{\frac{k_{s_1}}{k_{s_2}}}} \tag{226}$$

where k_{x_2} and k_t may be set to zero since they only refer to the frame invariance in time and in the x_2-direction.

The corresponding form for the mean flow (225) is

$$\bar{u}_1 = (k_{s_2}t + k_t)^{\frac{k_{s_1}}{k_{s_2}} - 1} \int_t \frac{\mathrm{d}f_1}{\mathrm{d}t}(k_{s_2}t + k_t)^{-\frac{k_{s_1}}{k_{s_2}}} dt + (k_{s_2}t + k_t)^{\frac{k_{s_1}}{k_{s_2}} - 1} \tilde{u}_1(\tilde{x}_{2_{alg}}) \tag{227}$$

The latter scaling law covers the two cases of a wake and mixing layer evolving in time.

5.2.1 Turbulent free mixing layer

The temporally evolving mixing layer describes the flow of two parallel streams with initially constant velocities which develop a free shear layers due to an prescribed velocity difference ΔU between the two streams. ΔU imposes an external velocity scale which is symmetry breaking for the scaling of velocity. In global form we find for the combined scaling symmetries taken from (218)/(219)

$$t^* = \mathrm{e}^{k_{s_2}} t \ , \quad x_2^* = \mathrm{e}^{k_{s_1}} x_2 \quad \text{and} \quad \bar{u}_1^* = \mathrm{e}^{k_{s_1} - k_{s_2}} \bar{u}_1 \ . \tag{228}$$

Due to ΔU we obtain from the latter groups

$$k_{s_1} = k_{s_2}. \tag{229}$$

Since the Euler and Navier-Stokes equations admit the symmetry of a longitudinal accelerating frame we have the symmetry due to $f(t)$. Considering the flow in a non-moving frame one obtains

$$f_1 = 0 . \tag{230}$$

We finally obtain from (227) if $k_t = 0$ and $k_{x_2} = 0$ is invoked

$$\bar{u}_1 = \tilde{u}_{MixL}\left(\frac{x_2}{t}\right) . \tag{231}$$

5.2.2 Turbulent wake flow

A turbulent wake flow describes the flow sufficient by far behind a small two-dimensional obstacle such as a wire. The mean velocity is as such that a constant mean velocity U_0 has a superimposed part $u_s(t)\tilde{u}_1(\delta)$ evolving in time and hence $\bar{u}_1 = U_0 + u_s(t)\tilde{u}_1(\delta)$. Comparing this with the general form of (227) $f(t)$ is uniquely defined as

$$f_1 = U_0(k_{s_2} - k_{s_1})t + c_1 , \tag{232}$$

where c_1 is arbitrary.

In contrast to the mixing layer the present case is dictated by the conservation of mean momentum in x_1-direction. Since the flow has no imposed pressure gradient we obtain from the Reynolds averaged mean momentum equation (78)

$$\frac{\partial}{\partial t}\int_{-\infty}^{\infty}(U_0 - \bar{u}_1)dx_2 = 0 , \tag{233}$$

where all remaining terms have cancelled out. The latter can readily be integrated,

$$\int_{-\infty}^{\infty}(U_0 - \bar{u}_1)dx_2 = \dot{m} , \tag{234}$$

where $\dot{m}$ is the mass flux deficit per unit depth. This is an additional constraint on the flow. Substituting (228) into (234) we find the two group parameter k_{s_1} and k_{s_2} are related to each other by

$$k_{s_2} = 2k_{s_1} . \tag{235}$$

As a result we obtain from (227) with $k_t = 0$ and $k_{x_2} = 0$

$$\bar{u}_1 = U_0 - \frac{1}{t^{\frac{1}{2}}}\hat{u}_{Wake}\left(\frac{x_2}{t^{\frac{1}{2}}}\right) . \tag{236}$$

From the latter we see that the velocity deficit decreases with time with the power of $1/2$ while the width of the wake increases with the same power.

6 Symmetries and Their Implications for Reynolds Averaged and Sub-grid Scale Models in Turbulence

Though the turbulent scaling laws derived in the previous chapters stress the importance of symmetries and give a deep insight into turbulent flows they are not applicable to engineering problems with their usually complex geometries. For this reason engineering problems still have to be tackled by semi-empirical turbulence models which are not derived from first principles. Particularly for the development and calibration of engineering turbulence models the new scaling laws derived from symmetry methods allow one to improve considerably existing models or may be a guideline to develop new models.

The primary premise of the present section is that a turbulence model should admit and reproduce all invariant solutions of the two- and multi-point correlation equations in order to describe turbulence properly.

Considering the development of Reynolds averaged turbulence models over the last hundred years we notice that from an engineering point of view with each new class of model a continuous model improvement was achieved. In view of the symmetry properties of these models we notice that each new class of models has more symmetry properties of the two-point correlation equations than the previous model. This property was usually not mentioned explicitly but was instead the result of physical intuition.

As an example we may consider old turbulence models which were very often not rotationally invariant or translationally invariant, since they were designed for special geometries such as wall-bounded shear flows. Only with the development of two-equation models, such as the K-ε model, there was major progress in this respect. Nevertheless, many two-equation models needed to use damping functions close to solid walls and these are in fact symmetry breaking with respect to translational and rotational invariance. A further important improvement was attained with the development of Reynolds stress transport models which admitted the essential symmetries of the two-point correlation equations. A similar development may be found in the evolution of sub-grid scale models in the context of large-eddy simulation.

A natural conclusion from this development leads to the above premise that a fundamental improvement of turbulence models may only be realized if the symmetry properties and the invariant solutions of the two- and multi-point equations are preserved.

In the following subsection we investigate two of the most common classes of turbulence models with respect to their transformation properties. We will derive necessary conditions for the presentation of symmetry properties. In subsection 6.1 we treat Reynolds averaged transport models and two-equation models such as the K-ε-model. In subsection 6.2 we investigate "sub-grid scale" (SGS) models for "large-eddy simulation" (LES).

6.1 Symmetries of Reynolds averaged turbulence models

It has been observed that certain turbulence models such as a the classical K-ε model exhibit non-physical behavior under certain flow conditions such as rotation or streamline curvature. The origin for this misbehavior can be traced back to certain symmetry properties of the model equations which do not conform to the symmetries of the large-scale two-point correlation equations.

In order to analyze the key problems of Reynolds averaged turbulence models with respect to their symmetry properties we investigate as an example the classical K-ε model of Hanjalić and Launder (1976) and the linear Reynolds stress transport model of Launder et al. (1975) (LRR).

The symmetry properties to be obeyed by the Reynolds averaged models are those admitted by the large-scale two-point correlation equation namely (92)-(95) and (114)/(115). In addition to this we may also consider 2DMFI given by (68) or rather in global form by (70). In Reynolds averaged form we have

$$\begin{aligned} \bar{T}_{2D}:\ t^* &= t\ ,\ \boldsymbol{x}^* = \mathbf{b}(t)\cdot\boldsymbol{x}\ ,\ \bar{\boldsymbol{u}}^* = \mathbf{b}(t)\cdot\bar{\boldsymbol{u}} + \dot{\mathbf{b}}(t)\cdot\boldsymbol{x}\ ,\\ \bar{p}^* &= \bar{p} + 2\,a_8 \int_Q (\bar{u}_1\,\mathrm{d}x_2 - \bar{u}_2\,\mathrm{d}x_1) + \tfrac{1}{2}a_8^2\,(x_1^2 + x_2^2)\ ,\\ \overline{\boldsymbol{u}'\boldsymbol{u}'}^* &= \mathbf{b}(t)\cdot\overline{\boldsymbol{u}'\boldsymbol{u}'}\cdot\mathbf{b}(t)\ , \end{aligned} \tag{237}$$

where $\mathbf{b}(t)$ is defined by

$$b_{ij}(t) = \begin{pmatrix} \cos(a_8 t) & -\sin(a_8 t) & 0\\ \sin(a_8 t) & \cos(a_8 t) & 0\\ 0 & 0 & 1 \end{pmatrix}\ . \tag{238}$$

Khor'kova and Verbovetsky (1995) were probably the first to compute all point symmetries of the K-ε model and the corresponding Lie algebra. From their calculation they concluded that the K-ε model admits all symmetries of inviscid fluid mechanics. These are essentially the symmetries of the large-scale two-point correlation equations as pointed out above.

Though the K-ε model apparently admits all necessary symmetries, we find that it is still not able to reproduce several of the invariant solutions given in subsection 4. This apparent contradiction may be illuminated by two examples of rotating shear flows. For this purpose we consider the K-ε model in its classical form given by

$$\frac{\bar{\mathrm{D}}K}{\bar{\mathrm{D}}t} = -\overline{u_k' u_l'}\frac{\partial \bar{u}_k}{\partial x_l} - \varepsilon + c_K \frac{\partial}{\partial x_l}\left(\nu_t \frac{\partial K}{\partial x_l}\right)\ , \tag{239}$$

$$\frac{\bar{\mathrm{D}}\varepsilon}{\bar{\mathrm{D}}t} = -c_{\varepsilon_1}\overline{u_k' u_l'}\frac{\partial \bar{u}_k}{\partial x_l}\frac{\varepsilon}{K} - c_{\varepsilon_2}\frac{\varepsilon^2}{K} + c_\varepsilon \frac{\partial}{\partial x_l}\left(\nu_t \frac{\partial \varepsilon}{\partial x_l}\right)\ , \tag{240}$$

$$\overline{u_i' u_j'} = \frac{2}{3}\delta_{ij}K - \nu_t 2\bar{S}_{ij} \quad \text{where} \quad \bar{S}_{ij} = \frac{1}{2}\left(\frac{\partial \bar{u}_i}{\partial x_j} + \frac{\partial \bar{u}_j}{\partial x_i}\right)\ , \quad \nu_t = C_\mu \frac{K^2}{\varepsilon}\ . \tag{241}$$

The model parameters are not important at this point and may be taken from Hanjalić and Launder (1976). The equation (239)-(241) have to be supplemented by the continuity equation (77) and the momentum equation (78). For all of the following considerations we introduce the limit $1/Re = 0$ and hence molecular viscosity will be neglected.

For the first test case of a plane shear flow the continuity equation is identical to zero and the momentum equation reduces to

$$0 = -\frac{\mathrm{d}\bar{p}}{\mathrm{d}x_1} + \frac{\mathrm{d}}{\mathrm{d}x_2}\left(\nu_t \frac{\mathrm{d}\bar{u}_1}{\mathrm{d}x_2}\right)\ . \tag{242}$$

Correspondingly the K- and ε-equations simplify to

$$0 = \nu_t \left(\frac{\mathrm{d}\bar{u}_1}{\mathrm{d}x_2}\right)^2 - \varepsilon + c_K \frac{\mathrm{d}}{\mathrm{d}x_2}\left(\nu_t \frac{\mathrm{d}K}{\mathrm{d}x_2}\right) \quad \text{and}$$
$$0 = c_{\varepsilon_1} \nu_t \left(\frac{\mathrm{d}\bar{u}_1}{\mathrm{d}x_2}\right)^2 \frac{\varepsilon}{K} - c_{\varepsilon_2}\frac{\varepsilon^2}{K} + c_\varepsilon \frac{\mathrm{d}}{\mathrm{d}x_2}\left(\nu_t \frac{\mathrm{d}\varepsilon}{\mathrm{d}x_2}\right) . \tag{243}$$

Without carrying out any symmetry analysis it is apparent that the rotation rate $\boldsymbol{\Omega}$ is not contained in any of the latter two equations. The equations (239)-(241) do not contain Coriolis terms for any type of flow and the momentum equations do not contain Coriolis terms for the special test case of a rotating channel flow. Hence it is apparent that the K-ε-model is not able to distinguish between a rotating flow and a flow in an inertial frame of reference. However, we have seen in subsection 4 in Figures 6 and 7 that the rotating channel flow exhibits characteristics very different from those of the non-rotating channel flow.

An improved derivation of the classical two-equation model, which overcomes this problem, at least partially, has been developed by Pope (1975) and further generalized by Gatski and Speziale (1993). Within this framework the Boussinesq model (241) for the Reynolds stress tensor $\overline{\boldsymbol{u}'\boldsymbol{u}'}$ and the mean strain tensor $\bar{S}$ has been extended by the inclusion of the mean rotation tensor

$$\overline{W}_{ij} = \frac{1}{2}\left(\frac{\partial \bar{u}_i}{\partial x_j} - \frac{\partial \bar{u}_j}{\partial x_i}\right) . \tag{244}$$

These extended two-equation models are able to account for strong streamline curvature and in particular they distinguish properly between an inertial and a non-inertial frame of reference, such as for rotating flows.

The problem may also be attacked from a symmetry point of view. The group analysis of (243) reveals beside translational invariance in the x_2-direction and Galilean invariance in the x_1-direction the two scaling groups

$$x_2^* = \mathrm{e}^{k_{s_1}} x_2 \ , \ \ \bar{u}_1^* = \mathrm{e}^{k_{s_1}-k_{s_2}} \bar{u}_1 \ , \ \ K^* = \mathrm{e}^{2(k_{s_1}-k_{s_2})} K \ , \ \ \varepsilon^* = \mathrm{e}^{2k_{s_1}-3k_{s_2}} \varepsilon \ . \tag{245}$$

Evidently we find all symmetries which are necessary to account for all the invariant solutions of plane parallel shear flows in subsection 4. However, since there is no mechanism that allows for a symmetry breaking of scaling of time due to Coriolis forces within the equation, a linear mean velocity profile according to (151) can not be obtained. As a conclusion the K-ε-model admits too many symmetries for the rotating channel case.

A similar problem with the K-ε model occurs for axisymmetric parallel shear flows with rotation, such as rotating turbulent pipe flow. For this test case also the model equations (239)-(241) have too many symmetries, which are not contained in the two-and multi-point equations for the large scales. In cylindrical coordinates the momentum equation in the azimuthal direction reduces to

$$0 = \frac{1}{r^2}\frac{\mathrm{d}}{\mathrm{d}r}\left(r^3 \nu_t \frac{\mathrm{d}}{\mathrm{d}r}\frac{\bar{u}_\phi}{r}\right) . \tag{246}$$

while the K- and ε-equations simplify to

$$0 = \nu_t \left[\left(r \frac{\mathrm{d}}{\mathrm{d}r} \frac{\bar{u}_\phi}{r} \right)^2 + \left(\frac{\mathrm{d}\bar{u}_z}{\mathrm{d}r} \right)^2 \right] - \varepsilon + c_K \frac{1}{r} \frac{\mathrm{d}}{\mathrm{d}r} \left(\nu_t r \frac{\mathrm{d}K}{\mathrm{d}r} \right) \quad \text{and}$$

$$0 = c_{\varepsilon_1} \nu_t \left[\left(r \frac{\mathrm{d}}{\mathrm{d}r} \frac{\bar{u}_\phi}{r} \right)^2 + \left(\frac{\mathrm{d}\bar{u}_z}{\mathrm{d}r} \right)^2 \right] \frac{\varepsilon}{K} - c_{\varepsilon_2} \frac{\varepsilon^2}{K} + c_\varepsilon \frac{1}{r} \frac{\mathrm{d}}{\mathrm{d}r} \left(\nu_t r \frac{\mathrm{d}\varepsilon}{\mathrm{d}r} \right) . \tag{247}$$

One may easily verify that the above equations admit all the necessary symmetries, as given in subsection 4.2, for a rotating turbulent pipe flow. These are Galilean invariance in the z-direction as well as both scaling symmetries. However a complete group analysis of the latter equations discloses an additional symmetry of the form:

$$r^* = r \ , \ \bar{u}_z^* = \bar{u}_z \ , \ \bar{u}_\phi^* = \bar{u}_\phi + b\, r \ , \ K^* = K \ , \ \varepsilon^* = \varepsilon \ , \tag{248}$$

where b represents the group parameter. This additional unphysical symmetry allows one to add a solid body rotation to the azimuthal velocity without any change to the remaining flow quantities. As an immediate result this leads to a linear azimuthal velocity in a rotating pipe flow, as has been reported by Hirai et al. (1988) who solved the K-ε-model for the rotating pipe case. The unphysical linear profile may readily be derived from the condition of invariant solution in (167) if, in addition to the classical symmetries, the above unphysical symmetry is added.

In contrast to the predictions based on the two-equation models such as the K-ε model for the rotating pipe flow, Hirai et al. (1988) found from numerical solutions of the Reynolds stress transport model of Launder et al. (1975) that, at least qualitatively, the experimental results are reproduced. The foundation of this is the symmetry structure of the model equations

$$\begin{aligned}
\frac{\bar{\mathrm{D}} \overline{u_i' u_j'}}{\bar{\mathrm{D}} t} = & - \overline{u_i' u_k'} \frac{\partial \bar{u}_j}{\partial x_k} - \overline{u_j' u_k'} \frac{\partial \bar{u}_i}{\partial x_k} \\
& + c_1 K \bar{S}_{ij} + c_2 K \left(b_{ik} \bar{S}_{jk} + b_{jk} \bar{S}_{ik} - \frac{2}{3} \delta_{ij} b_{kl} \bar{S}_{kl} \right) + c_3 K \left(b_{ik} \overline{W}_{jk} + b_{jk} \overline{W}_{ik} \right) \\
& - c_{\mathrm{R}} b_{ij} \varepsilon - \frac{2}{3} \delta_{ij} \varepsilon \\
& + c_s \frac{\partial}{\partial x_k} \frac{K}{\varepsilon} \left(\overline{u_i' u_l'} \frac{\partial \overline{u_j' u_k'}}{\partial x_l} + \overline{u_j' u_l'} \frac{\partial \overline{u_k' u_i'}}{\partial x_l} + \overline{u_k' u_l'} \frac{\partial \overline{u_i' u_j'}}{\partial x_l} \right) \\
& - 2\, \Omega_k \left[e_{kli} \overline{u_j' u_l'} + e_{klj} \overline{u_i' u_l'} \right] ,
\end{aligned} \tag{249}$$

$$\frac{\bar{\mathrm{D}} \varepsilon}{\bar{\mathrm{D}} t} = - c_{\varepsilon_1} \overline{u_k' u_l'} \frac{\partial \bar{u}_k}{\partial x_l} - c_{\varepsilon_2} \frac{\varepsilon^2}{K} + c_\varepsilon \frac{\partial}{\partial x_l} \left(\nu_t \frac{\partial \varepsilon}{\partial x_l} \right) , \tag{250}$$

where $\bar{S}_{ij}$ is defined in (241) and $\overline{W}_{ij}$ is given by the extended definition of (244)

$$\overline{W}_{ij} = \frac{1}{2} \left(\frac{\partial \bar{u}_i}{\partial x_j} - \frac{\partial \bar{u}_j}{\partial x_i} \right) + e_{mji} \Omega_m \ , \tag{251}$$

due to the system rotation.

In an inertial system ($\boldsymbol{\Omega} = 0$) the LRR-model admits all symmetries of the large-scale two-point correlation equations. In fact, one may easily show that not only linear pressure-strain models admit all proper symmetries but also all non-linear pressure-strain models. Only the introduction of wall-damping functions leads to an unphysical symmetry breaking. For this reason there have been many attempts to overcome wall treatment by wall-damping functions e.g. with the elliptic-relaxation model introduced by Durbin (1991).

A detailed symmetry analysis of the different flow cases in subsection 4 confirms that all proper symmetries are admitted and in fact also the necessary symmetry breaking is revealed by the system rotation.

The condition for an invariant solution in the case of the plane parallel shear flow is given by

$$\frac{\mathrm{d}x_2}{k_{s_1}x_2 + k_{x_2}} = \frac{\mathrm{d}\bar{u}_1}{(k_{s_1} - k_{s_2})\bar{u}_1 + k_{\bar{u}_1}} = \frac{\mathrm{d}\overline{u_{[i]}u_{[j]}}}{2(k_{s_1} - k_{s_2})\overline{u_{[i]}u_{[j]}}} = \frac{\mathrm{d}\varepsilon}{(2k_{s_1} - 3k_{s_2})\varepsilon} \,. \tag{252}$$

The essence of the analysis of the K-ε- and the LRR-model given above is that Reynolds-averaged turbulence models may not be invariant under time-dependent rotation. This problem becomes particularly obvious for flows with two or three homogeneous directions since in this case the model equations decouple from the mean momentum equations.

Accordingly we have to conclude that it is not sufficient to consider the symmetry properties of the entire system comprising the continuity, momentum and model equations for the statistical quantities. Instead we have to investigate the symmetry properties for those cases separately where the continuity and momentum equation decouple and only the model equations for the statistical quantities determine the flow. As a result we find that all model equations independent of each other must be sensitive to an non-inertial frame and hence may not be independent under time-dependent rotation.

The most important necessary conditions for Reynolds-averaged turbulence models may be summarized as follows:

a) All symmetries of the two- and multi-point correlation equations must be admitted by the model equations.

b) There should be no additional symmetries in the model equations even for reduced cases such as those admitting rotational symmetry.

c) The symmetry conditions a) and b) have to be admitted by each single model equation and independently by the momentum and continuity equation.

d) All invariant solutions implied by the two- and multi-point correlation equations must also be admitted by the model equations.

The last condition implies that by implementing the invariant solution into the model equation either a reduction of the independent variables to similarity variables must be achieved or an exact solution has to be obtained.

Within this subsection it has become apparent that it is difficult to develop model equations which admit all proper symmetries. In particular the two-equation models do not properly describe rotating flows such as the rotating channel or pipe flow. These deficiencies originate from inappropriate symmetry properties particularly the violation of condition b) and c).

Reynolds transport models have superior symmetry properties. In fact a few models been developed which not only comply with condition a) and b) but also with condition c). (see e.g. Oberlack, 1994, 1997 and Tagawa et al., 1991).

6.2 Symmetries in sub-grid scale-models in large-eddy simulation

In contrast to the Reynolds averaged equations, where statistical quantities are used, in large-eddy simulation of turbulence all flow quantities such as velocity $\hat{u}_i$ and pressure $\hat{p}$ are still random variables. Filtered quantities are denoted by "^" and will be defined according to

$$\hat{Z}(\boldsymbol{x},t) = \mathcal{T}[Z(\boldsymbol{x},t)] = \int_V G(\boldsymbol{x},\boldsymbol{y}) Z(\boldsymbol{y},t) \mathrm{d}^3 y \ . \tag{253}$$

Applying the latter operator onto the Navier-Stokes equations (53) and (54) and neglecting commutation issues for the moment we obtain

$$\frac{\partial \hat{u}_k}{\partial x_k} = 0 \tag{254}$$

and

$$\frac{\partial \hat{u}_i}{\partial t} + \hat{u}_k \frac{\partial \hat{u}_i}{\partial x_k} = -\frac{\partial \hat{p}}{\partial x_i} + \nu \frac{\partial^2 \hat{u}_i}{\partial x_k^2} - \frac{\partial \hat{\tau}_{ik}}{\partial x_k} \ . \tag{255}$$

The latter equation contains the sub-grid scale stress $\hat{\tau}_{ik}$ as an unclosed term, which is usually sub-divided into three parts

$$\hat{\tau}_{ik} = L_{ik} + C_{ik} + Q_{ik} \ , \tag{256}$$

defined by

$$L_{ik} = \widehat{\hat{u}_i \hat{u}_k} - \hat{u}_i \hat{u}_k \ , \quad C_{ik} = \widehat{u_i'' \hat{u}_k + \hat{u}_i u_k''} \quad \text{and} \quad Q_{ik} = \widehat{u_i'' u_k''} \ . \tag{257}$$

u_i'' and p'' are defined by

$$u_i'' = u_i - \hat{u}_i \quad \text{and} \quad p'' = p - \hat{p} \ . \tag{258}$$

They denote the sub-grid quantities which are not resolved for a given grid spacing Δ.

The unclosed terms L_{ik}, C_{ik} and Q_{ik} or rather $\hat{\tau}_{ik}$ are usually modelled as algebraic or differential-algebraic functionals of $\hat{u}_i$ and x_k. Transport equations for the sub-grid energy $\hat{\tau}_{kk}/2$ have also been suggested. An excellent overview on the existing models may be found in Ferziger (1996), Ferziger and Peric (1996) and Fureby et al. (1997). Due to its very common use we will restrict ourselves to algebraic or differential-algebraic closure models.

Since we intend to model important properties of the Euler- and Navier-Stokes equations we require all symmetries (61)-(66) and (70) to be admitted also by the filtered equations.

Since the sub-grid scale stresses are determined by the unresolved velocities their transformation properties need to be investigated. For this reason the filtering (253) will be applied to each of the symmetries (61)-(66). As a direct result we obtain the transformation properties of the filtered velocity and pressure. From the decomposition of the resolved and the sub-grid quantities in (258) we obtain the corresponding symmetry transformations. If in the following we implement the resulting symmetry transformation rules for the resolved and the sub-grid quantities into the equation for the sub-grid stresses (256) and (257) we obtain the symmetry properties for all terms in the filtered Navier-Stokes equations (254) and (255). The detailed transformation rules for each of the symmetries in (61)-(66) are given by the following list:

$$\begin{array}{lllll}\hat{T}_t: & t^* = t + a_1 \ , & \boldsymbol{x}^* = \boldsymbol{x} \ , & \hat{\boldsymbol{u}}^* = \hat{\boldsymbol{u}} \ , & \hat{p}^* = \hat{p}, \\ & \hat{\boldsymbol{\tau}}^* = \hat{\boldsymbol{\tau}} \ , & L^* = L \ , & C^* = C \ , & Q^* = Q,\end{array} \qquad (259)$$

$$\begin{array}{lllll}\hat{T}_{r_1} - \hat{T}_{r_3}: & t^* = t \ , & \boldsymbol{x}^* = \mathbf{a} \cdot \boldsymbol{x} \ , & \hat{\boldsymbol{u}}^* = \mathbf{a} \cdot \hat{\boldsymbol{u}} \ , & \hat{p}^* = \hat{p}, \\ & \hat{\boldsymbol{\tau}}^* = \mathbf{a} \cdot \hat{\boldsymbol{\tau}} \cdot \mathbf{a}^{\mathsf{T}}, & L^* = \mathbf{a} \cdot L \cdot \mathbf{a}^{\mathsf{T}}, & C^* = \mathbf{a} \cdot C \cdot \mathbf{a}^{\mathsf{T}}, & Q^* = \mathbf{a} \cdot Q \cdot \mathbf{a}^{\mathsf{T}},\end{array} \qquad (260)$$

$$\begin{array}{lllll}\hat{T}_{\hat{u}_1} - \hat{T}_{\hat{u}_3}: & t^* = t \ , & \boldsymbol{x}^* = \boldsymbol{x} + \boldsymbol{f}(t) \ , & \hat{\boldsymbol{u}}^* = \hat{\boldsymbol{u}} + \dfrac{\mathrm{d}\boldsymbol{f}}{\mathrm{d}t} \ , & \hat{p}^* = \hat{p} - \boldsymbol{x} \cdot \dfrac{\mathrm{d}^2\boldsymbol{f}}{\mathrm{d}t^2}, \\ & \boldsymbol{\tau}^* = \boldsymbol{\tau} \ , & L^* = L - & C^* = C + & Q^* = Q, \\ & & \dfrac{\mathrm{d}\boldsymbol{f}}{\mathrm{d}t}\hat{\boldsymbol{u}''} - \hat{\boldsymbol{u}''}\dfrac{\mathrm{d}\boldsymbol{f}}{\mathrm{d}t}, & \dfrac{\mathrm{d}\boldsymbol{f}}{\mathrm{d}t}\hat{\boldsymbol{u}''} + \hat{\boldsymbol{u}''}\dfrac{\mathrm{d}\boldsymbol{f}}{\mathrm{d}t}, & \end{array} \qquad (261)$$

$$\begin{array}{lllll}\hat{T}_{\hat{p}}: & t^* = t \ , & \boldsymbol{x}^* = \boldsymbol{x} \ , & \hat{\boldsymbol{u}}^* = \hat{\boldsymbol{u}} \ , & \hat{p}^* = \hat{p} + a_4 f_4(t), \\ & \hat{\boldsymbol{\tau}}^* = \hat{\boldsymbol{\tau}} \ , & L^* = L \ , & C^* = C \ , & Q^* = Q,\end{array} \qquad (262)$$

$$\begin{array}{lllll}\hat{T}_{s_1}: & t^* = t \ , & \boldsymbol{x}^* = \mathrm{e}^{a_2}\boldsymbol{x} \ , & \hat{\boldsymbol{u}}^* = \mathrm{e}^{a_2}\hat{\boldsymbol{u}} \ , & \hat{p}^* = \mathrm{e}^{2a_2}\hat{p}, \\ & \hat{\boldsymbol{\tau}}^* = \mathrm{e}^{2a_2}\hat{\boldsymbol{\tau}} \ , & L^* = \mathrm{e}^{2a_2}L \ , & C^* = \mathrm{e}^{2a_2}C \ , & Q^* = \mathrm{e}^{2a_2}Q,\end{array} \qquad (263)$$

$$\begin{array}{lllll}\hat{T}_{s_2}: & t^* = \mathrm{e}^{a_3}t \ , & \boldsymbol{x}^* = \boldsymbol{x} \ , & \hat{\boldsymbol{u}}^* = \mathrm{e}^{-a_3}\hat{\boldsymbol{u}} \ , & \hat{p}^* = \mathrm{e}^{-2a_3}\hat{p}, \\ & \hat{\boldsymbol{\tau}}^* = \mathrm{e}^{-2a_3}\hat{\boldsymbol{\tau}} \ , & L^* = \mathrm{e}^{-2a_3}L \ , & C^* = \mathrm{e}^{-2a_3}C \ , & Q^* = \mathrm{e}^{-2a_3}Q,\end{array} \qquad (264)$$

Time invariance $\hat{T}_t$ in (259) and pressure invariance $\hat{T}_{\hat{p}}$ in (262) are rather easily observed by simply ensuring that the models for the functionals L, C, Q or $\hat{\boldsymbol{\tau}}$ do not explicitly depend on t and $\hat{p}$. Any model in the literature is consistent with the two constraints (259) and (262).

Rotation invariance $\hat{T}_{r_1}$-$\hat{T}_{r_3}$ in (260) has an identical structure for both the large and sub-grid scale quantities. The transformation matrix $\mathbf{a}$ obeys the constraints defined below (70). In fact, any sub-grid scale model which is written in proper tensorial form obeys the symmetry (260). None of the existing SGS models violate this property.

Speziale (1985) was the first to observe that there are a variety of SGS models which are not Galilean invariant. In extension of this fact we find the generalized Galilean invariance leads to the transformation rule $\hat{T}_{\hat{u}_1}$-$\hat{T}_{\hat{u}_3}$ in (261). Apparently L and C are not form-invariant under (261) while the sum of all stress is invariant under a translatorial motion $\boldsymbol{f}$. Germano (1986) suggested a modified decomposition of $\hat{\boldsymbol{\tau}}$ in which each single term is Galilean invariant. However, any decomposition of $\hat{\boldsymbol{\tau}}$ is arbitrary so that one should only investigate $\hat{\boldsymbol{\tau}}$ on its transformation properties.

The necessity of Galilean invariant SGS models has been nicely demonstrated by Härtel and Kleiser (1997) by showing that its violation may in fact lead to negative dissipation.

The two scaling groups $\hat{T}_{s_1}$ and $\hat{T}_{s_2}$ in (263) and (264) are in fact only valid for the case $\nu = 0$ i.e. Euler's equation. On page 360 ff. we shall prove that a very large number of SGS models including the classical Smagorinsky model (Smagorinsky, 1963) violates (263) and (264) in the limit $\nu = 0$. The immediate consequence is that owing to this artificial symmetry breaking a variety of invariant solutions can not be obtained.

The last symmetry which we will investigate with respect to its transformation properties of SGS models is the 2DMFI which is given in global form in (70). Though this symmetry is strictly speaking only valid for 2D flows it is clearly important for flows which approach the 2D state such as flows very close to boundaries or more importantly for flows with strong rotation. The latter flows have a tendency to become two-dimensional in the limit of strong rotation in accordance with the Taylor-Proudman theorem.

According to the LES decomposition the symmetry (70) leads to

$$
\begin{aligned}
\hat{T}_{2D}: \quad t^* &= t \ , \ \boldsymbol{x}^* = \mathbf{b}(t)\cdot\boldsymbol{x} \ , \ \hat{\boldsymbol{u}}^* = \mathbf{b}(t)\cdot\hat{\boldsymbol{u}} + \dot{\mathbf{b}}(t)\cdot\boldsymbol{x} \ , \\
\hat{p}^* &= \hat{p} + 2\,a_8 \int_Q (\hat{u}_1\,\mathrm{d}x_2 - \hat{u}_2\,\mathrm{d}x_1) + \tfrac{1}{2}a_8^2\,(x_1^2 + x_2^2) \ , \\
\hat{\boldsymbol{\tau}}^* &= \mathbf{b}(t)\cdot\hat{\boldsymbol{\tau}}\cdot\mathbf{b}(t)^{\mathsf{T}} \ , \\
L^* &= \mathbf{b}(t)\cdot L\cdot\mathbf{b}(t)^{\mathsf{T}} - \left(\mathbf{b}(t)\cdot\hat{\boldsymbol{u''}}\right)\left(\dot{\mathbf{b}}(t)\cdot\boldsymbol{x}\right) - \left(\dot{\mathbf{b}}(t)\cdot\boldsymbol{x}\right)\left(\mathbf{b}(t)\cdot\hat{\boldsymbol{u''}}\right) \ , \\
C^* &= \mathbf{b}(t)\cdot C\cdot\mathbf{b}(t)^{\mathsf{T}} + \left(\mathbf{b}(t)\cdot\hat{\boldsymbol{u''}}\right)\left(\dot{\mathbf{b}}(t)\cdot\boldsymbol{x}\right) + \left(\dot{\mathbf{b}}(t)\cdot\boldsymbol{x}\right)\left(\mathbf{b}(t)\cdot\hat{\boldsymbol{u''}}\right) \ , \\
Q^* &= \mathbf{b}(t)\cdot Q\cdot\mathbf{b}(t)^{\mathsf{T}} \ .
\end{aligned}
\tag{265}
$$

where the rotation matrix $\mathbf{b}(t)$ is defined by

$$
b_{ij}(t) = \begin{pmatrix} \cos(a_8 t) & -\sin(a_8 t) & 0 \\ \sin(a_8 t) & \cos(a_8 t) & 0 \\ 0 & 0 & 1 \end{pmatrix} \ . \tag{266}
$$

Due to two-dimensionality all dependent variables are functions of x_1, x_2 and t only. The rotation axis is parallel to the axis of independence, here x_3. From (265) we find that the sum of all SGS stresses is invariant under a time-dependent rotation.

In order to examine the symmetry properties of certain SGS models in detail we shall investigate in the following certain very common models. All of these models are of the usual type

$$
\hat{\boldsymbol{\tau}} = \boldsymbol{\Xi}(\hat{\boldsymbol{u}}, \boldsymbol{x}) \ , \tag{267}
$$

where $\boldsymbol{\Xi}$ is a functional of the argument $\hat{\boldsymbol{u}}$ and $\boldsymbol{x}$. In order to close the filtered Navier-Stokes equations (255) $\hat{\boldsymbol{\tau}}$ is replaced by (267). For all of the above symmetries the equation (255) and the modelled equation should have the same properties.

Accordingly the model (267) should also be valid in any of the transformed coordinate systems. For the functional form (267) this means

$$
\hat{\boldsymbol{\tau}}^* = \boldsymbol{\Xi}(\hat{\boldsymbol{u}}^*, \boldsymbol{x}^*) \ , \tag{268}
$$

for any of the symmetries (259)-(265).

Practically all known models are correct with respect to time invariance (259), rotation invariance (260) and pressure invariance (262). Consequently we will not consider these symmetries here.

Speziale (1985) was the first to note that several SGS models are not Galilean invariant (e.g. Biringen and Reynolds, 1981, Moin and Kim, 1982, Bardina et al., 1980). Hence these models are also not invariant under the more general transformation (261).

The models to be investigated in the following are the classical Smagorinsky model (Smagorinsky, 1963), the "structure-functions" model by Métais and Lesieur (1992) and the "dynamic model" by Germano et al. (1990, 1991).

The Smagorinsky model assumes a pivotal role amongst the SGS models since it is the model most often used in the field of LES. Though it captures many technical flows with reasonable accuracy it essentially fails to properly describe near-wall flows without empirical wall functions. This problem is caused by the fact, that scaling invariance (263) is violated by the Smagorinsky model. In its classical form we have

$$\hat{\tau}_{ik} - \frac{1}{3}\delta_{ik}\hat{\tau}_{mm} = -C\Delta^2|\hat{S}|\hat{S}_{ik} \qquad \text{where} \qquad \hat{S}_{ik} = \frac{1}{2}\left(\frac{\partial \hat{u}_i}{\partial x_k} + \frac{\partial \hat{u}_k}{\partial x_i}\right) . \tag{269}$$

Δ is the filter length which is usually represented by the local mesh width. In order to see the disadvantageous properties of (269) which lead to the errors mentioned above, we may analyze the equivalence between (267) and (268) in the context of scale invariance (263). Writing the Smagorinsky model in the transformed "*"-coordinate system we have

$$\hat{\tau}^*_{ik} - \frac{1}{3}\delta_{ik}\hat{\tau}^*_{mm} = -C\Delta^2|\hat{S}^*|\hat{S}^*_{ik} \qquad \text{where} \qquad \hat{S}^*_{ik} = \frac{1}{2}\left(\frac{\partial \hat{u}^*_i}{\partial x^*_k} + \frac{\partial \hat{u}^*_k}{\partial x^*_i}\right) . \tag{270}$$

Imposing the known exact scale invariance (263) into the left hand side of (270) and in the same manner putting the scale invariance in the modelled right hand side of (270) we obtain

$$\begin{aligned} \mathrm{e}^{2(a_2-a_3)}\left(\hat{\tau}_{ik} - \frac{1}{3}\delta_{ik}\hat{\tau}_{mm}\right) &= -C\Delta^2|\hat{S}|\hat{S}_{ik}\mathrm{e}^{-2a_3} \\ \Rightarrow \quad \hat{\tau}_{ik} - \frac{1}{3}\delta_{ik}\hat{\tau}_{mm} &= -C\Delta^2|\hat{S}|\hat{S}_{ik}\mathrm{e}^{-2a_2} . \end{aligned} \tag{271}$$

Apparently the last expression is not form-invariant under the scaling transformation since a_2 is still a parameter in the equation. Hence we find $a_2 = 0$. The source of this problem is that there is a constant filter-length in the problem which is symmetry breaking with respect to space.

One may object that also Δ, similar to the spatial coordinate x, has to be scaled as $\Delta^* = \mathrm{e}^{a_2}\Delta$. As expected this *appears* to remedy the problem of the Smagorinsky model with respect to the scaling symmetry. However, Δ is not a free parameter, which is determined by the flow, but instead it is fixed a priori by the choice of the grid.

In order to better understand this deficiency of the Smagorinsky model we consider the invariant solution of the Navier-Stokes equations in the limit of $Re \to \infty$. In particular we employ the scaling groups X_{s_1} and X_{s_2} in (60) for the case of decaying turbulence. From the condition of an invariant solution (52) we find from the combination of the latter two scaling groups

$$\begin{aligned} &\frac{\mathrm{d}x_1}{a_2x_1} = \frac{\mathrm{d}x_2}{a_2x_2} = \frac{\mathrm{d}x_3}{a_2x_3} = \frac{\mathrm{d}t}{a_3t} \\ &\qquad = \frac{\mathrm{d}u_1}{(a_2-a_3)u_1} = \frac{\mathrm{d}u_2}{(a_2-a_3)u_1} = \frac{\mathrm{d}u_3}{(a_2-a_3)u_3} = \frac{\mathrm{d}p}{2(a_2-a_3)p} , \end{aligned} \tag{272}$$

where a_2 and a_3 are respectively the group parameters of scaling of space and time. From the solution of the latter equation we obtain the invariants $\check{\boldsymbol{u}}$ and $\check{p}$ which are taken as new variables

$$\boldsymbol{u} = t^{\frac{a_2}{a_3}-1}\check{\boldsymbol{u}}(\boldsymbol{\eta}), \quad p = t^{2\left(\frac{a_2}{a_3}-1\right)}\check{p}(\boldsymbol{\eta}) \; . \tag{273}$$

Employing them in (54) with $\nu = 0$ we obtain the reduced equation

$$\left(\frac{a_2}{a_3} - 1\right)\check{\boldsymbol{u}}_i - \frac{a_2}{a_3}\eta_k \frac{\partial \check{\boldsymbol{u}}_i}{\partial \eta_k} + \check{\boldsymbol{u}}_k \frac{\partial \check{\boldsymbol{u}}_i}{\partial \eta_k} = -\frac{\partial \check{p}}{\partial \eta_i} \; . \tag{274}$$

As expected the invariants (273) from the two scaling groups lead to a reduction of the equation of fluid motion.

If the modelled equations would also admit two scaling groups we should also find a corresponding reduction. However, using (273) in the filtered equations where the Smagorinsky model (269) has been employed we find

$$\left(\frac{a_2}{a_3} - 1\right)\hat{\check{\boldsymbol{u}}}_i - \frac{a_2}{a_3}\eta_k \frac{\partial \hat{\check{\boldsymbol{u}}}_i}{\partial \eta_k} + \hat{\check{\boldsymbol{u}}}_k \frac{\partial \hat{\check{\boldsymbol{u}}}_i}{\partial \eta_k} = -\frac{\partial \hat{\check{p}}}{\partial \eta_i} + \frac{\partial}{\partial \eta_k}\left(C\Delta^2|\hat{\check{S}}|\hat{\check{S}}_{ik}\right)t^{-2\frac{a_2}{a_3}} \; , \tag{275}$$

where $\check{\boldsymbol{u}}$ and $\check{p}$ have been replaced by $\hat{\check{\boldsymbol{u}}}$ and $\hat{\check{p}}$.

Obviously (275) is not independent of t. Hence, similarity can only be obtained for the restricted case

$$a_2 = 0. \tag{276}$$

As a result we find that the Smagorinsky model does not admit two scaling groups since the reduction which has been achieved in (274) does not work out in (275).

We have to conclude that the general form of the similarity reduction of (273) is violated by the Smagorinsky model. If in addition $\nu \neq 0$ is considered we find that any similarity reduction is destroyed. In fact, the latter problem of the Smagorinsky model is inherent to almost all exact solutions of the Euler and the Navier-Stokes equations.

It is crucial to understand that the above problem of the Smagorinsky model also "transfers" to all statistical equations which can be derived from the filtered Navier-Stokes equations extended by equation (269). In complete analogy to the two- and multi-point correlation equationsobtained from the Navier-Stokes equations one can derive two- and multi-point correlation equations from the filtered equations including a model. Investigating these equations with respect to their consistency with respect to certain turbulent scaling laws obtained in subsections 4 and 5 we find that e.g. the logarithmic law of the wall is violated. This is due to the fact that the combination of two scaling groups is needed for its derivation.

A model which violates both scaling symmetry (263) and 2DMFI (265) is the "structure-function" model of Métais and Lesieur (1992). The functional form of the model is given by

$$\hat{\tau}_{ik} - \frac{1}{3}\delta_{ik}\hat{\tau}_{mm} = C^{SF}\,\Delta I\{(\hat{\boldsymbol{u}}(\boldsymbol{x}+\boldsymbol{r}) - \hat{\boldsymbol{u}}(\boldsymbol{x}))^2\}^{1/2}\hat{S}_{ik} \; , \tag{277}$$

where C^{SF} is a constant and $I\{\cdot\}$ represents spatial averaging in the distance space $\boldsymbol{r}$. In analogy to the equivalence between (267) and (268) for any of the transformation rules (259)-(265) we require here also the equivalence between

$$\hat{\tau}^*_{ik} - \frac{1}{3}\delta_{ik}\hat{\tau}^*_{mm} = C^{SF}\Delta\, I\{(\hat{\boldsymbol{u}}^*(\boldsymbol{x}^* + \boldsymbol{r}^*) - \hat{\boldsymbol{u}}^*(\boldsymbol{x}))^2\}^{1/2}\hat{S}^*_{ik} \; . \tag{278}$$

and (277).

Investigating the 2DMFI symmetry we impose on both sides the transformation rules according to (265) yielding

$$\hat{\tau}_{ik} - \frac{1}{3}\delta_{ik}\hat{\tau}_{mm} = C^{SF}\Delta\, I\{(\hat{\boldsymbol{u}}_{(m)}(\boldsymbol{x}+\boldsymbol{r}) - \hat{\boldsymbol{u}}_{(m)}(\boldsymbol{x}) - e_{3l(m)}a_8 r_l)^2\}^{1\overline{2}}\hat{S}_{ik}. \tag{279}$$

It is apparent that the model (277) is not invariant under 2DMFI. As in our investigation of the Smagorinsky model we find that (277) also violates scaling properties.

In contrast to the latter two models the "dynamic model" by Germano et al. (1990, 1991) preserves all necessary symmetry properties (259)-(265). In its base form the model reads

$$\hat{\tau}_{ik} - \frac{1}{3}\delta_{ik}\hat{\tau}_{mm} = \frac{\widetilde{(\hat{u}_m\hat{u}_n - \tilde{\hat{u}}_m\tilde{\hat{u}}_n)\hat{S}_{mn}}}{\left(\frac{\tilde{\Delta}}{\Delta}\right)^2 |\tilde{\hat{S}}|\tilde{\hat{S}}_{mn}\tilde{\hat{S}}_{mn} - \widetilde{|\hat{S}|\hat{S}_{pq}}\tilde{\hat{S}}_{pq}}|\hat{S}|\hat{S}_{ik} \ . \tag{280}$$

All quantities denoted by "˜" signify test filter variables

$$\tilde{Z}(\boldsymbol{x},t) = \int_{V_{\tilde{\Delta}}} \tilde{G}(\boldsymbol{x},\boldsymbol{y})Z(\boldsymbol{y},t)\mathrm{d}^3y \ . \tag{281}$$

The definition of the test filter is in full analogy to (253) though the filter length is over a larger spatial domain such that

$$\tilde{\Delta} > \Delta \ . \tag{282}$$

Those quantities denoted by "ˆ" correspond to the resolved scales while in contrast "˜" signify variables which are obtained from explicit filtering of the resolved flow field.

The physical reason for the preservation of scaling symmetries in (280) is due to the explicit filtering. This procedure comes about with the introduction of an additional length-scale which in fact only appears as the ratio between $\tilde{\Delta}$ and Δ. This fraction is a non-dimensional expression which does not cause an artificial symmetry breaking.

It should be noted that, in addition to the scaling symmetry, all other symmetries (259)-(265) are also observed by the dynamic model. Since its first publication a number of modifications of the dynamic model were developed such as by Lilly (1992), Yoshizawa et al. (1996), Zang et al. (1993) which all preserve the proper symmetries.

References

Andreev, V. K., and Rodionov, A. A. (1988). Group analysis of the equations for plane flows of an ideal fluid in the lagrange variables. *Dokl. Akad. Nauk SSSR* 298(6):1358–1361.

Andreev, V. K., Kaptsov, O. V., Pukhnachov, V. V., and Rodionov, A. A. (1998). *Applications of Group Theoretical Methods in Hydrodynamics*. Kluwer Academic Press.

Bardina, J., Ferziger, J. H., and Reynolds, W. C. (1980). Improved sub-grid scale models for large eddy simulation. *AIAA-paper* 80(1357).

Barenblatt, G. I. (1993). Scaling laws for fully developed turbulent shear flows. part 1. basic hypotheses and analysis. *J. Fluid Mech.* 248:513–520.

Batchelor, G. K. (1946). The theory of axisymmetric turbulence. *Proc. Roy. Soc. A* 186:480–502.

Batchelor, G. K. (1967). *An Introduction to Fluid Dynamics*. Cambridge University Press.

Biringen, S., and Reynolds, W. C. (1981). Large-eddy simulation of the shear-free turbulent boundary layer. *J. Fluid Mech.* 103:53–63.

Birkhoff, G. (1954). Fourier synthesis of homogeneous turbulence. *Comm. Pure Appl. Math.* 7:19–44.

Bluman, G. W., and Kumei, S. (1989). *Symmetries and Differential Equations*. Applied Mathematical Sciences 81. Springer-Verlag.

Cantwell, B. J. (1978). Similarity transformations for the two-dimensional, unsteady, stream-function equation. *J. Fluid Mech.* 85:257–271.

Cantwell, B. J. (1997). Introduction to symmetry analysis. Course Notes prepared for AA 218 - Similitude in Engineering Mechanics. Stanford University.

Chandrasekhar, S. (1950). The theory of axisymmetric turbulence. *Phil. Trans. Roy. Soc. Lond. A* 242:557–577.

DeGraaff, D. B., Webster, D. R., and Eaton, J. K. (1999). The effect of Reynolds number on boundary layer turbulence. *Experimental Thermal and Fluid Science* 18(4):341–346.

Durbin, P. (1991). Near-wall turbulence closure modeling without 'damping functions'. *Theoret. Comput. Fluid Dyn.* 3:1–13.

El Telbany, M. M. M., and Reynolds, A. J. (1980). Velocity distributions in plane turbulent channel flows. *J. Fluid Mech.* 100:1–29.

Fernholz, H. H., Krause, E., Nockemann, M., and Schober, M. (1995). Comparative measurements in the canonical boundary layer at $\mathrm{Re}_{\delta_2} \leq 6 \times 10^4$ on the wall of the german-dutch windtunnel. *Phys. Fluids* 7(6):1275–1281.

Ferziger, J. H., and Peric, M. (1996). *Computational Methods for Fluid Dynamics*. Springer-Verlag.

Ferziger, J. H. (1996). Large eddy simulation. In Gatski, T. B., Hussaini, M. Y., and Lumley, J. L., eds., *Simulation and Modeling of Turbulent Flows*. Oxford University Press. 109–154.

Fischer, M., Durst, F., and Jovanović, J. (1999). Reynolds number dependence of near wall turbulent statistics in channel flows. In *Laser Techniques Applied to Fluid Mechanics (Selected Papers from the 9th International Symposium, Lissabon, Portugal, 13-16 Juli 1998)*. Springer-Verlag.

Fureby, C., Tabor, G., Weller, H. G., and Gosman, A. D. (1997). A comparative study of subgrid scale models in homogeneous isotropic turbulence. *Phys. Fluids* 9(5):1416–1429.

Gatski, T. B., and Speziale, C. G. (1993). On explicit algebraic stress models for complex turbulent flows. *J. Fluid Mech.* 254:59–78.

George, W. K., Castillo, L., and Knecht, P. (1996). The zero pressure-gradient turbulent boundary layer. Report TRL-153, Turbulence Research Laboratory, School of Engineering and Applied Sciences, SUNY Bufallo, NY.

Germano, M., Piomelli, U., Moin, P., and Cabot, W. (1990). A dynamic subgrid-scale eddy viscosity model. In *Proceedings of the Summer Program 1990*. Center for Turbulence Research.

Germano, M., Piomelli, U., Moin, P., and Cabot, W. (1991). A dynamic subgrid-scale eddy viscosity model. *Phys. Fluids A* 3:1760–1765.

Germano, M. (1986). A proposal for a redefinition of the turbulent stresses in the filtered Navier-Stokes equations. *Phys. Fluids* 29:2323–2324.

Greenspan, H. P. (1990). *The Theory of Rotating Fluids*. Breukelen.

Hanjalić, K., and Launder, B. E. (1976). Contribution towards a Reynolds stress closure for low Reynolds number turbulence. *J. Fluid Mech.* 74:593–610.

Härtel, C., and Kleiser, L. (1997). Galilean invariance and filtering dependence of near-wall grid-scale/subgrid-scale interactions in large-eddy simulation. *Phys. Fluids* 9:473–475.

Hirai, S., Takagi, T., and Matsumoto, M. (1988). Predictions of the laminarization phenomena in an axially rotating pipe flow. *J. Fluids Eng.* 110:424–430.

Ibragimov, N. H. (1995a). *CRC Handbook of Lie Group Analysis of Differential Equations*, volume 1: Symmetries, Exact Solutions, and Conservation Laws. CRC Press.

Ibragimov, N. H. (1995b). *CRC Handbook of Lie Group Analysis of Differential Equations*, volume 2: Applications in Engineering and Physical Sciences. CRC Press.

Ibragimov, N. H. (1996). *CRC Handbook of Lie Group Analysis of Differential Equations*, volume 3: New Trends in Theoretical Developments and Computational Methods. CRC Press.

Iida, O., and Kasagi, N. (1993). Redistribution of the Reynolds stresses and destruction of the turbulent heat flux in homogeneous decaying turbulence. In *9th. Symp. on Turb. Shear Flows, Kyoto, Japan*, 24.4.1–24.4.6.

Johnston, J. P., Halleen, R. M., and Lazius, D. K. (1972). Effects of spanwise rotation on the structure of two-dimensional fully developed turbulent channel flow. *J. Fluid Mech.* 56:533–557.

Khor'kova, N. G., and Verbovetsky, A. M. (1995). On symmetry subalgebras and conservation laws for the k-ε turbulence model and the Navier-Stokes equations. *Amer. Math. Soc. Transl.* 167(2):61–90.

Kikuyama, K., Murakami, M., Nishibori, K., and Maeda, K. (1983). Flow in axially rotating pipe. *Bulletin JSME* 26(214).

Kim, J., Moin, P., and Moser, R. (1987). Turbulence statistics in fully developed channel flow at low Reynolds number. *J. Fluid Mech.* 177:133–166.

Kolmogorov, A. N. (1941a). Decay of isotropic turbulence in incompressible viscous fluids. *Dokl. Akad. Nauk SSSR* A 31:538.

Kolmogorov, A. N. (1941b). Dissipation of energy in the locally isotropic turbulence. *C.R. Acad. Sci. SSSR* 32:16–18.

Kolmogorov, A. N. (1941c). The local structure of turbulence in incompressible viscous fluids for very large Reynolds numbers. *Dokl. Akad. Nauk SSSR* 30.

Kristoffersen, R., and Andersson, H. I. (1993). Direct simulations of low-Reynolds-number turbulent flow in a rotating channel. *J. Fluid Mech.* 256:163–197.

Launder, B. E., Reece, G. E., and Rodi, W. (1975). Progress in the development of Reynolds-stress turbulence closure. *J. Fluid Mech.* 68:537–566.

Lee, M. J., and Kim, J. (1991). The structure of turbulence in a simulated plane couette flow. In *8th Symp. Turb. Shear Flows, München*, 5.3.1–5.3.6.

Lilly, D. K. (1992). A proposed modification of the germano subgrid-scale closure method. *Phys. Fluids* 4:633–635.

Loitsyansky, L. (1939). Some basic laws of isotropic turbulent flow. Technical Report 440, Centr. Aero. Hydrodyn. Inst. Moscow. (Trans. NACA Tech. Memo. 1079).

Métais, O., and Lesieur, M. (1992). Spectral large eddy simulation of isotropic and stably-stratified turbulence. *J. Fluid Mech.* 239:157–194.

Mohamed, M. S., and LaRue, J. C. (1990). The decay power law in grid-generated turbulence. *J. Fluid Mech.* 219:195–214.

Moin, P., and Kim, J. (1982). Numerical investigation of turbulent channel flow. *J. Fluid Mech.* 118:341–377.

Niederschulte, G. L. (1996). *Turbulent Flow through a Rectangular Channel.* Dissertation, University of Illinois, Department of Theoretical and Applied Mechanics.

Oberlack, M., and Peters, N. (1993). Closure of the two-point correlation equation as a basis of Reynolds stress models. In So, R., Speziale, C., and Launder, B., eds., *Near-Wall Turbulent Flows*, 85–94. Elsevier Science Publisher.

Oberlack, M., Cabot, W., and Rogers, M. M. (1998). Group analysis, DNS and modeling of a turbulent channel flow with streamwise rotation. In Moin, P., ed., *Proceedings of the Center for Turbulence Summer Program 1998.* Center for Turbulence Research, Stanford University/NASA Ames, CA, USA. 221–242.

Oberlack, M. (1994). *Herleitung und Lösung einer Längenmass- und Dissipations-Tensorgleichung für turbulente Strömungen.* Dissertation, Inst. f. Techn. Mechanik, RWTH Aachen.

Oberlack, M. (1997). Non-isotropic dissipation in non-homogeneous turbulence. *J. Fluid Mech.* 350:351–374.

Oberlack, M. (1999). Similarity in non-rotating and rotating turbulent pipe flows. *J. Fluid Mech.* 379:1–22.

Oberlack, M. (2000a). On symmetries and invariant solutions of laminar and turbulent wall-bounded flows. *Zeitschrift für Angewandte Mathematik und Mechanik* 80(11-12):791–800.

Oberlack, M. (2000b). *Symmetrie, Invarianz und Selbstähnlichkeit in der Turbulenz.* Habilitation, Inst. f. Techn. Mechanik, RWTH Aachen.

Oberlack, M. (2001). On the decay exponent of isotropic turbulence. In *Proceedings of the Annual GAMM Meeting, ETH Zurich.*

Olver, P. J. (1986). *Applications of Lie Groups to Differential Equations.* Graduate Texts in Mathematics. Springer-Verlag.

Orlandi, P., and Fatica, M. (1997). Direct simulations of turbulent flow in a pipe rotating about its axis. *J. Fluid Mech.* 343:43–72.

Pope, S. B. (1975). A more general effective-viscosity hypothesis. *J. Fluid Mech.* 72:331–340.

Pukhnachev, V. V. (1972). Invariant solutions of Navier-Stokes equations describing motions with free boundary. *Dokl. Akad. Nauk* 202:302.

Reich, G. (1988). *Strömung und Wärmeübertragung in einem axial rotierenden Rohr.* Dissertation, Technische Universität Darmstadt.

Rogers, M. M., and Moin, P. (1987). The structure of the vorticity field in homogeneous turbulent flows. *J. Fluid Mech.* 243:33–66.

Rotta, J. C. (1972). *Turbulente Strömungen.* Teubner, Stuttgart.

Saddoughi, S. G., and Veeravalli, S. V. (1994). Local isotropy in turbulent boundary layers at high Reynolds number. *J. Fluid Mech.* 268:333–372.

Saffman, P. G. (1967). The large-scale structure of homogeneous turbulence. *J. Fluid Mech.* 27:581–593.

Schlichting, H. (1982). *Grenzschicht-Theorie.* Verlag G.Braun, Karlsruhe, 8. edition.

Smagorinsky, J. (1963). General circulation experiments with the primitive equations. *Mon. Weath. Rev.* 91:99–164.

Speziale, C. G. (1981). Some interesting properties of two-dimensional turbulence. *Phys. Fluids A* 28(8):1425–1427.

Speziale, C. G. (1985). Galilean invariance of subgrid-scale stress models in the large-eddy simulation of turbulence. *J. Fluid Mech.* 156:55–62.

Tagawa, M., Nagano, Y., and Tsuji, T. (1991). Turbulence model for the dissipation components of Reynolds stresses. In *8th Symp. Turb. Shear Flows, München*, 29.3.1–29.3.6.

von Kármán, T., and Howarth, L. (1938). On the statistical theory of isotropic turbulence. *Proc. Roy. Soc.* A 164:192–215.

Wagner, C., and Friedrich, R. (1998). On the turbulence structure in solid and permeable pipes. *Int. J. Heat Fluid Flow* 19:459–469.

Yoshizawa, A., Tsubokura, M., Kobayashi, T., and Taniguchi, N. (1996). Modeling of the dynamic subgrid-scale viscosity in large eddy simulation. *Phys. Fluids* 8:2254–2256.

Zagarola, M. V., Smits, A. J., Orszag, S. A., and V., Y. (1997). Scaling of the mean velocity profile for turbulent pipe flow. *Phys. Rev. Letters* 78(2).

Zagarola, M. V. (1996). *Mean-flow Scaling of Turbulent Pipe Flow*. Dissertation, Princeton University.

Zang, Y., Street, R. L., and Koseff, J. R. (1993). A dynamic mixed subgrid-scale model and its application to turbulent recirculating flows. *Phys. Fluids* 8:3186–3196.

Index

Printed in the United States
By Bookmasters